FLUORIDE
IN DRINKING WATER

A SCIENTIFIC REVIEW OF EPA'S STANDARDS

Committee on Fluoride in Drinking Water

Board on Environmental Studies and Toxicology

Division on Earth and Life Studies

NATIONAL RESEARCH COUNCIL
OF THE NATIONAL ACADEMIES

THE NATIONAL ACADEMIES PRESS
Washington, D.C.
www.nap.edu

THE NATIONAL ACADEMIES PRESS   500 Fifth Street, NW   Washington, DC 20001

NOTICE: The project that is the subject of this report was approved by the Governing Board of the National Research Council, whose members are drawn from the councils of the National Academy of Sciences, the National Academy of Engineering, and the Institute of Medicine. The members of the committee responsible for the report were chosen for their special competences and with regard for appropriate balance.

This project was supported by Contract No. 68-C-03-013 between the National Academy of Sciences and the U.S. Environmental Protection Agency. Any opinions, findings, conclusions, or recommendations expressed in this publication are those of the author(s) and do not necessarily reflect the view of the organizations or agencies that provided support for this project.

International Standard Book Number 10   0-309-10128-X (Book)
International Standard Book Number 13   978-0-309-10128-8 (Book)
International Standard Book Number 10   0-309-65796-2 (PDF)
International Standard Book Number 13   978-0-309-65796-9 (PDF)
Library of Congress Control Number 2006933237

Additional copies of this report are available from

The National Academies Press
500 Fifth Street, NW
Box 285
Washington, DC 20055

800-624-6242
202-334-3313 (in the Washington metropolitan area)
http://www.nap.edu

Copyright 2006 by the National Academy of Sciences. All rights reserved.

Printed in the United States of America

THE NATIONAL ACADEMIES
Advisers to the Nation on Science, Engineering, and Medicine

The **National Academy of Sciences** is a private, nonprofit, self-perpetuating society of distinguished scholars engaged in scientific and engineering research, dedicated to the furtherance of science and technology and to their use for the general welfare. Upon the authority of the charter granted to it by the Congress in 1863, the Academy has a mandate that requires it to advise the federal government on scientific and technical matters. Dr. Ralph J. Cicerone is president of the National Academy of Sciences.

The **National Academy of Engineering** was established in 1964, under the charter of the National Academy of Sciences, as a parallel organization of outstanding engineers. It is autonomous in its administration and in the selection of its members, sharing with the National Academy of Sciences the responsibility for advising the federal government. The National Academy of Engineering also sponsors engineering programs aimed at meeting national needs, encourages education and research, and recognizes the superior achievements of engineers. Dr. Wm. A. Wulf is president of the National Academy of Engineering.

The **Institute of Medicine** was established in 1970 by the National Academy of Sciences to secure the services of eminent members of appropriate professions in the examination of policy matters pertaining to the health of the public. The Institute acts under the responsibility given to the National Academy of Sciences by its congressional charter to be an adviser to the federal government and, upon its own initiative, to identify issues of medical care, research, and education. Dr. Harvey V. Fineberg is president of the Institute of Medicine.

The **National Research Council** was organized by the National Academy of Sciences in 1916 to associate the broad community of science and technology with the Academy's purposes of furthering knowledge and advising the federal government. Functioning in accordance with general policies determined by the Academy, the Council has become the principal operating agency of both the National Academy of Sciences and the National Academy of Engineering in providing services to the government, the public, and the scientific and engineering communities. The Council is administered jointly by both Academies and the Institute of Medicine. Dr. Ralph J. Cicerone and Dr. Wm. A. Wulf are chair and vice chair, respectively, of the National Research Council.

www.national-academies.org

COMMITTEE ON FLUORIDE IN DRINKING WATER

Members

JOHN DOULL (*Chair*), University of Kansas Medical Center, Kansas City
KIM BOEKELHEIDE, Brown University, Providence, RI
BARBARA G. FARISHIAN, Washington, DC
ROBERT L. ISAACSON, Binghamton University, Binghamton, NY
JUDITH B. KLOTZ, University of Medicine and Dentistry of New Jersey, Piscataway
JAYANTH V. KUMAR, New York State Department of Health, Albany
HARDY LIMEBACK, University of Toronto, Ontario, Canada
CHARLES POOLE, University of North Carolina at Chapel Hill, Chapel Hill
J. EDWARD PUZAS, University of Rochester, Rochester, NY
NU-MAY RUBY REED, California Environmental Protection Agency, Sacramento
KATHLEEN M. THIESSEN, SENES Oak Ridge, Inc., Oak Ridge, TN
THOMAS F. WEBSTER, Boston University School of Public Health, Boston, MA

Staff

SUSAN N. J. MARTEL, Project Director
CAY BUTLER, Editor
TAMARA DAWSON, Senior Program Assistant

Sponsor

U.S. ENVIRONMENTAL PROTECTION AGENCY

COMMITTEE ON TOXICOLOGY

Members

WILLIAM E. HALPERIN (*Chair*), New Jersey Medical School, Newark
LAWRENCE S. BETTS, Eastern Virginia Medical School, Norfolk
EDWARD C. BISHOP, Parsons Corporation, Pasadena, CA
JAMES V. BRUCKNER, University of Georgia, Athens
GARY P. CARLSON, Purdue University, West Lafayette, IN
MARION EHRICH, Virginia Tech, Blacksburg
SIDNEY GREEN, Howard University, Washington, DC
MERYL KAROL, University of Pittsburgh, Pittsburgh, PA
JAMES MCDOUGAL, Wright State University School of Medicine, Dayton, OH
ROGER MCINTOSH, Science Applications International Corporation, Baltimore, MD
GERALD WOGAN, Massachusetts Institute of Technology, Cambridge

Staff

KULBIR S. BAKSHI, Program Director
EILEEN N. ABT, Senior Program Officer for Risk Analysis
ELLEN K. MANTUS, Senior Program Officer
SUSAN N. J. MARTEL, Senior Program Officer
AIDA NEEL, Program Associate
TAMARA DAWSON, Senior Program Assistant
RADIAH ROSE, Senior Editorial Assistant
ALEXANDRA STUPPLE, Senior Editorial Assistant
SAMMY BARDLEY, Librarian

BOARD ON ENVIRONMENTAL STUDIES AND TOXICOLOGY[1]

Members

JONATHAN M. SAMET (*Chair*), Johns Hopkins University, Baltimore, MD
RAMÓN ALVAREZ, Environmental Defense, Austin, TX
JOHN M. BALBUS, Environmental Defense, Washington, DC
THOMAS BURKE, Johns Hopkins University, Baltimore, MD
DALLAS BURTRAW, Resources for the Future, Washington, DC
JAMES S. BUS, Dow Chemical Company, Midland, MI
COSTEL D. DENSON, University of Delaware, Newark
E. DONALD ELLIOTT, Willkie Farr & Gallagher LLP, Washington, DC
J. PAUL GILMAN, Oak Ridge National Laboratory, Oak Ridge, TN
SHERRI W. GOODMAN, Center for Naval Analyses, Alexandria, VA
JUDITH A. GRAHAM, American Chemistry Council, Arlington, VA
DANIEL S. GREENBAUM, Health Effects Institute, Cambridge, MA
WILLIAM P. HORN, Birch, Horton, Bittner and Cherot, Washington, DC
ROBERT HUGGETT, Michigan State University (emeritus), East Lansing
JAMES H. JOHNSON, JR., Howard University, Washington, DC
JUDITH L. MEYER, University of Georgia, Athens
PATRICK Y. O'BRIEN, ChevronTexaco Energy Technology Company, Richmond, CA
DOROTHY E. PATTON, International Life Sciences Institute, Washington, DC
STEWARD T.A. PICKETT, Institute of Ecosystem Studies, Millbrook, NY
DANNY D. REIBLE, University of Texas, Austin
JOSEPH V. RODRICKS, ENVIRON International Corporation, Arlington, VA
ARMISTEAD G. RUSSELL, Georgia Institute of Technology, Atlanta
LISA SPEER, Natural Resources Defense Council, New York, NY
KIMBERLY M. THOMPSON, Massachusetts Institute of Technology, Cambridge
MONICA G. TURNER, University of Wisconsin, Madison
MARK J. UTELL, University of Rochester Medical Center, Rochester, NY
CHRIS G. WHIPPLE, ENVIRON International Corporation, Emeryville, CA
LAUREN ZEISE, California Environmental Protection Agency, Oakland

Senior Staff

JAMES J. REISA, Director
DAVID J. POLICANSKY, Scholar
RAYMOND A. WASSEL, Senior Program Officer for Environmental Sciences and Engineering

---

[1]This study was planned, overseen, and supported by the Board on Environmental Studies and Toxicology.

KULBIR BAKSHI, Senior Program Officer for Toxicology
EILEEN N. ABT, Senior Program Officer for Risk Analysis
K. JOHN HOLMES, Senior Program Officer
ELLEN K. MANTUS, Senior Program Officer
SUSAN N.J. MARTEL, Senior Program Officer
SUZANNE VAN DRUNICK, Senior Program Officer
RUTH E. CROSSGROVE, Senior Editor

OTHER REPORTS OF THE
BOARD ON ENVIRONMENTAL STUDIES AND TOXICOLOGY

Assessing the Human Health Risks of Trichloroethylene: Key Scientific Issues (2006)
New Source Review for Stationary Sources of Air Pollution (2006)
Human Biomonitoring for Environmental Chemicals (2006)
Health Risks from Dioxin and Related Compounds: Evaluation of the EPA Reassessment (2006)
State and Federal Standards for Mobile-Source Emissions (2006)
Superfund and Mining Megasites—Lessons from the Coeur d'Alene River Basin (2005)
Health Implications of Perchlorate Ingestion (2005)
Air Quality Management in the United States (2004)
Endangered and Threatened Species of the Platte River (2004)
Atlantic Salmon in Maine (2004)
Endangered and Threatened Fishes in the Klamath River Basin (2004)
Cumulative Environmental Effects of Alaska North Slope Oil and Gas Development (2003)
Estimating the Public Health Benefits of Proposed Air Pollution Regulations (2002)
Biosolids Applied to Land: Advancing Standards and Practices (2002)
The Airliner Cabin Environment and Health of Passengers and Crew (2002)
Arsenic in Drinking Water: 2001 Update (2001)
Evaluating Vehicle Emissions Inspection and Maintenance Programs (2001)
Compensating for Wetland Losses Under the Clean Water Act (2001)
A Risk-Management Strategy for PCB-Contaminated Sediments (2001)
Acute Exposure Guideline Levels for Selected Airborne Chemicals (4 volumes, 2000-2004)
Toxicological Effects of Methylmercury (2000)
Strengthening Science at the U.S. Environmental Protection Agency (2000)
Scientific Frontiers in Developmental Toxicology and Risk Assessment (2000)
Ecological Indicators for the Nation (2000)
Waste Incineration and Public Health (1999)
Hormonally Active Agents in the Environment (1999)
Research Priorities for Airborne Particulate Matter (4 volumes, 1998-2004)
The National Research Council's Committee on Toxicology: The First 50 Years (1997)
Carcinogens and Anticarcinogens in the Human Diet (1996)

Upstream: Salmon and Society in the Pacific Northwest (1996)
Science and the Endangered Species Act (1995)
Wetlands: Characteristics and Boundaries (1995)
Biologic Markers (5 volumes, 1989-1995)
Review of EPA's Environmental Monitoring and Assessment Program (3 volumes, 1994-1995)
Science and Judgment in Risk Assessment (1994)
Pesticides in the Diets of Infants and Children (1993)
Dolphins and the Tuna Industry (1992)
Science and the National Parks (1992)
Human Exposure Assessment for Airborne Pollutants (1991)
Rethinking the Ozone Problem in Urban and Regional Air Pollution (1991)
Decline of the Sea Turtles (1990)

*Copies of these reports may be ordered from
the National Academies Press
(800) 624-6242 or (202) 334-3313
www.nap.edu*

OTHER REPORTS OF THE COMMITTEE ON TOXICOLOGY

Review of the Department of Defense Research Program on Low-Level Exposures to Chemical Warfare Agents (2005)
Emergency and Continuous Exposure Guidance Levels for Selected Submarine Contaminants, Volume 1 (2004)
Spacecraft Water Exposure Guidelines for Selected Contaminants, Volume 1 (2004)
Toxicologic Assessment of Jet-Propulsion Fuel 8 (2003)
Review of Submarine Escape Action Levels for Selected Chemicals (2002)
Standing Operating Procedures for Developing Acute Exposure Guideline Levels for Hazardous Chemicals (2001)
Evaluating Chemical and Other Agent Exposures for Reproductive and Developmental Toxicity (2001)
Acute Exposure Guideline Levels for Selected Airborne Contaminants, Volume 1 (2000), Volume 2 (2002), Volume 3 (2003), Volume 4 (2004)
Review of the US Navy's Human Health Risk Assessment of the Naval Air Facility at Atsugi, Japan (2000)
Methods for Developing Spacecraft Water Exposure Guidelines (2000)
Review of the U.S. Navy Environmental Health Center's Health-Hazard Assessment Process (2000)
Review of the U.S. Navy's Exposure Standard for Manufactured Vitreous Fibers (2000)
Re-Evaluation of Drinking-Water Guidelines for Diisopropyl Methylphosphonate (2000)
Submarine Exposure Guidance Levels for Selected Hydrofluorocarbons: HFC-236fa, HFC-23, and HFC-404a (2000)
Review of the U.S. Army's Health Risk Assessments for Oral Exposure to Six Chemical-Warfare Agents (1999)
Toxicity of Military Smokes and Obscurants, Volume 1(1997), Volume 2 (1999), Volume 3 (1999)
Assessment of Exposure-Response Functions for Rocket-Emission Toxicants (1998)
Toxicity of Alternatives to Chlorofluorocarbons: HFC-134a and HCFC-123 (1996)
Permissible Exposure Levels for Selected Military Fuel Vapors (1996)
Spacecraft Maximum Allowable Concentrations for Selected Airborne Contaminants, Volume 1 (1994), Volume 2 (1996), Volume 3 (1996), Volume 4 (2000)

Preface

In 1986, the U.S. Environmental Protection Agency (EPA) established a maximum-contaminant-level goal (MCLG) of 4 milligrams per liter (mg/L) and a secondary maximum contaminant level (SMCL) of 2 mg/L for fluoride in drinking water. These exposure values are not recommendations for the artificial fluoridation of drinking water, but are guidelines for areas in the United States that are contaminated or have high concentrations of naturally occurring fluoride. The goal of the MCLG is to establish an exposure guideline to prevent adverse health effects in the general population, and the goal of the SMCL is to reduce the occurrence of adverse cosmetic consequences from exposure to fluoride. Both the MCLG and the SMCL are nonenforceable guidelines.

The regulatory standard for drinking water is the maximum contaminant level (MCL), which is set as close to the MCLG as possible, with the use of the best technology available. For fluoride, the MCL is the same as the MCLG of 4 mg/L. In 1993, a previous committee of the National Research Council (NRC) reviewed the health effects of ingested fluoride and EPA's MCL. It concluded that the MCL was an appropriate interim standard, but that further research was needed to fill data gaps on total exposures to fluoride and its toxicity. Because new research on fluoride is now available and because the Safe Drinking Water Act requires periodic reassessment of regulations for drinking water contaminants, EPA requested that the NRC evaluate the adequacy of its MCLG and SMCL for fluoride to protect public health. In response to EPA's request, the NRC convened the Committee on Fluoride in Drinking Water, which prepared this report. The committee was charged to review toxicologic, epidemiologic, and clinical data on fluoride, particularly data published since 1993, and exposure data on orally ingested fluoride from drinking water and other sources. Biographical information on the committee members is provided in Appendix A.

This report presents the committee's review of the scientific basis of EPA's MCLG and SMCL for fluoride, and their adequacy for protecting children and others from adverse health effects. The committee considers the relative contribution of various sources of fluoride (e.g., drinking water, food, dental hygiene products) to total exposure, and identifies data gaps and makes recommendations for future research relevant to setting the MCLG and SMCL for fluoride. Addressing questions of economics, risk-benefit assessment, or water-treatment technology was not part of the committee's charge.

This report has been reviewed in draft form by individuals chosen for their diverse perspectives and technical expertise, in accordance with procedures approved by the NRC's Report Review Committee. The purpose of this independent review is to provide candid and critical comments that will assist the institution in making its published report as sound as possible and to ensure that the report meets institutional standards for objectivity, evidence, and responsiveness to the study charge. The review comments and draft manuscript remain confidential to protect the integrity of the deliberative process. We wish to thank the following individuals for their review of this report: Kenneth Cantor, National Cancer Institute; Caswell Evans, Jr., University of Illinois at Chicago; Michael Gallo, University of Medicine and Dentistry of New Jersey; Mari Golub, California Environmental Protection Agency; Philippe Grandjean, University of Southern Denmark; David Hoel, Medical University of South Carolina; James Lamb, The Weinberg Group Inc.; Betty Olson, University of California at Irvine; Elizabeth Platz, Johns Hopkins Bloomberg School of Public Health; George Stookey, Indiana University School of Dentistry; Charles Turner, University of Indiana; Robert Utiger, Harvard Institute of Medicine; Gary Whitford, Medical College of Georgia; and Gerald Wogan, Massachusetts Institute of Technology.

Although the reviewers listed above have provided many constructive comments and suggestions, they were not asked to endorse the conclusions or recommendations, nor did they see the final draft of the report before its release. The review of this report was overseen by John C. Bailar, University of Chicago, and Gilbert S. Omenn, University of Michigan Medical School. Appointed by the NRC, they were responsible for making certain that an independent examination of this report was carried out in accordance with institutional procedures and that all review comments were carefully considered. Responsibility for the final content of this report rests entirely with the authoring committee and the institution.

The committee gratefully acknowledges the individuals who made presentations to the committee at its public meetings. They include Paul Connett, St. Lawrence University; Joyce Donohue, EPA; Steve Levy, University of Iowa; William Maas, Centers for Disease Control and Prevention; Edward Ohanian, EPA; Charles Turner, Indiana University; and Gary Whitford, University of Georgia. The committee also wishes to thank Thomas Burke, Johns Hopkins University; Michael Morris, University of Michigan; Bernard Wagner, Wagner and Associates; and Lauren Zeise, California Environmental Protection Agency, who served as consultants to the committee.

The committee is grateful for the assistance of the NRC staff in preparing the report. It particularly wishes to acknowledge the outstanding staff support from project director Susan Martel. We are grateful for her persistence and patience in keeping us focused and moving ahead on the task and her expertise and skill in reconciling the differing viewpoints of committee members. Other staff members who contributed to this effort are James Reisa, director of the Board on Environmental Studies and Toxicology; Kulbir Bakshi, program director for the Committee on Toxicology; Cay Butler, editor; Mirsada Karalic-Loncarevic, research associate; Jennifer Saunders, research associate; and Tamara Dawson, senior project assistant.

Finally, I would like to thank all the members of the committee for their efforts throughout the development of this report.

John Doull, M.D., Ph.D., *Chair*
Committee on Fluoride in Drinking Water

Contents

SUMMARY   1

1   INTRODUCTION   13
Fluoride in Drinking Water, 14
History of EPA's Regulation of Fluoride, 16
Committee's Task, 18
Committee's Approach, 21
Structure of the Report, 22

2   MEASURES OF EXPOSURE TO FLUORIDE IN THE UNITED STATES   23
Sources of Fluoride Exposure, 24
Recent Estimates of Total Fluoride Exposure, 54
Total Exposure to Fluoride, 55
Summary of Exposure Assessment, 64
Biomarkers of Exposure, Effect, and Susceptibility, 69
Findings, 81
Recommendations, 87

3   PHARMACOKINETICS OF FLUORIDE   89
Overview of Fluoride Chemistry, Units, and Measurement, 89
Short Review of Fluoride Pharmacokinetics: Absorption, Distribution, and Elimination, 90
Pharmacokinetic Models, 92
Fluoride Concentrations in Human Bone Versus Water Concentration, 93

Fluoride Concentrations in Bones after Clinical Studies, 96
Comparative Pharmacokinetics of Rats and Humans, 98
Organofluorine Compounds, 99
Factors Modifying Pharmacokinetics and Their Implications
  for Potentially Susceptible Populations, 99
Findings, 101
Research Recommendations, 101

4   EFFECTS OF FLUORIDE ON TEETH   103
    Enamel Fluorosis, 103
    Other Dental Effects, 126
    Findings, 127
    Recommendations, 130

5   MUSCULOSKELETAL EFFECTS   131
    Chemistry of Fluoride As It Relates to Mineralizing Tissues, 131
    Effect of Fluoride on Cell Function, 133
    Effects of Fluoride on Human Skeletal Metabolism, 138
    Effect of Fluoride on Chondrocyte Metabolism and Arthritis, 177
    Findings, 178
    Recommendations, 180

6   REPRODUCTIVE AND DEVELOPMENTAL EFFECTS OF FLUORIDE   181
    Reproductive Effects, 181
    Developmental Effects, 193
    Findings, 204
    Recommendations, 204

7   NEUROTOXICITY AND NEUROBEHAVIORAL EFFECTS   205
    Human Studies, 205
    Animal Studies, 214
    Neurochemical Effects and Mechanisms, 218
    Findings, 220
    Recommendations, 222

8   EFFECTS ON THE ENDOCRINE SYSTEM   224
    Thyroid Follicular Cells, 224
    Thyroid Parafollicular Cells, 236
    Parathyroid Glands, 238
    Pineal Gland, 252
    Other Endocrine Organs, 256
    Summary, 260
    Recommendations, 266

9	EFFECTS ON THE GASTROINTESTINAL, RENAL, HEPATIC, AND IMMUNE SYSTEMS GI System, 268 The Renal System, 280 Hepatic System, 292 Immune System, 293 Findings, 295 Recommendations, 302	268
10	GENOTOXICITY AND CARCINOGENICITY Genotoxicity, 304 Carcinogenicity, 316 EPA Guidelines and Practice in Setting MCLGs Regarding Carcinogenicity, 334 Findings, 335 Recommendations, 338	304
11	DRINKING WATER STANDARDS FOR FLUORIDE Current Methods for Setting Standards for Drinking Water, 340 New Risk Assessment Considerations, 342 Fluoride Standards, 345 Findings and Recommendations, 352	340

REFERENCES 354

Appendixes

A Biographical Information on the Committee on Fluoride in Drinking Water, 411
B Measures of Exposure to Fluoride in the United States: Supplementary Information, 416
C Ecologic and Partially Ecologic Studies in Epidemiology, 439
D Comparative Pharmacokinetics of Rats and Humans, 442
E Detailed Information on Endocrine Studies of Fluoride, 447

FLUORIDE
IN DRINKING WATER

Summary

Under the Safe Drinking Water Act, the U.S. Environmental Protection Agency (EPA) is required to establish exposure standards for contaminants in public drinking-water systems that might cause any adverse effects on human health. These standards include the maximum contaminant level goal (MCLG), the maximum contaminant level (MCL), and the secondary maximum contaminant level (SMCL). The MCLG is a health goal set at a concentration at which no adverse health effects are expected to occur and the margins of safety are judged "adequate." The MCL is the enforceable standard that is set as close to the MCLG as possible, taking into consideration other factors, such as treatment technology and costs. For some contaminants, EPA also establishes an SMCL, which is a guideline for managing drinking water for aesthetic, cosmetic, or technical effects.

Fluoride is one of the drinking-water contaminants regulated by EPA. In 1986, EPA established an MCLG and MCL for fluoride at a concentration of 4 milligrams per liter (mg/L) and an SMCL of 2 mg/L. These guidelines are restrictions on the total amount of fluoride allowed in drinking water. Because fluoride is well known for its use in the prevention of dental caries, it is important to make the distinction here that EPA's drinking-water guidelines are not recommendations about adding fluoride to drinking water to protect the public from dental caries. Guidelines for that purpose (0.7 to 1.2 mg/L) were established by the U.S. Public Health Service more than 40 years ago. Instead, EPA's guidelines are maximum allowable concentrations in drinking water intended to prevent toxic or other adverse effects that could result from exposure to fluoride.

In the early 1990s at the request of EPA, the National Research Council

(NRC) independently reviewed the health effects of ingested fluoride and the scientific basis for EPA's MCL. It concluded that the MCL was an appropriate interim standard but that further research was needed to fill data gaps on total exposure to fluoride and its toxicity. Because new research on fluoride is now available and because the Safe Drinking Water Act requires periodic reassessment of regulations for drinking-water contaminants, EPA requested that the NRC again evaluate the adequacy of its MCLG and SMCL for fluoride to protect public health.

COMMITTEE'S TASK

In response to EPA's request, the NRC convened the Committee on Fluoride in Drinking Water, which prepared this report. The committee was charged to review toxicologic, epidemiologic, and clinical data on fluoride—particularly data published since the NRC's previous (1993) report—and exposure data on orally ingested fluoride from drinking water and other sources. On the basis of its review, the committee was asked to evaluate independently the scientific basis of EPA's MCLG of 4 mg/L and SMCL of 2 mg/L in drinking water and the adequacy of those guidelines to protect children and others from adverse health effects. The committee was asked to consider the relative contribution of various fluoride sources (e.g., drinking water, food, dental-hygiene products) to total exposure. The committee was also asked to identify data gaps and to make recommendations for future research relevant to setting the MCLG and SMCL for fluoride. Addressing questions of artificial fluoridation, economics, risk-benefit assessment, and water-treatment technology was not part of the committee's charge.

THE COMMITTEE'S EVALUATION

To accomplish its task, the committee reviewed a large body of research on fluoride, focusing primarily on studies generated since the early 1990s, including information on exposure; pharmacokinetics; adverse effects on various organ systems; and genotoxic and carcinogenic potential. The collective evidence from in vitro assays, animal research, human studies, and mechanistic information was used to assess whether multiple lines of evidence indicate human health risks. The committee only considered adverse effects that might result from exposure to fluoride; it did not evaluate health risk from lack of exposure to fluoride or fluoride's efficacy in preventing dental caries.

After reviewing the collective evidence, including studies conducted since the early 1990s, the committee concluded unanimously that the present MCLG of 4 mg/L for fluoride should be lowered. Exposure at the MCLG clearly puts children at risk of developing severe enamel fluorosis, a condition that is associated with enamel loss and pitting. In addition, the majority of the committee concluded that the MCLG is not likely to be protective against bone fractures. The basis for these conclusions is expanded upon below.

Exposure to Fluoride

The major sources of exposure to fluoride are drinking water, food, dental products, and pesticides. The biggest contributor to exposure for most people in the United States is drinking water. Estimates from 1992 indicate that approximately 1.4 million people in the United States had drinking water with natural fluoride concentrations of 2.0-3.9 mg/L, and just over 200,000 people had concentrations equal to or exceeding 4 mg/L (the presented MCL). In 2000, it was estimated that approximately 162 million people had artificially fluoridated water (0.7-1.2 mg/L).

Food sources contain various concentrations of fluoride and are the second largest contributor to exposure. Beverages contribute most to estimated fluoride intake, even when excluding contributions from local tap water. The greatest source of nondietary fluoride is dental products, primarily toothpastes. The public is also exposed to fluoride from background air and from certain pesticide residues. Other sources include certain pharmaceuticals and consumer products.

Highly exposed subpopulations include individuals who have high concentrations of fluoride in drinking water, who drink unusually large volumes of water, or who are exposed to other important sources of fluoride. Some subpopulations consume much greater quantities of water than the 2 L per day that EPA assumes for adults, including outdoor workers, athletes, and people with certain medical conditions, such as diabetes insipidus. On a per-body-weight basis, infants and young children have approximately three to four times greater exposure than do adults. Dental-care products are also a special consideration for children, because many tend to use more toothpaste than is advised, their swallowing control is not as well developed as that of adults, and many children under the care of a dentist undergo fluoride treatments.

Overall, the committee found that the contribution to total fluoride exposure from fluoride in drinking water in the average person, depending on age, is 57% to 90% at 2 mg/L and 72% to 94% at 4 mg/L. For high-water-intake individuals, the drinking-water contribution is 86% to 96% at 2 mg/L and 92% to 98% at 4 mg/L. Among individuals with an average water-intake rate, infants and children have the greatest total exposure to fluoride, ranging from 0.079 to 0.258 mg/kg/day at 4 mg/L and 0.046 to 0.144 mg/kg/day at 2 mg/L in drinking water. For high-water-intake individuals exposed to fluoride at 4 mg/L, total exposure ranges from 0.294

mg/kg/day for adults to 0.634 mg/kg/day for children. The corresponding intake range at 2 mg/L is 0.154 to 0.334 mg/kg/day for adults and children, respectively.

Dental Effects

Enamel fluorosis is a dose-related mottling of enamel that can range from mild discoloration of the tooth surface to severe staining and pitting. The condition is permanent after it develops in children during tooth formation, a period ranging from birth until about the age of 8. Whether to consider enamel fluorosis, particularly the moderate to severe forms, to be an adverse health effect or a cosmetic effect has been the subject of debate for decades. In previous assessments, all forms of enamel fluorosis, including the severest form, have been judged to be aesthetically displeasing but not adverse to health. This view has been based largely on the absence of direct evidence that severe enamel fluorosis results in tooth loss; loss of tooth function; or psychological, behavioral, or social problems.

Severe enamel fluorosis is characterized by dark yellow to brown staining and discrete and confluent pitting, which constitutes enamel loss. The committee finds the rationale for considering severe enamel fluorosis only a cosmetic effect to be much weaker for discrete and confluent pitting than for staining. One of the functions of tooth enamel is to protect the dentin and, ultimately, the pulp from decay and infection. Severe enamel fluorosis compromises that health-protective function by causing structural damage to the tooth. The damage to teeth caused by severe enamel fluorosis is a toxic effect that is consistent with prevailing risk assessment definitions of adverse health effects. This view is supported by the clinical practice of filling enamel pits in patients with severe enamel fluorosis and restoring the affected teeth. Moreover, the plausible hypothesis concerning elevated frequency of caries in persons with severe enamel fluorosis has been accepted by some authorities, and the available evidence is mixed but generally supportive.

Severe enamel fluorosis occurs at an appreciable frequency, approximately 10% on average, among children in U.S. communities with water fluoride concentrations at or near the current MCLG of 4 mg/L. Thus, the MCLG is not adequately protective against this condition.

Two of the 12 members of the committee did not agree that severe enamel fluorosis should now be considered an adverse health effect. They agreed that it is an adverse dental effect but found that no new evidence has emerged to suggest a link between severe enamel fluorosis, as experienced in the United States, and a person's ability to function. They judged that demonstration of enamel defects alone from fluorosis is not sufficient to change the prevailing opinion that severe enamel fluorosis is an adverse cosmetic effect. Despite their disagreement on characterization of the condition, these two members concurred with the committee's conclusion that the MCLG should prevent the occurrence of this unwanted condition.

Enamel fluorosis is also of concern from an aesthetic standpoint because it discolors or results in staining of teeth. No data indicate that staining alone affects tooth function or susceptibility to caries, but a few studies have shown that tooth mottling affects aesthetic perception of facial attractiveness. It is difficult to draw conclusions from these studies, largely because perception of the condition and facial attractiveness are subjective and culturally influenced. The committee finds that it is reasonable to assume that some individuals will find *moderate* enamel fluorosis on front teeth to be detrimental to their appearance and that it could affect their overall sense of well-being. However, the available data are not adequate to categorize moderate enamel fluorosis as an adverse health effect on the basis of structural or psychological effects.

Since 1993, there have been no new studies of enamel fluorosis in U.S. communities with fluoride at 2 mg/L in drinking water. Earlier studies indicated that the prevalence of moderate enamel fluorosis at that concentration could be as high as 15%. Because enamel fluorosis has different distribution patterns among teeth, depending on when exposure occurred during tooth development and on enamel thickness, and because current indexes for categorizing enamel fluorosis do not differentiate between mottling of anterior and posterior teeth, the committee was not able to determine what percentage of moderate cases might be of cosmetic concern.

Musculoskeletal Effects

Concerns about fluoride's effects on the musculoskeletal system historically have been and continue to be focused on skeletal fluorosis and bone fracture. Fluoride is readily incorporated into the crystalline structure of bone and will accumulate over time. Since the previous 1993 NRC review of fluoride, two pharmacokinetic models were developed to predict bone concentrations from chronic exposure to fluoride. Predictions based on these models were used in the committee's assessments below.

Skeletal Fluorosis

Skeletal fluorosis is a bone and joint condition associated with prolonged exposure to high concentrations of fluoride. Fluoride increases bone density and appears to exacerbate the growth of osteophytes present in the bone and joints, resulting in joint stiffness and pain. The condition is categorized into one of four stages: a preclinical stage and three clinical stages that increase in severity. The most severe stage (clinical stage III) historically has been referred to as the "crippling" stage. At stage II, mobility is not significantly affected, but it is characterized by chronic joint pain, arthritic symptoms, slight calcification of ligaments, and osteosclerosis of the cancellous bones. Whether EPA's MCLG of 4 mg/L protects against these precursors to more serious mobility problems is unclear.

Few clinical cases of skeletal fluorosis in healthy U.S. populations have been reported in recent decades, and the committee did not find any recent studies to evaluate the prevalence of the condition in populations exposed to fluoride at the MCLG. Thus, to answer the question of whether EPA's MCLG protects the general public from stage II and stage III skeletal fluorosis, the committee compared pharmacokinetic model predictions of bone fluoride concentrations and historical data on iliac-crest bone fluoride concentrations associated with the different stages of skeletal fluorosis. The models estimated that bone fluoride concentrations resulting from lifetime exposure to fluoride in drinking water at 2 mg/L (4,000 to 5,000 mg/kg ash) or 4 mg/L (10,000 to 12,000 mg/kg ash) fall within or exceed the ranges historically associated with stage II and stage III skeletal fluorosis (4,300 to 9,200 mg/kg ash and 4,200 to 12,700 mg/kg ash, respectively). However, this comparison alone is insufficient for determining whether stage II or III skeletal fluorosis is a risk for populations exposed to fluoride at 4 mg/L, because bone fluoride concentrations and the levels at which skeletal fluorosis occurs vary widely. On the basis of the existing epidemiologic literature, stage III skeletal fluorosis appears to be a rare condition in the United Sates; furthermore, the committee could not determine whether stage II skeletal fluorosis is occurring in U.S. residents who drink water with fluoride at 4 mg/L. Thus, more research is needed to clarify the relationship between fluoride ingestion, fluoride concentrations in bone, and stage of skeletal fluorosis before any conclusions can be drawn.

Bone Fractures

Several epidemiologic studies of fluoride and bone fractures have been published since the 1993 NRC review. The committee focused its review on observational studies of populations exposed to drinking water containing fluoride at 2 to 4 mg/L or greater and on clinical trials of fluoride (20-34 mg/day) as a treatment for osteoporosis. Several strong observational studies indicated an increased risk of bone fracture in populations exposed to fluoride at 4 mg/L, and the results of other studies were qualitatively consistent with that finding. The one study using serum fluoride concentrations found no appreciable relationship to fractures. Because serum fluoride concentrations may not be a good measure of bone fluoride concentrations or long-term exposure, the ability to show an association might have been diminished in that study. A meta-analysis of randomized clinical trials reported an elevated risk of new nonvertebral fractures and a slightly decreased risk of vertebral fractures after 4 years of fluoride treatment. An increased risk of bone fracture was found among a subset of the trials that the committee found most informative for assessing long-term exposure. Although the duration and concentrations of exposure to fluoride differed between the observational studies and the clinical trials, bone fluoride content was similar (6,200 to more than 11,000 mg/kg ash in observational studies and 5,400 to 12,000 mg/kg ash in clinical trials).

Fracture risk and bone strength have been studied in animal models. The weight of evidence indicates that, although fluoride might increase bone volume, there is less strength per unit volume. Studies of rats indicate that bone strength begins to decline when fluoride in bone ash reaches 6,000 to 7,000 mg/kg. However, more research is needed to address uncertainties associated with extrapolating data on bone strength and fractures from animals to humans. Important species differences in fluoride uptake, bone remodeling, and growth must be considered. Biochemical and physiological data indicate a biologically plausible mechanism by which fluoride could weaken bone. In this case, the physiological effect of fluoride on bone quality and risk of fracture observed in animal studies is consistent with the human evidence.

Overall, there was consensus among the committee that there is scientific evidence that under certain conditions fluoride can weaken bone and increase the risk of fractures. The majority of the committee concluded that lifetime exposure to fluoride at drinking-water concentrations of 4 mg/L or higher is likely to increase fracture rates in the population, compared with exposure to 1 mg/L, particularly in some demographic subgroups that are prone to accumulate fluoride into their bones (e.g., people with renal disease). However, 3 of the 12 members judged that the evidence only supports a conclusion that the MCLG *might not* be protective against bone fracture. Those members judged that more evidence is needed to conclude that bone fractures occur at an appreciable frequency in human populations exposed to fluoride at 4 mg/L and that the MCLG is not *likely* to be protective.

There were few studies to assess fracture risk in populations exposed to fluoride at 2 mg/L in drinking water. The best available study, from Finland, suggested an increased rate of hip fracture in populations exposed to fluoride at concentrations above 1.5 mg/L. However, this study alone is not sufficient to judge fracture risk for people exposed to fluoride at 2 mg/L. Thus, no conclusions could be drawn about fracture risk or safety at 2 mg/L.

Reproductive and Developmental Effects

A large number of reproductive and developmental studies in animals have been conducted and published since the 1993 NRC report, and the overall quality of that database has improved significantly. Those studies indicated that adverse reproductive and developmental outcomes occur only at very high concentrations that are unlikely to be encountered by U.S. populations. A few human studies suggested that high concentrations of fluoride exposure might be associated with alterations in reproductive hormones, effects on fertility, and developmental outcomes, but design limitations make those studies insufficient for risk evaluation.

Neurotoxicity and Neurobehavioral Effects

Animal and human studies of fluoride have been published reporting adverse cognitive and behavioral effects. A few epidemiologic studies of Chinese populations have reported IQ deficits in children exposed to fluoride at 2.5 to 4 mg/L in drinking water. Although the studies lacked sufficient detail for the committee to fully assess their quality and relevance to U.S. populations, the consistency of the results appears significant enough to warrant additional research on the effects of fluoride on intelligence.

A few animal studies have reported alterations in the behavior of rodents after treatment with fluoride, but the committee did not find the changes to be substantial in magnitude. More compelling were studies on molecular, cellular, and anatomical changes in the nervous system found after fluoride exposure, suggesting that functional changes could occur. These changes might be subtle or seen only under certain physiological or environmental conditions. More research is needed to clarify the effect of fluoride on brain chemistry and function.

Endocrine Effects

The chief endocrine effects of fluoride exposures in experimental animals and in humans include decreased thyroid function, increased calcitonin activity, increased parathyroid hormone activity, secondary hyperparathyroidism, impaired glucose tolerance, and possible effects on timing of sexual maturity. Some of these effects are associated with fluoride intake that is achievable at fluoride concentrations in drinking water of 4 mg/L or less, especially for young children or for individuals with high water intake. Many of the effects could be considered subclinical effects, meaning that they are not adverse health effects. However, recent work on borderline hormonal imbalances and endocrine-disrupting chemicals indicated that adverse health effects, or increased risks for developing adverse effects, might be associated with seemingly mild imbalances or perturbations in hormone concentrations. Further research is needed to explore these possibilities.

Effects on Other Organ Systems

The committee also considered effects on the gastrointestinal system, kidneys, liver, and immune system. There were no human studies on drinking water containing fluoride at 4 mg/L in which gastrointestinal, renal, hepatic, or immune effects were carefully documented. Case reports and in vitro and animal studies indicated that exposure to fluoride at concentrations greater than 4 mg/L can be irritating to the gastrointestinal system, affect renal tissues and function, and alter hepatic and immunologic parameters. Such effects are unlikely to be a risk for the average individual exposed to fluoride at 4 mg/L in drinking water. However, a potentially susceptible subpopulation comprises individuals with renal impairments who retain more fluoride than healthy people do.

Genotoxicity and Carcinogenicity

Many assays have been performed to assess the genotoxicity of fluoride. Since the 1993 NRC review, the most significant additions to the database are in vivo assays in human populations and, to a lesser extent, in vitro assays with human cell lines and in vivo experiments with rodents. The results of the in vivo human studies are mixed. The results of in vitro tests are also conflicting and do not contribute significantly to the interpretation of the existing database. Evidence on the cytogenetic effects of fluoride at environmental concentrations is contradictory.

Whether fluoride might be associated with bone cancer has been a subject of debate. Bone is the most plausible site for cancer associated with fluoride because of its deposition into bone and its mitogenic effects on bone cells in culture. In a 1990 cancer bioassay, the overall incidence of osteosarcoma in male rats exposed to different amounts of fluoride in drinking water showed a positive dose-response trend. In a 1992 study, no increase in osteosarcoma was reported in male rats, but most of the committee judged the study to have insufficient power to counter the evidence for the trend found in the 1990 bioassay.

Several epidemiologic investigations of the relation between fluoride and cancer have been performed since the 1993 evaluation, including both individual-based and ecologic studies. Several studies had significant methodological limitations that made it difficult to draw conclusions. Overall, the results are mixed, with some studies reporting a positive association and others no association.

On the basis of the committee's collective consideration of data from humans, genotoxicity assays, and studies of mechanisms of action in cell systems (e.g., bone cells in vitro), the evidence on the potential of fluoride to initiate or promote cancers, particularly of the bone, is tentative and mixed. Assessing whether fluoride constitutes a risk factor for osteosarcoma is complicated by the rarity of the disease and the difficulty of characterizing biologic dose because of the ubiquity of population exposure to fluoride and the difficulty of acquiring bone samples in nonaffected individuals.

A relatively large hospital-based case-control study of osteosarcoma and fluoride exposure is under way at the Harvard School of Dental Medicine and is expected to be published in 2006. That study will be an important addition to the fluoride database, because it will have exposure information on residence histories, water consumption, and assays of bone and toenails. The results of that study should help to identify what future research will be most useful in elucidating fluoride's carcinogenic potential.

DRINKING-WATER STANDARDS

Maximum-Contaminant-Level Goal

In light of the collective evidence on various health end points and total exposure to fluoride, the committee concludes that EPA's MCLG of 4 mg/L should be lowered. Lowering the MCLG will prevent children from developing severe enamel fluorosis and will reduce the lifetime accumulation of fluoride into bone that the majority of the committee concludes is likely to put individuals at increased risk of bone fracture and possibly skeletal fluorosis, which are particular concerns for subpopulations that are prone to accumulating fluoride in their bones.

To develop an MCLG that is protective against severe enamel fluorosis, clinical stage II skeletal fluorosis, and bone fractures, EPA should update the risk assessment of fluoride to include new data on health risks and better estimates of total exposure (relative source contribution) for individuals. EPA should use current approaches for quantifying risk, considering susceptible subpopulations, and characterizing uncertainties and variability.

Secondary Maximum Contaminant Level

The prevalence of severe enamel fluorosis is very low (near zero) at fluoride concentrations below 2 mg/L. From a cosmetic standpoint, the SMCL does not completely prevent the occurrence of moderate enamel fluorosis. EPA has indicated that the SMCL was intended to reduce the severity and occurrence of the condition to 15% or less of the exposed population. The available data indicate that fewer than 15% of children will experience moderate enamel fluorosis of aesthetic concern (discoloration of the front teeth) at that concentration. However, the degree to which moderate enamel fluorosis might go beyond a cosmetic effect to create an adverse psychological effect or an adverse effect on social functioning is not known.

OTHER PUBLIC HEALTH ISSUES

The committee's conclusions regarding the potential for adverse effects from fluoride at 2 to 4 mg/L in drinking water do not address the lower exposures commonly experienced by most U.S. citizens. Fluoridation is widely practiced in the United States to protect against the development of dental caries; fluoride is added to public water supplies at 0.7 to 1.2 mg/L. The charge to the committee did not include an examination of the benefits and risks that might occur at these lower concentrations of fluoride in drinking water.

RESEARCH NEEDS

As noted above, gaps in the information on fluoride prevented the committee from making some judgments about the safety or the risks of fluoride at concentrations of 2 to 4 mg/L. The following research will be useful for filling those gaps and guiding revisions to the MCLG and SMCL for fluoride.

- Exposure assessment
  — Improved assessment of exposure to fluoride from all sources is needed for a variety of populations (e.g., different socioeconomic conditions). To the extent possible, exposures should be characterized for individuals rather than communities, and epidemiologic studies should group individuals by exposure level rather than by source of exposure, location of residence, or fluoride concentration in drinking water. Intakes or exposures should be characterized with and without normalization for body weight. Fluoride should be included in nationwide biomonitoring surveys and nutritional studies; in particular, analysis of fluoride in blood and urine samples taken in these surveys would be valuable.
- Pharmacokinetic studies
  — The concentrations of fluoride in human bone as a function of exposure concentration, exposure duration, age, sex, and health status should be studied. Such studies would be greatly aided by noninvasive means of measuring bone fluoride. Information is particularly needed on fluoride plasma and bone concentrations in people with small-to-moderate changes in renal function as well as in those with serious renal deficiency.
  — Improved and readily available pharmacokinetic models should be developed. Additional cross-species pharmacokinetic comparisons would help to validate such models.
- Studies of enamel fluorosis
  — Additional studies, including longitudinal studies, should be done in U.S. communities with water fluoride concentrations greater than 1 mg/L.

These studies should focus on moderate and severe enamel fluorosis in relation to caries and in relation to psychological, behavioral, and social effects among affected children, their parents, and affected children after they become adults.

— Methods should be developed and validated to objectively assess enamel fluorosis. Consideration should be given to distinguishing between staining or mottling of the anterior teeth and of the posterior teeth so that aesthetic consequences can be more easily assessed.

— More research is needed on the relation between fluoride exposure and dentin fluorosis and delayed tooth eruption patterns.

- Bone studies

— A systematic study of clinical stage II and stage III skeletal fluorosis should be conducted to clarify the relationship between fluoride ingestion, fluoride concentration in bone, and clinical symptoms.

— More studies of communities with drinking water containing fluoride at 2 mg/L or more are needed to assess potential bone fracture risk at these higher concentrations. Quantitative measures of fracture, such as radiologic assessment of vertebral body collapse, should be used instead of self-reported fractures or hospital records. Moreover, if possible, bone fluoride concentrations should be measured in long-term residents.

- Other health effects

— Carefully conducted studies of exposure to fluoride and emerging health parameters of interest (e.g., endocrine effects and brain function) should be performed in populations in the United States exposed to various concentrations of fluoride. It is important that exposures be appropriately documented.

1

Introduction

Under the Safe Drinking Water Act, the U.S. Environmental Protection Agency (EPA) is required to establish the concentrations of contaminants that are permitted in public drinking-water systems. A public water system is defined by EPA as a "system for the provision to the public of water for human consumption through pipes or other constructed conveyances, if such system has at least fifteen service connections or regularly serves at least twenty-five individuals" (63 Fed. Reg. 41940 [1998]). Section 1412 of the act, as amended in 1986, requires EPA to publish maximum-contaminant-level goals (MCLGs) and promulgate national primary drinking-water regulations (maximum contaminant levels [MCLs]) for contaminants in drinking water that might cause any adverse effect on human health and that are known or expected to occur in public water systems. MCLGs are health goals set at concentrations at which no known or expected adverse health effects occur and the margins of safety are adequate. MCLGs are not regulatory requirements but are used by EPA as a basis for establishing MCLs. MCLs are enforceable standards to be set as close as possible to the MCLG with use of the best technology available. For some contaminants, EPA also establishes secondary maximum contaminant levels (SMCLs), which are nonenforceable guidelines for managing drinking water for aesthetic, cosmetic, or technical effects related to public acceptance of drinking water.

Fluoride is one of the natural contaminants found in public drinking water supplies regulated by EPA. In 1986, an MCLG of 4 milligrams per liter (mg/L) and an SMCL of 2 mg/L were established for fluoride, and an MCL of 4 mg/L was promulgated. It is important to make the distinction that EPA's standards are guidelines for restricting the amount of naturally occurring fluoride in drinking water; they are not recommendations about the practice of adding fluoride to public drinking-water systems (see below). In this report, the National Research Council's (NRC's) Committee on Fluoride in Drinking Water reviews the nature of the human health risks from fluoride, estimates exposures to the general public from drinking water and other sources, and provides an assessment of the adequacy of the MCLG for protecting public health from adverse health effects from fluoride and of the SMCL for protecting against cosmetic effects. Assessing the efficacy of fluoride in preventing dental caries is not covered in this report.

This chapter briefly reviews the sources of fluoride in drinking water, states the task the committee addressed, sets forth the committee's activities and deliberative process in developing the report, and describes the organization of the report.

FLUORIDE IN DRINKING WATER

Fluoride may be found in drinking water as a natural contaminant or as an additive intended to provide public health protection from dental caries (artificial water fluoridation). EPA's drinking water standards are restrictions on the amount of naturally occurring fluoride allowed in public water systems, and are not recommendations about the practice of water fluoridation. Recommendations for water fluoridation were established by the U.S. Public Health Service, and different considerations were factored into how those guidelines were established.

Natural

Fluoride occurs naturally in public water systems as a result of runoff from weathering of fluoride-containing rocks and soils and leaching from soil into groundwater. Atmospheric deposition of fluoride-containing emissions from coal-fired power plants and other industrial sources also contributes to amounts found in water, either by direct deposition or by deposition to soil and subsequent runoff into water. Of the approximately 10 million people with naturally fluoridated public water supplies in 1992, around 6.7 million had fluoride concentrations less than or equal to 1.2 mg/L (CDC 1993). Approximately 1.4 million had natural fluoride concentrations between 1.3 and 1.9 mg/L, 1.4 million had between 2.0 and 3.9 mg/L, and 200,000 had concentrations equal to or exceeding 4.0 mg/L. Exceptionally high concentrations of fluoride in drinking water are found in areas of Colorado (11.2 mg/L), Oklahoma (12.0 mg/L), New Mexico (13.0 mg/L), and Idaho (15.9 mg/L).

Areas of the United States with concentrations of fluoride in drinking water greater than 1.3 mg/L are all naturally contaminated. As discussed below, a narrow concentration range of 0.7 to 1.2 mg/L is recommended when decisions are made to intentionally add fluoride into water systems. This lower range also occurs naturally in some areas of the United States. Information on the fluoride content of public water supplies is available from local water suppliers and local, county, or state health departments.

Artificial

Since 1945, fluoride has been added to many public drinking-water supplies as a public-health practice to control dental caries. The "optimal" concentration of fluoride in drinking water for the United States for the prevention of dental caries has been set at 0.7 to 1.2 mg/L, depending on the mean temperature of the locality (0.7 mg/L for areas with warm climates, where water consumption is expected to be high, and 1.2 mg/L for cool climates, where water consumption is low) (PHS 1991). The optimal range was determined by selecting concentrations that would maximize caries prevention and limit enamel fluorosis, a dose-related mottling of teeth that can range from mild discoloration of the surface to severe staining and pitting. Decisions about fluoridating a public drinking-water supply are made by state or local authorities. CDC (2002a) estimates that approximately 162 million people (65.8% of the population served by public water systems) received optimally fluoridated water in 2000.

The practice of fluoridating water supplies has been the subject of controversy since it began (see reviews by Nesin 1956; Wollan 1968; McClure 1970; Marier 1977; Hileman 1988). Opponents have questioned the motivation for and the safety of the practice; some object to it because it is viewed as being imposed on them by the states and as an infringement on their freedom of choice (Hileman 1988; Cross and Carton 2003). Others claim that fluoride causes various adverse health effects and question whether the dental benefits outweigh the risks (Colquhoun 1997). Another issue of controversy is the safety of the chemicals used to fluoridate water. The most commonly used additives are silicofluorides, not the fluoride salts used in dental products (such as sodium fluoride and stannous fluoride). Silicofluorides are one of the by-products from the manufacture of phosphate fertilizers. The toxicity database on silicofluorides is sparse and questions have been raised about the assumption that they completely dissociate in water and, therefore, have toxicity similar to the fluoride salts tested in laboratory studies and used in consumer products (Coplan and Masters 2001).

It also has been maintained that, because of individual variations in exposure to fluoride, it is difficult to ensure that the right individual dose to protect against dental caries is provided through large-scale water fluoridation. In addition, a body of information has developed that indicates the major anticaries benefit of fluoride is topical and not systemic (Zero et al. 1992; Rölla and Ekstrand 1996; Featherstone 1999; Limeback 1999a; Clarkson and McLoughlin 2000; CDC 2001; Fejerskov 2004). Thus, it has been argued that water fluoridation might not be the most effective way to protect the public from dental caries.

Public health agencies have long disputed these claims. Dental caries is a common childhood disease. It is caused by bacteria that colonize on tooth surfaces, where they ferment sugars and other carbohydrates, generating lactic acid and other acids that decay tooth enamel and form a cavity. If the cavity penetrates to the dentin (the tooth component under the enamel), the dental pulp can become infected, causing toothaches. If left untreated, pulp infection can lead to abscess, destruction of bone, and systemic infection (Cawson et al. 1982; USDHHS 2000). Various sources have concluded that water fluoridation has been an effective method for preventing dental decay (Newbrun 1989; Ripa 1993; Horowitz 1996; CDC 2001; Truman et al. 2002). Water fluoridation is supported by the Centers for Disease Control and Prevention (CDC) as one of the 10 great public health achievements in the United States, because of its role in reducing tooth decay in children and tooth loss in adults (CDC 1999). Each U.S. Surgeon General has endorsed water fluoridation over the decades it has been practiced, emphasizing that "[a] significant advantage of water fluoridation is that all residents of a community can enjoy its protective benefit. . . . A person's income level or ability to receive dental care is not a barrier to receiving fluoridation's health benefits" (Carmona 2004).

As noted earlier, this report does not evaluate nor make judgments about the benefits, safety, or efficacy of artificial water fluoridation. That practice is reviewed only in terms of being a source of exposure to fluoride.

HISTORY OF EPA'S REGULATION OF FLUORIDE

In 1975, EPA proposed an interim primary drinking-water regulation for fluoride of 1.4-2.4 mg/L. That range was twice the "optimal" range of 0.7-1.2 mg/L recommended by the U.S. Public Health Service for water fluoridation. EPA's interim guideline was selected to prevent the occurrence of objectionable enamel fluorosis, mottling of teeth that can be classified as mild, moderate, or severe. In general, mild cases involve the development of white opaque areas in the enamel of the teeth, moderate cases involve visible brown staining, and severe cases include yellow to brown staining and pitting and cracking of the enamel (NRC 1993). EPA considered objectionable enamel fluorosis to involve moderate to severe cases with dark stains and pitting of the teeth.

The history of EPA's regulation of fluoride is documented in 50 Fed. Reg. 20164 (1985). In 1981, the state of South Carolina petitioned EPA to exclude fluoride from the primary drinking-water regulations and to set only an SMCL. South Carolina contended that enamel fluorosis should be considered a cosmetic effect and not an adverse health effect. The American Medical Association, the American Dental Association, the Association of State and Territorial Dental Directors, and the Association of State and Territorial Health Officials supported the petition. After reviewing the issue, the U.S. Public Health Service concluded there was no evidence that fluoride in public water supplies has any adverse effects on dental health, as measured by loss of teeth or tooth function. U.S. Surgeon General C. Everett Koop supported that position. The National Drinking Water Advisory Council (NDWAC) recommended that enamel fluorosis should be the basis for a secondary drinking-water regulation. Of the health effects considered to be adverse, NDWAC found osteosclerosis (increased bone density) to be the most relevant end point for establishing a primary regulation.

EPA asked the U.S. Surgeon General to review the available data on the nondental effects of fluoride and to determine the concentrations at which adverse health effects would occur and an appropriate margin of safety to protect public health. A scientific committee convened by the surgeon general concluded that exposure to fluoride at 5.0 to 8.0 mg/L was associated with radiologic evidence of osteosclerosis. Osteosclerosis was considered to be not an adverse health effect but an indication of osseous changes that would be prevented if the maximum content of fluoride in drinking water did not exceed 4 mg/L. The committee further concluded that there was no scientific documentation of adverse health effects at 8 mg/L and lower; thus, 4 mg/L would provide a margin of safety. In 1984, the surgeon general concluded that osteosclerosis is not an adverse health effect and that crippling skeletal fluorosis was the most relevant adverse health effect when considering exposure to fluoride from public drinking-water supplies. He continued to support limiting fluoride concentrations to 2 mg/L to avoid objectionable enamel fluorosis (50 Fed. Reg. 20164 [1985]).

In 1984, NDWAC took up the issue of whether psychological and behavioral effects from objectionable enamel fluorosis should be considered adverse. The council concluded that the cosmetic effects of enamel fluorosis could lead to psychological and behavioral problems that affect the overall well-being of the individual. EPA and the National Institute of Mental Health convened an ad hoc panel of behavioral scientists to further evaluate the potential psychological effects of objectionable enamel fluorosis. The panel concluded that "individuals who have suffered impaired dental appearance as a result of moderate or severe fluorosis are probably at increased risk for psychological and behavioral problems or difficulties" (R. E. Kleck, unpublished report, Nov. 17, 1984, as cited in 50 Fed. Reg. 20164 [1985]). NDWAC recommended that the primary drinking-water guideline for fluoride be set at 2 mg/L (50 Fed. Reg. 20164 [1985]).

On the basis of its review of the available data and consideration of the recommendations of various advisory bodies, EPA set an MCLG of 4 mg/L on the basis of crippling skeletal fluorosis (50 Fed. Reg. 47,142 [1985]). That value was calculated from an estimated lowest-observed-adverse-effect level of 20 mg/day for crippling skeletal fluorosis, the assumption that adult water intake is 2 L per day, and the application of a safety factor of 2.5. This factor was selected by EPA to establish an MCLG that was in agreement with a recommendation from the U.S. Surgeon General. In 1986, the MCL for fluoride was promulgated to be the same as the MCLG of 4 mg/L (51 Fed. Reg. 11,396 [1986]).

EPA also established an SMCL for fluoride of 2 mg/L to prevent objectionable enamel fluorosis in a significant portion of the population (51 Fed. Reg. 11,396 [1986]). To set that guideline, EPA reviewed data on the incidence of moderate and severe enamel fluorosis and found that, at a fluoride concentration of 2 mg/L, the incidence of moderate fluorosis ranged from 0% to 15%. Severe cases appeared to be observed only at concentrations above 2.5 mg/L. Thus, 2 mg/L was considered adequate for preventing enamel fluorosis that would be cosmetically objectionable. EPA established the SMCL as an upper boundary guideline for areas that have high concentrations of naturally occurring fluoride. EPA does not regulate or promote the addition of fluoride to drinking water. If fluoride in a community water system exceeds the SMCL but not the MCL, a notice about potential risk of enamel fluorosis must be sent to all customers served by the system (40 CFR 141.208[2005]).

In the early 1990s, the NRC was asked to independently review the health effects of ingested fluoride and EPA's MCL. The NRC (1993) found EPA's MCL of 4 mg/L to be an appropriate interim standard. Its report identified inconsistencies in the fluoride toxicity database and gaps in knowledge. Accordingly, the NRC recommended research in the areas of fluoride intake, enamel fluorosis, bone strength and fractures, and carcinogenicity. A list of the specific recommendations from that report is provided in Box 1-1.

COMMITTEE'S TASK

The Safe Drinking Water Act requires that EPA periodically review existing standards for water contaminants. Because of that requirement and new research on fluoride, EPA's Office of Water requested that the NRC reevaluate the adequacy of the MCLG and SMCL for fluoride to protect public health. The NRC assigned this task to the standing Committee on Toxicology, and convened the Committee on Fluoride in Drinking Water. The committee was asked to review toxicologic, epidemiologic, and clinical data, particularly data published since 1993, and exposure data on orally ingested fluoride from drinking water and other sources (e.g., food, tooth-

BOX 1-1
Recommendations from NRC (1993) Report

Intake, Metabolism, and Disposition of Fluoride
- Determine and compare intake of fluoride from all sources, including fluoride-containing dental products, in communities with fluoridated and nonfluoridated water. That information would improve our understanding of trends in dental caries, enamel fluorosis, and possibly other disorders or diseases.
- Determine the effects of factors that affect human acid-base balance and urinary pH on the metabolic characteristics, balance, and tissue concentrations of fluoride.
- Determine the metabolic characteristics of fluoride in infants, young children, and the elderly.
- Determine prospectively the metabolic characteristics of fluoride in patients with progressive renal disease.
- Using preparative and analytical methods now available, determine soft-tissue fluoride concentrations and their relation to plasma fluoride concentrations. Consider the relation of tissue concentrations to variables of interest, including past fluoride exposure and age.
- Identify the compounds that compose the "organic fluoride pool" in human plasma and determine their sources, metabolic characteristics, fate, and biological importance.

Enamel Fluorosis
- Identify sources of fluoride during the critical stages of tooth development in childhood and evaluate the contribution of each source to enamel fluorosis.
- Conduct studies on the relation between water fluoride concentrations and enamel fluorosis in various climatic zones.
- Determine the lowest concentration of fluoride in toothpaste that produces acceptable cariostasis.
- Conduct studies on the contribution of ingested fluoride and fluoride applied topically to teeth to prevent caries.

Bone Fracture
- Conduct a workshop to evaluate the advantages and disadvantages of the various doses, treatments, laboratory animal models, weight-bearing versus non-weight-bearing bones, and testing methods for bone strength that can be used to determine the effects of fluoride on bone.
- Conduct additional studies of hip and other fractures in geo-

continued

> **BOX 1-1
> Continued**
>
> graphic areas with high and low fluoride concentration in drinking water and make use of individual information about water consumption. These studies also should collect individual information on bone fluoride concentrations and intake of fluoride from all sources, as well as reproductive history, past and current hormonal status, intake of dietary and supplemental calcium and other cations, bone density, and other factors that might influence the risk of hip fracture.
>
> **Carcinogenicity**
> - Conduct one or more highly focused, carefully designed analytical studies (case control or cohort) of the cancer sites that are most highly suspect, based on data from animal studies and the few suggestions of a carcinogenic effect reported in the epidemiologic literature. Such studies should be designed to gather information on individual study subjects so that adjustments can be made for the potential confounding effects of other risk factors in analyses of individuals. Information on fluoride exposure from sources other than water must be obtained, and estimates of exposure from drinking water should be as accurate as possible. In addition, analysis of fluoride in bone samples from patients and controls would be valuable in inferring total lifetime exposures to fluoride. Among the disease outcomes that warrant separate study are osteosarcomas and cancers of the buccal cavity, kidney, and bones and joints.

paste, dental rinses). On the basis of those reviews, the committee was asked to evaluate independently the scientific basis of EPA's MCLG of 4 mg/L and SMCL of 2 mg/L in drinking water and the adequacy of those guidelines to protect children and others from adverse health effects. The committee was asked to consider the relative contribution of various fluoride sources (e.g., food, dental-hygiene products) to total exposure. The committee also was asked to identify data gaps and make recommendations for future research relevant to setting the MCLG and SMCL for fluoride. Addressing questions of economics, risk-benefit assessment, and water-treatment technology was not part of the committee's charge.

The committee is aware that some readers expect this report to make a determination about whether public drinking-water supplies should be fluoridated. That expectation goes beyond the committee's charge. As noted above, the MCLG and SMCL are guidelines for areas where fluoride concentrations are naturally high. They are designed with the intent to protect the public from adverse health effects related to fluoride exposure and not as guidelines to provide health benefits.

COMMITTEE'S APPROACH

To accomplish its task, the committee held six meetings between August 2003 and June 2005. The first two meetings involved data-gathering sessions that were open to the public. The committee heard presentations from EPA, CDC, individuals involved in fluoride research, fluoridation supporters, and antifluoridation proponents. The committee also reviewed a large body of written material on fluoride, primarily focusing on research that was completed after publication of the 1993 NRC report. The available data included numerous research articles, literature reviews, position papers, and unpublished data submitted by various sources, including the public. Each paper and submission was evaluated case by case on its own merits.

Unless otherwise noted, the term fluoride is used in this report to refer to the inorganic, ionic form. Most of the nonepidemiologic studies reviewed involved exposure to a specified fluoride compound, usually sodium fluoride. Various units of measure are used to express exposure to fluoride in terms of exposure concentrations and internal dose (see Table 1-1 and Chapter 3). To the extent possible, the committee has tried to use units that allow for easy comparisons.

In this report, the committee updates information on the issues considered in the 1993 review—namely, data on pharmacokinetics; dental effects; skeletal effects; reproductive and developmental effects; neurological and behavioral effects; endocrine effects; gastrointestinal, renal, hepatic, and immune effects; genotoxicity; and carcinogenicity. More inclusive reviews are provided on effects to the endocrine and central nervous systems, because the previous NRC review did not give those effects as much attention. The committee used a general weight-of-evidence approach to evaluate the literature, which involved assessing whether multiple lines of evidence

TABLE 1-1 Units Commonly Used for Measuring Fluoride

Medium	Unit	Equivalent
Water	1 ppm	1 mg/L
Plasma	1 µmol/L	0.019 mg/L
Bone ash	1 ppm	1 mg/kg
	1%	10,000 mg/kg

ABBREVIATIONS: mg/kg, milligrams per kilogram; mg/L, milligrams per liter; µmol/L, micromoles per liter; ppm, parts per million.

indicate a human health risk. This included an evaluation of in vitro assays, animal research, and human studies (conducted in the United States and other countries). Positive and negative results were considered, as well as mechanistic and nonmechanistic information. The collective evidence was considered in perspective with exposures likely to occur from fluoride in drinking water at the MCLG or SMCL.

In evaluating the effects of fluoride, consideration is given to the exposure associated with the effects in terms of dose and time. Dose is a simple variable (such as mg/kg/day), and time is a complex variable because it involves not only the frequency and duration of exposure but also the persistence of the agent in the system (kinetics) and the effect produced by the agent (dynamics). Whether the key rate-limiting events responsible for the adverse effect are occurring in the kinetic or in the dynamic pathway is important in understanding the toxicity of a chemical and in directing future research (see Rozman and Doull 2000). The committee also attempts to characterize fluoride exposures from various sources to different subgroups within the general population and to identify subpopulations that might be particularly susceptible to the effects of fluoride.

STRUCTURE OF THE REPORT

The remainder of this report is organized into 10 chapters. Chapter 2 characterizes the general public's exposure to fluoride from drinking water and other sources. Chapter 3 provides a description of the chemistry of fluoride and pharmacokinetic information that was considered in evaluating the toxicity data on fluoride. In Chapters 4-9, the committee evaluates the scientific literature on adverse effects of fluoride on teeth, the musculoskeletal system, reproduction and development, the nervous system, the endocrine system, the gastrointestinal system, the kidneys, the liver, and the immune system. Chapter 10 evaluates the genotoxic and carcinogenic potential of fluoride. Finally, Chapter 11 provides an assessment of the most significant health risks from fluoride in drinking water and its implications for the adequacy of EPA's MCLG and SMCL for protecting the public.

2

Measures of Exposure to Fluoride in the United States

The major sources of internal exposure of individuals to fluorides are the diet (food, water, beverages) and fluoride-containing dental products (toothpaste, fluoride supplements). Internal exposure to fluorides also can occur from inhalation (cigarette smoke, industrial emissions), dermal absorption (from chemicals or pharmaceuticals), ingestion or parenteral administration of fluoride-containing drugs, and ingestion of fluoride-containing soil. Information on the pharmacokinetics of fluoride are provided in Chapter 3.

The National Research Council's (NRC's) 1993 review of the health effects of ingested fluoride reported estimates of average daily fluoride intake from the diet of 0.04-0.07 milligrams per kilogram (mg/kg) of body weight for young children in an area with fluoridated water (fluoride concentration in drinking water, 0.7-1.2 mg per liter [L]; NRC 1993). Dietary intake of fluoride by adults in an area with fluoridated water was variously estimated to be between 1.2 and 2.2 mg/day (0.02-0.03 mg/kg for a 70-kg adult). The fluoride intake from toothpaste or mouth rinse by children with good control of swallowing, assuming twice-a-day use, was estimated to equal the intake from food, water, and beverages. The review acknowledged that "substantially" higher intakes of fluoride from consumption of fluoridated water would result for individuals such as outdoor laborers in warm climates or people with high-urine-output disorders, but these intakes were not quantified. Similarly, children and others with poor control of swallowing could have intakes of fluoride from dental products that exceed the dietary intakes, but these intakes also were not quantified. Other factors cited as affecting individual fluoride intakes include changes in the guidelines for fluoride supplementation and use of bottled water or home water purification systems rather than fluoridated municipal water. The NRC (1993) recommended further research to "determine and compare the intake of fluoride from all sources, including fluoride-containing dental products, in fluoridated and nonfluoridated communities."

This chapter provides a review of the available information on fluoride exposures in the United States, including sources of fluoride exposure, intakes from various fluoride sources, and factors that could affect individual exposures to fluorides. Population subgroups with especially high exposures are discussed. The major emphasis of this chapter is on chronic exposure rather than acute exposure. The use of biomarkers as alternative approaches to estimation of actual individual exposures is also discussed.

In practice, most fluorine added to drinking water is in the form of fluosilicic acid (fluorosilicic acid, H_2SiF_6) or the sodium salt (sodium fluosilicate, Na_2SiF_6), collectively referred to as fluorosilicates (CDC 1993); for some smaller water systems, fluoride is added as sodium fluoride (NaF). Fluoride in toothpaste and other dental products is usually present as sodium fluoride (NaF), stannous fluoride (SnF_2), or disodium monofluorophosphate (Na_2PO_3F). Fluorine-containing pesticides and pharmaceuticals also contribute to total fluorine exposures and are considered separately. Fluoride in food and drinking water usually is considered in terms of total fluorine content, assumed to be present entirely as fluoride ion (F^-). Information on exposures to fluorosilicates and aluminofluorides is also included.

SOURCES OF FLUORIDE EXPOSURE

Drinking Water

General Population

The major dietary source of fluoride for most people in the United States is fluoridated municipal (community) drinking water, including water consumed directly, food and beverages prepared at home or in restaurants from municipal drinking water, and commercial beverages and processed foods originating from fluoridated municipalities. On a mean per capita basis, community (public or municipal) water constitutes 75% of the total water ingested in the United States; bottled water constitutes 13%, and other sources (e.g., wells and cisterns) constitute 10% (EPA 2000a). Municipal water sources that are not considered "fluoridated" could contain low concentrations of naturally occurring fluoride, as could bottled water and private wells, depending on the sources.

An estimated 162 million people in the United States (65.8% of the population served by public water systems) received "optimally fluoridated"[1] water in 2000 (CDC 2002a). This represents an increase from 144 million (62.1%) in 1992. The total number of people served by public water systems in the United States is estimated to be 246 million; an estimated 35 million people obtain water from other sources such as private wells (CDC 2002a,b). The U.S. Environmental Protection Agency (EPA) limits the fluoride that can be present in public drinking-water supplies to 4 mg/L (maximum contaminant level, or MCL) to protect against crippling skeletal fluorosis, with a secondary maximum contaminant level (SMCL) of 2 mg/L to protect against objectionable enamel fluorosis (40CFR 141.62(b)[2001], 40CFR 143.3[2001]).

Of the 144 million people with fluoridated public water supplies in 1992, approximately 10 million (7%) received naturally fluoridated water, the rest had artificially fluoridated water (CDC 2002c). Of the population with artificially fluoridated water in 1992, more than two-thirds had a water fluoride concentration of 1.0 mg/L, with almost one-quarter having lower concentrations and about 5% having concentrations up to 1.2 mg/L (CDC 1993; see Appendix B).

Of the approximately 10 million people with naturally fluoridated public water supplies in 1992, approximately 67% had fluoride concentrations ≤ 1.2 mg/L (CDC 1993; see Appendix B). Approximately 14% had fluoride concentrations between 1.3 and 1.9 mg/L and another 14% had between 2.0 and 3.9 mg/L; 2% (just over 200,000 persons) had natural fluoride concentrations equal to or exceeding 4.0 mg/L.[2] Water supplies that exceeded 4.0 mg/L ranged as high as 11.2 mg/L in Colorado, 12.0 mg/L in Oklahoma, 13.0 mg/L in New Mexico, and 15.9 mg/L in Idaho (see Appendix B, Table B-3).[3] States with the largest populations receiving water supplies with fluoride at ≥ 4.0 mg/L included Virginia (18,726 persons, up to 6.3 mg/L), Oklahoma (18,895 persons, up to 12.0 mg/L), Texas (36,863 persons, up to 8.8 mg/L), and South Carolina (105,618 persons, up to 5.9 mg/L).

Little information is available on the fluoride content of private water sources, but the variability can reasonably be expected to be high and to

---

[1]The term optimally fluoridated water means a fluoride level of 0.7-1.2 mg/L; water fluoride levels are based on the average maximum daily air temperature of the area (see Appendix B).

[2]More recently (2000), CDC has estimated that 850,000 people are served by public water supplies containing fluoride in excess of 2 mg/L; of these, 152,000 people receive water containing fluoride in excess of 4 mg/L (unpublished data from CDC as reported in EPA 2003a). Based on analytical data from 16 states, EPA (2003a) estimates that 1.5-3.3 million people nationally are served by public water supplies with fluoride concentrations exceeding 2 mg/L; of these 118,000-301,000 people receive water with fluoride concentrations greater than 4 mg/L.

[3]High-fluoride municipal waters are generally found in regions that have high fluoride concentrations in the groundwater or in surface waters. ATSDR (2003) has reviewed fluoride concentrations in environmental media, including groundwater and surface water. Fleischer (1962) and Fleischer et al. (1974) reported fluoride concentrations in groundwater by county for the coterminous United States.

depend on the region of the country. Fluoride measured in well water in one study in Iowa ranged from 0.06 to 7.22 mg/L (mean, 0.45 mg/L); home-filtered well water contained 0.02-1.00 mg/L (mean, 0.32 mg/L; Van Winkle et al. 1995). Hudak (1999) determined median fluoride concentrations for 237 of 254 Texas counties (values were not determined for counties with fewer than five observations). Of the 237 counties, 84 have median groundwater fluoride concentrations exceeding 1 mg/L; of these, 25 counties exceed 2 mg/L and five exceed 4 mg/L. Residents in these areas (or similar areas in other states) who use groundwater from private wells are likely to exceed current guidelines for fluoride intake.

Duperon et al. (1995) pointed out that fluoride concentrations reported by local water suppliers can be substantially different from concentrations measured in water samples obtained in homes. Use of home water filtration or purification systems can reduce the fluoride concentration in community water by 13% to 99%, depending on the type of system (Duperon et al. 1995; Van Winkle et al. 1995; Jobson et al. 2000). Distillation or reverse osmosis can remove nearly all the fluoride. The extent of use of home water filtration or purification systems nationally is not known but obviously would affect the fluoride intake for people using such systems. Van Winkle et al. (1995) reported that 11% of their study population (in Iowa) used some type of home filtration either for well water or for public water.

Fluoride concentrations in bottled water[4] are regulated by law to a maximum of 1.4-2.4 mg/L if no fluoride is added and a maximum of 0.8-1.7 mg/L if fluoride is added (the applicable value within the range depends on the annual average of maximum daily air temperatures at the location of retail sale; 21CFR 165.110[2003]). Maximum fluoride concentrations for imported bottled water are 1.4 mg/L if no fluoride is added and 0.8 mg/L if fluoride is added (21CFR 165.110[2003]). Fluoride concentrations are required on labels in the United States only if fluoride is added. Fluoride concentrations listed on labels or in chemical analyses available on the Internet for various brands range from 0 to 3.6 mg/L (Bartels et al. 2000; Johnson and DeBiase 2003; Bottled Water Web 2004); of those without added fluoride, most are below 0.6 mg/L. Most brands appear to list fluoride content only if they are specifically advertising the fact that their water is fluoridated; fluoride concentrations of these brands range from 0.5 to 0.8 mg/L (for "nursery" or "infant" water) up to 1.0 mg/L. Several reports indicate

---

[4]The term "bottled water" applies to water intended for human consumption, containing no added ingredients besides fluoride or appropriate antimicrobial agents; the regulations apply to bottled water, drinking water, artesian water, artesian well water, groundwater, mineral water, purified water, demineralized water, deionized water, distilled water, reverse osmosis water, purified drinking water, demineralized drinking water, deionized drinking water, distilled drinking water, reverse osmosis drinking water, sparkling water, spring water, and well water (21CFR 165.110[2003]).

that fluoride concentrations obtained from the manufacturer or stated on labels for bottled waters might not be accurate (Weinberger 1991; Toumba et al. 1994; Bartels et al. 2000; Lalumandier and Ayers 2000; Johnson and DeBiase 2003; Zohouri et al. 2003).

Measured fluoride concentrations in bottled water sold in the United States have varied from 0 to 1.36 mg/L (Nowak and Nowak 1989; Chan et al. 1990; Stannard et al. 1990; Van Winkle et al. 1995; Bartels et al. 2000; Lalumandier and Ayers 2000; Johnson and DeBiase 2003). Van Winkle et al. (1995) reported a mean of 0.18 mg/L for 78 commercial bottled waters in Iowa. Johnson and DeBiase (2003) more recently reported values ranging from 0 to 1.2 mg/L for 65 bottled waters purchased in West Virginia, with 57 brands having values below 0.6 mg/L. Measured fluoride concentrations in bottled waters in other countries have similar ranges: 0.05-4.8 mg/L in Canada (Weinberger 1991), 0.10-0.80 mg/L in the United Kingdom (Toumba et al. 1994), and 0.01-0.37 mg/L more recently in the United Kingdom (Zohouri et al. 2003).[5] Bartels et al. (2000) found significant variation in fluoride concentrations among samples of the same brand with different bottling dates purchased in the same city. In general, distilled and purified (reverse osmosis) waters contain very low concentrations of fluoride; drinking water (often from a municipal tap) and spring water vary with their source, as do mineral waters, which can be very low or very high in fluoride. Most spring water sold in the United States probably has a low fluoride content (<0.3 mg/L). Typical fluoride concentrations in various types of drinking water in the United States are summarized in Table 2-1.

Average per capita ingestion of community or municipal water is estimated to be 927 mL/day (EPA 2000a; see Appendix B[6]). The estimated 90th percentile of the per capita ingestion of community water from that survey is 2.016 L/day. Estimated intakes by those actually consuming community water (excluding people with zero ingestion of community water) are higher, with a mean of 1.0 L/day and a 90th percentile of 2.069 L/day (EPA 2000a). Thus, if national estimates of water intake (see Appendix B)

---

[5]The European Commission has set a maximum limit of 5.0 mg/L for fluoride in natural mineral waters, effective January 1, 2008 (EC 2003). In addition, natural mineral waters with a fluoride concentration exceeding 1.5 mg/L must be labeled with the words "contains more than 1.5 mg/L of fluoride: not suitable for regular consumption by infants and children under 7 years of age," and for all natural mineral waters, the actual fluoride content is to be listed on the label. England has essentially the same requirements (TSO 2004), applicable to all bottled waters (natural mineral waters, spring water, and bottled drinking water).

[6]As described more fully in Appendix B, the values from EPA (2000a) are from a short-term survey of more than 15,000 individuals in the United States. Although these values are considered reasonable indicators both of typical water consumption and of the likely range of water consumption on a long-term basis, they should not be used by themselves to predict the number of individuals or percentage of the population that consumes a given amount of water on a long-term basis.

TABLE 2-1 Typical Fluoride Concentrations of Major Types of Drinking Water in the United States

Source	Range, mg/L[a]
Municipal water (fluoridated)	0.7-1.2
Municipal water (naturally fluoridated)	0.7-4.0+
Municipal water (nonfluoridated)	<0.7
Well water	0-7+
Bottled water from municipal source	0-1.2
Spring water	0-1.4 (usually <0.3)
Bottled "infant" or "nursery" water	0.5-0.8
Bottled water with added fluoride[b]	0.8-1.0
Distilled or purified water	<0.15

[a]See text for relevant references.
[b]Other than "infant" or "nursery" water.

are assumed to be valid for the part of the population with fluoridated water supplies, the intake of fluoride for a person with average consumption of community water (1 L/day) in a fluoridated area ranges from 0.7 to 1.2 mg/day, depending on the area. A person with consumption of community water equivalent to the 90th percentile in that survey (2.069 L/day) would have a fluoride intake between 1.4 and 2.5 mg/day, from community water alone. Table 2-2 provides examples of fluoride intake by typical and high consumers of municipal water by age group.

The estimates of water consumption described in Appendix B are in keeping with recently published "adequate intake" values for total water consumption (including drinking water, all beverages, and moisture in food; IOM 2004; see Appendix B, Table B-10). Note that these estimates are national values; the range of values for optimal fluoridation was intended to account for expected regional differences in water consumption due to regional temperature differences (see Appendix B). A separate study based on the same data used by EPA (2000a) found no strong or consistent association between water intake and month or season (Heller et al. 1999). Another recent study of American children aged 1-10 years also found no significant relationship between water consumption and mean temperature in modern conditions (perhaps due to artificial temperature regulation) and suggested that the temperature-related guidelines for fluoride concentrations in drinking water be reevaluated (Sohn et al. 2001).

Actual intakes of fluoride from drinking water by individuals depend on their individual water intakes, the source or sources of that water, and the use of home water purification or filtration systems. As described earlier, fluoride concentrations in community water might vary from their reported concentrations; fluoride content of bottled water also varies considerably with brand or source, with packaging date for a given brand, and from

TABLE 2-2 Examples of Fluoride Intake from Consumption of Community (Municipal) Water by People Living in Fluoridated Areas[a]

	Typical Consumers[b]				High Consumers[c]			
	Water Consumption		Fluoride Intake[d]		Water Consumption		Fluoride Intake[d]	
	mL/day	mL/kg/day	mg/day	mg/kg/day	mL/day	mL/kg/day	mg/day	mg/kg/day
U.S. population (total)	1,000	17	0.7-1.2	0.012-0.020	2,100	33	1.5-2.5	0.023-0.040
All infants (<1 year)[e]	500	60	0.35-0.6	0.042-0.072	950	120	0.67-1.1	0.084-0.14
Children 1-2 years	350	26	0.25-0.42	0.018-0.031	700	53	0.49-0.84	0.037-0.064
Children 3-5 years	450	23	0.32-0.54	0.016-0.028	940	45	0.66-1.1	0.032-0.054
Children 6-12 years	500	16	0.35-0.6	0.011-0.019	1,000	33	0.7-1.2	0.023-0.040
Youths 13-19 years	800	12	0.56-0.96	0.0084-0.014	1,700	26	1.2-2.0	0.018-0.031
Adults 20-49 years	1,100	16	0.77-1.3	0.011-0.019	2,200	32	1.5-2.6	0.022-0.038
Adults 50+ years	1,200	17	0.84-1.4	0.012-0.020	2,300	32	1.6-2.8	0.022-0.038
Females 13-49 years[f]	980	15	0.69-1.2	0.011-0.018	2,050	32	1.4-2.5	0.022-0.038

[a]Based on consumption data described in Appendix B for people actually consuming community (municipal) water.
[b]Based on a typical consumption rate of community (municipal) water for the age group.
[c]Based on a reasonably high (but not upper bound) consumption rate of community (municipal) water for the age group; some individual exposures could be higher.
[d]Based on fluoride concentrations of 0.7-1.2 mg/L.
[e]Includes any infant, nursing or nonnursing, who consumes at least some community water; these infants may be fed primarily breast milk, ready-to-feed formula (to which no water is normally added), or formula prepared from concentrate (which requires addition of water).
[f]Women of childbearing age.

information (if any) given on the labels or provided by the manufacturer. Private water sources (e.g., wells and cisterns) probably are even more variable in fluoride content, with some regions of the country being especially high and others very low. A number of authors have pointed out the difficulty doctors and dentists face in ascertaining individual fluoride intakes, just from drinking water (from all sources), for the purpose of prescribing appropriate fluoride supplementation (Nowak and Nowak 1989; Chan et al. 1990; Stannard et al. 1990; Levy and Shavlick 1991; Weinberger 1991; Dillenberg et al. 1992; Jones and Berg 1992; Levy and Muchow 1992; Toumba et al. 1994; Duperon et al. 1995; Van Winkle et al. 1995; Heller et al. 1999; Bartels et al. 2000; Lalumandier and Ayers 2000; Johnson and DeBiase 2003; Zohouri et al. 2003).

High Intake Population Subgroups

EPA, in its report to Congress on sensitive subpopulations (EPA 2000b), defines sensitive subpopulations in terms of either their response (more severe response or a response to a lower dose) or their exposure (greater exposure than the general population). Hence, it is appropriate to consider those population subgroups whose water intake is likely to be substantially above the national average for the corresponding sex and age group. These subgroups include people with high activity levels (e.g., athletes, workers with physically demanding duties, military personnel); people living in very hot or dry climates, especially outdoor workers; pregnant or lactating women; and people with health conditions that affect water intake. Such health conditions include diabetes mellitus, especially if untreated or poorly controlled; disorders of water and sodium metabolism, such as diabetes insipidus; renal problems resulting in reduced clearance of fluoride; and short-term conditions requiring rapid rehydration, such as gastrointestinal upsets or food poisoning (EPA 2000a). (While the population sample described in Appendix B [Water Ingestion and Fluoride Intakes] included some of these individuals, the study did not attempt to estimate means or distributions of intake for these specific subgroups.)

As shown in Appendix B (Tables B-4 to B-9), some members of the U.S. population could have intakes from community water sources of as much as 4.5-5 L/day (as high as 80 mL/kg/day for adults). Some infants have intakes of community water exceeding 200 mL/kg/day. Heller et al. (1999), using the same data set as EPA (2000a), reported that 21 of 14,640 people (of all ages) had water intakes over 6 standard deviations from the mean (greater than 249 mL/kg/day). Whyte et al. (2005) describe an adult woman who consistently consumed 1-2 gallons (3.8-7.6 L) of fluid per day (instant tea made with well water); no specific reason for her high fluid consumption is given.

Fluid requirements of athletes, workers, and military personnel depend on the nature and intensity of the activity, the duration of the activity, and the ambient temperature and humidity. Total sweat losses for athletes in various sports can range from 200 to 300 mL/hour to 2,000 mL/hour or more (Convertino et al. 1996; Horswill 1998; Cox et al. 2002; Coyle 2004). Most recommendations on fluid consumption for athletes are concerned with matching fluid replacement to fluid losses during the training session or competition to minimize the detrimental effects of dehydration on athletic performance (Convertino et al. 1996; Horswill 1998; Coris et al. 2004; Coyle 2004). Depending on the nature of the sport or training session, the ease of providing fluid, and the comfort of the athlete with respect to content of the gastrointestinal tract, fluid intake during exercise is often only a fraction (e.g., one-half) of the volume lost, and losses of 2% of body weight or more might occur during an exercise session in spite of fluid consumption during the session (Convertino et al. 1996; Cox et al. 2002; Coris et al. 2004; Coyle 2004).

Total daily fluid consumption by athletes generally is not reported; for many athletes, it is probably on the order of 5% of body weight (50 mL/kg/day) or more to compensate for urinary and respiratory losses as well as sweat losses. For example, Crossman (2003) described a professionally prepared diet plan for a major league baseball player that includes 26 cups (6.2 L) of water or sports drink on a workout day and 19 cups (4.5 L) on an off-day; this is in addition to 9-11 cups (2.1-2.6 L) of milk, fruit juice, and sports drink with meals and scheduled snacks (total fluid intake of 6.8-8.8 L/day, or 52-67 mL/kg/day for a 132-kg player[7]). While some players and teams probably use bottled or distilled water, most (especially at the amateur and interscholastic levels) probably use local tap water; also, sports drinks might be prepared (commercially or by individuals) with tap water.

The U.S. Army's policy on fluid replacement for warm-weather training calls for 0.5-1 quart/hour (0.47-0.95 L/hour), depending on the temperature, humidity, and type of work (Kolka et al. 2003; USASMA 2003). In addition, fluid intake is not to exceed 1.5 quarts/hour (1.4 liter/hour) or 12 quarts/day (11.4 L/day). The Army's planning factor for individual tap water consumption ranges from 1.5 gallons/day (5.7 L/day) for temperate conditions to 3.0 gallons/day (11.4 L/day) for hot conditions (U.S. Army 1983). Hourly intake can range from 0.21 to 0.65 L depending on the temperature (McNall and Schlegel 1968), and daily intake among physically active individuals can range from 6 to 11 L (U.S. Army 1983, cited by EPA 1997). Nonmilitary outdoor workers in hot or dry climates probably would have similar needs.

---

[7]The player's weight was obtained from the 2003 roster of the Cleveland Indians baseball team (http://cleveland.indians.mlb.com).

Water intakes for pregnant and lactating women are listed separately in Appendix B (Tables B-4 to B-9). Total water intake for pregnant women does not differ greatly from that for all adult females (Table B-9), while total water consumption by lactating women is generally higher. For the highest consumers among lactating women, consumption rates approximate those for athletes and workers (50-70 mL/kg/day).

Diabetes mellitus and diabetes insipidus are both characterized by high water intakes and urine volumes, among other things (Beers and Berkow 1999; Eisenbarth et al. 2002; Robinson and Verbalis 2002; Belchetz and Hammond 2003). People with untreated or poorly controlled diabetes mellitus would be expected to have substantially higher fluid intakes than nondiabetic members of the population. The American Diabetes Association (2004) estimates that 18.2 million people in the United States (6.3% of the population) have diabetes mellitus and that 5.2 million of these are not aware they have the disease. Other estimates range from 16 to 20 million people in the United States, with up to 50% undiagnosed (Brownlee et al. 2002; Buse et al. 2002).

Diabetes insipidus, or polyuria, is defined as passage of large volumes of urine, in excess of about 2 L/m^2/day (approximately 150 mL/kg/day at birth, 110 mL/kg/day at 2 years, and 40 mL/kg/day in older children and adults) (Baylis and Cheetham 1998; Cheetham and Baylis 2002). Diabetes insipidus includes several types of disease distinguished by cause, including both familial and acquired disorders (Baylis and Cheetham 1998; Cheetham and Baylis 2002; Robinson and Verbalis 2002). Water is considered a therapeutic agent for diabetes insipidus (Beers and Berkow 1999; Robinson and Verbalis 2002); in addition, some kinds of diabetes insipidus can be treated by addressing an underlying cause or by administering vasopressin (antidiuretic hormone) or other agents to reduce polyuria to a tolerable level. The Diabetes Insipidus Foundation (2004) estimates the number of diabetes insipidus patients in the United States at between 40,000 and 80,000.

Someone initially presenting with central or vasopressin-sensitive diabetes insipidus might ingest "enormous" quantities of fluid and may produce 3-30 L of very dilute urine per day (Beers and Berkow 1999) or up to 400 mL/kg/day (Baylis and Cheetham 1998). Most patients with central diabetes insipidus have urine volumes of 6-12 L/day (Robinson and Verbalis 2002). Patients with primary polydipsia might ingest and excrete up to 6 L of fluid per day (Beers and Berkow 1999). Pivonello et al. (1998) listed water intakes of 5.5-8.6 L/day for six adults with diabetes insipidus who did not take vasopressin and 1.4-2.5 L/day for 12 adults who used a vasopressin analogue. An estimated 20% to 40% of patients on lithium therapy have a urine volume > 2.5 L/day, and up to 12% have frank nephrogenic diabetes insipidus characterized by a urine volume > 3 L/day (Mukhopadhyay et al. 2001).

Five papers described enamel fluorosis in association with diabetes insipidus or polydipsia (Table 2-3). Two of the papers described cases of enamel fluorosis in the United States resulting from fluoride concentrations of 1, 1.7, or 2.6 mg/L in drinking water (Juncos and Donadio 1972; Greenberg et al. 1974). The two individuals drinking water with fluoride at 1.7 and 2.6 mg/L also had roentgenographic bone changes consistent with "systemic fluorosis"[8] (Juncos and Donadio 1972). These patients and four other renal patients in the U.S. "in whom fluoride may have been the cause of detectable clinical and roentgenographic effects" were also reported by Johnson et al. (1979); most of the patients had urine volumes exceeding 3 L/day and drinking water with fluoride concentrations around 1.7-3 mg/L.

Moderate and severe enamel fluorosis have been reported in diabetes insipidus patients in other countries with drinking water containing fluoride at 0.5 mg/L (Klein 1975) or 1 mg/L (Seow and Thomsett 1994), and severe enamel fluorosis with skeletal fluorosis has been reported with fluoride at 3.4 mg/L (Mehta et al. 1998). Greenberg et al. (1974) recommended that children with any disorder that gives rise to polydipsia and polyuria[9] be supplied a portion of their water from a nonfluoridated source.

Table 2-4 provides examples of fluoride intake by members of several population subgroups characterized by above-average water consumption (athletes and workers, patients with diabetes mellitus or diabetes insipidus). It should be recognized that, for some groups of people with high water intakes (e.g., those with a disease condition or those playing indoor sports such as basketball or hockey), there probably will be little correlation of water intake with outdoor temperature—such individuals in northern states would consume approximately the same amounts of water as their counterparts in southern states. However, fluoridation still varies from state to state (Appendix B), so that some individuals could consume up to 1.7 times as much as others for the same water intake (1.2 versus 0.7 mg/L).

Background Food

Measured fluoride in samples of human breast milk is very low. Dabeka et al. (1986) found detectable concentrations in only 92 of 210 samples (44%) obtained in Canada, with fluoride ranging from <0.004 to 0.097 mg/L. The mean concentration in milk from mothers in fluoridated

---

[8]These two individuals also had impaired renal function, which could have increased their retention of fluoride (see Chapter 3).

[9]Greenberg et al. (1974) listed "central diabetes insipidus, psychogenic water ingestion, renal medullary disease, including hypercalcemic nephropathy, hypokalemic nephropathy and anatomic and vascular disturbances and those diseases causing solute diuresis" as disorders associated with "excessive" consumption of water and therefore the possibility of "fluoride toxicity in a community with acceptable fluoride concentration."

TABLE 2-3 Case Reports of Fluorosis in Association with Diabetes Insipidus or Polydipsia

Study Subjects	Exposure Conditions	Comments	Reference
(a) 18-year-old boy, 57.4 kg (b) 17-year-old girl, 45.65 kg (United States)	(a) "high" intake of well water containing fluoride at 2.6 mg/L since early childhood; current intake, 7.6 L/day (0.34 mg/kg/day) (b) "high" intake of water containing fluoride at 1.7 mg/L since infancy; current intake, 4 L/day (0.15 mg/kg/day)	Enamel fluorosis and roentgenographic bone changes consistent with "systemic fluorosis," attributed to the combination of renal insufficiency and polydipsia (the latter resulting from the renal disease); reported by the Mayo Clinic	Juncos and Donadio 1972
2 boys (ages 10 and 11) with familial nephrogenic diabetes insipidus (United States)	Fluoridated communities in the U.S. (1 mg/L); one child since birth, one since age 4; fluid intake ranged from 2.6 to 6 times normal daily intake for age (approximately 1.25-3 L/day at time of study)	Enamel fluorosis; fluoride concentrations in deciduous teeth (enamel layer 50-100 μm from surface) 3-6 times those in controls (normal boys aged 10-14 residing in an area with fluoride at 1 mg/L)	Greenberg et al. 1974
Mother and four children with familial pituitary diabetes insipidus (Israel)	Water had "lower than accepted" fluoride content (0.5 mg/L); water consumption by mother and two teenage daughters (none used vasopressin) was 10-15 L/day each; two younger children treated for diabetes insipidus from ages 3 and 5	Enamel fluorosis in all four children: severe in the older two who were not treated for diabetes insipidus, milder in the two younger children who were treated for diabetes insipidus. Mother also had diabetes insipidus and fluorosis; she had grown up in Kurdistan with an unknown water fluoride content	Klein 1975
Six cases of familial pituitary diabetes insipidus (Australia)	Children had average water intake of 8-10 L/day; two of the children lived in fluoridated areas (1 mg/L)	Moderate (one child) or severe (one child) enamel fluorosis in the two children who lived in fluoridated areas	Seow and Thomsett 1994
Two brothers with pituitary diabetes insipidus (ages 17 and 7) (India)	Well water with fluoride at 3.4 mg/L	Severe enamel fluorosis, skeletal deformities, and radiological evidence of skeletal fluorosis	Mehta et al. 1998

TABLE 2-4 Examples of Fluoride Intake from Drinking Water by Members of Selected Population Subgroups Living in Fluoridated Areas[a]

Population Subgroup (Weight)	Typical Consumers[b]				High Consumers[c]			
	Water Consumption		Fluoride Intake[d]		Water Consumption		Fluoride Intake[d]	
	mL/day	mL/kg/day	mg/day	mg/kg/day	mL/day	mL/kg/day	mg/day	mg/kg/day
Athletes, workers, military (50 kg)	2,500	50	1.8-3.0	0.035-0.06	3,500	70	2.5-4.2	0.049-0.084
Athletes, workers, military (70 kg)	3,500	50	2.5-4.2	0.035-0.06	4,900	70	3.4-5.9	0.049-0.084
Athletes, workers, military (100 kg)	5,000	50	3.5-6.0	0.035-0.06	7,000	70	4.9-8.4	0.049-0.084
Athletes and workers (120 kg)	6,000	50	4.2-7.2	0.035-0.06	8,400	70	5.9-10	0.049-0.084
DM patients (20 kg)	1,000	50	0.7-1.2	0.035-0.06	2,000	100	1.4-2.4	0.07-0.12
DM patients (70 kg)	3,500	50	2.5-4.2	0.035-0.06	4,900	70	3.4-5.9	0.049-0.084
NDI patients (20 kg)	1,000	50	0.7-1.2	0.035-0.06	3,000	150	2.1-3.6	0.11-0.18
NDI patients (70 kg)	3,500	50	2.5-4.2	0.035-0.06	10,500	150	7.4-13	0.11-0.18

[a] Assumes all drinking water is from fluoridated community (municipal) sources.
[b] Based on a typical consumption rate for the population subgroup.
[c] Based on a reasonably high (but not upper bound) consumption rate for the population subgroup; some individual exposures could be higher.
[d] Based on fluoride concentrations of 0.7-1.2 mg/L.
ABBREVIATIONS: DM, diabetes mellitus; NDI, nephrogenic diabetes insipidus.

communities (1 mg/L in the water) was 0.0098 mg/L; in nonfluoridated communities, the mean was 0.0044 mg/L). Fluoride concentrations were correlated with the presence of fluoride in the mother's drinking water. Spak et al. (1983) reported mean fluoride concentrations in colostrum of 0.0053 mg/L (0.28 µM/L) in an area in Sweden with fluoride at 0.2 mg/L in drinking water and 0.0068 mg/L (0.36 µM/L) in an area with fluoride at 1.0 mg/L in the drinking water; in the fluoridated area, the mean fluoride concentration in mature milk was 0.007 mg/L (0.37 µM/L). No statistically significant difference in milk fluoride concentration between the two areas was found.

Hossny et al. (2003) reported fluoride concentrations in breast milk of 60 mothers in Cairo, Egypt, ranging from 0.002 to 0.01 mg/L [0.1-0.6 µM/L; median, 0.0032 mg/L (0.17 µM/L); mean, 0.0046 mg/L (0.24 µM/L)]. Cairo is considered nonfluoridated, with a reported water fluoride concentration of 0.3 mg/L (Hossny et al. 2003). Opinya et al. (1991) found higher fluoride concentrations in mothers' milk (mean, 0.033 mg/L; range, 0.011-0.073 mg/L), but her study population was made up of mothers in Kenya with an average daily fluoride intake of 22.1 mg. However, even at very high fluoride intakes by mothers, breast milk still contains very low concentrations of fluoride compared with other dietary fluoride sources. No significant correlation was established between the fluoride in milk and the intake of fluoride in the Kenyan study (Opinya et al. 1991).

Cows' milk likewise contains very low fluoride concentrations, compared with other dietary sources such as drinking water. Dairy milk samples measured in Houston contained fluoride at 0.007 to 0.068 mg/L (average, 0.03 mg/L) (Liu et al. 1995). Milk samples in 11 Canadian cities contained 0.007-0.086 mg/L (average, 0.041 mg/L) (Dabeka and McKenzie 1987). A sample of soy milk contained much more fluoride than a sample of dairy milk, with a measured concentration of 0.491 mg/L (Liu et al. 1995).

Infant formulas vary in fluoride content, depending on the type of formula and the water with which it is prepared. Dabeka and McKenzie (1987) reported mean fluoride concentrations in ready-to-use formulas of 0.23 mg/L for formulas manufactured in the United States and 0.90 mg/L for formulas manufactured in Canada. Van Winkle et al. (1995) analyzed 64 infant formulas, 47 milk-based and 17 soy-based. For milk-based formulas, mean fluoride concentrations were 0.17 mg/L for ready-to-feed, 0.12 mg/L for liquid concentrates reconstituted with distilled water, and 0.14 mg/L for powdered concentrates reconstituted with distilled water. Mean fluoride concentrations for soy-based formulas were 0.30, 0.24, and 0.24 mg/L for ready-to-feed, liquid concentrates, and powdered concentrates, respectively (the latter two were reconstituted with distilled water). Obviously, the fluoride concentration in home-prepared formula depends on the fluoride concentrations in both the formula concentrate and the home drinking water. Fomon et al. (2000) have recommended using low-fluoride water to dilute infant formulas.

Heilman et al. (1997) found 0.01 to 8.38 µg of fluoride per g of prepared infant foods. The highest concentrations were found in chicken (1.05-8.38 µg/g); other meats varied from 0.01 µg/g (veal) to 0.66 µg/g (turkey). Other foods—fruits, desserts, vegetables, mixed foods, and cereals—ranged from 0.01 to 0.63 µg/g. The fluoride concentrations in most foods are attributable primarily to the water used in processing (Heilman et al. 1997); fluoride in chicken is due to processing methods (mechanical deboning) that leave skin and residual bone particles in the meat (Heilman et al. 1997; Fein and Cerklewski 2001). An infant consuming 2 oz (about 60 g) of chicken daily at 8 µg of fluoride per g would have an intake of about 0.48 mg (Heilman et al. 1997).

Tea can contain considerable amounts of fluoride, depending on the type of tea and its source. Tea plants take up fluoride from soil along with aluminum (Shu et al. 2003; Wong et al. 2003). Leaf tea, including black tea and green tea, is made from the buds and young leaves of the tea plant, the black tea with a fermentation process, and the green tea without. Oolong tea is intermediate between black and green tea. Brick tea, considered a low-quality tea, is made from old (mature) leaves and sometimes branches and fruits of the tea plant (Shu et al. 2003; Wong et al. 2003). Fluoride accumulates mostly in the leaves of the tea plant, especially the mature or fallen leaves. Measured fluoride concentrations in tea leaves range from 170 to 878 mg/kg in different types of tea, with brick tea generally having 2-4 times as much fluoride as leaf tea (Wong et al. 2003). Commercial tea brands in Sichuan Province of China ranged from 49 to 105 mg/kg dry weight for green teas and 590 to 708 mg/kg dry weight for brick teas (Shu et al. 2003). Infusions of Chinese leaf tea (15 kinds) made with distilled water have been shown to have fluoride at 0.6-1.9 mg/L (Wong et al. 2003). Brick teas, which are not common in the United States, contain 4.8-7.3 mg/L; consumption of brick teas has been associated with fluorosis in some countries (Wong et al. 2003).

Chan and Koh (1996) measured fluoride contents of 0.34-3.71 mg/L (mean, 1.50 mg/L) in caffeinated tea infusions (made with distilled, deionized water), 1.01-5.20 mg/L (mean, 3.19 mg/L) in decaffeinated tea infusions, and 0.02-0.15 mg/L (mean, 0.05 mg/L) in herbal tea infusions, based on 44 brands of tea available in the United States (Houston area). Whyte et al. (2005) reported fluoride concentrations of 1.0-6.5 mg/L in commercial teas (caffeinated and decaffeinated) obtained in St. Louis (prepared with distilled water according to label directions). Warren et al. (1996) found fluoride contents of 0.10-0.58 mg/L in various kinds and brands of coffee sold in the United States (Houston area), with a slightly lower mean for decaffeinated (0.14 mg/L) than for caffeinated (0.17 mg/L) coffee. Instant coffee had a mean fluoride content of 0.30 mg/L (all coffees tested were prepared with deionized distilled water). Fluoride concentrations of 0.03 mg/L (fruit tea) to 3.35 mg/L (black tea) were reported for iced-tea products sold in Germany primarily by international companies (Behrendt et al. 2002).

In practice, fluoride content in tea or coffee as consumed will be higher if the beverage is made with fluoridated water; however, for the present purposes, the contribution from water for beverages prepared at home is included in the estimated intakes from drinking water, discussed earlier. Those estimates did not include commercially available beverages such as fruit juices (not including water used to reconstitute frozen juices), juice-flavored drinks, iced-tea beverages, carbonated soft drinks, and alcoholic beverages. Kiritsy et al. (1996) reported fluoride concentrations in juices and juice-flavored drinks of 0.02-2.8 mg/L (mean, 0.56 mg/L) for 532 different drinks (including five teas) purchased in Iowa City (although many drinks represented national or international distribution); frozen-concentrated beverages were reconstituted with distilled water before analysis. White grape juices had the highest mean fluoride concentration (1.45 mg/L); upper limits on most kinds of juices exceeded 1.50 mg/L. Stannard et al. (1991) previously reported fluoride concentrations from 0.15 to 6.80 mg/L in a variety of juices originating from a number of locations in the United States. The variability in fluoride concentrations is due primarily to variability in fluoride concentrations in the water used in manufacturing the product (Kiritsy et al. 1996). The high fluoride content of grape juices (and grapes, raisins, and wines), even when little or no manufacturing water is involved, is thought to be due to a pesticide (cryolite) used in grape growing (Stannard et al. 1991; Kiritsy et al. 1996; Burgstahler and Robinson 1997).

Heilman et al. (1999) found fluoride concentrations from 0.02 to 1.28 mg/L (mean, 0.72 mg/L) in 332 carbonated beverages from 17 production sites, all purchased in Iowa. In general, these concentrations reflect that of the water used in manufacturing. Estimated mean intakes from the analyzed beverages were 0.36 mg/day for 2- to 3-year-old children and 0.60 mg/day for 7- to 10-year-olds (Heilman et al. 1999). Pang et al. (1992) estimated mean daily fluoride intakes from beverages (excluding milk and water) for children of 0.36, 0.54, and 0.60 mg, for ages 2-3, 4-6, and 7-10, respectively; daily total fluid intake ranged from 970 to 1,240 mL, and daily beverage consumption ranged from 585 to 756 mL.

Burgstahler and Robinson (1997) reported fluoride contents of 0.23-2.80 mg/L in California wines, with 7 of 19 samples testing above 1 mg/L; the fluoride in wine and in California grapes (0.83-5.20 mg/kg; mean, 2.71 mg/kg) was attributed to the use of cryolite (Na_3AlF_6) as a pesticide in the vineyards. Martínez et al. (1998) reported fluoride concentrations from 0.03 to 0.68 mg/L in wines from the Canary Islands; most fluoride concentrations in the wines were in the range of 0.10-0.35 mg/L. A maximum legal threshold of 1 mg/L for the fluoride concentration in wine has been established by the Office International de la Vigne et du Vin (OIV 1990; cited by Martínez et al. 1998). Warnakulasuriya et al. (2002) reported mean fluoride concentrations of 0.08-0.71 mg/L in beers available in Great Britain; one Irish beer contained fluoride at 1.12 mg/L. Examples of fluoride intakes that could be expected in heavy drinkers (8-12 drinks per day) are given in Table 2-5.

R.D. Jackson et al. (2002) reported mean fluoride contents from 0.12 µg/g (fruits) to 0.49 µg/g (grain products) in a variety of noncooked, nonreconstituted foods (excluding foods prepared with water). Fluoride contents in commercial beverages (excluding reconstituted and fountain beverages) averaged 0.55 µg/g; those in milk and milk products averaged 0.31 µg/g. In the same study, fluoride contents in water, reconstituted beverages, and cooked vegetables and grain products (cereals, pastas, soups) differed significantly between two towns in Indiana, one with a water fluoride content of 0.2 mg/L and one with an optimally fluoridated water supply (1.0 mg/L). Bottled fruit drinks, water, and carbonated beverages purchased in the two towns did not differ significantly. The mean daily fluoride ingestion for children 3-5 years old from food and beverages (including those prepared with community water) was estimated to be 0.454 mg in the low-fluoride town and 0.536 mg in the fluoridated town.

Dabeka and McKenzie (1995) reported mean fluoride contents in various food categories in Winnipeg, ranging up to 2.1 µg/g for fish, 0.61 µg/g for soup, and 1.15 µg/g for beverages; the highest single items were cooked veal (1.2 µg/g), canned fish (4.6 µg/g), shellfish (3.4 µg/g), cooked wheat cereal (1.0 µg/g), and tea (5.0 µg/g). Estimated dietary intakes (including fluoridated tap water) varied from 0.35 mg/day for children aged 1-4 to 3.0 mg/day for 40- to 64-year-old males. Over all ages and both sexes, the esti-

TABLE 2-5 Examples of Fluoride Intakes by Heavy Drinkers from Alcoholic Beverages Alone

Beverage	Fluoride Concentration, mg/L	Fluoride Intake, mg/day	
		8 drinks per day	12 drinks per day
Beer (12-oz. cans or bottles)	0.5	1.4	2.1
	1.0	2.8	4.3
Wine (5-oz. glasses)	0.3	0.35	0.53
	1.0	1.2	1.8
Mixed drinks (1.5 oz. liquor + 6.5 oz. mixer and ice)	0.7[a]	1.1	1.6
	1.0[a]	1.5	2.3

[a]In carbonated soda and ice.

mated average dietary intake of fluoride was 1.76 mg/day; the food category contributing most to the estimated intake was beverages (80%).

Rojas-Sanchez et al. (1999) estimated fluoride intakes for children (aged 16-40 months) in three communities in Indiana, including a low-fluoride community, a "halo" community (not fluoridated, but in the distribution area of a fluoridated community), and a fluoridated community. For fluoride in food, the mean intakes were 0.116-0.146 mg/day, with no significant difference between communities. Intake from beverages was estimated to be 0.103, 0.257, and 0.396 mg/day for the low-, halo, and high-fluoride communities; differences between the towns were statistically significant.

Apart from drinking water (direct and indirect consumption, as described earlier), the most important foods in terms of potential contribution to individual fluoride exposures are infant formula, commercial beverages such as juice and soft drinks, grapes and grape products, teas, and processed chicken (Table 2-6). Grapes and grape products, teas, and processed chicken can be high in fluoride apart from any contribution from preparation or process water. Commercial beverages and infant formulas, however, greatly depend on the fluoride content of the water used in their preparation or manufacture (apart from water used in their in-home preparation); due to widespread distribution, such items could have similar fluoride concentrations in most communities, on average.

TABLE 2-6 Summary of Typical Fluoride Concentrations of Selected Food and Beverages in the United States

Source	Range, mg/L	Range, mg/kg
Human breast milk		
Fluoridated area (1 mg/L)	0.007-0.01	—
Nonfluoridated area	0.004	—
Cow's milk	≤0.07	—
Soy milk	0.5	—
Milk-based infant formula[a]	≤0.2	—
Soy-based infant formula[a]	0.2-0.3	—
Infant food—chicken	—	1-8
Infant food—other	—	0.01-0.7
Tea[a]	0.3-5	—
Herbal tea[a]	0.02-0.15	—
Coffee[a]	0.1-0.6	—
Grape juice[a]	≤3	—
Other juices and juice drinks[a]	≤1.5	—
Grapes	—	0.8-5
Carbonated beverages	0.02-1.3	—
Wine	0.2-3	—
Beer	0.08-1	—

[a]Not including contribution from local tap water.

Because of the wide variability in fluoride content in items such as tea, commercial beverages and juices, infant formula, and processed chicken, and the possibility of a substantial contribution to an individual's total fluoride intake, a number of authors have suggested that such fluoride sources be considered in evaluating an individual's need for fluoride supplementation (Clovis and Hargreaves 1988; Stannard et al. 1991; Chan and Koh 1996; Kiritsy et al. 1996; Warren et al. 1996; Heilman et al. 1997, 1999; Levy and Guha-Chowdhury 1999), especially for individuals who regularly consume large amounts of a single product (Stannard et al. 1991; Kiritsy et al. 1996). Several authors also point out the difficulty in evaluating individual fluoride intake, given the wide variability of fluoride content among similar items (depending on point of origin, etc.), the wide distribution of many products, and the lack of label or package information about fluoride content for most products (Stannard et al. 1991; Chan and Koh 1996; Behrendt et al. 2002).

Dental Products and Supplements

Fluoridated dental products include dentifrices (toothpastes, powders, liquids, and other preparations for cleaning teeth) for home use and various gels and other topical applications for use in dental offices. More than 90% of children ages 2-16 years surveyed in 1983 or 1986 used fluoride toothpaste (Wagener et al. 1992). Of these children, as many as 15% to 20% in some age groups also used fluoride supplements or mouth rinses (Wagener et al. 1992). Using the same 1986 survey data, Nourjah et al. (1994) reported that most children younger than 2 years of age used fluoride dentifrices.

Most toothpaste sold in the United States contains fluoride (Newbrun 1992), usually 1,000-1,100 parts per million (ppm) (0.1-0.11%).[10] The amount of fluoride actually swallowed by an individual depends on the amount of toothpaste used, the swallowing control of the person (especially for young children), and the frequency of toothpaste use. Ophaug et al. (1980, 1985) estimated the intake of fluoride by small children (2-4 years) to be 0.125-0.3 mg per brushing; a 2-year-old child brushing twice daily would ingest nearly as much fluoride from the toothpaste as from food and fluoridated drinking water combined (Ophaug et al. 1985). Levy and Zarei-M (1991) reported estimates of 0.12-0.38 mg of fluoride ingested per brushing. Burt (1992) and Newbrun (1992) reported estimates of 0.27

---

[10]Equivalent to 1-1.1 mg fluoride ion per gram of toothpaste. This may be expressed in various ways on the package, e.g., as 0.24% or 0.243% sodium fluoride (NaF), 0.76% or 0.8% monofluorophosphate (Na_2PO_3F), or 0.15% w/v fluoride (1.5 mg fluoride ion per cubic centimeter of toothpaste).

mg/day for a preschool child brushing twice daily with standard-strength (1,000 ppm) toothpaste.

Levy (1993, 1994) and Levy et al. (1995a) reviewed a number of studies of the amount of toothpaste people of various ages ingest. Amounts of toothpaste used per brushing range from 0.2 to 5 g, with means around 0.4-2 g, depending on the age of the person. The estimated mean percentage of toothpaste ingested ranges from 3% in adults to 65% in 2-year-olds. Children who did not rinse after toothbrushing ingested 75% more toothpaste than those who rinsed. Perhaps 20% of children have fluoride intakes from toothpaste several times greater than the mean values, and some children probably get more than the recommended amount of fluoride from toothpaste alone, apart from food and beverages (Levy 1993, 1994). Mean intakes of toothpaste by adults were measured at 0.04 g per brushing (0.04 mg of fluoride per brushing for toothpaste with 0.1% fluoride), with the 90th percentile at 0.12 g of toothpaste (0.12 mg of fluoride) per brushing (Barnhart et al. 1974).

Lewis and Limeback (1996) estimated the daily intake of fluoride from dentifrice (products for home use) to be 0.02-0.06, 0.008-0.02, 0.0025, and 0.001 mg/kg, for ages 7 months to 4 years, 5-11 years, 12-19 years, and 20+ years, respectively. Rojas-Sanchez et al. (1999) estimated fluoride intake from dentifrice at between 0.42 and 0.58 mg/day in children aged 16-40 months in three communities in Indiana. Children tend to use more toothpaste when provided special "children's" toothpaste than when given adult toothpaste (Levy et al. 1992; Adair et al. 1997), and many children do not rinse or spit after brushing (Naccache et al. 1992; Adair et al. 1997).

Estimates of typical fluoride ingestion from toothpaste are given by age group in Table 2-7; these estimates are for typical rather than high or upper-bound intakes, and many individuals could have substantially higher intakes. A number of papers have suggested approaches to decreasing children's intake of fluoride from toothpaste, including decreasing the fluoride content in

TABLE 2-7 Estimated Typical Fluoride Intakes from Toothpaste[a]

Age Group, years	Fluoride Intake, mg/day	Age Group, years	Fluoride Intake, mg/day
Infants < 0.5[b]	0	Youth 13-19	0.2
Infants 0.5-1	0.1	Adults 20-49	0.1
Children 1-2	0.15	Adults 50+	0.1
Children 3-5	0.25	Females 13-49[c]	0.1
Children 6-12	0.3		

[a]Based on information reviewed by Levy et al. (1995a). Estimates assume two brushings per day with fluoride toothpaste (0.1% fluoride) and moderate rinsing.

[b]Assumes no brushing before 6 months of age.

[c]Women of childbearing age.

children's toothpaste, discouraging the use of fluoride toothpaste by children less than 2 years old, avoiding flavored children's toothpastes, encouraging the use of very small amounts of toothpaste, encouraging rinsing and expectorating (rather than swallowing) after brushing, and recommending careful parental supervision (e.g., Szpunar and Burt 1990; Levy and Zarei-M 1991; Simard et al. 1991; Burt 1992; Levy et al. 1992, 1993, 1997, 2000; Naccache et al. 1992; Newbrun 1992; Levy 1993, 1994; Bentley et al. 1999; Rojas-Sanchez et al. 1999; Warren and Levy 1999; Fomon et al. 2000).

Topical applications of fluoride in a professional setting can lead to ingestion of 1.3-31.2 mg (Levy and Zarei-M 1991). Substantial ingestion of fluoride also has been demonstrated from the use of fluoride mouth rinse and self-applied topical fluoride gel (Levy and Zarei-M 1991). Heath et al. (2001) reported that 0.3-6.1 mg of fluoride (5-29% of total applied) was ingested by young adults who used gels containing 0.62-62.5 mg of fluoride.

Levy et al. (2003a) found that two-thirds of children had at least one fluoride treatment by age 6 and that children with dental caries were more likely to have had such a treatment. Their explanation is that professional application of topical fluoride is used mostly for children with moderate to high risk for caries. In contrast, Eklund et al. (2000), in a survey of insurance claims for more than 15,000 Michigan children treated by 1,556 different dentists, found no association between the frequency of use of topical fluoride (professionally applied) and restorative care. Although these were largely low-risk children, for whom routine use of professionally applied fluoride is not recommended, two-thirds received topical fluoride at nearly every office visit. The authors recommended that the effectiveness of professionally applied topical fluoride products in modern clinical practice be evaluated.

Exposures from topical fluorides during professional treatment are unlikely to be significant contributors to chronic fluoride exposures because they are used only a few times per year. However, they could be important with respect to short-term or peak exposures.

Heath et al. (2001) found that retention of fluoride ion in saliva after the use of dentifrice (toothpaste, mouthrinse, or gel) was proportional to the quantity used, at least for young adults. They were concerned with maximizing the retention in saliva to maximize the topical benefit of the fluoride. Sjögren and Melin (2001) were also concerned about enhancing the retention of fluoride in saliva and recommend minimal rinsing after toothbrushing. However, fluoride in saliva eventually will be ingested, so enhancing the retention of fluoride in saliva after dentifrice use also enhances the ingestion of fluoride from the dentifrice.

Fluoride supplements (NaF tablets, drops, lozenges, and rinses) are intended for prescriptions for children in low-fluoride areas; dosages generally range from 0.25 to 1.0 mg of fluoride/day (Levy 1994; Warren and Levy

1999). Appropriate dosages should be based on age, risk factors (e.g., high risk for caries), and ingestion of fluoride from other sources (Dillenberg et al. 1992; Jones and Berg 1992; Levy and Muchow 1992; Levy 1994; Warren and Levy 1999). Although compliance is often considered to be a problem, inappropriate use of fluoride supplements has also been identified as a risk factor for enamel fluorosis (Dillenberg et al. 1992; Levy and Muchow 1992; Levy 1994; Pendrys and Morse 1995; Warren and Levy 1999).

The dietary fluoride supplement schedule in the United States, as revised in 1994 by the American Dental Association, now calls for no supplements for children less than 6 months old and none for any child whose water contains at least 0.6 mg/L (Record et al. 2000; ADA 2005; Table 2-8). Further changes in recommendations for fluoride supplements have been suggested (Fomon and Ekstrand 1999; Newbrun 1999; Fomon et al. 2000), including dosages based on individual body weight rather than age (Adair 1999) and the use of lozenges to be sucked rather than tablets to be swallowed (Newbrun 1999), although others disagree (Moss 1999). The Canadian recommendations for fluoride supplementation include an algorithm for determining the appropriateness for a given child and then a schedule of doses; no supplementation is recommended for children whose water contains at least 0.3 mg/L or who are less than 6 months old (Limeback et al. 1998; Limeback 1999b).

Fluoride in Air

Fluoride (either as hydrogen fluoride, particulate fluorides, or fluorine gas) is released to the atmosphere by natural sources such as volcanoes[11] and by a number of anthropogenic sources. In North America, anthropogenic sources of airborne fluoride include coal combustion by electrical utilities and other entities, aluminum production plants, phosphate fertilizer plants, chemical production facilities, steel mills, magnesium plants, and manufacturers of brick and structural clay (reviewed by ATSDR 2003). Estimated airborne releases of hydrogen fluoride in the United States in 2001 were 67.4 million pounds (30.6 million kg; TRI 2003), of which at least 80% was attributed to electrical utilities (ATSDR 2003). Airborne releases of fluorine gas totaled about 9,000 pounds or 4,100 kg (TRI 2003). Anthropogenic hydrogen fluoride emissions in Canada in the mid-1990s were estimated at 5,400 metric tons (5.4 million kg or 11.9 million pounds), of which 75% was attributed to primary aluminum producers (CEPA 1996).

---

[11]Volcanic activity historically has been a major contributor of HF and other contaminants to the atmosphere in some parts of the world, with some volcanoes emitting 5 tons of HF per day (Nicaragua) or as much as 15 million tons during a several month eruption (Iceland) (Durand and Grattan 2001; Grattan et al. 2003; Stone 2004).

TABLE 2-8 Dietary Fluoride Supplement Schedule of 1994

	Fluoride Concentration in Drinking Water, mg/L		
Age	< 0.3	0.3-0.6	> 0.6
Birth to 6 months	None	None	None
6 months to 3 years	0.25 mg/day	None	None
3-6 years	0.50 mg/day	0.25 mg/day	None
6-16 years	1.0 mg/day	0.50 mg/day	None

SOURCE: ADA 2005. Reprinted with permission; copyright 2005, American Dental Association.

Measured fluoride concentrations in air in the United States and Canada typically range from 0.01 to 1.65 µg/m^3, with most of it (75%) present as hydrogen fluoride (CEPA 1996). The highest concentrations (>1 µg/m^3) correspond to urban locations or areas in the vicinity of industrial operations. Historically, concentrations ranging from 2.5 to 14,000 µg/m^3 have been reported near industrial operations in various countries (reviewed by EPA 1988). Ernst et al. (1986) reported an average concentration of airborne fluoride of about 600 µg/m^3 during the 1981 growing season in a rural inhabited area (Cornwall Island) on the U.S.-Canadian border directly downwind from an aluminum smelter. Hydrogen fluoride is listed as a hazardous air pollutant in the Clean Air Act Amendments of 1990 (reviewed by ATSDR 2003), and as such, its emissions are subject to control based on "maximum achievable control technology" emission standards. Such standards are already in effect for fluoride emissions from primary and secondary aluminum production, phosphoric acid manufacture and phosphate fertilizer production, and hydrogen fluoride production (ATSDR 2003).

For most individuals in the United States, exposure to airborne fluoride is expected to be low compared with ingested fluoride (EPA 1988); exceptions include people in heavily industrialized areas or having occupational exposure. Assuming inhalation rates of 10 m^3/day for children and 20 m^3/day for adults, fluoride exposures from inhalation in rural areas (<0.2 µg/m^3 fluoride) would be less than 2 µg/day (0.0001-0.0002 mg/kg/day) for a child and 4 µg/day (0.00006 mg/kg/day) for an adult. In urban areas (<2 µg/m^3), fluoride exposures would be less than 20 µg/day (0.0001-0.002 mg/kg/day) for a child and 40 µg/day (0.0006 mg/kg/day) for an adult. Lewis and Limeback (1996) used an estimate of 0.01 µg/kg/day (0.00001 mg/kg/day) for inhaled fluoride for Canadians; this would equal 0.1 µg/day for a 10-kg child or 0.7 µg/day for a 70-kg adult.

Occupational exposure at the Occupational Safety and Health Administration (OSHA) exposure limit of 2.5 mg/m^3 would result in a fluoride intake of 16.8 mg/day for an 8-hour working day (0.24 mg/kg/day for a

70-kg person) (ATSDR 2003). Heavy cigarette smoking could contribute as much as 0.8 mg of fluoride per day to an individual (0.01 mg/kg/day for a 70-kg person) (EPA 1988).

Fluoride in Soil

Fluoride in soil could be a source of inadvertent ingestion exposure, primarily for children. Typical fluoride concentrations in soil in the United States range from very low (<10 ppm) to as high as 3% to 7% in areas with high concentrations of fluorine-containing minerals (reviewed by ATSDR 2003). Mean or typical concentrations in the United States are on the order of 300-430 ppm. Soil fluoride content may be higher in some areas due to use of fluoride-containing phosphate fertilizers or to deposition of airborne fluoride released from industrial operations.

Estimated values for inadvertent soil ingestion by children (excluding those with pica) are 100 mg/day (mean) and 400 mg/day (upper bound) (EPA 1997); the estimated mean value for soil ingestion by adults is 50 mg/day (EPA (1997). For a typical fluoride concentration in soil of 400 ppm, therefore, estimated intakes of fluoride by children would be 0.04 (mean) to 0.16 mg/day (upper bound) and by adults, 0.02 mg/day. For a 20-kg child, the mass-normalized intake would be 0.002-0.008 mg/kg/day; for a 70-kg adult, the corresponding value would be 0.0003 mg/kg/day. Erdal and Buchanan (2005) estimated intakes of 0.0025 and 0.01 mg/kg/day for children (3-5 years), for mean and reasonable maximum exposures, respectively, based on a fluoride concentration in soil of 430 ppm. In their estimates, fluoride intake from soil was 5-9 times lower than that from fluoridated drinking water.

For children with pica (a condition characterized by consumption of nonfood items such as dirt or clay), an estimated value for soil ingestion is 10 g/day (EPA 1997). For a 20-kg child with pica, the fluoride intake from soil containing fluoride at 400 ppm would be 4 mg/day or 0.2 mg/kg/day. Although pica in general is not uncommon among children, the prevalence is not known (EPA 1997). Pica behavior specifically with respect to soil or dirt appears to be relatively rare but is known to occur (EPA 1997); however, fluoride intake from soil for a child with pica could be a significant contributor to total fluoride intake. For most children and for adults, fluoride intake from soil probably would be important only in situations in which the soil fluoride content is high, whether naturally or due to industrial pollution.

Pesticides

Cryolite and sulfuryl fluoride are the two pesticides that are regulated for their contribution to the residue of inorganic fluoride in foods. For food use pesticides, EPA establishes a tolerance for each commodity to which a pesticide is allowed to be applied. Tolerance is the maximum amount of pesticide allowed to be present in or on foods. In the environment, cryolite breaks down to fluoride, which is the basis for the safety evaluation of cryolite and synthetic cryolite pesticides (EPA 1996a). Fluoride ions are also degradation products of sulfuryl fluoride (EPA 1992). Thus, the recent evaluation of the dietary risk of sulfuryl fluoride use on food takes into account the additional exposure to fluoride from cryolite (EPA 2004). Sulfuryl fluoride is also regulated as a compound with its own toxicologic characteristics.

Cryolite, sodium hexafluoroaluminate (Na_3AlF_6), is a broad spectrum insecticide that has been registered for use in the United States since 1957. Currently, it is used on many food (tree fruits, berries, and vegetables) and feed crops, and on nonfood ornamental plants (EPA 1996a). The respective fluoride ion concentrations from a 200 ppm aqueous synthetic cryolite (97.3% pure) at pH 5, 7, and 9 are estimated at 16.8, 40.0, and 47.0 ppm (approximately 15.5%, 37%, and 43% of the total available fluorine) (EPA 1996a). A list of tolerances for the insecticidal fluorine compounds cryolite and synthetic cryolite is published in the Code of Federal Regulations (40 CFR § 180.145(a, b, c) [2004]). Current tolerances for all commodities are at 7 ppm.

Sulfuryl fluoride (SO_2F_2), is a structural fumigant registered for use in the United States since 1959 for the control of insects and vertebrate pests. As of January 2004, EPA published a list of tolerances for sulfuryl fluoride use as a post-harvest fumigant for grains, field corn, nuts, and dried fruits (69 Fed. Reg. 3240 [2004]; 40 CFR 180.575(a) [2004]). The calculated exposure threshold at the drinking-water MCL of 4 mg/L was used as the basis for assessing the human health risk associated with these decisions (EPA 2004).

Concerns were raised that foods stored in the freezer during sulfuryl fluoride residential fumigation might retain significant amounts of fluoride residue. Scheffrahn et al. (1989) reported that unsealed freezer foods contained fluoride at as high as 89.7 ppm (flour, at 6,803 mg-hour/L rate of sulfuryl fluoride application) while no fluoride residue was detected (0.8 ppm limit of detection) in foods that were sealed with polyethylene film. A later study reported fluoride residue above 1 ppm in food with higher fat contents (e.g., 5.643 ppm in margarine) or that was improperly sealed (e.g., 7.66 ppm in a reclosed peanut butter PETE [polyethylene terephthalate] jar) (Scheffrahn et al. 1992).

Dietary exposure for a food item is calculated as the product of its consumption multiplied by the concentration of the residue of concern. The total daily dietary exposure for an individual is the sum of exposure from all food items consumed in a day. A chronic dietary exposure assessment of fluoride was recently conducted for supporting the establishment of tolerances for the post-harvest use of sulfuryl fluoride. EPA (2004) used the Dietary Exposure Evaluation Model (DEEM-FCID), a computation program, to estimate the inorganic fluoride exposure from cryolite, sulfuryl fluoride, and the background concentration of fluoride in foods. DEEM-FCID (Exponent, Inc) uses the food consumption data from the 1994-1996 and 1998 Continuing Survey of Food Intakes by Individuals (CSFII) conducted by the U.S. Department of Agriculture (USDA). The 1994-1996 database consists of food intake diaries of more than 15,000 individuals nationwide on two nonconsecutive days. A total of 4,253 children from birth to 9 years of age are included in the survey. To ensure that the eating pattern of young children is adequately represented in the database, an additional survey was conducted in 1998 of 5,559 children 0-9 years of age. The latter survey was designed to be compatible with the CSFII 1994-1996 data so that the two sets of data can be pooled to increase the sample size for children. The Food Commodity Intake Database (FCID) is jointly developed by EPA and USDA for the purpose of estimating dietary exposure from pesticide residues in foods. It is a translated version of the CSFII data that expresses the intake of consumed foods in terms of food commodities (e.g., translating apple pie into its ingredients, such as apples, flour, sugar, etc.) (EPA 2000c).

All foods and food forms (e.g., grapes—fresh, cooked, juice, canned, raisins, wine) with existing tolerances for cryolite and sulfuryl fluoride were included in the recent EPA fluoride dietary exposure analysis (EPA 2004). For the analysis of fluoride exposure from cryolite, residue data taken from monitoring surveys, field studies, and at tolerance were adjusted to reflect changes in concentration during food processing (e.g., mixing in milling, dehydration, and food preparation). For the fluoride exposure from post-harvest treatment with sulfuryl fluoride, the measured residues are used without further adjustment except for applying drawdown factors in grain mixing (EPA 2004). In estimating fluoride exposure from both cryolite- and sulfuryl fluoride-treated foods, residue concentrations were adjusted for the percentage of crop treated with these pesticides based on the information from market share and agricultural statistics on pesticide use.

Fluoride exposures from a total of 543 forms of foods (e.g., plant-based, bovine, poultry, egg, tea) containing fluoride were also estimated as the background food exposure. Residue data were taken from surveys and residue trials (EPA 2004). No adjustments were made to account for residue concentration through processing or dehydration. Theoretically, the exposure from some processed foods (e.g., dried fruits) could potentially be higher than if their residue concentrations were assumed to be the same as in the fresh commodities (e.g., higher exposure from higher residue in dried fruits than assuming same residue concentration for both dried and fresh fruits.) However, these considerations are apparently offset by the use of higher residue concentrations for many commodities (e.g., using the highest values from a range of survey data, the highest value as surrogate for when data are not available, assuming residue in dried fruits and tree nuts at one-half the limit of quantification when residue is not detected) such that the overall dietary exposure was considered overestimated (EPA 2004). The dietary fluoride exposure thus estimated ranged from 0.0003 to 0.0031 mg/kg/day from cryolite, 0.0003 to 0.0013 mg/kg/day from sulfuryl fluoride, and 0.005 to 0.0175 mg/kg/day from background concentration in foods (EPA 2004). Fine-tuning the dietary exposure analysis using the comprehensive National Fluoride Database recently published by USDA (2004) for many foods also indicates that the total background food exposure would not be significantly different from the analysis by EPA, except for the fluoride intake from tea. A closer examination of the residue profile used by EPA (2004) for background food exposure analysis reveals that 5 ppm, presumably a high-end fluoride concentration in brewed tea, was entered in the residue profile that called for fluoride concentration in powdered or dried tea. According to the USDA survey database (2004), the highest detected fluoride residue in instant tea powder is 898.72 ppm. The corrected exposure estimate is presented in the section "Total Exposure to Fluoride" later in this chapter.

Fluorinated Organic Chemicals

Many pharmaceuticals, consumer products, and pesticides contain organic fluorine (e.g., $-CF_3$, $-SCF_3$, $-OCF_3$). Unlike chlorine, bromine, and iodine, organic fluorine is not as easily displaced from the alkyl carbon and is much more lipophilic than the hydrogen substitutes (Daniels and Jorgensen 1977; PHS 1991). The lipophilic nature of the trifluoromethyl group contribute to the enhanced biological activity of some pharmaceutical chemicals.

The toxicity of fluorinated organic chemicals usually is related to their molecular characteristics rather than to the fluoride ions metabolically displaced. Fluorinated organic chemicals go through various degrees of biotransformation before elimination. The metabolic transformation is minimal for some chemicals. For example, the urinary excretion of ciprofloxacin (fluoroquinolone antibacterial agent) consists mainly of the unchanged parent compound or its fluorine-containing metabolites (desethylene-, sulfo-, oxo-, and N-formyl ciprofloxacin) (Bergan 1989). Nevertheless, Pradhan et al. (1995) reported an increased serum fluoride concentration from 4 µM (0.076 ppm) to 11 µM (0.21 ppm) in 19 children from India (8 months to 13 years old) within 12 hours after the initial oral dose of ciprofloxacin at 15-25 mg/kg. The presumed steady state (day 7 of repeated dosing) 24-hour urinary fluoride concentration was 15.5% higher than the predosing concentration (59 µM versus 51 µM; or, 1.12 ppm versus 0.97 ppm). Another example of limited contribution to serum fluoride concentration from pharmaceuticals was reported for flecainide, an antiarrhythmic drug. The peak serum fluoride concentration ranged from 0.0248 to 0.0517 ppm (1.3 to 2.7 µM) in six healthy subjects (26-54 years old, three males, and three females) 4.5 hours after receiving a single oral dose of 100 mg of flecainide acetate (Rimoli et al. 1991). One to two weeks before the study, the subjects were given a poor fluoride diet, used toothpaste without fluoride, and had low fluoride (0.08 mg/L) in their drinking water.

Other fluoride-containing organic chemicals go through more extensive metabolism that results in greater increased bioavailability of fluoride ion. Elevated serum fluoride concentrations from fluorinated anesthetics have been extensively studied because of the potential nephrotoxicity of methoxyflurane in association with elevated serum fluoride concentrations beyond a presumed toxicity benchmark of 50 µM (Cousins and Mazze 1973; Mazze et al. 1977). A collection of data on peak serum fluoride ion concentrations from exposures to halothane, enflurane, isoflurane, and sevoflurane is given in Appendix B. These data serve to illustrate a wide range of peak concentrations associated with various use conditions (e.g., length of use, minimum alveolar concentration per hour), biological variations (e.g., age, gender, obesity, smoking), and chemical-specific characteristics (e.g., biotransformation pattern and rates). It is not clear how these episodically elevated serum fluoride ion concentrations contribute to potential adverse effects of long-term sustained exposure to inorganic fluoride from other media, such as drinking water, foods, and dental-care products.

Elevated free fluoride ion (< 2% of administered dose) also was detected in the plasma and urine of some patients after intravenous administration of fluorouracil (Hull et al. 1988). Nevertheless, the major forms of urinary excretion were still the unchanged parent compound and its fluorine-containing metabolites (dihydrofluorouracil, α-fluoro-β-ureidopropanoic acid, α-fluoro-β-alanine). The extent of dermal absorption of topical fluorouracil cream varies with skin condition, product formulation, and the conditions of use. Levy et al. (2001a) reported less than 3% systemic fluorouracil absorption in patients treated with 0.5% or 5% cream for actinic keratosis.

A group of widely used consumer products is the fluorinated telomers and polytetrafluoroethylene, or Teflon. EPA is in the process of evaluating the environmental exposure to low concentrations of perfluorooctanoic acid (PFOA) and its principal salts that are used in manufacturing fluoropolymers or as their breakdown products (EPA 2003b). PFOA is persistent in the environment. It is readily absorbed through oral and inhalation exposure and is eliminated in urine and feces without apparent biotransformation (EPA 2003b; Kudo and Kawashima 2003). Unchanged plasma and urine fluoride concentrations in rats that received intraperitoneal injections of

PFOA also indicated a lack of defluorination (Vanden Heuvel et al. 1991). (See Chapter 3 for more discussion of PFOA.)

Aluminofluorides, Beryllofluorides, and Fluorosilicates

Aluminofluorides and Beryllofluorides

Complexes of aluminum and fluoride (aluminofluorides, most often AlF_3 or AlF_4^-) or beryllium and fluoride (beryllofluorides, usually as BeF_3^-) occur when the two elements are present in the same environment (Strunecka and Patocka 2002). Fluoroaluminate complexes are the most common forms in which fluoride can enter the environment. Eight percent of the earth's crust is composed of aluminum; it is the most abundant metal and the third most abundant element on earth (Liptrot 1974). The most common form for the inorganic salt of aluminum and fluoride is cryolite (Na_3AlF_6). In fact, of the more than 60 metals on the periodic chart, Al^{3+} binds fluoride most strongly (Martin 1988). With the increasing prevalence of acid rain, metal ions such as aluminum become more soluble and enter our day-to-day environment; the opportunity for bioactive forms of AlF to exist has increased in the past 100 years. Human exposure to aluminofluorides can occur when a person ingests both a fluoride source (e.g., fluoride in drinking water) and an aluminum source; sources of human exposure to aluminum include drinking water, tea, food residues, infant formula, aluminum-containing antacids or medications, deodorants, cosmetics, and glassware (ATSDR 1999; Strunecka and Patocka 2002; Li 2003; Shu et al. 2003; Wong et al. 2003). Aluminum in drinking water comes both from the alum used as a flocculant or coagulant in water treatment and from leaching of aluminum into natural water by acid rain (ATSDR 1999; Li 2003). Exposure specifically to aluminofluoride complexes is not the issue so much as the fact that humans are routinely exposed to both elements. Human exposure to beryllium occurs primarily in occupational settings, in the vicinity of industrial operations that process or use beryllium, and near sites of beryllium disposal (ATSDR 2002).

Aluminofluoride and beryllofluoride complexes appear to act as analogues of phosphate groups—for example, the terminal phosphate of guanidine triphosphate (GTP) or adenosine triphosphate (ATP) (Chabre 1990; Antonny and Chabre 1992; Caverzasio et al. 1998; Façanha and Okorokova-Façanha 2002; Strunecka and Patocka 2002; Li 2003). Thus, aluminofluorides might influence the activity of a variety of phosphatases, phosphorylases, and kinases, as well as the G proteins involved in biological signaling systems, by inappropriately stimulating or inhibiting normal function of the protein (Yatani and Brown 1991; Caverzasio et al. 1998; Façanha and Okorokova-Façanha 2002; Strunecka and Patocka 2002; Li

2003). Aluminofluoride complexes have been reported to increase the concentrations of second messenger molecules (e.g., free cytosolic Ca^{2+}, inositol 1,4,5-trisphosphate, and cyclic AMP) for many bodily systems (Sternweis and Gilman 1982; Strunecka et al. 2002; Li 2003). The increased toxicity of beryllium in the presence of fluoride and vice versa was noted as early as 1949 (Stokinger et al. 1949). For further discussion of aluminofluorides, see Chapters 5 and 7.

Further research should include characterization of both the exposure conditions and the physiological conditions (for fluoride and for aluminum or beryllium) under which aluminofluoride and beryllofluoride complexes can be expected to occur in humans as well as the biological effects that could result.

Fluorosilicates

Most fluoride in drinking water is added in the form of fluosilicic acid (fluorosilicic acid, H_2SiF_6) or the sodium salt (sodium fluorosilicate, Na_2SiF_6), collectively referred to as fluorosilicates (CDC 1993). Of approximately 10,000 fluoridated water systems included in the CDC's 1992 fluoridation census, 75% of them (accounting for 90% of the people served) used fluorosilicates. This widespread use of silicofluorides has raised concerns on at least two levels. First, some authors have reported an association between the use of silicofluorides in community water and elevated blood concentrations of lead in children (Masters and Coplan 1999; Masters et al. 2000); this association is attributed to increased uptake of lead (from whatever source) due to incompletely dissociated silicofluorides remaining in the drinking water (Masters and Coplan 1999; Masters et al. 2000) or to increased leaching of lead into drinking water in systems that use chloramines (instead of chlorine as a disinfectant) and silicofluorides (Allegood 2005; Clabby 2005; Maas et al. 2005).[12,13] Macek et al. (2006) have also compared blood lead concentrations in children by method of water fluoridation; they stated that their analysis did not support an association between blood lead concentrations and silicofluorides, but also could not refute it,

---

[12]In common practice, chloramines are produced with an excess of ammonia, which appears to react with silicofluorides to produce an ammonium-fluorosilicate intermediate which facilitates lead dissolution from plumbing components (Maas et al. 2005).

[13]Another possible explanation for increased blood lead concentrations which has not been examined is the effect of fluoride intake on calcium metabolism; a review by Goyer (1995) indicates that higher blood and tissue concentrations of lead occur when the diet is low in calcium. Increased fluoride exposure appears to increase the dietary requirement for calcium (see Chapter 8); in addition, the substitution of tap-water based beverages (e.g., soft drinks or reconstituted juices) for dairy products would result in both increased fluoride intake and decreased calcium intake.

especially for children living in older housing. Second, essentially no studies have compared the toxicity of silicofluorides with that of sodium fluoride, based on the assumption that the silicofluorides will have dissociated to free fluoride before consumption (see also Chapter 7).

Use of more sophisticated analytical techniques such as nuclear magnetic resonance has failed to detect any silicon- and fluorine-containing species other than hexafluorosilicate ion (SiF_6^{2-}) (Urbansky 2002; Morris 2004). In drinking water at approximately neutral pH and typical fluoride concentrations, all the silicofluoride appears to be dissociated entirely to silicic acid [$Si(OH)_4$], fluoride ion, and HF (Urbansky 2002; Morris 2004); any intermediate species either exist at extremely low concentrations or are highly transient. SiF_6^{2-} would be present only under conditions of low pH (pH < 5; Urbansky 2002; Morris 2004) and high fluoride concentration (above 16 mg/L according to Urbansky [2002]; at least 1 g/L to reach detectable levels of SiF_6^{2-}, according to Morris [2004]). Urbansky (2002) also stated that the silica contribution from the fluoridating agent is usually trivial compared with native silica in the water; therefore, addition of any fluoridating agent (or the presence of natural fluoride) could result in the presence of SiF_6^{2-} in any water if other conditions (low pH and high total fluoride concentration) are met. Both Urbansky (2002) and Morris (2004) indicate that other substances in the water, especially metal cations, might form complexes with fluoride, which, depending on pH and other factors, could influence the amount of fluoride actually present as free fluoride ion. For example, P.J. Jackson et al. (2002) have calculated that at pH 7, in the presence of aluminum, 97.46% of a total fluoride concentration of 1 mg/L is present as fluoride ion, but at pH 6, only 21.35% of the total fluoride is present as fluoride ion, the rest being present in various aluminum fluoride species (primarily AlF_2^+ and AlF_3). Calculations were not reported for pH < 6.

Further research should include analysis of the concentrations of fluoride and various fluoride species or complexes present in tap water, using a range of water samples (e.g., of different hardness and mineral content). In addition, given the expected presence of fluoride ion (from any fluoridation source) and silica (native to the water) in any fluoridated tap water, it would be useful to examine what happens when that tap water is used to make acidic beverages or products (commercially or in homes), especially fruit juice from concentrate, tea, and soft drinks. Although neither Urbansky (2002) nor Morris (2004) discusses such beverages, both indicate that at pH < 5, SiF_6^{2-} would be present, so it seems reasonable to expect that some SiF_6^{2-} would be present in acidic beverages but not in the tap water used to prepare the beverages. Consumption rates of these beverages are high for many people, and therefore the possibility of biological effects of SiF_6^{2-}, as opposed to free fluoride ion, should be examined.

RECENT ESTIMATES OF TOTAL FLUORIDE EXPOSURE

A number of authors have reviewed fluoride intake from water, food and beverages, and dental products, especially for children (NRC 1993; Levy 1994; Levy et al. 1995a,b,c; Lewis and Limeback 1996; Levy et al. 2001b). Heller et al. (1999, 2000) estimated that a typical infant less than 1 year old who drinks fluoridated water containing fluoride at 1 mg/L would ingest approximately 0.08 mg/kg/day from water alone. Shulman et al. (1995) also calculated fluoride intake from water, obtaining an estimate of 0.08 mg/kg/day for infants (7-9 months of age), with a linearly declining intake with age to 0.034 mg/kg/day for ages 12.5-13 years.

Levy et al. (1995b,c; 2001b) have estimated the intake of fluoride by infants and children at various ages based on questionnaires completed by the parents in a longitudinal study. For water from all sources (direct, mixed with formula, etc.), the intake of fluoride by infants (Levy et al. 1995b) ranged from 0 (all ages examined) to as high as 1.73 mg/day (9 months old). Infants fed formula prepared from powdered or liquid concentrate had fluoride intakes just from water in the formula of up to 1.57 mg/day. The sample included 124 infants at 6 weeks old and 77 by 9 months old. Thirty-two percent of the infants at 6 weeks and 23% at age 3 months reportedly had no water consumption (being fed either breast milk or ready-to-feed formula without added water). Mean fluoride intakes for the various age groups ranged from 0.29 to 0.38 mg/day; however, these values include the children who consumed no water, and so are not necessarily applicable for other populations. For the same children, mean fluoride intakes from water, fluoride supplement (if used), and dentifrice (if used) ranged from 0.32 to 0.38 mg/day (Levy et al. 1995c); the maximum fluoride intakes ranged from 1.24 (6 weeks old) to 1.73 mg/day (9 months old). Ten percent of the infants at 3 months old exceeded an intake of 1.06 mg/day.

For a larger group of children (about 12,000 at 3 months and 500 by 36 months of age; Levy et al. 2001b), mean fluoride intakes from water, supplements, and dentifrice combined ranged from 0.360 mg/day (12 months old) to 0.634 mg/day (36 months old). The 90th percentiles ranged from 0.775 mg/day (16 months old) to 1.180 mg/day (32 months old). Maximum intakes ranged from 1.894 mg/day (16 months old) to 7.904 mg/day (9 months old) and were attributable only to water (consumption of well water with 5-6 mg/L fluoride; about 1% of the children had water sources containing more than 2 mg/L fluoride). For ages 1.5-9 months, approximately 40% of the infants exceeded a mass-normalized intake level for fluoride of 0.07 mg/kg/day; for ages 12-36 months, about 10-17% exceeded that level (Levy et al. 2001b).

Levy et al. (2003b) reported substantial variation in total fluoride intake among children aged 36-72 months, with some individual intakes greatly exceeding the means. The mean intake per unit of body weight declined with age from 0.05 to 0.06 mg/kg/day at 36 months to 0.03-0.04 mg/kg/day at 72 months; 90th percentile values declined from about 0.10 mg/kg/day to about 0.06 mg/kg/day (Levy et al. 2003b). Singer et al. (1985) reported mean estimated total fluoride intakes of 1.85 mg/day for 15- to 19-year-old males (based on a market-basket survey and a diet of 2,800 calories per day) in a fluoridated area (>0.7 mg/L) and 0.86 mg/day in nonfluoridated areas (<0.3 mg/L). Beverages and drinking water contributed approximately 75% of the total fluoride intake.

Lewis and Limeback (1996) estimated total daily fluoride intakes of 0.014-0.093 mg/kg for formula-fed infants and 0.0005-0.0026 mg/kg for breast-fed infants (up to 6 months). For children aged 7 months to 4 years, the estimated daily intakes from food, water, and household products (primarily dentifrice) were 0.087-0.160 mg/kg in fluoridated areas and 0.045-0.096 mg/kg in nonfluoridated areas. Daily intakes for other age groups were 0.049-0.079, 0.033-0.045, and 0.047-0.058 mg/kg for ages 5-11, 12-19, and 20+ in fluoridated areas, and 0.026-0.044, 0.017-0.021, and 0.032-0.036 mg/kg for the same age groups in nonfluoridated areas.

Rojas-Sanchez et al. (1999) estimated mean total daily fluoride intakes from foods, beverages, and dentifrice by 16- to 40-month-old children to be 0.767 mg (0.056 mg/kg) in a nonfluoridated community and 0.965 mg (0.070-0.073 mg/kg) in both a fluoridated community and a "halo" community. The higher mean dentifrice intake in the halo community than in the fluoridated community compensated for the lower dietary intake of fluoride in the halo community. Between 45% and 57% of children in the communities with higher daily fluoride intake exceeded the "upper estimated threshold limit" of 0.07 mg/kg, even without including any fluoride intake from supplements, mouth rinses, or gels in the study.

Erdal and Buchanan (2005), using a risk assessment approach based on EPA practices, estimated the cumulative (all sources combined) daily fluoride intake by infants (<1-year-old) in fluoridated areas to be 0.11 and 0.20 mg/kg for "central tendency" and "reasonable maximum exposure" conditions, respectively. For infants in nonfluoridated areas, the corresponding intakes were 0.08 and 0.11 mg/kg. For children aged 3-5, the estimated intakes were 0.06 and 0.23 mg/kg in fluoridated areas and 0.06 and 0.21 in nonfluoridated areas.

TOTAL EXPOSURE TO FLUORIDE

A systematic estimation of fluoride exposure from pesticides, background food, air, toothpaste, fluoride supplement, and drinking water is presented in this section. The estimated typical or average chronic exposures to inorganic fluoride from nonwater sources are presented in Table 2-9.

TABLE 2-9 Total Estimated Chronic Inorganic Fluoride Exposure from Nonwater Sources

Population Subgroups	Average Inorganic Fluoride Exposure, mg/kg/day							
	Sulfuryl Fluoride[a]	Cryolite[a]	Background Food[a]	Toothpaste[b]	Air[a]	Total Nonwater	Supplement[c]
All infants (<1 year)	0.0005	0.0009	0.0096	0	0.0019	0.0129	0.0357
Nursing	0.0003	0.0004	0.0046	0	0.0019	0.0078[d]	0.0357
Nonnursing	0.0006	0.0012	0.0114	0	0.0019	0.0151	0.0357
Children 1-2 years	0.0013	0.0031	0.0210	0.0115	0.0020	0.0389	0.0192
Children 3-5 years	0.0012	0.0020	0.0181	0.0114	0.0012	0.0339	0.0227
Children 6-12 years	0.0007	0.0008	0.0123	0.0075	0.0007	0.0219	0.0250
Youth 13-19 years	0.0004	0.0003	0.0097	0.0033	0.0007	0.0144	0.0167
Adults 20-49 years	0.0003	0.0004	0.0114	0.0014	0.0006	0.0141	0
Adults 50+ years	0.0003	0.0005	0.0102	0.0014	0.0006	0.0130	0
Females 13-49 years[e]	0.0003	0.0005	0.0107	0.0016	0.0006	0.0137	0

[a]Based on the exposure assessment by EPA (2004). Background food exposures are corrected for the contribution from powdered or dried tea at 987.72 ppm instead of 5 ppm used in EPA analysis.
[b]Based on Levy et al. (1995a), assuming two brushings per day with fluoride toothpaste (0.1% F) and moderate rinsing. The estimated exposures are: 0 mg/day for infants; 0.15 mg/day for 1-2 years; 0.25 mg/day for 3-5 years; 0.3 mg/day for 6-12 years; 0.2 mg/day for 13-19 years; 0.1 mg/day for all adults and females 13-49 years. The calculated exposure in mg/kg/day is based on the body weights from EPA (2004). For most age groups, these doses are lower than the purported maximum of 0.3 mg/day used for all age groups by EPA (2004).
[c]Based on ADA (2005) schedule (Table 2-8) and body weights from EPA (2004). Note that the age groups here do not correspond exactly to those listed by ADA (2005). The estimated exposures are: 0.25 mg/day for infant and 1-2 years; 0.5 mg/day for 3-5 years, and 1 mg/day for 6-12 years and 13-19 years.
[d]Includes the estimated 0.0006 mg/kg/day from breast milk. Using the higher estimated breast-milk exposure from a fluoridated area (approximately 0.0014 mg/kg/day) results in 0.0086 mg/kg/day for total nonwater exposure.
[e]Women of childbearing age.

The exposures from pesticides (sulfuryl fluoride and cryolite), background food, and air are from a recent exposure assessment by EPA (2004). The background food exposure is corrected for the contribution from powdered or dried tea by using the appropriate residue concentration of 897.72 ppm for instant tea powder instead of the 5 ppm for brewed tea used in the EPA (2004) analysis. It should be noted that the exposure from foods treated with sulfuryl fluoride is not applicable before its registration for post-harvest fumigation in 2004. The exposure from toothpaste is based on Levy et al. (1995a; see Table 2-7). The use of fluoride-containing toothpaste is assumed not to occur during the first year of life. Fluoride supplements are considered separately in Table 2-9 and are not included in the "total nonwater" column. Children 1-2 years old have the highest exposures from all nonwater source components. The two highest nonwater exposure groups are children 1-2 and 3-5 years old, at 0.0389 and 0.0339 mg/kg/day, respectively (Table 2-9). These doses are approximately 2.5-3 times those of adult exposures.

The estimated exposures from drinking water are presented in Table 2-10, using the DEEM-FCID model (version 2.03, Exponent Inc.). The water consumption data are based on the FCID translated from the CSFII 1994-1996 and 1998 surveys and represent an update to the information presented in Appendix B. The food forms for water coded as "direct, tap"; "direct, source nonspecified"; "indirect, tap"; and "indirect, source nonspecified" are assumed to be from local tap water sources. The sum of these four categories constitutes 66-77% of the total daily water intake. The remaining 23-34% is designated as nontap, which includes four food forms coded as "direct, bottled"; "direct, others"; "indirect, bottled"; and

TABLE 2-10 Estimated Chronic (Average) Inorganic Fluoride Exposure (mg/kg/day) from Drinking Water (All Sources)[a]

Population Subgroups	Fluoride Concentrations in Tap Water (fixed nontap water at 0.5 mg/L)				
	0 mg/L	0.5 mg/L	1.0 mg/L	2.0 mg/L	4.0 mg/L
All infants (<1 year)	0.0120	0.0345	0.0576	0.1040	0.1958
Nursing	0.0050	0.0130	0.0210	0.0370	0.0700
Nonnursing	0.0140	0.0430	0.0714	0.1290	0.2430
Children 1-2 years	0.0039	0.0157	0.0274	0.0510	0.0982
Children 3-5 years	0.0036	0.0146	0.0257	0.0480	0.0920
Children 6-12 years	0.0024	0.0101	0.0178	0.0330	0.0639
Youth 13-19 years	0.0018	0.0076	0.0134	0.0250	0.0484
Adults 20-49 years	0.0024	0.0098	0.0173	0.0320	0.0620
Adults 50+ years	0.0023	0.0104	0.0184	0.0340	0.0664
Females 13-49 years[b]	0.0025	0.0098	0.0171	0.0320	0.0609

[a]Estimated from DEEM-FCID model (version 2.03, Exponent Inc.). The water consumption data are based on diaries from the CSFII 1994-1996 and 1998 surveys that are transformed into food forms by the Food Commodity Intake Database (FCID). The food forms coded as "direct, tap"; "direct, source nonspecified"; "indirect, tap"; and "indirect, source nonspecified" are assumed to be from tap water sources.

[b]Women of childbearing age.

"indirect, others". Fluoride exposures from drinking water (Table 2-10) are estimated for different concentrations of fluoride in the local tap water (0, 0.5, 1.0, 2.0, or 4.0 mg/L), while assuming a fixed 0.5 mg/L for all nontap sources (e.g., bottled water). The assumption for nontap water concentration is based on the most recent 6-year national public water system compliance monitoring from a 16-state cross section that represents approximately 41,000 public water systems, showing average fluoride concentrations of 0.482 mg/L in groundwater and 0.506 mg/L in surface water (EPA 2003a). The reported best estimates for exceeding 1.2, 2, and 4 mg/L in surface-water source systems are 9.37%, 1.11%, and 0.0491%, respectively; for groundwater source systems, the respective estimates are 8.54%, 3.05%, and 0.55%. Table 2-10 shows that nonnursing infants have the highest exposure from drinking water. The estimated daily drinking-water exposures at tap-water concentrations of 1, 2, and 4 mg/L are 0.0714, 0.129, and 0.243 mg/kg, respectively. These values are approximately 2.6 times those for children 1-2 and 3-5 years old and 4 times the exposure of adults.

The estimated total fluoride exposures aggregated from all sources are presented in Table 2-11. These values represent the sum of exposures from Table 2-9 and 2-10, assuming fluoride supplements might be given to infants and children up to 19 years old in low-fluoride tap-water scenarios (0 and 0.5 mg/L). Table 2-11 shows that, when tap water contains fluoride, nonnursing infants have the highest total exposure. They are 0.087, 0.144, and 0.258 mg/kg/day in tap water at 1, 2, and 4 mg/L, respectively. At 4 mg/L, the total exposure for nonnursing infants is approximately twice the exposure for children 1-2 and 3-5 years old and 3.4 times the exposure for adults.

The relative source contributions to the total exposure in Table 2-11 for scenarios with 1, 2, and 4 mg/L in tap water are illustrated in Figures 2-1, 2-2, and 2-3, respectively. Numerical values for the 1-, 2-, and 4-mg/L scenarios are given later in the summary tables (Tables 2-13, 2-14, and 2-15). Under the assumptions for estimating the exposure, the contribution from pesticides plus fluoride in the air is within 4% to 10% for all population subgroups at 1 mg/L in tap water, 3-7% at 2 mg/L in tap water, and 1-5% at 4 mg/L in tap water. The contributions from the remaining sources also vary with different tap-water concentrations. For nonnursing infants, who represent the highest total exposure group even without any exposure from toothpaste, the contribution from drinking water is 83% for 1 mg/L in tap water (Figure 2-1). As the tap-water concentration increases to 2 and 4 mg/L, the relative drinking-water contribution increases to 90% and 94%, respectively (Figures 2-2 and 2-3). The proportion of the contribution from all sources also varies in children 1-2 and 3-5 years old. At 1 mg/L, the drinking-water contribution is approximately 42%, while the contributions from toothpaste and background food are sizable, approximately 18% and

TABLE 2-11 Total Estimated (Average) Chronic Inorganic Fluoride Exposure (mg/kg/day) from All Sources, Assuming Nontap Water at a Fixed Concentration[a]

Population Subgroups	Concentration in Tap Water (fixed nontap water at 0.5 mg/L)						
	With Fluoride Supplement		Without Fluoride Supplement				
	0 mg/L	0.5 mg/L	0 mg/L	0.5 mg/L	1 mg/L	2 mg/L	4 mg/L
All infants (<1 year)	0.061	0.083	0.025	0.047	0.070	0.117	0.209
Nursing[b]	0.049	0.057	0.013	0.021	0.030	0.046	0.079
Nonnursing	0.065	0.094	0.029	0.058	0.087	0.144	0.258
Children 1-2 years	0.062	0.074	0.043	0.055	0.066	0.090	0.137
Children 3-5 years	0.060	0.071	0.038	0.049	0.060	0.082	0.126
Children 6-12 years	0.049	0.057	0.024	0.032	0.040	0.055	0.086
Youth 13-19 years	0.033	0.039	0.016	0.022	0.028	0.039	0.063
Adults 20-49 years	0.017	0.024	0.017	0.024	0.031	0.046	0.076
Adults 50+ years	0.015	0.023	0.015	0.023	0.031	0.047	0.079
Females 13-49 years[c]	0.016	0.024	0.016	0.024	0.031	0.046	0.075

[a]The estimated exposures from fluoride supplements and total nonwater sources (including pesticides, background food, air, and toothpaste) are from Table 2-9. The estimated exposures from drinking water are from Table 2-10. For nonfluoridated areas (tap water at 0 and 0.5 mg/L), the total exposures are calculated both with and without fluoride supplements.

[b]The higher total nonwater exposure of 0.0086 mg/kg/day that includes breast milk from a fluoridated area (footnote in Table 2-9) is used to calculate the exposure estimates for the "without supplement" groups that are exposed to fluoride in water at 1, 2, and 4 mg/L.

[c]Women of childbearing age.

31%, respectively (Figure 2-1). At 2 mg/L, the drinking-water contribution is raised to approximately 57%, while the contributions from toothpaste and background food are reduced to 13% and 23%, respectively (Figure 2-2). At 4 mg/L, the relative contribution of drinking water continues to increase to approximately 72%, while the contribution from toothpaste and background food are further reduced to approximately 9% and 15%, respectively (Figure 2-3). As age increases toward adulthood (20+ years), the contribution from toothpaste is reduced to approximately 5% at 1 mg/L, 3-4% at 2 mg/L, and 2% at 4 mg/L. Correspondingly, the contribution from drinking water increases to approximately 57% at 1 mg/L, 70% at 2 mg/L, and 82% at 4 mg/L.

Data presented in Tables 2-9 to 2-11 are estimates of typical exposures, while the actual exposure for an individual could be lower or higher. There are inherent uncertainties in estimating chronic exposure based on the 2-day CSFII surveys. The DEEM-FCID model assumes that the average

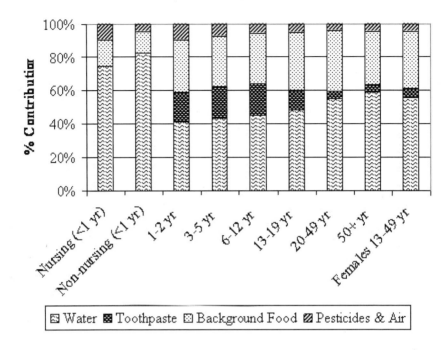

FIGURE 2-1 Source contribution to total inorganic fluoride exposure, including fluoride at 1 mg/L in tap water. The estimated chronic inorganic fluoride exposures from the various routes are presented in Tables 2-9 and 2-10. No fluoride supplement is included for any population subgroup. The total exposures as presented in Table 2-11 for the population subgroups are: 0.030 mg/kg/day (nursing infants), 0.087 mg/kg/day (non-nursing infants), 0.066 mg/kg/day (1-2 years old), 0.060 mg/kg/day (3-5 years old), 0.040 mg/kg/day (6-12 years old), 0.028 mg/kg/day (13-19 years old), and 0.031 mg/kg/day for adults (20 to 50+ years old) and women of childbearing age (13-49 years old).

intake from the cross-sectional survey represents the longitudinal average for a given population. Thus, the chronic exposures of those who have persistently high intake rates, especially for food items that contain high concentrations of fluoride (e.g., tea), are likely to be underestimated. For example, at an average fluoride concentration of 3.3 mg/L for brewed tea and 0.86 mg/L for iced tea (USDA 2004), the tea component in the background food presented in Table 2-9 represents an average daily consumption of one-half cup of brewed tea or 2 cups of iced tea. A habitual tea drinker, especially for brewed tea, can be expected to significantly exceed these con-

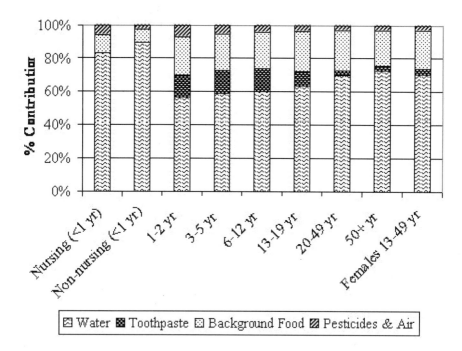

FIGURE 2-2 Source contribution to total inorganic fluoride exposure, including fluoride at 2 mg/L fluoride in tap water. The estimated chronic inorganic fluoride exposures from the various routes are presented in Tables 2-9 and 2-10. No fluoride supplement is included for any population subgroup. The total exposures as presented in Table 2-11 for the population subgroups are: 0.046 mg/kg/day (nursing infants), 0.144 mg/kg/day (non-nursing infants), 0.090 mg/kg/day (1-2 years old), 0.082 mg/kg/day (3-5 years old), 0.055 mg/kg/day (6-12 years old), 0.039 mg/kg/day (13-19 years old), and 0.046-0.047 mg/kg/day for adults (20-50+ years old) and women of childbearing age (13-49 years old).

sumption rates. Other groups of people who are expected to have exposures higher than those calculated here include infants given fluoride toothpaste before age 1, anyone who uses toothpaste more than twice per day or who swallows excessive amounts of toothpaste, children inappropriately given fluoride supplements in a fluoridated area, children in an area with high fluoride concentrations in soil, and children with pica who consume large amounts of soil.

The exposure estimates presented in this chapter for non-drinking-water routes are based on the potential profile of fluoride residue concentrations

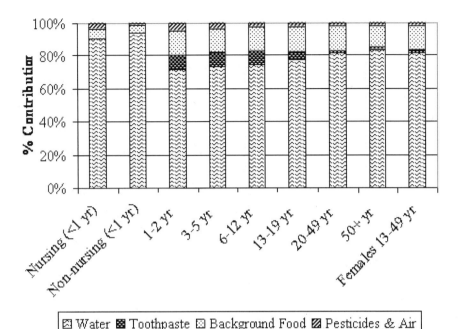

FIGURE 2-3 Source contribution to total inorganic fluoride exposure, including fluoride at 4 mg/L in tap water. The estimated chronic inorganic fluoride exposures from the various routes are presented in Tables 2-9 and 2-10. No fluoride supplement is included for any population subgroup. The total exposures as presented in Table 2-11 for the population subgroups are: 0.079 mg/kg/day (nursing infants), 0.258 mg/kg/day (nonnursing infants), 0.137 mg/kg/day (1-2 years old), 0.126 mg/kg/day (3-5 years old), 0.086 mg/kg/day (6-12 years old), 0.063 mg/kg/day (13-19 years old), 0.075-0.079 mg/kg/day for adults (20-50+ years old) and women of childbearing age (13-49 years old).

in the current exposure media. They likely do not reflect the concentration of past exposure scenarios, particularly for routes that show changes in time (e.g., pesticide use practices). Any new and significant source of fluoride exposure, such as commodities approved for sulfuryl fluoride fumigation application beyond April 2005, is expected to alter the percentage of drinking water contribution as presented in this chapter.

Different assumptions for the drinking-water concentration alone also can result in slightly different estimates. For example, values in Table 2-11 are derived from assuming that the nontap water has a fixed fluoride concentration of 0.5 mg/L, while tap-water concentration varies up to 4 mg/L. Table 2-12 provides alternative calculations of total exposure by assuming

TABLE 2-12 Total Estimated (Average) Chronic Inorganic Fluoride Exposure (mg/kg/day) from All Sources, Assuming the Same Specified Fluoride Concentration for Both Tap and Nontap Waters[a]

	Concentration in All Water					
	1 mg/L	2 mg/L	4 mg/L	1 mg/L	2 mg/L	4 mg/L
Population Subgroups	Modeled water intake[b]			EPA default water intake[c]		
All infants (<1 year)	0.082	0.151	0.289	0.113	0.213	0.413
Nursing	0.034	0.060	0.111	0.109	0.209	0.409
Nonnursing	0.100	0.186	0.357	0.115	0.215	0.415
Children 1-2 years	0.070	0.102	0.164	0.139	0.239	0.439
Children 3-5 years	0.063	0.093	0.151	NA	NA	NA
Children 6-12 years	0.042	0.062	0.103	NA	NA	NA
Youth 13-19 years	0.030	0.045	0.075	NA	NA	NA
Adults 20-49 years	0.034	0.053	0.093	0.043	0.071	0.128
Adults 50+ years	0.034	0.054	0.096	0.042	0.070	0.127
Females 13-49 years[d]	0.033	0.053	0.092	0.042	0.071	0.128

[a]The estimated exposures from nonwater sources (including pesticides, background food, air, and toothpaste) are from Table 2-9. No fluoride supplement is included in the total fluoride exposure estimates.
[b]The component of drinking-water exposure is estimated from DEEM-FCID.
[c]The EPA default daily water intake rate is 1 L for a 10-kg child and 2 L for a 70-kg adult. NA: not applicable based on EPA's default body weight.
[d]Women of childbearing age.

that all sources of drinking water (both tap and nontap water) contain the same specified fluoride concentration. Within this assumption, the drinking-water component can be estimated from either the DEEM-FCID model or the default drinking-water intake rate currently used by EPA for establishing the MCL (1 L/day for a 10-kg child and 2 L/day for a 70-kg adult).

Some uncertainties exist regarding the extent the FCID database may include all processed waters (e.g., soft drinks and soups). Thus, the exposure using EPA's defaults as presented in Table 2-12 can serve as a bounding estimate from the water contribution. The difference in the total fluoride exposure calculated from the two water intake methods (i.e., EPA defaults versus FCID modeled) varies with different population subgroups shown in Table 2-12. In general, as the drinking-water contribution to the total exposure becomes more prominent at higher drinking-water concentration, the differences in total exposure approach the differences in drinking-water intake rates of the two methods. Using EPA's default adult water intake rate of 28.6 mL/kg/day (based on 2 L/day for a 70 kg adult) results in approximately 32-39% higher total exposure than the model estimates. This approximates the 38-45% lower model estimate of total water intake rate (i.e., 19.7 mL/kg/day for 20-49 year olds, 20.7 mL/kg/day for 50+ year olds). Using EPA's default water intake rate for a child results in approximately 16% higher total exposure than the model estimates for nonnursing infants at 4 mg/L drinking water. This reflects closely the difference in the total water intake between the default 100 mL/kg/day (based on 1 L/day for a 10 kg child) and the DEEM-FCID estimate of 85.5 mL/kg/day for this population group. Similarly, for nursing infants, the 3.7-fold higher total exposure at 4 mg/L from using the EPA's default of 100 mL/kg/day also reflects their significantly lower model estimate of total water intake (i.e., 25.6 mL/kg/day). Two additional simple conceptual observations can be made to relate data presented in Table 2-12 to those in Tables 2-9 and 2-11. By using a fixed rate of water intake for infants and children 1-2 years old, the difference in their total exposure is due to the contribution from all nonwater sources as presented in Table 2-9. The difference between model estimates presented in Table 2-11 (last 3 columns) by varying concentrations for tap water alone (with fixed nontap water at 0.5 mg/L) and estimates using one fluoride concentration for both tap and nontap waters in Table 2-12 (first 3 columns) reflects the contribution from the nontap-water component.

The fluoride exposure estimates presented thus far, regardless of the various assumptions (e.g., the same versus different fluoride concentrations in tap and nontap water) and different water intake rates (e.g., EPA default versus estimates from FCID database of the CSFII surveys), do not include those who have sustained high water intake rates as noted previously (athletes, workers, and individuals with diabetes mellitus or nephrogenic diabetes insipidus (see Table 2-4). The high-end exposures for these high-water-consumption population subgroups are included in the summaries below.

SUMMARY OF EXPOSURE ASSESSMENT

The estimated aggregated total fluoride exposures from pesticides, background food, air, toothpaste, and drinking water are summarized for drinking water fluoride concentrations of 1 mg/L (Table 2-13), 2 mg/L (Table 2-14), and 4 mg/L (Table 2-15). Two sets of exposures are presented using different approaches to estimate the exposure from drinking water. One is estimated by modeling water intakes based on FCID data and assuming a fixed nontap water concentration of 0.5 mg/L. The other is estimated using EPA default drinking-water intake rates (i.e., 1 L/day for a 10 kg child, 2 L/day for a 70 kg adult) and assuming the same concentration for tap and nontap waters. Both sets of estimates include the same fluoride exposure from nonwater sources. The total exposure from the latter approach is higher than the model estimates due to the higher default drinking water intake rates and the assumption that nontap waters contain the same concentration of fluoride residue as the tap water.

TABLE 2-13 Contributions to Total Fluoride Chronic Exposure at 1 mg/L in Drinking Water

Population Subgroups	Total Exposure, mg/kg/day	% Contribution to Total Exposure			
		Pesticides and Air	Background Food	Tooth-paste	Drinking Water
Modeled average water consumer					
(Tap water at 1 mg/L, nontap water at 0.5 mg/L; Table 2-11)					
All infants (<1 year)	0.070	4.7	13.6	0	81.7
Nursing	0.030	8.9	15.6	0	70.8
Nonnursing	0.087	4.3	13.2	0	82.5
Children 1-2 years	0.066	9.7	31.7	17.4	41.3
Children 3-5 years	0.060	7.4	30.4	19.1	43.1
Children 6-12 years	0.040	5.4	30.9	18.9	44.8
Youth 13-19 years	0.028	4.9	34.8	12.0	48.3
Adults 20-49 years	0.031	4.0	36.3	4.6	55.1
Adults 50+ years	0.031	4.4	32.4	4.6	58.7
Females 13-49 years[a]	0.031	4.4	34.7	5.3	55.6
EPA default water intake, all water at 1 mg/L					
(1 L/day for 10-kg child; 2 L/day for 70-kg adult; Table 2-12)					
All infants (<1 year)	0.113	2.9	8.5	0	88.6
Nursing	0.109	2.4	4.3	0	92.0
Nonnursing	0.115	3.2	9.9	0	86.9
Children 1-2 years	0.139	4.6	15.1	8.3	72.0
Adults 20-49 years	0.043	3.0	26.7	3.3	67.0
High end of high water intake individuals all water at 1 mg/L					
(based on intake rates in Table 2-4)					
Athletes and workers	0.084	1.5	13.5	1.7	83.3
DM patients (3-5 years)	0.134	3.3	13.5	8.5	74.7
DM patients (adults)	0.084	1.5	13.5	1.7	83.3
NDI patients (3-5 years)	0.184	2.4	9.9	6.2	81.6
NDI patients (adults)	0.164	0.8	6.9	0.9	91.4

[a]Women of childbearing age.

ABBREVIATIONS: DM, diabetes mellitus; NDI, nephrogenic diabetes insipidus.

Although each of these exposure estimates have areas of uncertainty, the average total daily fluoride exposure is expected to fall between them. For the modeling estimates, there are inherent uncertainties in modeling long-term intake rates based on the cross-sectional CSFII dietary survey data. Thus, the exposure from any dietary component, water or other foods, could be underestimated for individuals who have habitually higher intake rates (e.g., water, tea). Specific to the water component, there are also uncertainties regarding the extent the FCID database may include all processed waters (e.g., soft drinks and soups). On the other hand, the EPA

TABLE 2-14 Contributions to Total Fluoride Chronic Exposure at 2 mg/L in Drinking Water

Population Subgroups	Total Exposure, mg/kg/day	% Contribution to Total Exposure			
		Pesticides and Air	Background Food	Tooth-paste	Drinking Water
Modeled average water consumer					
(Tap water at 2 mg/L, nontap water at 0.5 mg/L; Table 2-11)					
All infants (<1 year)	0.117	2.8	8.2	0	89.0
Nursing	0.046	5.8	10.1	0	81.0
Nonnursing	0.144	2.6	7.9	0	89.5
Children 1-2 years	0.090	7.1	23.3	12.8	56.7
Children 3-5 years	0.082	5.4	22.1	13.9	58.6
Children 6-12 years	0.055	3.9	22.4	13.7	60.1
Youth 13-19 years	0.039	3.5	24.5	8.5	63.5
Adults 20-49 years	0.046	2.8	24.7	3.1	69.4
Adults 50+ years	0.047	2.9	21.7	3.0	72.4
Females 13-49 years[a]	0.046	3.0	23.4	3.6	70.1
EPA default water intake, all water at 1 mg/L					
(2 L/day for 10-kg child; 2 L/day for 70-kg adult; Table 2-12)					
All infants (<1 year)	0.213	1.6	4.5	0	93.9
Nursing	0.209	1.3	2.2	0	95.8
Nonnursing	0.215	1.7	5.3	0	93.0
Children 1-2 years	0.239	2.7	8.8	4.8	83.7
Adults 20-49 years	0.071	1.8	16.0	2.0	80.2
High end of high water intake individuals all water at 2 mg/L					
(based on intake rates in Table 2-4)					
Athletes and workers	0.154	0.8	7.4	0.9	90.9
DM patients (3-5 years)	0.234	1.9	7.7	4.9	85.5
DM patients (adults)	0.154	0.8	7.4	0.9	90.9
NDI patients (3-5 years)	0.334	1.3	5.4	3.4	89.9
NDI patients (adults)	0.314	0.4	3.6	0.5	95.5

[a]Women of childbearing age.

ABBREVIATIONS: DM, diabetes mellitus; NDI, nephrogenic diabetes insipidus.

default water intake rate is likely higher than the average rate for certain population subgroups (e.g., nursing infants).

The estimates presented in Tables 2-13, 2-14, and 2-15 show that on a per body weight basis, the exposures are generally higher for young children than for the adults. By assuming that the nontap water concentration is fixed at 0.5 mg/L, nonnursing infants have the highest model-estimated average total daily fluoride exposure: 0.087, 0.144, and 0.258 mg/kg/day when tap-water concentrations of fluoride are 1, 2, and 4 mg/L, respectively (Table

TABLE 2-15 Contributions to Total Fluoride Chronic Exposure at 4 mg/L in Drinking Water

Population Subgroups	Total Exposure, mg/kg/day	% Contribution to Total Exposure			
		Pesticides and Air	Background Food	Tooth-paste	Drinking Water
Modeled average water consumer (Tap water at 4 mg/L, nontap water at 0.5 mg/L; Table 2-11)					
All infants (<1 year)	0.209	1.6	4.6	0	93.9
Nursing	0.079	3.3	5.9	0	89.0
Nonnursing	0.258	1.4	4.4	0	94.1
Children 1-2 years	0.137	4.7	15.3	8.4	71.6
Children 3-5 years	0.126	3.5	14.4	9.0	73.1
Children 6-12 years	0.086	2.5	14.3	8.7	74.5
Youth 13-19 years	0.063	2.2	15.4	5.3	77.1
Adults 20-49 years	0.076	1.7	15.0	1.9	81.5
Adults 50+ years	0.079	1.7	12.8	1.8	83.7
Females 13-49 years[a]	0.075	1.8	14.3	2.2	81.7
EPA default water intake all water at 4 mg/L (1 L/day for 10-kg child; 2 L/day for 70-kg adult; Table 2-12)					
All infants (<1 year)	0.413	0.8	2.3	0	96.9
Nursing	0.409	0.6	1.1	0	97.9
Nonnursing	0.415	0.9	2.8	0	96.4
Children 1-2 years	0.439	1.5	4.8	2.6	91.1
Adults 20-49 years	0.128	1.0	8.9	1.1	89.0
High end of high water intake individuals, all water at 4 mg/L (based on intake rates in Table 2-4)					
Athletes and workers	0.294	0.4	3.9	0.5	95.2
DM patients (3-5 years)	0.434	1.0	4.2	2.6	92.2
DM patients (adults)	0.294	0.4	3.9	0.5	95.2
NDI patients (3-5 years)	0.634	0.7	2.9	1.8	94.7
NDI patients (adults)	0.614	0.2	1.9	0.2	97.7

[a]Women of childbearing age.

ABBREVIATIONS: DM, diabetes mellitus; NDI, nephrogenic diabetes insipidus

2-11, and Tables 2-13, 2-14, and 2-15). The major contributing factor is their much higher model-estimated drinking-water exposure than other age groups (Table 2-10). The total exposures of nonnursing infants are approximately 2.8-3.4 times that of adults. By holding the exposure from drinking water at a constant with the EPA default water intake rates, children 1-2 years old have slightly higher total exposure than the nonnursing infants, reflecting the higher exposure from nonwater sources (Table 2-9). The estimated total fluoride exposures for children 1-2 years old are 0.139, 0.239, and 0.439 mg/kg/day for 1, 2, and 4 mg/L of fluoride in drinking water, respectively (Tables 2-13, 2-14, 2-15). These exposures are approximately 3.4 times that of adults. The estimated total exposure for children 1-2 years old and adults at 4 mg/L fluoride in drinking water is approximately two times the exposure at 2 mg/L and three times the exposure at 1 mg/L.

The estimated total daily fluoride exposures for three population subgroups with significantly high water intake rates are included in Tables 2-13, 2-14, and 2-15. The matching age groups for data presented in Table 2-4 are: adults ≥ 20 years old for the athletes and workers, and both children 3-5 years old (default body weight of 22 kg) and adults for individuals with diabetes mellitus and nephrogenic diabetes insipidus. In estimating the total exposure, the high-end water intake rates from Table 2-4 are used to calculate the exposure from drinking water. The total exposures for adult athletes and workers are 0.084, 0.154, and 0.294 mg/kg/day at 1, 2, and 4 mg/L of fluoride in water, respectively. These doses are approximately two times those of the adults with a default water intake rate of 2 L/day. For individuals with nephrogenic diabetes insipidus, the respective total fluoride exposures for children (3-5 years old) and adults are 0.184 and 0.164 mg/kg/day at 1 mg/L, 0.334 and 0.314 mg/kg/day at 2 mg/L, and 0.634 and 0.614 mg/kg/day at 4 mg/L. Compared to the exposure of children 1-2 years old, who have the highest total exposure among all age groups of the general population (i.e., 0.139-0.439 mg/kg/day at 1-4 mg/L, assuming EPA's 100 mL/kg/day default water intake rate for children), the highest estimated total exposure among these high water intake individuals (i.e., 0.184-0.634 mg/kg/day for children 3-5 years old with nephrogenic diabetes insipidus, assuming 150 mL/kg/day high-end water intake rate) are 32-44% higher.

The relative contributions from each source of exposure are also presented in Tables 2-13, 2-14, and 2-15. For an average individual, the model-estimated drinking-water contribution to the total fluoride exposure is 41-83% at 1 mg/L in tap water, 57-90% at 2 mg/L, and 72-94% at 4mg/L in tap water (see also Figures 2-1, 2-2, and 2-3). Assuming that all drinking-water sources (tap and nontap) contain the same fluoride concentration and using the EPA default drinking-water intake rates, the drinking-water contribution is 67-92% at 1 mg/L, 80-96% at 2 mg/L, and 89-98% at 4 mg/L. The drinking-water contributions for the high water intake individuals among adult athletes and workers, and individuals with diabetes mellitus and nephrogenic diabetes insipidus, are 75-91% at 1 mg/L, 86-96% at 2 mg/L, and 92-98% at 4 mg/L.

As noted earlier, these estimates were based on the information that was available to the committee as of April 2005. Any new and significant sources of fluoride exposure are expected to alter the percentage of drinking-water contribution as presented in this chapter. However, water will still be the most significant source of exposure.

BIOMARKERS OF EXPOSURE, EFFECT, AND SUSCEPTIBILITY

Biological markers, or biomarkers, are broadly defined as indicators of variation in cellular or biochemical components or processes, structure, or function that are measurable in biological systems or samples (NRC 1989a). Biomarkers often are categorized by whether they indicate exposure to an agent, an effect of exposure, or susceptibility to the effects of exposure (NRC 1989a). Vine (1994) described categories of biological markers in terms of internal dose, biologically effective dose, early response, and disease, plus susceptibility factors that modify the effects of the exposure. Factors that must be considered in selecting a biomarker for a given study include the objectives of the study, the availability and specificity of potential markers, the feasibility of measuring the markers (including the invasiveness of the necessary techniques and the amount of biological specimen needed), the time to appearance and the persistence of the markers in biological media, the variability of marker concentrations within and between individuals, and aspects (e.g., cost, sensitivity, reliability) related to storage and analysis of the samples (Vine 1994). ATSDR (2003) recently reviewed biomarkers of exposure and effect for fluoride.

Biomarkers of exposure to fluoride consist of measured fluoride concentrations in biological tissues or fluids that can be used as indices of an individual's exposure to fluoride. For fluoride, concentrations in a number of tissues and fluids, including teeth, bones, nails, hair, urine, blood or plasma, saliva, and breast milk, have been used to estimate exposures (Vine 1994; Whitford et al. 1994; ATSDR 2003). Table 2-16 gives examples of measurements in humans together with the associated estimates of exposure. The Centers for Disease Control and Prevention (CDC 2003, 2005) has measured a number of chemicals in blood or urine of members of the U.S. population, but thus far fluoride has not been included in their survey.

Fluoride concentrations in bodily fluids (e.g., urine, plasma, serum, saliva) are probably most suitable for evaluating recent or current fluoride exposures or fluoride balance (intake minus excretion), although some sources indicate that samples obtained from fasting persons may be useful for estimating chronic fluoride intake or bone fluoride concentrations (e.g., Ericsson et al. 1973; Waterhouse et al. 1980). Examples of the association between estimated fluoride intakes (or mass-normalized intakes) and measured fluoride concentrations in urine, plasma, and serum for individuals and groups are shown in Figures 2-4, 2-5, 2-6, and 2-7. Note that in most cases, the variation in fluoride intake is not sufficient to explain the variation in the measured fluoride concentrations. A number of parameters affect individual fluoride uptake, retention, and excretion (Chapter 3) (Whitford 1996). In addition, a significant decrease in fluoride exposure might not be

TABLE 2-16 Summary of Selected Biomarkers for Fluoride Exposure in Humans

Fluoride Exposure	Number of Persons	Fluoride Concentration	Reference
Urine			
1.2-2.2 mg/day	5	0.8-1.2 mg/day	Teotia et al. 1978
2.5-3.8 mg/day[a]	2	1.2-2.2 mg/day	(Figure 2-4)
8.7-9.2 mg/day	3	3.2-5.8 mg/day	
21.0-28.0 mg/day	2	10.0-11.0 mg/day	
48.0-52.0 mg/day	2	15.0-18.5 mg/day	
1.0 mg/L in drinking water	17	1.5 (0.2) mg/L	Bachinskii et al. 1985
		1.9 (0.3) mg/day	(Figure 2-6)
2.3 mg/L in drinking water	30	2.4 (0.2) mg/L	
		2.7 (0.2) mg/day	
0.09 (range, 0.06-0.11) mg/L in drinking water	45	0.15 (0.07) mg/L[b]	Schamschula et al. 1985 (Figure 2-6)
0.82 (range, 0.5-1.1) mg/L in drinking water	53	0.62 (0.26) mg/L[b]	
1.91 (range, 1.6-3.1) mg/L in drinking water	41	1.24 (0.52) mg/L[b]	
0.32 mg/L in drinking water	100	0.77 (0.49) mg/L[b]	Czarnowski et al. 1999
1.69 mg/L in drinking water	111	1.93 (0.82) mg/L[b]	(Figure 2-6)
2.74 mg/L in drinking water	89	2.89 (1.39) mg/L[b]	
About 3 mg/day	1	2.30-2.87 mg/day	Whitford et al. 1999a
About 6 mg/day	1	4.40-5.13 mg/day	
7.35 (1.72) mg/day[b]	50	9.45 (4.11) mg/L[b]	Gupta et al. 2001
11.97 (1.8) mg/day[b]	50	15.9 (9.98) mg/L[b]	(Figure 2-7)
14.45 (3.19) mg/day[a]	50	17.78 (7.77) mg/L[a]	
32.56 (9.33) mg/day[a]	50	14.56 (7.88) mg/L[a]	
0.93 (0.39) mg/day[b] [0.053 (0.021) mg/kg/day[b]]	11	0.91 (0.45) mg/L[b]	Haftenberger et al. 2001 (Figure 2-5)
1.190 (0.772) mg/day from all sources[b]	20	0.481 (0.241) mg/day[b]	Pessan et al. 2005
Plasma			
1.2-2.2 mg/day	5	0.020-0.038 mg/L	Teotia et al. 1978
2.5-3.8 mg/day	2	0.036-0.12 mg/L	(Figure 2-4)
8.7-9.2 mg/day	3	0.15-0.18 mg/L	
21.0-28.0 mg/day	2	0.11-0.17 mg/L	
48.0-52.0 mg/day	2	0.14-0.26 mg/L	
Serum			
1.0 mg/L in drinking water	17	0.21 (0.01) mg/L	Bachinskii et al. 1985
2.3 mg/L in drinking water	30	0.25 (0.01) mg/L	(Figure 2-6)
7.35 (1.72) mg/day[b]	50	0.79 (0.21) mg/L[b]	Gupta et al. 2001
11.97 (1.8) mg/day[b]	50	1.10 (0.58) mg/L[b]	(Figure 2-7)
14.45 (3.19) mg/day[b]	50	1.10 (0.17) mg/L[b]	
32.56 (9.33) mg/day[b]	50	1.07 (0.17) mg/L[b]	

TABLE 2-16 Continued

Fluoride Exposure	Number of Persons	Fluoride Concentration	Reference
0.3 mg/L in drinking water:			Hossny et al. 2003
Breastfed infants	48	0.0042 (0.0027) mg/L[b]	
All infants (4 weeks-2 years)	97	0.0051 (0.0030) mg/L[b]	
Preschoolers (2-6 years)	100	0.011 (0.0049) mg/L[b]	
Primary schoolers (6-12 years)	99	0.010 (0.0042) mg/L[b]	
Saliva			
0.09 (range, 0.06-0.11) mg/L in drinking water	45	6.25 (2.44) µg/L[b]	Schamschula et al. 1985
0.82 (range, 0.5-1.1) mg/L in drinking water	53	11.23 (4.29) µg/L[b]	
1.91 (range, 1.6-3.1) mg/L in drinking water	41	15.87 (6.01) µg/L[b]	
0.1 mg/L in drinking water	27	1.9-55.1 µg/L	Oliveby et al. 1990
1.2 mg/L in drinking water	27	1.9-144 µg/L	Oliveby et al. 1990
Plaque			
0.09 (range, 0.06-0.11) mg/L in drinking water	45	5.04 (4.60) ppm[b]	Schamschula et al. 1985
0.82 (range, 0.5-1.1) mg/L in drinking water	53	8.47 (9.69) ppm[b]	
1.91 (range, 1.6-3.1) mg/L in drinking water	41	19.6 (19.3) ppm[b]	
Hair			
0.09 (range, 0.06-0.11) mg/L in drinking water	45	0.18 (0.07) µg/g[b]	Schamschula et al. 1985
0.82 (range, 0.5-1.1) mg/L in drinking water	53	0.23 (0.11) µg/g[b]	
1.91 (range, 1.6-3.1) mg/L in drinking water	41	0.40 (0.25) µg/g[b]	
0.27 mg/L in drinking water and 2.8 µg/m^3 in air	59	1.35 (0.95) µg/g[b]	Hac et al. 1997
0.32 mg/L in drinking water	53	4.13 (2.24) µg/g[b]	Czarnowski et al. 1999
1.69 mg/L in drinking water	111	10.25 (6.63) µg/g[b]	
2.74 mg/L in drinking water	84	14.51 (6.29) µg/g[b]	
Breast milk			
0.2 mg/L in drinking water	47	0.0053 mg/L (colostrum)	Spak et al. 1983

continued

TABLE 2-16 Continued

Fluoride Exposure	Number of Persons	Fluoride Concentration	Reference
1.0 mg/L in drinking water	79	0.0068 mg/L (colostrum)	
1.0 mg/L in drinking water	17	0.007 mg/L (mature milk)	
Nonfluoridated community	32	0.0044 mg/L	Dabeka et al. 1986
1 mg/L in drinking water	112	0.0098 mg/L	
22.1 mg/day (mean)	27	0.011-0.073 mg/L	Opinya et al. 1991
0.3 mg/L in drinking water	60	0.0046 (0.0025) mg/L[b]	Hossny et al. 2003
Fingernails			
0.09 (range, 0.06-0.11) mg/L in drinking water	45	0.79 (0.26) ppm[b]	Schamschula et al. 1985
0.82 (range, 0.5-1.1) mg/L in drinking water	53	1.31 (0.49) ppm[b]	
1.91 (range, 1.6-3.1) mg/L in drinking water	41	2.31 (1.14) ppm[b]	
About 3 mg/day	1	1.94-3.05 mg/kg	Whitford et al. 1999a
About 6 mg/day (after 3.5 months)	1	4.52-5.38 mg/kg	
0.1 mg/L in drinking water	10	0.75-3.53 mg/kg	
1.6 mg/L in drinking water	6	2.28-7.53 mg/kg	
2.3 mg/L in drinking water	9	4.00-13.18 mg/kg	
0.7-1.0 mg/L in drinking water, without fluoride dentifrice	10	2.3-7.3 mg/kg	Corrêa Rodrigues et al. 2004
0.7-1.0 mg/L in drinking water, with fluoride dentifrice (after 4 months)	10	10.1 mg/kg (peak)	
0.004 ± 0.003 mg/kg/day	15	0.42-6.11 μg/g	Levy et al. 2004
0.029 ± 0.029 mg/kg/day	15	0.87-7.06 μg/g	
Toenails			
0.09 mg/L in drinking water		4.2 ppm	Feskanich et al. 1998
1.0 mg/L in drinking water		6.4 ppm	
3 mg/day	1	1.41-1.60 mg/kg	Whitford et al. 1999a
0.7-1.0 mg/L in drinking water, without fluoride dentifrice	10	2.5-5.6 mg/kg	Corrêa Rodrigues et al. 2004
0.7-1.0 mg/L in drinking water, with fluoride dentifrice (after 4 months)	10	9.2 mg/kg (peak)	
0.004 ± 0.003 mg/kg/day	15	0.08-3.89 μg/g	Levy et al. 2004
0.029 ± 0.029 mg/kg/day	15	0.81-6.38 μg/g	
Teeth			
Normal	NA	190-300 ppm (total ash)	Roholm 1937

TABLE 2-16 Continued

Fluoride Exposure	Number of Persons	Fluoride Concentration	Reference
Cryolite workers	5	1,100-5,300 ppm (total ash)	
Enamel (0.44-0.48 μm depth)			
0.09 (range, 0.06-0.11) mg/L in drinking water	45	1,549 (728) ppm[b]	Schamschula et al. 1985
0.82 (range, 0.5-1.1) mg/L in drinking water	53	2,511 (1,044) ppm[b]	
1.91 (range, 1.6-3.1) mg/L in drinking water	41	3,792 (1,362) ppm[b]	
Enamel (2.44-2.55 μm depth)			
0.09 (range, 0.06-0.11) mg/L in drinking water	45	641 (336) ppm[b]	Schamschula et al. 1985
0.82 (range, 0.5-1.1) mg/L in drinking water	53	1,435 (502) ppm[b]	
1.91 (range, 1.6-3.1) mg/L in drinking water	41	2,107 (741) ppm[b]	
Enamel			
0.7 or 1.0 mg/L in drinking water	30	0-192 μg/g	Vieira et al. 2005
Dentin			
0.7 or 1.0 mg/L in drinking water	30	59-374 μg/g	Vieira et al. 2005
Bones			
Normal	NA	480-2,100 ppm in bone ash (ribs)	Roholm 1937
Cryolite workers	2	9,900 and 11,200 ppm in bone ash (ribs) ranges (ppm in bone ash, various bone types, 3,100-9,900 and 8,100-13,100 in the 2 individuals	
0.1-0.4 mg/L in drinking water	33	326-2,390 ppm in bone ash[c]	Zipkin et al. 1958
1.0 mg/L in drinking water	5	1,610-4,920 ppm in bone ash[d]	
2.6 mg/L in drinking water	27	1,560-10,800 ppm in bone ash[e]	
4.0 mg/L in drinking water	4	4,780-11,000 ppm in bone ash[f]	

continued

TABLE 2-16 Continued

Fluoride Exposure	Number of Persons	Fluoride Concentration	Reference
< 0.2 mg/L in drinking water since infancy	8	1,379 (179) ppm in bone ash[g]	Eble et al. 1992
1 mg/L in drinking water at least 23 years or since infancy	9	1,775 (313) ppm in bone ash[g]	
0.27 mg/L in drinking water and 2.8 μg/m³ in air	59	625.7 (346.5) ppm[b,h]	Hac et al. 1997
0.7 or 1.0 mg/L in drinking water	30	0-396 ppm[i]	Vieira et al. 2005

[a]Previous exposure of 30-38 mg/day, 2-5 years before study.
[b]Mean and standard deviation.
[c]Reported as 0.019-0.119% in bone, with ash content of 43.2-68.4%.
[d]Reported as 0.100-0.238% in bone, with ash content of 45.9-62.2%.
[e]Reported as 0.092-0.548% in bone, with ash content of 32.7-66.7%.
[f]Reported as 0.261-0.564% in bone, with ash content of 44.3-62.8%.
[g]Mean and standard error of the mean.
[h]Reported as μg fluoride per gram bone; appears to be dry weight of bone, not bone ash.
[i]Measured by Instrumental Neutron Activation Analysis; appears to be wet weight of bone.

ABBREVIATION: NA, not available.

reflected immediately in urine or plasma, presumably because of remobilization of fluoride from resorbed bone.[14]

Concentrations of salivary fluoride (as excreted by the glands) are typically about two-thirds of the plasma fluoride concentration and independent of the salivary flow rate (Rölla and Ekstrand 1996); fluoride in the mouth from dietary intake or dentifrices also affects the concentrations measured in whole saliva. Significantly higher concentrations of fluoride were found in whole saliva and plaque following use of a fluoridated dentifrice versus a nonfluoridated dentifrice by children residing in an area with low fluoride (<0.1 mg/L) in drinking water. Concentrations were 15 times higher in whole saliva and 3 times higher in plaque, on average, 1 hour after use of the dentifrice (Whitford et al. 2005). Whitford et al. (1999b) found that whole-saliva fluoride concentrations in 5- to 10-year-old children were not signifi-

---

[14]For example, following defluoridation of a town's water supply from 8 mg/L to around 1.3 mg/L (mean daily fluoride content over 113 weeks), urinary fluoride concentrations in males fell from means of 6.5 (children) and 7.7 (adults) mg/L before defluoridation to 4.9 and 5.1 mg/L, respectively, after 1 week, 3.5 and 3.4 mg/L, respectively, after 39 weeks, and 2.2 and 2.5 mg/L, respectively, after 113 weeks (Likins et al. 1956). An estimate of current fluoride intake (as opposed to fluoride balance) from a urine sample during this period would probably have been an overestimate.

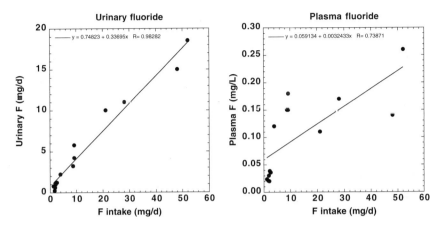

FIGURE 2-4 Urinary fluoride excretion (left) and fasting plasma fluoride concentration (right) as functions of current daily fluoride intake for individual adults (nine males, five females) aged 18-58 years. Data from Teotia et al. 1978.

cantly related to those in either plasma or parotid ductal saliva. However, fluoride concentrations in parotid ductal saliva were strongly correlated to the plasma fluoride concentrations ($r = 0.916$), with a saliva-to-plasma fluoride concentration ratio of 0.80 (SE = 0.03, range from 0.61 to 1.07). For three-quarters of the study population (13 of 17), the fluoride concentration in parotid ductal saliva could be used to estimate plasma fluoride concentrations within 20% or less, and the largest difference was 32%.

Measured fluoride concentrations in human breast milk have been correlated with the mother's fluoride intake in some studies (Dabeka et al. 1986) and not well correlated in other studies (Spak et al. 1983; Opinya et al. 1991). In general, measurements of fluoride in breast milk would be of limited use in exposure estimation because of the very low concentrations even in cases of high fluoride intake, lack of a consistent correlation with the mother's fluoride intake, and limitation of use to those members of a population who are lactating at the time of sampling.

Schamschula et al. (1985) found increasing concentrations of fluoride in urine, nails, hair, and saliva with increasing water fluoride concentration in a sample of Hungarian children, but fluoride contents were not directly proportional to the water fluoride content. Although means were significantly different between groups, there was sufficient variability among individuals within groups that individual values between groups overlapped. Feskanich et al. (1998) used toenail fluoride as an indicator of long-term

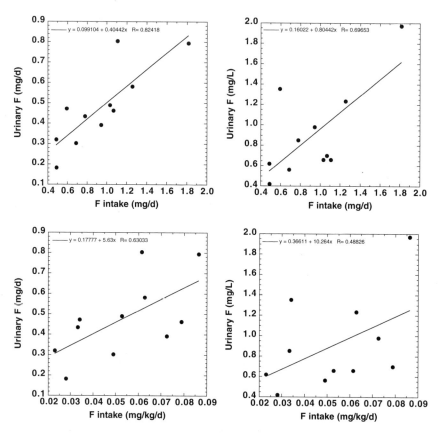

FIGURE 2-5 Urinary fluoride excretion (left) and concentration (right) as functions of current daily fluoride intake (top) or body-weight normalized intake (bottom) for individual children (six boys, five girls) aged 3-6 years. Data from Haftenberger et al. 2001.

fluoride intake and considered it to be a better long-term marker than plasma concentrations.

Whitford et al. (1999a) found a direct relationship between fluoride concentrations in drinking water and fluoride concentrations in fingernail clippings from 6- to 7-year-old children with no known fluoride exposure other than from drinking water. In nail samples from one adult, Whitford et al. (1999a) also found that an increase in fluoride intake was reflected in fingernail fluoride concentrations approximately 3.5 months later and that toenails had significantly lower fluoride concentrations than fingernails. Levy et al. (2004) also found higher fluoride concentrations in fingernails

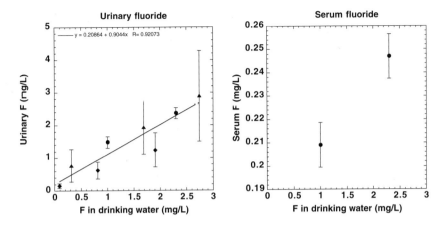

FIGURE 2-6 Urinary (left) and serum (right) fluoride concentrations as functions of fluoride concentration in drinking water. Dark symbols indicate means of groups; vertical lines indicate 1 standard deviation from the mean. Data from Bachinskii et al. (1985; circles), Schamschula et al. (1985; diamonds), and Czarnowski et al. (1999; triangles). Data from Bachinskii et al. represent 47 adults (ages 19-59); data from Schamschula et al. represent children aged 14 years; and data from Czarnowski et al. represent adults (ages 24-77, mean age 50).

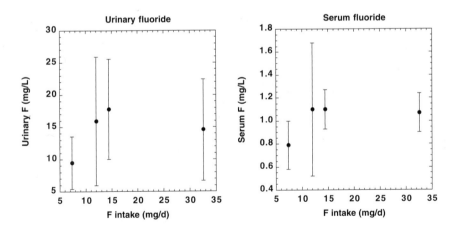

FIGURE 2-7 Urinary (left) and serum (right) fluoride concentrations as functions of estimated daily fluoride intake (data from Gupta et al. 2001). Dark circles indicate means of groups of 50 children (ages 6-12); vertical lines indicate 1 standard deviation from the mean.

than in toenails in 2- to 6-year old children and showed a correlation between nail concentrations and dietary fluoride intake (exclusive of fluoride in toothpaste). Plasma fluoride in these children was not correlated with fluoride in fingernails, toenails, diet, or drinking water.

In contrast, Corrêa Rodrigues et al. (2004), in samples from 2- to 3-year-old children, found no significant differences in fluoride concentrations between fingernails and toenails collected at the same time. An increase in fluoride intake in these children was reflected in nail samples approximately 4 months later (Corrêa Rodrigues et al. 2004). Most likely, differences in "lag times" and differences between fingernails and toenails in the same individual reflect differences in growth rates of the nails due to factors such as age or differences in blood flow. McDonnell et al. (2004) found a wide variation in growth rates of thumbnails of 2- and 3-year-old children; age, gender, and fluoride exposure had no effect on the growth rates. However, it was emphasized that, for any study in which it is of interest to estimate the timing of a fluoride exposure based on measurements of fluoride in nails, the growth rate of the nails should be measured for each individual.

Czarnowski et al. (1999) found correlations between water fluoride concentrations and urinary fluoride, fluoride in hair, and bone mineral density measured in 300 people in the Gdánsk region of Poland. For workers with occupational exposure to airborne fluoride (largely HF), Czarnowski and Krechniak (1990) found good correlation among groups of workers between fluoride concentrations in urine and nails ($r = 0.99$); correlation between concentrations in urine and hair or hair and nails was also positive but not as good ($r = 0.77$ and 0.70, respectively). For individual values, positive correlation was found only between concentrations in urine and nails ($r = 0.73$). It was not possible to establish correlations between fluoride concentrations in biological media and air (Czarnowski and Krechniak 1990).

Measuring the fluoride content of teeth and bones can give an indication of chronic or cumulative fluoride exposure, although after cessation of fluoride exposure, bone fluoride concentrations slowly decrease because of resorption of bone. In addition, bone turnover results in the accumulation of various concentrations of fluoride in different bone types and sites (Selwitz 1994). Dentin has also been suggested as a reasonably accurate marker for long-term exposure (Selwitz 1994), although Vieira et al. (2005) found no correlation between bone fluoride and either enamel or dentin fluoride in persons with exposure to 0.07 or 1.0 mg/L fluoride in drinking water.

Roholm (1937) reported that the fluoride content in normal teeth varied from 190 to 300 ppm (0.19 to 0.30 mg/g) in the total ash, with 5-7 times as much fluoride in the dentin as in the enamel. Fluoride content in the total ash of teeth from five cryolite workers (employed 8-10 years; three with osteosclerosis) contained 1,100-5,300 ppm (1.1-5.3 mg/g), with the most carious teeth containing the most fluoride. Roholm (1937) also reported normal bone fluoride concentrations of 480-2,100 ppm in bone ash (0.48-2.1 mg/g bone ash in ribs), with concentrations between 3,100 and 13,100 ppm in bone ash (3.1 and 13.1 mg/g bone ash; varying with type of bone) in two cryolite workers. Hodge and Smith (1965), summarizing several reports, listed mean concentrations of bone fluoride in normal individuals between 450 and 1,200 ppm in bone ash and in people "suffering excessive exposure" to fluorides between 7,500 and 20,830 ppm in bone ash. More recently, Eble et al. (1992) have reported fluoride concentrations in bone ash ranging from 378 ppm (16-year old with <0.2 mg/L fluoride in drinking water since infancy) to 3,708 ppm (79-year old with fluoridated water). A 46-year old female with chronic renal failure had a fluoride concentration in bone ash of 3,253 ppm (Eble et al. 1992).

The data of Zipkin et al. (1958) shows a good relationship between drinking-water fluoride and the mean percentage of fluoride in bone (iliac crest, rib, and vertebra) for adults in areas of various fluoride concentrations in drinking water. However, the ranges (Table 2-16; see also Chapter 3, Figure 3-1) suggest that variability among individuals within groups could be large, probably reflecting variability in individual fluoride intakes, duration of exposure, and age. A major disadvantage of measuring bone fluoride is the invasiveness of bone sampling in live individuals. Although easier to do, x-ray screening for increased bone density should be done only when the need for information justifies the radiation dose involved; in addition, bone density might not be related solely to fluoride exposure or to bone fluoride content.

The two most important biomarkers of effect for fluoride are considered to be enamel fluorosis and skeletal fluorosis (ATSDR 2003); these are discussed more fully in Chapters 4 and 5. Enamel fluorosis is characterized by mottling and erosion of the enamel of the teeth and is associated with elevated fluoride intakes during the childhood years when the teeth are developing. According to the U.S. Public Health Service (PHS 1991), both the percent prevalence and the increasing severity of enamel fluorosis are associated with increasing fluoride concentration in drinking water (and presumably actual fluoride intake). For "optimally" fluoridated water (0.7-1.2 mg/L), 22% of children examined in the 1980s showed some fluorosis (mostly very mild or mild); at water fluoride concentrations above 2.3 mg/L, more than 70% of children showed fluorosis (PHS 1991; NRC 1993). Some children developed fluorosis even at the lowest fluoride concentrations (<0.4 mg/L), suggesting that either fluoride intakes are variable within a population with the same water supply or there is variability in the susceptibility to fluorosis within populations (or both). Baelum et al. (1987) indicated that 0.03 mg/kg/day might not be protective against enamel fluorosis, and Fejerskov et al. (1987) stated that the borderline dose above which enamel fluorosis might develop could be as low as 0.03 mg/kg/day.

DenBesten (1994) described the limitations of using enamel fluorosis as a biomarker of exposure: enamel fluorosis is useful only for children less than about 7 years old when the exposure occurred; the incidence and degree of fluorosis vary with the timing, duration, and concentration; and there appear to be variations in individual response. Selwitz (1994), summarizing a workshop on the assessment of fluoride accumulation, also indicated that variability in response (incidence and severity of enamel fluorosis) to fluoride exposure may result from physiological differences among individuals and that enamel fluorosis is not an adequate biomarker for fluoride accumulation or potentially adverse health effects beyond the period of tooth formation. Selwitz (1994) did suggest that enamel fluorosis could be used as a biomarker of fluoride exposure in young children within a community over time.

Skeletal fluorosis (see also Chapter 5) is characterized by increased bone mass, increased radiographic density of the bones, and a range of skeletal and joint symptoms; preclinical skeletal fluorosis is associated with fluoride concentrations of 3,500-5,500 ppm in bone ash and clinical stages I, II, and III with concentrations of 6,000-7,000, 7,500-9,000, and >8,400, respectively (PHS 1991), although other sources indicate lower concentrations of bone fluoride in some cases of skeletal fluoride (see Chapter 5). According to the Institute of Medicine, "Most epidemiological research has indicated that an intake of at least 10 mg/day [of fluoride] for 10 or more years is needed to produce clinical signs of the milder forms of [skeletal fluorosis]" (IOM 1997). However, the National Research Council (NRC 1993) indicated that crippling (as opposed to mild) skeletal fluorosis "might occur in people who have ingested 10-20 mg of fluoride per day for 10-20 years." A previous NRC report (NRC 1977) stated that a retention of 2 mg of fluoride per day (corresponding approximately to a daily intake of 4-5 mg) "would mean that an average individual would experience skeletal fluorosis after 40 yr, based on an accumulation of 10,000 ppm fluoride in bone ash." Studies in other countries indicate that skeletal fluorosis might be in part a marker of susceptibility as well as exposure, with factors such as dietary calcium deficiency involved in addition to fluoride intake (Pettifor et al. 1989; Teotia et al. 1998).

Hodge and Smith (1965) summarized a number of studies of skeletal fluorosis, including two that indicated affected individuals in the United States with water supplies containing fluoride at 4.8 or 8 mg/L. They also stated categorically that "crippling fluorosis has never been seen in the United States." The individuals with endemic fluorosis at 4.8 mg/L are referred to elsewhere as having "radiographic osteosclerosis, but no evidence of skeletal fluorosis" (PHS 1991). In combination with high fluid intake and large amounts of tea, "the lowest drinking-water concentration of fluoride associated with symptomatic skeletal fluorosis that has been reported to date is 3 ppm, outside of countries such as India" (NRC 1977).

Both the PHS (1991) and the NRC (1993) indicated that only five cases of crippling skeletal fluorosis have been reported in the literature in the United States (including one case in a recent immigrant from an area with fluoride in the drinking water at 3.9 mg/L) (PHS 1991). These individuals were said to have water supplies ranging from 3.9 to 8.0 mg/L (water fluoride content given for one of the individuals is actually less than 3.9 mg/L) (PHS 1991). Two of the individuals had intakes of up to 6 L/day of water containing fluoride at 2.4-3.5 or 4.0-7.8 mg/L (PHS 1991; NRC 1993); this corresponds to fluoride intakes of up to 14.4-21 or 24-47 mg/day.

Several cases of skeletal fluorosis reported in the United States are summarized in Table 2-17. These reports indicate that a fluoride concentration of 7-8 mg/L for 7 years is sufficient to bring about skeletal fluorosis (Felsenfeld and Roberts 1991), but skeletal fluorosis may occur at much lower fluoride concentrations in cases of renal insufficiency (Juncos and Donadio 1972; Johnson et al. 1979). People who consume instant tea are at increased risk of developing skeletal fluorosis, especially if they drink large volumes, use extra-strength preparations, or use fluoridated or fluoride-contaminated water (Whyte et al. 2005).

In summary, selecting appropriate biomarkers for a given fluoride study depends on a number of factors, as listed above. A major consideration is the time period of interest for the study (e.g., current or recent exposures versus exposures in childhood versus cumulative exposures) and whether the intent is to demonstrate differences among groups or to characterize exposures of specific individuals. Many of the areas for further research identified by a 1994 workshop (Whitford et al. 1994) are still relevant for improving the assessment of fluoride exposures.

FINDINGS

Table 2-18 summarizes various published perspectives on the significance of given concentrations of fluoride exposure. Historically, a daily intake of 4-5 mg by an adult (0.057-0.071 mg/kg for a 70-kg adult) was considered a "health hazard" (McClure et al. 1945, cited by Singer et al. 1985). However, the Institute of Medicine (IOM 1997) now lists 10 mg/day as a "tolerable upper intake" for children > 8 years old and adults, although that intake has also been associated with the possibility of mild (IOM 1997) or even crippling (NRC 1993) skeletal fluorosis.

The recommended optimal fluoride intake for children to maximize caries prevention and minimize the occurrence of enamel fluorosis is often stated as being 0.05-0.07 mg/kg/day (Levy 1994; Heller et al. 1999, 2000). Burt (1992) attempted to track down the origin of the estimate of 0.05-0.07

TABLE 2-17 Case Reports of Skeletal Fluorosis in the United States

Study Subjects	Exposure Conditions	Comments	Reference
(a) 18-year-old boy, 57.4 kg (b) 17-year-old girl, 45.65 kg	(a) "high" intake of well water containing fluoride at 2.6 mg/L since early childhood; current intake, 7.6 L/day (0.34 mg/kg/day) (b) "high" intake of water containing fluoride at 1.7 mg/L since infancy; current intake, 4 L/day (0.15 mg/kg/day)	Enamel fluorosis and roentgenographic bone changes consistent with "systemic fluorosis," attributed to the combination of renal insufficiency and polydipsia (the latter resulting from the renal disease); reported by the Mayo Clinic	Juncos and Donadio 1972
Six renal patients seen at the Mayo Clinic over a several year period (includes the two patients reported by Juncos and Donadio)	Drinking water with 1.7-3 mg/L fluoride; water consumption not stated, but urine volumes of "most" of the patients exceeded 3 L/day	Fluoride "may have been the cause of detectable clinical and roentgenographic effects" Five of the patients had renal disease of at least 15 years duration before skeletal symptoms developed	Johnson et al. 1979
54-year-old woman in Oklahoma	Well water with fluoride concentration of 7.3-8.2 mg/L (382-429 μmol/L); duration of residence at that location, 7 years; prior to that she had used municipal water at less than 2 mg/L fluoride; water consumption not reported, but considered likely to be "increased" due to hot summers	Osteosclerosis, elevated serum alkaline phosphatase, stiffness of knees and hips (2 years duration), kyphosis Renal insufficiency was not a factor	Felsenfeld and Roberts 1991
52-year-old woman in Missouri	Daily consumption of 1-2 gallons (3.8-7.6 L) per day of double-strength instant tea made with unfiltered well water (2.8 mg/L fluoride in the well water) for close to 10 years; estimated fluoride intake of 37-74 mg/day (11-22 mg/day from well water and 26-52 mg/day from tea)	Osteosclerosis, increased bone mineral density, bone and joint pains Intake of fluoride from well water alone was considered sufficient to cause mild skeletal fluorosis No mention of any renal disease	Whyte et al. 2005

MEASURES OF EXPOSURE TO FLUORIDE IN THE UNITED STATES 83

TABLE 2-18 Summary of Current and Historical Perspectives on Fluoride Exposure

Exposure, mg/kg/day	Description	Reference
0.0014	"Adequate intake" for children < 6 months old[a] (0.01 mg/day)	IOM 1997; ADA 2005
0.01-0.04	Average daily dietary fluoride intake for children 0-2 years old residing in nonfluoridated areas (< 0.4 mg/L)	IOM 1997[b]
0.017-0.031	Average daily intake by adults in a fluoridated area (1.2-2.2 mg/day)[c]	NRC 1993
0.017-0.054	Lower end of "safe and adequate daily dietary intake" for children 0-10 years[d] (0.1-1.5 mg/day)	NRC 1989b
0.019-0.033	Lower end of "safe and adequate daily dietary intake" for children ≥ 10 years and adults[d] (1.5 mg/day)	NRC 1989b
0.02-0.10	Average daily dietary fluoride intake for children 1-9 years residing in fluoridated areas (0.7-1.1 mg/L)	McClure 1943[e]
0.038-0.069	Upper end of "safe and adequate daily dietary intake" for children ≥ 10 years and adults[d] (2.5-4.0 mg/day)	NRC 1989b
0.04-0.07	Average daily intake by children in a fluoridated area	NRC 1993
0.05	"Adequate intake" for all ages above 6 months old[a,f]	IOM 1997; ADA 2005
0.05	ATSDR's minimal risk level[g] (chronic duration, based on increased rate of bone fractures)[h]	ATSDR 2003
0.05-0.13	Average daily dietary fluoride intake for children 0-2 years old residing in fluoridated areas (0.7-1.1 mg/L)	IOM 1997[b]
0.05-0.07	"Optimal" intake to maximize caries prevention and minimize the occurrence of enamel fluorosis	Levy 1994; Heller et al. 1999, 2000
0.05-0.07	"Useful upper limit for fluoride intake in children"	Burt 1992
0.057-0.071	"Health hazard" for adults (4-5 mg/day)[c]	McClure et al. 1945
0.057	EPA's SMCL (2 mg/l; adult intake)[i]	40CFR 143.3[2001]
0.06	EPA's reference dose[j] (based on protection of children from objectionable enamel fluorosis)[k]	EPA 1989
0.083-0.13	Upper end of "safe and adequate daily dietary intake" for children 0-10 years old[d] (0.5-2.5 mg/day)	NRC 1989b
0.10	"Tolerable upper intake"[l] for ages 0-8[a] (0.7-2.2 mg/day)	IOM 1997; ADA 2005
0.10	EPA's SMCL (2 mg/L; child intake)[m]	40CFR 143.3 [2001]
0.11	EPA's MCLG and MCL (4 mg/L; adult intake)[n]	40CFR 141.62(b)[2001]
0.13-0.18	"Tolerable upper intake"[o] for ages ≥ 14[a] (10 mg/day)	IOM 1997; ADA 2005
0.2	EPA's MCLG and MCL (4 mg/L; child intake)[p]	40CFR 141.62(b)[2001]

continued

TABLE 2-18 Continued

Exposure, mg/kg/day	Description	Reference
0.25	"Tolerable upper intake"[o] for ages 9-13[a] (10 mg/day)	IOM 1997; ADA 2005

[a]Based on intakes and average body weights listed by IOM (1997) and ADA (2005); see Table B-17 in Appendix B.
[b]Summaries of papers published between 1979 and 1988 (IOM 1997).
[c]Based on a 70-kg adult.
[d]Based on intakes and median weights listed by NRC (1989b); see Table B-16 in Appendix B.
[e]Summarized by IOM (1997).
[f]Range, 0.045-0.056 mg/kg/day.
[g]A minimal risk level (MRL) is an estimate of the daily human exposure to a hazardous substance that is likely to be without appreciable risk of adverse noncancer health effects over a specified duration of exposure (ATSDR 2003).
[h]The ATSDR (2003) states that an intermediate-duration MRL derived from a study of thyroid effects in rats would have been lower (more protective) than the chronic-duration MRL of 0.05, but the value of that MRL is not given.
[i]Based on intake of 2 L/day by a 70-kg adult of water containing fluoride at 2 mg/L.
[j]Reference dose (RfD) is an estimate (with uncertainty spanning perhaps an order of magnitude) of a daily oral exposure to the human population (including sensitive subgroups) that is likely to be without an appreciable risk of deleterious effects during a lifetime (EPA 1989).
[k]Based on a fluoride concentration of 1 mg/L in drinking water; the RfD for fluoride contains no uncertainty factor or modifying factor, although RfDs for other substances contain uncertainty factors to account for things such as variability within the human population (EPA 2003b).
[l]Based on moderate enamel fluorosis (IOM 1997).
[m]Based on intake of 1 L/day by a 20-kg child of water containing fluoride at 2 mg/L.
[n]Based on intake of 2 L/day by a 70-kg adult of water containing fluoride at 4 mg/L.
[o]Based on skeletal fluorosis for adults and children ≥ age 9 (IOM 1997).
[p]Based on intake of 1 L/day by a 20-kg child of water containing fluoride at 4 mg/L.

mg/kg/day as an optimum intake of fluoride but was unable to find it. He interpreted the available evidence as suggesting that 0.05-0.07 mg/kg/day (from all sources) "remains a useful upper limit for fluoride intake in children" (see also NRC 1993).

Figure 2-8 shows the average intake of fluoride from all sources estimated in this report (Table 2-11), with 1 mg/L in drinking water; Figure 2-9 shows the average intake of fluoride from drinking water alone (Table 2-10), given a fluoride concentration at the MCLG/MCL (4 mg/L). For comparison purposes, an intake of 0.05-0.07 mg/kg/day is indicated on the graphs.

Based on EPA's estimates of community water consumption by consumers with an average intake (EPA 2000a), if that water is fluoridated, children

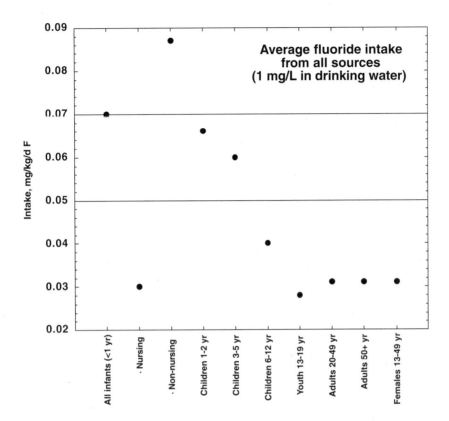

FIGURE 2-8 Estimated average intake of fluoride from all sources, at 1 mg/L in drinking water (based on Table 2-11). Horizontal lines indicate an intake of 0.05-0.07 mg/kg/day.

less than 6 months old have an intake at or above 0.05-0.07 mg/kg/day (see Appendix B, Table B-10). Children from 6 months to 1 year old have similar intakes if their water is fluoridated at 1 or 1.2 mg/L. No other age groups have that intake at ordinary fluoride concentrations; all age groups reach or exceed that intake with water at 4 mg/L. For individuals with higher-than-average intake of community water, intakes for the youngest children (<1 year) might exceed 0.05-0.07 mg/kg/day at all concentrations of water fluoridation (see Appendix B, Tables B-11, B-12, and B-13); for fluoride concentrations corresponding to the SMCL (2 mg/L) or MCL (4 mg/L), an intake of 0.05-0.07 mg/kg/day is reached or exceeded by all age groups. Note that the estimates in Appendix B include only the fluoride contribution from

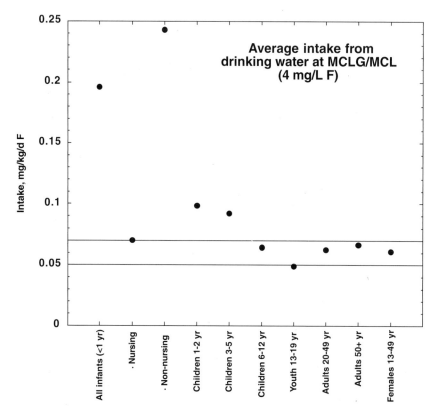

FIGURE 2-9 Estimated average intake of fluoride from drinking water alone, based on a fluoride concentration of 4 mg/L (MCLGl/MCL; based on Table 2-10). Horizontal lines indicate an intake of 0.05-0.07 mg/kg/day.

community water (drinking water, plus beverages and foods prepared with community water at home or in local eating establishments); if contributions from food, tea, commercial beverages, toothpastes, and other sources are added, total intakes by individuals will increase accordingly.

Estimates of total exposure (typical or average) shown in Table 2-11 indicate that all children through age 12 who take fluoride supplements (assuming low water fluoride) will reach or exceed 0.05-0.07 mg/kg/day. For children not on supplements, nonnursing infants with fluoride in tap water at ≥0.5 mg/L will exceed 0.05-0.07 mg/kg/day for typical exposures. Also, children through 5 years old (≥0.5 mg/L in tap water), children 6-12 years old (≥2 mg/L in tap water), and teenagers and adults (≥4 mg/L in tap water) will exceed 0.05-0.07 mg/kg/day with typical or average fluoride exposures in terms of water consumption and toothpaste ingestion.

A number of researchers have pointed out both the importance of evaluating individual fluoride intake from all sources and the difficulties associated with doing so, given the variability of fluoride content in various foods and beverages and the variability of individual intakes of the specific items (Clovis and Hargreaves 1988; Nowak and Nowak 1989; Chan et al. 1990; Stannard et al. 1990, 1991; Weinberger 1991; Toumba et al. 1994; Duperon et al. 1995; Van Winkle et al. 1995, Chan and Koh 1996; Kiritsy et al. 1996; Warren et al. 1996; Heilman et al. 1997, 1999; Heller et al. 1999; Levy and Guha-Chowdhury 1999; Lalumandier and Ayers 2000). However, as shown in Figure 2-1, for typical individuals, the single most important contributor to fluoride exposures (approaching 50% or more) is fluoridated water and other beverages and foods prepared or manufactured with fluoridated water.

RECOMMENDATIONS

- Fluoride should be included in nationwide biomonitoring surveys and nutritional studies (e.g., CDC's National Health and Nutrition Examination Survey and affiliated studies). In particular, analysis of fluoride in blood and urine samples taken in these surveys would be valuable.
- National data on fluoridation (e.g., CDC 1993) should be updated on a regular basis.
- Probabilistic analysis should be performed for the uncertainty in estimates of individual and group exposures and for population distributions of exposure (e.g., variability with respect to long-term water consumption). This would permit estimation of the number of people exposed at various concentrations, identification of population subgroups at unusual risk for high exposures, identification or confirmation of those fluoride sources with the greatest impact on individual or population exposures, and identification or characterization of fluoride sources that are significant contributors to total exposure for certain population subgroups.
- To assist in estimating individual fluoride exposure from ingestion, manufacturers and producers should provide information on the fluoride content of commercial foods and beverages.
- To permit better characterization of current exposures from airborne fluorides, ambient concentrations of airborne hydrogen fluoride and particulates should be reported on national and regional scales, especially for areas of known air pollution or known sources of airborne fluorides. Additional information on fluoride concentrations in soils in residential and recreational areas near industrial fluoride sources also should be obtained.
- Additional studies on the relationship between individual fluoride exposures and measurements of fluoride in tissues (especially bone and nails) and bodily fluids (especially serum and urine) should be conducted. Such

studies should determine both absolute intakes (mg/day) and body-weight normalized intakes (mg/kg/day).

- Assumptions about the influence of environmental factors, particularly temperature, on water consumption should be reevaluated in light of current lifestyle practices (e.g., greater availability of air conditioning, participation in indoor sports).
- Better characterization of exposure to fluoride is needed in epidemiology studies investigating potential effects. Important exposure aspects of such studies would include the following:
  - collecting data on general dietary status and dietary factors that could influence exposure or effects, such as calcium, iodine, and aluminum intakes
  - characterizing and grouping individuals by estimated (total) exposure, rather than by source of exposure, location of residence, fluoride concentration in drinking water, or other surrogates
  - reporting intakes or exposures with and without normalization for body weight (e.g., mg/day and mg/kg/day)
  - addressing uncertainties associated with exposure, including uncertainties in measurements of fluoride concentrations in bodily fluids and tissues
  - reporting data in terms of individual correlations between intake and effect, differences in subgroups, and differences in percentages of individuals showing an effect and not just differences in group or population means.
- Further analysis should be done of the concentrations of fluoride and various fluoride species or complexes (especially fluorosilicates and aluminofluorides) present in tap water, using a range of water samples (e.g., of different hardness and mineral content). Research also should include characterizing any changes in speciation that occur when tap water is used for various purposes—for example, to make acidic beverages.
- The possibility of biological effects of SiF_6^{2-}, as opposed to free fluoride ion, should be examined.
- The biological effects of aluminofluoride complexes should be researched further, including the conditions (exposure conditions and physiological conditions) under which the complexes can be expected to occur and to have biological effects.

3

Pharmacokinetics of Fluoride

This chapter updates pharmacokinetic information on fluoride developed since the earlier National Research Council review (NRC 1993). Particular attention is given to several potentially important issues for evaluation of the U.S. Environmental Protection Agency (EPA) maximum-contaminant-level goal (MCLG), including the accumulation of fluoride in bone, pharmacokinetic modeling, cross-species extrapolation, and susceptible populations. Consideration of biomarkers is provided in Chapter 2.

OVERVIEW OF FLUORIDE CHEMISTRY, UNITS, AND MEASUREMENT

Fluoride is the ionic form of fluorine, the most electronegative element. Water in the United States is typically fluoridated with fluorosilicates or sodium fluoride. In water at approximately neutral pH, fluorosilicates appear to entirely dissociate, producing fluoride ion, hydrofluoric acid (HF), and silicic acid (Si(OH)4). Fluoride reversibly forms HF in water. It also complexes with aluminum. See Chapter 2 for additional discussion of fluorosilicates and aluminum fluoride complexes.

Inorganic fluoride takes two primary forms in body fluids: fluoride ion and HF. Organofluorine compounds, and their potential relationship to inorganic fluoride, are discussed in Chapter 2 and later in this chapter.

A number of different units are commonly used to measure fluoride concentrations in water and biological samples (Table 3-1). Because the atomic weight of fluorine is 19, 1 μmol/L is equal to 0.019 milligrams per liter (mg/L). Bone ash is typically about 56% of wet bone by weight (Rao

TABLE 3-1 Commonly Used Units for Measuring Fluoride

Medium	Unit	Equivalent
Water	1 ppm	1 mg/L
Plasma	1 µmol/L	0.019 mg/L
Bone ash	1 ppm	1 mg/kg

et al. 1995), so 1,000 milligrams per kilogram (mg/kg) of fluoride in bone ash is equivalent to about 560 mg/kg wet weight.

Fluoride concentrations in body fluids typically are measured with a fluoride-specific electrode, an instrument that cannot reliably measure concentrations below about 0.019 mg/L and tends to overpredict at lower concentrations. As many people living in areas with artificially fluoridated water have plasma concentrations in this range, studies that rely on fluoride electrodes alone might tend to overpredict concentrations in plasma and body fluids. The hexamethyldisiloxane diffusion method provides a way around this problem by concentrating the fluoride in samples before analysis (reviewed by Whitford 1996).

SHORT REVIEW OF FLUORIDE PHARMACOKINETICS: ABSORPTION, DISTRIBUTION, AND ELIMINATION

A comprehensive review of fluoride pharmacokinetics is provided by Whitford (1996), and this section presents a brief overview of that information. The pharmacokinetics of fluoride are primarily governed by pH and storage in bone. HF diffuses across cell membranes far more easily than fluoride ion. Because HF is a weak acid with a pKa of 3.4, more of the fluoride is in the form of HF when pH is lower. Consequently, pH—and factors that affect it—play an important role in the absorption, distribution, and excretion of fluoride. Fluoride is readily incorporated into calcified tissues, such as bone and teeth, substituting for hydroxyls in hydroxyapatite crystals. Fluoride exchanges between body fluids and bone, both at the surface layer of bone (a short-term process) and in areas undergoing bone remodeling (a longer-term process). Most of the fluoride in the body, about 99%, is contained in bone.

Fluoride is well absorbed in the alimentary tract, typically 70% to 90%. For sodium fluoride and other very soluble forms, nearly 100% is absorbed. Fluoride absorption is reduced by increased stomach pH and increased concentrations of calcium, magnesium, and aluminum. At high concentrations, those metals form relatively insoluble fluoride salts. A recent study comparing hard and soft water found little difference in fluoride bioavailability in healthy young volunteers (Maguire et al. 2004). Fluoride can increase the uptake of aluminum into bone (Ahn et al. 1995) and brain (Varner et al. 1998).

Fluoride concentrations in plasma, extracellular fluid, and intracellular fluid are in approximate equilibrium. The concentrations in the water of most tissues are thought to be 40% to 90% of plasma concentrations, but there are several important exceptions. Tissue fluid/plasma (T/P) ratios exceed one for the kidney because of high concentrations in the renal tubules. T/P ratios can exceed one in tissues with calcium deposits, such as the placenta near the end of pregnancy. The pineal gland, a calcifying organ that lies near the center of the brain but outside the blood-brain barrier, has been found to accumulate fluoride (Luke 2001). Fluoride concentrations in adipose tissue and brain are generally thought to be about 20% of plasma or less (Whitford 1996). The blood-brain barrier is thought to reduce fluoride transfer, at least in short-term experiments (Whitford 1996). It is possible that brain T/P ratios are higher for exposure before development of the blood-brain barrier.

Most tissue measurements are based on short-term exposures of healthy adult animals. Similar T/P ratios have been found for liver and kidney in some chronic animal experiments (Dunipace et al. 1995), but not all organs have been examined. The literature contains some unexplained exceptions to these T/P generalizations (Mullenix et al. 1995; Inkielewicz and Krechniak 2003). Mullenix et al. (1995) reported atypically high, dose-dependent T/P ratios for the rat brain: more than 20 for control animals and about 3 for animals exposed to fluoride at 125 mg/L in drinking water for 20 weeks. Because these T/P ratios for brain are much higher than earlier results, Whitford (1996) speculated that the results of Mullenix et al. were due to analytical error. Additional measurements of fluoride tissue concentrations after chronic dosing are needed.

Fluoride is cleared from plasma through two primary mechanisms: uptake by bone and excretion in urine. Plasma clearance by the two routes is approximately equal in healthy adult humans. (Plasma clearance is the volume of plasma from which fluoride is removed per unit time. The rate of removal equals the clearance times the plasma fluoride concentration. Clearances are additive.) The relative clearance by bone is larger in young animals and children because of their growing skeletal systems. "In contrast to the compact nature of mature bone, the crystallites of developing bone are small in size, large in number and heavily hydrated. Thus, they afford a relatively enormous surface area for reactions involving fluoride" (Whitford 1996, p. 94). Experimental work in growing dogs demonstrates that extrarenal clearance, almost entirely uptake by bone, is inversely related to age. Renal clearance depends on pH and glomerular filtration rate. At low pH, more HF is formed, promoting reabsorption. Excretion of previously absorbed fluoride from the body is almost entirely via urine. Fluoride not absorbed by the gut is found in feces. High concentrations of calcium in contents of the gastrointestinal tract can cause net excretion of fluoride.

Fluoride is rapidly absorbed from the gastrointestinal tract, with a half-life of about 30 minutes. After a single dose, plasma concentrations rise to a peak and then fall as the fluoride is cleared by the renal system and bone, decreasing back to (short-term) baseline with a half-life of several hours. Fluoride concentrations in plasma are not homeostatically controlled (Whitford 1996). Chronic dosing leads to accumulation in bone and plasma (although it might not always be detectable in plasma.) Subsequent decreases in exposure cause fluoride to move back out of bone into body fluids, becoming subject to the same kinetics as newly absorbed fluoride. A study of Swiss aluminum workers found that fluoride bone concentrations decreased by 50% after 20 years. The average bone ash concentration in the workers was about 6,400 mg/kg at the end of exposure, estimated via regression (Baud et al. 1978). The bone concentration found in these workers is similar to that found in long-term consumers of drinking water containing fluoride in the range of 2-4 mg/L (discussed later in this chapter). Twenty years might not represent a true half-life. Recent pharmacokinetic models (see below) are nonlinear, suggesting that elimination rates might be concentration dependent.

PHARMACOKINETIC MODELS

Pharmacokinetic models can be useful for integrating research results and making predictions. Two important fluoride models have been published since the 1993 NRC review. Turner et al. (1993) modeled bone concentrations in healthy adult humans. They assumed a nonlinear function relating the concentrations of fluoride in newly formed bone to plasma/extracellular fluids. The relationship is close to linear until bone ash concentrations reach about 10,000 mg/kg; above that concentration the curve levels off. (Based on the chemical structure of fluorapatite, $Ca_{10}(PO_4)_6F_2$, the theoretical limit on bone fluoride concentration is 37,700 mg/kg.) The model was relatively successful at predicting fluoride bone concentrations due to chronic exposure compared with experimental data—for example, the human bone measurements of Zipkin et al. (1958). Bone fluoride concentrations were predicted to increase approximately linearly as a function of water concentration, at least up to 4 mg/L. The most sophisticated model to date (Rao et al. 1995) extended this work with a physiologically based pharmacokinetic (PBPK) model. Among other features, it models change in body weight, plasma clearance, and bone uptake as a function of sex and age, allowing predictions for lifetime exposures. It can model both rats and humans, making it useful for comparing these species. Predicted bone concentrations were comparable with data from several studies of humans, including the study by Zipkin et al. (1958), and two rat carcinogenicity studies (Maurer et al. 1990; Bucher et al. 1991). Both models predicted increasing fluoride concentrations in bone with length of chronic exposure. None of these studies presented results for plasma.

Both models also performed well in predicting bone concentrations of fluoride resulting from osteoporosis treatment, involving about 25 mg of fluoride per day for up to 6 years. This suggests that the models can adequately predict the results of both long-term lower exposures (drinking water) and shorter-term, higher exposures (treatment regimes) by changing exposure assumptions.

The PBPK model of Rao et al. (1995) could be used in several ways, including (1) predicting bone concentrations in people after lifetime exposures to assumed water concentrations or other exposure scenarios, and (2) comparing plasma and bone fluoride concentrations in rats and humans with the same exposure. The Rao model is quite complicated and relies on several numerical functions not provided in the paper. The Turner model is more limited in scope, unable to compare species or take sex- and age-related effects into account, but it is much simpler. Not enough detail on either model was available to replicate them nor was the committee able to obtain operational versions of the models.

FLUORIDE CONCENTRATIONS IN HUMAN BONE VERSUS WATER CONCENTRATION

Remarkably few data are available for studying the association between fluoride in human bone and low-dose chronic exposure via drinking water. Although there are a number of cross-sectional studies comparing bone concentrations with water concentrations, very few contain estimates of length of exposure. Most studies are autopsies, as bone samples can be difficult to obtain from healthy living subjects. Among studies examining exposure to fluoride at 4 mg/L, Zipkin et al. (1958) provided the only data set that included exposure durations. The results of that study were also modeled by Turner et al. (1993) and Rao et al. (1995). Sixty-three of the 69 subjects, aged 26 to 90, died suddenly, primarily due to trauma, cardiovascular disease, and cerebrovascular causes; three had renal disease. The authors recorded concentrations of fluoride in drinking water and bone as well as sex, age, and years of residence. Compared with today, many other sources of fluoride exposure were uncommon or did not exist. The average residence time for the whole study was 31 years, 34 years for the 2.6-mg/L group and 21 years for the 4-mg/L group. Exposure took place for most people as adults. No estimates of water consumption are provided: water concentration serves as an ecologic measure of exposure.

Table 3-2 summarizes data on fluoride content of the iliac crest, the

TABLE 3-2 Fluoride in Bone Due to Chronic Water Exposure[a]

Water Concentration, mg/L	Average Iliac Crest Concentration, mg/kg Ash
0.1	665 ± 224 (n = 17)
1	2,249 ± 506 (n = 4)
2.6	4,496 ± 2,015 (n = 25)
4	6,870 ± 1,629 (n = 4)
Total	3,203 (n = 50)

[a]Fifty-three subjects had data for the iliac crest; 3 from the 0.2 and 0.3 mg/L groups are omitted because they were also exposed to fluoridated water for 2 to 4 years.

SOURCE: Zipkin et al. 1958.

bone modeled by Turner et al. and Rao et al. Zipkin et al. concluded that average bone fluoride concentrations were linearly related to water concentration. (As discussed in Appendix C, this analysis is fully ecologic). The committee regressed individual-level bone concentrations versus water concentrations (a group measure of exposure) and individual-level covariates such as age. (This analysis is partially ecologic.) Figure 3-1 plots bone versus water concentrations and the result of simple regression with no covariates. (Note the apparent heteroscedasticity.) The model was improved

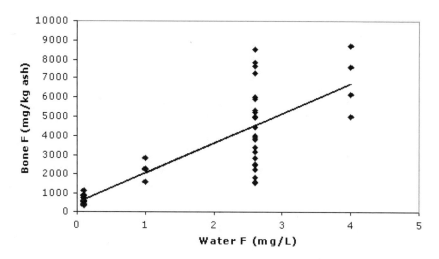

FIGURE 3-1 Illiac crest data from Zipkin et al. (1958). Crude regression results: $y = 517 + 1,549x$; ($r^2 = 0.66$); slope = 1,549 (95% confidence interval [CI] = 1,227, 1,872).

by including residence years and sex; age had little additional impact and was omitted in the final model (Table 3-3).

Several cross-sectional studies have found an association between fluoride bone concentrations and age (Jackson and Weidmann 1958; Kuo and Stamm 1974; Parkins et al. 1974; Charen et al. 1979; Alhava et al. 1980; Eble et al. 1992; Richards et al. 1994; Torra et al. 1998). Jackson and Weidmann (1958) were unusual in finding a leveling off at an older age. But most studies did not have information on length of exposure, a variable often correlated with age (R = 0.41 in the Zipkin data set). Because of the potential for rapid fluoride uptake by bones during childhood, the committee modeled exposure before puberty with an indicator variable, but this added little to the model. Very few data are available on bone fluoride concentrations in children. Most studies do not distinguish between trabecular and cortical bone, although the former have higher fluoride concentrations (Eble et al. 1992).

The model in Table 3-3 indicates that fluoride bone concentrations increased with fluoride water concentrations and residence time; females tended to have higher concentrations than males. These results need to be interpreted with caution. Some subjects had renal disease, which can sometimes increase fluoride concentrations (see discussion below), potentially reducing the generalizability of the results to a healthier population. The committee's analysis is partially ecologic (Appendix C). However, the Turner and Rao pharmacokinetic models also predict that fluoride bone concentrations increase with water concentration and duration of chronic exposure.

What bone fluoride concentration occurs after 70 years of exposure to water at 4 mg/L? The multiple regression model predicts about 8,100 mg/kg ash for females, within the range of the data set used to construct the model but near its maximum. Few people studied by Zipkin et al. were exposed for 70 years and only four were exposed at 4 mg/L. Fluoride is taken up by bone more rapidly during growth than in adulthood. This phenomenon, not addressed by the regression model, could cause the model to underpredict. Only the model of Rao et al. was constructed to examine lifetime exposure. Assuming 70 years of exposure at 4 mg/L in water, Rao et al. predicted fluoride concentrations of 10,000 to 12,000 mg/kg in bone ash for females. Even

TABLE 3-3 Multiple Regression Results for Zipkin Data

	Coefficient	95% CI	P value
Intercept	−556 mg/kg	(−1,512, 401)	0.25
Water fluoride	1,527	(1,224, 1,831)	2.7×10^{-13}
Residence, years	26.5 mg/kg/year	(7.48, 45.5)	0.007
Sex (M = 0)	663 mg/kg	(−148, 1,475)	0.11

higher values would be predicted if other sources of fluoride exposure were included. This prediction lies beyond the range of the human data used to check the model, but it represents the current best estimate. In making this prediction, the authors appear to have assumed consumption of 1 L of water per day up to age 10 and 2 L/day thereafter. Higher water consumption rates (e.g., 5 L/day) would further increase bone concentrations of fluoride but by less than fivefold because of the nonlinear kinetics.

Unfortunately, Rao et al. did not publish predictions for 2 mg/L. The regression model of Table 3-3 predicts about 5,000 mg/kg ash for females after 70 years of exposure. This value exceeds the mean value (4,500 mg/kg) observed at 2.6 mg/L in the Zipkin study, primarily because of the assumed longer time of residence. As this estimate is based on regression modeling of the Zipkin data, it may underestimate predictions based on pharmacokinetic modeling or additional sources of exposure. The committee located only a few other studies that measured bone fluoride at similar water concentrations. A British study found bone concentrations of about 5,700 mg/kg ash in people chronically exposed to water with fluoride at 1.9 mg/L; these people are also thought to be exposed to fluoride in tea (Jackson and Weidmann 1958; see Turner et al 1993 for unit conversions). In an area of rural Finland with fluoride in drinking water exceeding 1.5 mg/L, the average bone concentrations from 57 autopsies were 3,490 mg/kg ash in females and 2,830 mg/kg ash in males (Arnala et al. 1985). Most had lived their whole lives in the same place, most were over 50, and 7 had impaired renal function. For 16, fluoride concentrations were measured in the water sources (2.6 ± 1.4 mg/L); bone concentrations were 4,910 ± 2,250 mg/kg ash. In a later study of the same area of Finland, the mean bone concentration in 18 hip fracture patients was 3,720 ± 2,390 mg/kg, assumed to be ash (Arnala et al. 1986). The mean age was 79, 14 were female, 3 had diabetes, and 1 had elevated serum creatinine; residence time was not specified. For people exposed to fluoride at 2 mg/L in drinking water for a lifetime, the committee concludes that average bone concentration can be expected to be in the range of 4,000 to 5,000 mg/kg ash. Considerable variation around the average is expected.

FLUORIDE CONCENTRATIONS IN BONES AFTER CLINICAL STUDIES

A number of clinical studies measured bone fluoride concentrations after therapeutic treatment (van Kesteren et al. 1982; Boivin et al. 1988; Bayley et al. 1990; Gutteridge et al. 1990; Orcel et al. 1990; Boivin et al. 1993; Søgaard et al. 1994; Lundy et al. 1995). Figure 3-2 summarizes these data, plotting fluoride concentrations in bone ash after treatment versus total exposure from the studies. The weighted least squares (WLS) regression line weighted points according to the number of participants in each trial (see Appendix C). Note that the two points farthest above the regression line (Bayley et al. 1990; Lundy et al. 1995) were from studies carried out in Toronto and Minnesota, presumably fluoridated areas; most (possibly all) of the other studies were conducted in European countries that do not fluoridate water. The two points farthest below the line delivered fluoride in a form designed to reduce bioavailability (Boivin et al. 1988; Turner et al. 1993). This analysis is ecologic, plotting average bone concentrations versus total exposure. However, analysis of individual-level data in two studies (van Kesteren et al. 1982; Gutteridge et al. 1990) provides similar results.

Because the pharmacokinetics of fluoride are nonlinear, we would not necessarily expect people with the same cumulative exposure to have the same bone fluoride concentrations. Indeed, the model may overpredict bone concentrations for long-term exposure to lower fluoride concentrations via water. Figure 3-2 also shows the average bone ash concentrations measured by Zipkin et al. for fluoride at 4 mg/L plotted against estimated total exposure. The latter was estimated assuming consumption of 1.51 L of water per day (Turner et al. 1993) and 21 years of exposure to fluoride in the 4-mg/L area. (The Zipkin study reported residence time and water concentrations but not water consumption.) While not completely out of range, the bone concentration is lower than expected based on the regression for the clinical data. Analysis of Turner's pharmacokinetic model (Turner et al. 1993) suggests that short-term (months to years), high-dose exposures

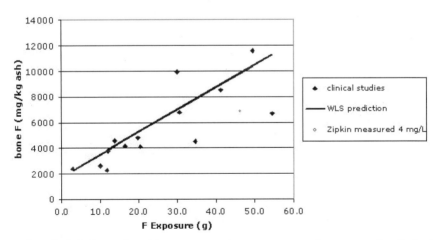

FIGURE 3-2 Bone fluoride concentrations versus total exposure in clinical trials. For comparison, the average bone concentration found by Zipkin et al. (1958) among subjects drinking water with fluoride at 4 mg/L is provided.

may produce higher bone fluoride concentrations than long-term (decades), low-dose exposures. More time means more bone resorption, allowing a greater fraction of the total fluoride dose to be excreted. Additional research on this topic would be useful.

More detailed information on fluoride's effects on bone cells and bone formation is presented in Chapter 5.

COMPARATIVE PHARMACOKINETICS OF RATS AND HUMANS

Among animal species, fluoride toxicology has been studied most extensively in rats. When extrapolating from rats to humans, it is useful to consider their relative pharmacokinetics. There are at least two ways to do this. Bone, tissue, or plasma concentrations may provide an appropriate biomarker of internal exposure for some effects. Alternatively, one can compare plasma, tissue, and bone concentrations in rats and humans given the same dose.

Our knowledge of the comparative pharmacokinetics of fluoride is primarily limited to short-term studies of a small number of mammals. Using estimates of plasma, renal, and extrarenal fluoride clearances scaled to body weight, Whitford et al. (1991) concluded that dogs were the best pharmacokinetic model for humans, based on studies of healthy young adults. In contrast, renal clearance in rats (age 12 weeks) was more than three times larger than in humans; rat extrarenal clearance was about twice as large (Whitford et al. 1991). Unlike in humans, rat bones do not undergo Haversian remodeling (remodeling along channels within the bone). Fluoride uptake by the bones of adult rats should be minimal (Turner et al. 1995).

Comparisons between species—and within species for different experiments—are complicated by several factors. With chronic exposure, fluoride bone concentrations tend to increase over time. The amount of calcium in the diet affects the amount of fluoride absorbed. The dose of fluoride can depend on the concentration of fluoride in water, water consumption, and the amount of fluoride in the diet. If fluoride concentration is kept constant in water, dose can vary as the animal ages. Species age at different rates, and age affects pharmacokinetics, especially bone development and kidney function.

Evidence suggests that rats require higher chronic exposure than humans to achieve the same plasma and bone fluoride concentrations. It has been suggested that rats might require water concentrations about five times larger than humans to reach the same plasma concentration (Dunipace et al. 1995). For bone, Turner et al. (1992) estimated that "humans incorporate fluoride ~18 times more readily than rats when the rats are on a normal calcium diet." This comparison was also based on water concentrations. In Appendix D, this issue is briefly reviewed. The factor for plasma is uncertain, in part because it could change with age or duration of dose. It might be more appropriate to compare exposures than water concentration. Bone comparisons are also uncertain but appear to support a rat-to-human conversion factor for older rats and humans of at least an order of magnitude.

ORGANOFLUORINE COMPOUNDS

Two types of fluorine are found in human plasma: inorganic and organic. Up to now, this chapter has discussed the inorganic form. Remarkably, the amount of organic fluoride in serum is generally greater than the amount of inorganic fluoride (Whitford 1996). Interest in organofluorine compounds has grown tremendously in the last decade. Two compounds (and their salts) dominate recent biological research: perfluorooctanesulfonate (PFOS; $C_8F_{17}SO_3^-$) and perfluorooctanoate (PFOA; $C_7F_{15}COO^-$). Both are straight-chain compounds with fluorine substituted for aliphatic hydrogens. These compounds are biologically stable with long half-lives, on the order of years, in humans. Relatively little is known about the routes of human exposure. A recent study of American Red Cross adult blood donors found median serum concentrations of 35 µg/L of PFOS and 5 µg/L of PFOA (Olsen et al. 2003).

Defluorination of PFOA has not been detected in rat experiments (Vanden Heuvel et al. 1991; Kudo and Kawashima 2003). Given the stability of PFOA and PFOS, they do not appear to be important sources of inorganic fluoride, although more research is needed, particularly for PFOS. Degradation of other fluorocarbons might produce fluoride ion. Perfluorooctanesulfonyl fluoride (POSF, $C_8F_{17}SO_2F$) is used as a starting material for manufacturing polymers and surfactants. Residual POSF in products "may degrade or metabolize, to an undeterminate degree" to PFOS (Olsen et al. 2004, p. 1600). Certain anesthetics release fluoride ion during use (see Chapter 2).

FACTORS MODIFYING PHARMACOKINETICS AND THEIR IMPLICATIONS FOR POTENTIALLY SUSCEPTIBLE POPULATIONS

Changes in chronic exposure to fluoride will tend to alter plasma and bone fluoride concentrations. A number of factors can modify the pharmacokinetics, providing another way to change fluoride tissue concentrations.

Fluoride clearance tends to increase with urinary pH. One proposed mechanism is decreased reabsorption in the renal tubule, easily crossed by HF and nearly impermeable to fluoride ion. Increasing urinary pH thus tends to decrease fluoride retention. As a result, fluoride retention might be affected by environments or conditions that chronically affect urinary pH, including diet, drugs, altitude, and certain diseases (e.g., chronic obstructive pulmonary disease) (reviewed by Whitford 1996).

Because of their growing skeleton, infants and children clear relatively larger amounts of fluoride into bones than adults (Ekstrand et al. 1994; Whitford 1999). As discussed earlier, fluoride plasma and bone concentrations tend to increase with age. Although this trend is partly due to accumulation over time, decreased renal clearance and differences in bone resorption (preferential removal of cystallites with little or no fluoride in the elderly have been hypothesized to play a role.

Because the kidney is the major route of excretion, increased plasma and bone fluoride concentrations are not surprising in patients with kidney disease. Plasma fluoride concentrations are clearly elevated in patients with severely compromised kidney function, reduced glomerular filtration rates of around 20% of normal, as measured via creatinine clearance or serum creatinine concentrations (Hanhijärvi 1974, 1982; Parsons et al. 1975; Schiffl and Binswanger 1980; Waterhouse et al. 1980; Hanhijärvi and Penttilä 1981). Kuo and Stamm (1975) found no association. However, elevated serum concentrations were found in renal patients with normal serum creatinine (Hanhijärvi 1982).

Only a few studies have examined fluoride concentrations in bone in renal patients. Call et al. (1965) found doubled bone fluoride concentrations in five patients with chronic, severe kidney disease. Juncos and Donadio (1972) diagnosed systemic fluorosis (but did not measure bone fluoride concentrations) in two patients with reduced renal function and exposure to drinking water with fluoride at 1.7 and 2.6 mg/L. Four renal patients with severe skeletal changes or bone pain had elevated serum and bone fluoride concentrations; the bone concentrations ranged from about 5,500 to 11,000 mg/kg (Johnson et al. 1979). Fluoride bone concentrations more than doubled in four patients with severe, chronic pyelonephritis (Hefti and Marthaler 1981). Arnala et al. (1985) reported elevated bone concentrations (roughly 50%) in six people with "slightly impaired renal function" from a fluoridated area. Bone fluoride concentrations were significantly increased in dialysis patients compared with normal controls (Cohen-Solal et al. 2002). In rats with surgically induced renal deficiency (80% nephrectomy), glomerular filtration rate decreased by 68%. After 6 months of fluoride treatment, bone fluoride concentrations approximately doubled (Turner et al. 1996).

Hanhijärvi and Penttilä (1981) reported elevated serum fluoride in patients with cardiac failure. Fluoride concentrations were positively related to serum creatinine, although the concentrations of the latter did not indicate renal insufficiency. During cardiac failure, the body tries to maintain blood flow to the heart and brain.

Although some studies report no difference in plasma fluoride concentrations between men and women (e.g., Torra et al. 1998), others found greater rates of increase with age in females (Husdan et al. 1976; Hanhijärvi et al. 1981). Enhanced release of fluoride in postmenopausal women is one possible explanation. Similar to our regression results of the Zipkin data, some studies have found a tendency toward elevated bone fluoride concentrations in women (Arnala et al. 1985; Richards et al. 1994). A Finnish study reported that bone fluoride concentrations increased more rapidly with age in women than in men (Alhava et al. 1980). This variability might be due to several factors, including individual differences in water consumption and pharmacokinetics.

In sum, although the data are sparse, severe renal insufficiency appears to increase bone fluoride concentrations, perhaps as much as twofold. The elderly are at increased risk of high bone fluoride concentrations due to accumulation over time; although less clear, decreased renal function and gender may be important.

FINDINGS

- Bone fluoride concentrations increase with both magnitude and length of exposure. Empirical data suggest substantial variations in bone fluoride concentrations at any given water concentration.
- On the basis of pharmacokinetic modeling, the current best estimate for bone fluoride concentrations after 70 years of exposure to fluoride at 4 mg/L in water is 10,000 to 12,000 mg/kg in bone ash. Higher values would be predicted for people consuming large amounts of water (>2 L/day) or for those with additional sources of exposure. Less information was available for estimating bone concentrations from lifetime exposure to fluoride in water at 2 mg/L. The committee estimates average bone concentrations of 4,000 to 5,000 mg/kg ash.
- Groups likely to have increased bone fluoride concentrations include the elderly and people with severe renal insufficiency.
- Pharmacokinetics should be taken into account when comparing effects of fluoride in different species. Limited evidence suggests that rats require higher chronic exposures than humans to achieve the same plasma and bone concentrations.

RESEARCH RECOMMENDATIONS

- Additional research is needed on fluoride concentrations in human bone as a function of magnitude and duration of exposure, age, gender, and health status. Such studies would be greatly aided by noninvasive means of measuring bone fluoride. As discussed in other chapters of this report, some soft tissue effects may be associated with fluoride exposure. Most measurements of fluoride in soft tissues are based on short-term exposures and some atypically high values have been reported. Thus, more studies are needed on fluoride concentrations in soft tissues (e.g., brain, thyroid, kidney) following chronic exposure.

- Research is needed on fluoride plasma and bone concentrations in people with small to moderate changes in renal function as well as patients with serious renal deficiency. Other potentially sensitive populations should be evaluated, including the elderly, postmenopausal women, and people with altered acid-base balance.
- Improved and readily available pharmacokinetic models should be developed.
- Additional studies comparing pharmacokinetics across species are needed.
- More work is needed on the potential for release of fluoride by the metabolism of organofluorines.

4

Effects of Fluoride on Teeth

In this chapter, the committee reviews research on the occurrence of enamel fluorosis at different concentrations of fluoride in drinking water, with emphasis on severe enamel fluorosis and water fluoride concentrations at or near the current maximum contaminant level goal (MCLG) of 4 mg/L and the secondary maximum contaminant level (SMCL) of 2 mg/L. Evidence on dental caries in relation to severe enamel fluorosis, aesthetic and psychological effects of enamel fluorosis, and effects of fluoride on dentin fluorosis and delayed tooth eruption is reviewed as well. Evidence on caries prevention at water concentrations below the SMCL of 2 mg/L is not reviewed. Strengths and limitations of study methods, including issues pertaining to diagnosis and measurement, are considered.

ENAMEL FLUOROSIS

Fluoride has a great affinity for the developing enamel because tooth apatite crystals have the capacity to bind and integrate fluoride ion into the crystal lattice (Robinson et al. 1996). Excessive intake of fluoride during enamel development can lead to enamel fluorosis, a condition of the dental hard tissues in which the enamel covering of the teeth fails to crystallize properly, leading to defects that range from barely discernable markings to brown stains and surface pitting. This section provides an overview of the clinical and histopathological manifestations of enamel fluorosis, diagnostic issues, indexes used to characterize the condition, and possible mechanisms.

Clinical and Histological Features

Enamel fluorosis is a mottling of the tooth surface that is attributed to fluoride exposure during tooth formation. The process of enamel maturation consists of an increase in mineralization within the developing tooth and concurrent loss of early-secreted matrix proteins. Exposure to fluoride during maturation causes a dose-related disruption of enamel mineralization resulting in widening gaps in its crystalline structure, excessive retention of enamel proteins, and increased porosity. These effects are thought to be due to fluoride's effect on the breakdown rates of matrix proteins and on the rate at which the by-products from that degradation are withdrawn from the maturing enamel (Aoba and Fejerskov 2002).

Clinically, mild forms of enamel fluorosis are evidenced by white horizontal striations on the tooth surface or opaque patches, usually located on the incisal edges of anterior teeth or cusp tips of posterior teeth. Opaque areas are visible in tangential reflected light but not in normal light. These lesions appear histopathologically as hypomineralization of the subsurface covered by a well-mineralized outer enamel surface (Thylstrup and Fejerskov 1978). In mild fluorosis, the enamel is usually smooth to the point of an explorer, but not in moderate and severe cases of the condition (Newbrun 1986). In moderate to severe forms of fluorosis, porosity increases and lesions extend toward the inner enamel. After the tooth erupts, its porous areas may flake off, leaving enamel defects where debris and bacteria can be trapped. The opaque areas can become stained yellow to brown, with more severe structural damage possible, primarily in the form of pitting of the tooth surface.

Enamel in the transitional or early maturation stage of development is the most susceptible to fluorosis (DenBesten and Thariani 1992). For most children, the first 6 to 8 years of life appear to be the critical period of risk. In the Ikeno district of Japan, where a water supply containing fluoride at 7.8 mg/L was inadvertently used for 12 years, no enamel fluorosis was seen in any child who was age 7 years or older at the start of this period or younger than 11 months old at the end of it (Ishii and Suckling 1991). For anterior teeth, which are of the most aesthetic concern, the risk period appears to be the first 3 years of life (Evans and Stamm 1991; Ishii and Suckling 1991; Levy et al. 2002a). Although it is possible for enamel fluorosis to occur when teeth are exposed during enamel maturation alone, it is unclear whether it will occur if fluoride exposure takes place only at the stage of enamel-matrix secretion. Fejerskov et al. (1994) noted that fluoride uptake into mature enamel is possible only as a result of concomitant enamel dissolution, such as caries development. Because the severity of fluorosis is related to the duration, timing, and dose of fluoride intake, cumulative exposure during the entire maturation stage, not merely during critical periods of certain types of tooth development, is probably the most important exposure measure to consider when assessing the risk of fluorosis (DenBesten 1999).

Mechanisms

Dental enamel is formed by matrix-mediated biomineralization. Crystallites of hydroxyapatite $(Ca_{10}(PO_4)_6(OH)_2)$ form a complex protein matrix that serves as a nucleation site (Newbrun 1986). The matrix consists primarily of amelogenin, proteins synthesized by secretory ameloblasts that have a functional role in establishing and maintaining the spacing between enamel crystallites. Full mineralization of enamel occurs when amelogenin fragments are removed from the extracellular space. The improper mineralization that occurs with enamel fluorosis is thought to be due to inhibition of the matrix proteinases responsible for removing amelogenin fragments. The delay in removal impairs crystal growth and makes the enamel more porous (Bronckers et al. 2002). DenBesten et al. (2002) showed that rats exposed to fluoride in drinking water at 50 or 100 mg/L had lower total proteinase activity per unit of protein than control rats. Fluoride apparently interferes with protease activities by decreasing free Ca^{2+} concentrations in the mineralizing milieu (Aoba and Fejerskov 2002).

Matsuo et al. (1998) investigated the mechanism of enamel fluorosis in rats administered sodium fluoride (NaF) at 20 mg/kg by subcutaneous injections for 4 days or at 240 mg/L in drinking water for 4 weeks. They found that fluoride alters intracellular transport in the secretory ameloblasts and suggested that G proteins play a role in the transport disturbance. They found different immunoblotting-and-pertussis-toxin-sensitive G proteins on the rough endoplasmic reticulum and Golgi membranes of the germ cells of rats' incisor teeth.

Health Issues and Clinical Treatment

Whether to consider enamel fluorosis, particularly the moderate to severe forms, an adverse cosmetic effect or an adverse health effect has been the subject of debate for decades. Some early literature suggests that the clinical course of caries could be compromised by untreated severe enamel fluorosis. Smith and Smith (1940, pp.1050-1051) observed, "There is ample evidence that mottled teeth, though they be somewhat more resistant to the onset of decay, are structurally weak, and that unfortunately when decay does set in, the result is often disastrous. Caries once started evidently spreads rapidly. Steps taken to repair the cavities in many cases were unsuccessful, the tooth breaking away when attempts were made to anchor the fillings, so that extraction was the only course." Gruebbel (1952, p.153) expressed a similar viewpoint: "Severe mottling is as destructive to teeth as is dental caries. Therefore, when the concentration is excessive, defluorination or a new water supply should be recommended. The need for removing excessive amounts of fluorides calls attention to the peculiar situation in public health practice in which a chemical substance is added to water in some localities to prevent a disease and the same chemical substance is removed in other localities to prevent another disease." Dean advised that when the average child in a community has mild fluorosis (0.6 on his scale, described in the next section), ". . . it begins to constitute a public health problem warranting increasing consideration" (Dean 1942, p. 29).

There appears to be general acceptance in today's dental literature that enamel fluorosis is a toxic effect of fluoride intake that, in its severest forms, can produce adverse effects on dental health, such as tooth function and caries experience. For example:

- "The most severe forms of fluorosis manifest as heavily stained, pitted, and friable enamel that can result in loss of dental function" (Burt and Eklund 1999).
- "In more severely fluorosed teeth, the enamel is pitted and discolored and is prone to fracture and wear" (ATSDR 2003, p. 19).
- "The degree of porosity (hypermineralization) of such teeth results in a diminished physical strength of the enamel, and parts of the superficial enamel may break away . . . In the most severe forms of dental fluorosis, the extent and degree of porosity within the enamel are so severe that most of the outermost enamel will be chipped off immediately following eruption" (Fejerskov et al. 1990, p. 694).
- "With increasing severity, the subsurface enamel all along the tooth becomes increasingly porous . . . the more severe forms are subject to extensive mechanical breakdown of the surface" (Aoba and Fejerskov 2002, p. 159).
- "With more severe forms of fluorosis, caries risk increases because of pitting and loss of the outer enamel" (Levy 2003, p. 286).
- " . . . the most severe forms of dental fluorosis might be more than a cosmetic defect if enough fluorotic enamel is fractured and lost to cause pain, adversely affect food choices, compromise chewing efficiency, and require complex dental treatment" (NRC 1993, p. 48).

Severe enamel fluorosis is treated to prevent further enamel loss and to address the cosmetic appearance of teeth. Treatments include bleaching, microabrasion, and the application of veneers or crowns. Bleaching and microabrasion are typically used with the mild to moderate forms of enamel fluorosis. Bleaching is the least invasive procedure, but does not eliminate the dark stains associated with severe enamel fluorosis. Microabrasion involves the controlled abrasion of enamel to remove superficial stains.

This technique has been reported to be minimally invasive and successful in treating single-line or patched opacities, but was not effective in treating defects that extend deeper into the enamel (Wong and Winter 2002). Train et al. (1996) found that while microabrasion improved the appearance of all degrees of enamel fluorosis, severely fluorosed teeth exhibited more defective surfaces following treatment. Pits and fissures can be filled with flowable composites. Partial veneers, composite veneers, and crowns provide the best aesthetic results for very severe enamel fluorosis, but are the most invasive treatments. Crowns are usually used as a last resort because they can be a threat to tooth vitality (Christensen 2005). The procedure requires the further removal of tooth enamel to allow for bonding of the crown, and sometimes requires replacement within a few years. The more invasive treatments should be used only in the most severe cases of enamel fluorosis.

Ascertaining Enamel Fluorosis

Enamel Fluorosis Indexes

The three main indexes used to grade enamel fluorosis in research are Dean's index, the Thylstrup-Fejerskov index (TFI), and the tooth surface index of fluorosis (TSIF). A particularly useful review of the characteristics, strengths, and limitations of these indexes is given by Rozier (1994).

Dean's index (Table 4-1) uses a 6-point ordinal scale, ranging from normal to severe, to classify individuals with regard to enamel fluorosis (Dean 1942). Scores are assigned on the basis of the two worst-affected teeth and are derived from an assessment of the whole tooth rather than the worst-affected tooth surface. Although Dean's index is considered adequate for a broad definition of prevalence and trends, it suffers from limited sensitivity for analytical research in several ways. Because a person is assigned to a fluorosis category on the basis of only two severely affected teeth, the score may not discriminate between those individuals who have more affected teeth from those with only a few affected teeth. In addition, as the teeth most frequently affected by enamel fluorosis are posterior teeth and not the aesthetically important anterior teeth, Dean's index may misclassify individuals with respect to aesthetic effects (Griffin et al. 2002). As a score assigned at the level of the person, Dean's index enables the computation of prevalence estimates but does not permit an analysis of the effects of changes in exposure during the development of different teeth. Finally, with only one category for severe fluorosis, Dean's index does not discriminate between staining and pitting or between discrete and confluent pitting. In fact, Dean revised the index in 1942 to create the version in use today, which combines the original "moderately severe" and "severe" categories. Despite its limitations, Dean's index is by far the most widely used measure of enamel

TABLE 4-1 Clinical Criteria for Dean's Enamel Fluorosis Index

Diagnosis	Criteria
Normal (0)	The enamel represents the usually translucent semivitriform type of structure. The surface is smooth, glossy, and usually a pale creamy white color.
Questionable (0.5)	The enamel discloses slight aberrations from the translucency of normal enamel, ranging from a few white flecks to occasional white spots. This classification is utilized when a definite diagnosis of the mildest form of fluorosis is not warranted and a classification of "normal" is not justified.
Very mild (1)	Small, opaque, paper white area scattered irregularly over the tooth but not involving as much as approximately 25% of the tooth surface. Frequently included in this classification are teeth showing no more than 1 to 2 mm of white opacity at the tip of the summit of the cusps of the bicuspids or second molars.
Mild (2)	The white opaque areas in the enamel of the teeth are more extensive but do not involve as much as 50% of the tooth.
Moderate (3)	All enamel surfaces of the teeth are affected, and surfaces subject to attrition show marked wear. Brown stain is frequently a disfiguring feature.
Severe (4)	All enamel surfaces are affected and hypoplasia is so marked that the general form of the tooth may be altered. The major diagnostic sign of this classification is the discrete or confluent pitting. Brown stains are widespread and teeth often present a corroded appearance.

SOURCE: Dean 1942. Reprinted with permission; copyright 1942, American Association for the Advancement of Science.

fluorosis in the research literature. As a consequence, any comprehensive review of the literature must rely upon it.

The TFI (Table 4-2), which classifies the facial surface of each tooth on a 10-point scale (0 to 9), provides more criteria and categories for characterizing mild and severe forms of fluorosis than Dean's index allows (Thylstrup and Fejerskov 1978). At the upper end of the severity scale, the TFI usefully distinguishes among marked discoloration without pitting (score 4); discrete or focal pitting (score 5); and degrees of confluent pitting, enamel loss, and tooth deformation (scores 6-9). The TFI has been shown to be a valid indication of the fluoride content of fluorotic enamel. Most investigators combine TFI scores of 5 and higher, all of which include pitting, to form a category of severe enamel fluorosis.

The TSIF (Table 4-3) ascribes a fluorosis score on an 8-point scale (0 to 7) to each unrestored surface of each tooth (Horowitz et al. 1984). At the higher end of the scale, there is a greater range of criteria for characterization of effects. A TSIF score of 5 is the lowest classification on this scale that involves enamel pitting. Although some researchers combine scores 5-7

TABLE 4-2 Clinical Criteria and Scoring for the Thylstrup and Fejerskov Index (TFI) of Enamel Fluorosis

Score	Criteria
0	Normal translucency of enamel remains after prolonged air-drying.
1	Narrow white lines corresponding to the perikymata.
2	Smooth surfaces: More pronounced lines of opacity that follow the perikymata. Occasionally confluence of adjacent lines. Occlusal surfaces: Scattered areas of opacity < 2 mm in diameter and pronounced opacity of cuspal ridges.
3	Smooth surfaces: Merging and irregular cloudy areas of opacity. Accentuated drawing of perikymata often visible between opacities. Occlusal surfaces: Confluent areas of marked opacity. Worn areas appear almost normal but usually circumscribed by a rim of opaque enamel.
4	Smooth surfaces: The entire surface exhibits marked opacity or appears chalky white. Parts of surface exposed to attrition appear less affected. Occlusal surfaces: Entire surface exhibits marked opacity. Attrition is often pronounced shortly after eruption.
5	Smooth and occlusal surfaces: Entire surface displays marked opacity with focal loss of outermost enamel (pits) < 2 mm in diameter.
6	Smooth surfaces: Pits are regularly arranged in horizontal bands < 2 mm in vertical extension. Occlusal surfaces: Confluent areas < 3 mm in diameter exhibit loss of enamel. Marked attrition.
7	Smooth surfaces: Loss of outermost enamel in irregular areas involving less than half of entire surface. Occlusal surfaces: Changes in morphology caused by merging pits and marked attrition.
8	Smooth and occlusal surfaces: Loss of outermost enamel involving more than half of surface.
9	Smooth and occlusal surfaces: Loss of main part of enamel with change in anatomic appearance of surface. Cervical rim of almost unaffected enamel is often noted.

SOURCE: Thylstrup and Fejerskov 1978. Reprinted with permission; copyright 1978, Community Dentistry and Oral Epidemiology.

to classify severe enamel fluorosis, others extend their highest category of severity to include score 4, which includes staining but not pitting.

Other fluorosis indexes, such as those developed by Siddiqui (1955) and Al-Alousi et al. (1975), are used less frequently in research and almost never in the United States. The developmental defects of enamel (DDE) index was designed as a general classification scheme for enamel defects (FDI 1982; Clarkson and O'Mullane 1989). As it emphasizes aesthetic concerns and is not based on etiologic considerations, it is not technically an index of enamel fluorosis. The fluorosis risk index (FRI) was developed specifically for use in case-control studies (Pendrys 1990), very few of which have been conducted.

TABLE 4-3 Clinical Criteria and Scoring for the Tooth Surface Index of Fluorosis (TSIF)

Score	Criteria
0	Enamel shows no evidence of fluorosis.
1	Enamel shows definite evidence of fluorosis—namely, areas with parchment-white color that total less than one-third of the visible enamel surface. This category includes fluorosis confined only to incisal edges of anterior teeth and cusp tips of posterior teeth ("snowcapping").
2	Parchment-white fluorosis totals at least one-third, but less than two-thirds, of the visible surface.
3	Parchment-white fluorosis totals at least two-thirds of the visible surface.
4	Enamel shows staining in conjunction with any of the preceding levels of fluorosis. Staining is defined as an area of definite discoloration that may range from light to very dark brown.
5	Discrete pitting of the enamel exists, unaccompanied by evidence of staining of intact enamel. A pit is defined as a definite physical defect in the enamel surface with a rough floor that is surrounded by a wall of intact enamel. The pitted area is usually stained or differs in color from the surrounding enamel.
6	Both discrete pitting and staining of the intact enamel exist.
7	Confluent pitting of the enamel surface exists. Large areas of enamel may be missing and the anatomy of the tooth may be altered. Dark-brown stain is usually present.

SOURCE: Horowitz et al. 1984. Reprinted with permission; copyright 1984, American Dental Association.

A major difference among the three principal enamel fluorosis indexes is the level at which the scores are recorded: the level of the person on Dean's index, the level of the tooth on the TFI, and the level of the tooth surface on the TSIF. As the tooth-level scores for Dean's index are usually recorded but not reported, it is impossible to break the reported person-level scores down to the tooth or tooth-surface level. Similarly, the tooth level TFI scores cannot be broken down to the level of the tooth surface. In contrast, it is possible to combine TFI scores up to the person level and to combine TSIF scores up to the tooth or person levels.

Because the person-level Dean's index is the oldest and still the most widely used enamel fluorosis index, researchers using the TFI or TSIF sometimes, though rarely, aggregate scores on those scales up to the person level for comparability. When this is done, the most severe one or two teeth or tooth surfaces are typically used. As a consequence, the prevalence of a given level of enamel fluorosis severity (other than "normal" or "unaffected") will tend to be lowest if expressed as a proportion of all tooth surfaces, intermediate in magnitude if expressed as a proportion of all teeth, and highest if expressed as a proportion of all persons in a given sample. Prevalence estimates at the person level are reviewed by the committee later in this chapter. When the interest is in aesthetic concerns about milder forms of fluorosis, the person level and tooth level have disadvantages, as the affected teeth may be located in the posterior part of the mouth and thus less visible under ordinary (nonclinical) circumstances. For the severest forms, in contrast, the considerations are reversed. It is more informative to know the proportion of a population who have any teeth with dark staining and pitting than the proportion of all teeth or of all tooth surfaces that have these most severe manifestations of enamel fluorosis.

Diagnostic Issues

The 1993 National Research Council (NRC) report found that the accuracy of clinical diagnosis of fluorotic lesions, especially those of the mild form, has been plagued by the fact that not all white or light yellow opacities in dental enamel are caused by fluoride. The ascertainment of severe enamel fluorosis, in contrast, is much more secure. This is especially true in studies of children in communities with relatively high water fluoride concentrations in the United States and similar locales, where there are few if any alternative explanations for dark yellow to brown staining and pitting of the enamel of recently erupted permanent teeth.

Some studies in the international literature have reported severe mottling of the teeth that could not be attributed to fluoride exposure. For example, Whitford (1996) was unable to explain a high prevalence of severe lesions resembling fluorosis in individuals in Morrococha, Peru, on the basis of exposure to fluoride in water, food, or dental products. Yoder et al. (1998) found severe dental mottling in a population in Tanzania with negligible fluoride in the water (<0.2 mg/L). They noted that urinary fluoride concentrations in affected subjects from that area were not consistent with concentrations found in subjects from a high-fluoride area who had severe enamel fluorosis. Mottling unrelated to fluoride has been suggested to be due to malnutrition, metabolic disorders, exposure to certain dietary trace elements, widespread introduction of tea drinking among children at very early ages, or physical trauma to the tooth (Curzon and Spector 1977; Cutress and Suckling 1990).

A genetic condition called amelogeneis imperfecta causes enamel defects that can be mistaken for enamel fluorosis (Seow 1993); the hypoplastic lesions of this condition have a deficiency in the quantity of enamel with grooves and pits on the surface. Hypocalcified lesions have low mineralization, appear pigmented, and have softened and easily detachable enamel. Hypomaturation conditions are evident as opaque and porous enamel. The prevalence of amelogeneis imperfecta ranges from approximately 1 in 700 to 1 in 14,000, depending on the population studied (Seow 1993).

Angmar-Mansson and Whitford (1990) reported that acute and chronic exposures to hypobaric hypoxia that occurs at high altitudes are associated with bilaterally symmetrical and diffuse disturbances in enamel mineralization that might be mistaken for fluorosis. More recently, Rweneyonyi et al. (1999) reported higher prevalences of severe enamel fluorosis at higher altitudes than at lower altitudes in Ugandan populations with the same water fluoride levels.

Some evidence from animal studies indicates that genetics might contribute to susceptibility to enamel fluorosis (Everett et al. 2002). It has also been proposed that use of the antibiotic amoxicillin during infancy might contribute to the development of enamel fluorosis of the primary teeth (Hong et al. 2004).

A number of review articles evaluate the strengths and deficiencies of the various indexes used to diagnose and characterize the degree of enamel fluorosis (Clarkson 1989; Ellwood et al. 1994; Kingman 1994; Rozier 1994). In general, the following observations may be made:

- The various indexes use different examination techniques, classification criteria, and ways of reporting data. All indexes are based on subjective assessment, and little information is available on their validity or comparability. Prevalence data obtained from these indexes also can vary considerably because of differences in study protocols and case definitions. Nevertheless, the American Dental Association (2005) considers severe and even moderate fluorosis "typically easy to detect."
- Examiner reliability is an important consideration in evaluation studies. Systematic interexaminer variability has been reported (Burt et al. 2003). Rozier (1994) noted that only about half the studies available in 1994 provided evidence that examiner reliability was evaluated. Although almost all of those assessments were conducted in populations in which severe enamel fluorosis was very rare, they showed an acceptable level of agreement.
- Agreement among examiners tends to be lower when enamel fluorosis is recorded at the level of the tooth or tooth surface than when it is recorded at the person level.

Prevalence of Severe Enamel Fluorosis in Relation to Water Fluoride Concentrations

In many reviews and individual studies, all levels of enamel fluorosis severity are grouped together. This approach is less problematic at comparatively low levels of fluoride intake, where all or almost all of the cases are mild or moderate in severity. At higher intake levels, such as those typically found in communities with water fluoride concentrations at the current MCLG of 4 mg/L or the current SMCL of 2 mg/L, it is more informative to report results for the different levels of fluorosis severity. Those reviews in which severity distinctions have been drawn, such as NRC (1993) and IOM (1997), have tended to combine moderate and severe fluorosis into a single category. The present report focuses more specifically on the severe forms.

The committee compiled prevalence estimates at the person level for severe enamel fluorosis in relation to water fluoride levels from studies around the world. The starting points were the estimates provided in EPA's documentation supporting the MCLG (50 Fcd. Reg. 20164 [1985]) and Appendix C6 of McDonagh et al. (2000a). To these were added results from 24 additional studies (Venkateswarlu et al. 1952; Forsman 1974; Retief et al. 1979; Rozier and Dudney 1981; Subbareddy and Tewari 1985; Haimanot et al. 1987; Kaur et al. 1987; Mann et al. 1987, 1990; Szpunar and Burt 1988; Thaper et al. 1989; Jackson et al. 1995; Cortes et al. 1996; Akpata et al. 1997; Gopalakrishnan et al. 1999; Kumar and Swango 1999; Menon and Indushekar 1999; Rwenyonyi et al. 1999; Sampaio and Arneberg 1999; Awadia et al. 2000; Alarcón-Herrera et al. 2001; Grobler et al. 2001; Ermiş et al. 2003; Wondwossen et al. 2004). Results were excluded if they were for fluorosis indexes other Dean's index, the TFI, the TSIF, or modifications thereof (e.g., Goward 1982; Nunn et al. 1992); for all fluorosis or for moderate and severe fluorosis combined (e.g., Warnakulasuriya et al. 1992; Mella et al. 1994; Alonge et al. 2000; Burt et al. 2003); for primary or deciduous teeth as opposed to permanent teeth (e.g., McInnes et al. 1982); for different teeth separately with no results at the person level or for all teeth combined (e.g., Opinya et al. 1991); for unbounded upper categories of water fluoride for which no mean or median value was given (e.g., > 1.2 mg/L in Heller et al. [1997], > 2 mg/L in Ray et al. [1982], > 2.5 mg/L in Angelillo et al. [1999]); for bounded but extremely wide water fluoride ranges (e.g., 0.8 to 4.3 mg/L in Haimanot et al. [1987], 0.7 to 4.0 in Beltran-Aguilar et al. [2002], 0.3 to 2.2 mg/L in Wondwossen et al. [2004]). For narrower bounded categories, the midrange water fluoride level was used. Results from studies of children and teenagers (age 20 years or younger) were tallied separately from results for adults. Severe enamel fluorosis was classified as the "severe" classification in Dean's index and, depending on the groupings created by the original invesgtigators, TFI scores of 4-9 or 5-9 and TSIF scores of 4-7 or 5-7. Because of the wide variability in methods and populations, and the lack of independence when a given study provided more than one result, the estimates were not subjected to formal statistical analyses. Instead, plots of the prevalence estimates in relation to water fluoride concentration were examined for the presence of any clear and obvious patterns or trends.

Figure 4-1 shows 94 prevalence estimates from studies in the United States. Despite the wide range of research methods, fluorosis indexes, water fluoride measurement methods, and population characteristics in these studies conducted over a period spanning half a century, a clear trend is evident.

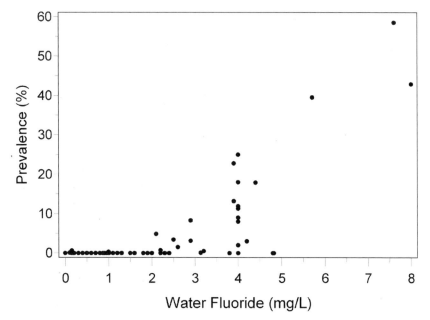

FIGURE 4-1 Prevalence of severe enamel fluorosis at the person level by water fluoride concentration, permanent teeth, age < 20 years, U.S. communities.

The prevalence of severe enamel fluorosis is close to zero in communities at all water fluoride concentrations below 2 mg/L. Above 2 mg/L, the prevalence rises sharply. The shape of this curve differs dramatically from the linear trend observed when all levels of fluorosis severity are combined and related to either the water fluoride concentration (Dean 1942) or the estimated daily dose in milligrams per kilogram (Fejerskov et al. 1990).

Not shown in Figure 4-1 are a prevalence of 54% in a community with a water fluoride concentration of 14 mg/L (50 Fed. Reg. 20164 [1985]) and results from two studies of adults. One, with an age range of 20-44 years, reported prevalences of zero at <0.1 mg/L and 2% at 2.5 mg/L (Russell and Elvove 1951). In the other, with an age range of 27-65 years, the prevalences were zero at 0.7 mg/L and 76% at 3.5 mg/L (Eklund et al. 1987). These results are broadly consistent with those in Figure 4-1.

Strongly supporting evidence comes from a series of surveys conducted by researchers at the National Institute of Dental Health (Selwitz et al. 1995, 1998). In these studies using the TSIF, scores were reported only at the tooth-surface level (Figure 4-2). As with the person-level prevalence estimates (Figure 4-1), an approximate population threshold for severe enamel fluorosis is evident at water concentrations below 2 mg/L.

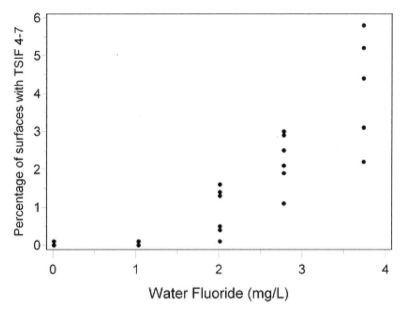

FIGURE 4-2 Percentage of tooth surfaces with severe enamel fluorosis (TSIF scores 4-7) by water fluoride concentration, permanent teeth, ages 8-10 and 13-16 years, U.S. communities, 1980, 1985 and 1990. (Some samples of children at a given water fluoride concentration had identical percentages of tooth surfaces with TSIF scores 4-7.) SOURCE: Selwitz et al. 1995, 1998.

Figure 4-3 shows 143 prevalence estimates from studies of children outside the United States. Not shown are results for three Ethiopian communities with extremely high water fluoride concentrations of 26, 34 and 36 mg/L and prevalences of 18%, 48% and 25%, respectively (Haimanot et al. 1987). Although a positive association may be discernible, it is much less obvious than in the U.S. studies. There is little evidence of an approximate population threshold as in the results in U.S. communities (Figure 4-1). In many regions around the world, water intake among children whose permanent teeth are forming can be much more variable than in the United States, susceptibility may differ more widely, sources of fluoride intake other than the community water supply may be more prevalent, or the ascertainment of severe enamel fluorosis may be more often compromised by other determinants of dental discoloration and pitting.

One question is whether the most severe forms of enamel fluorosis, specifically those involving confluent pitting, occur at water concentrations in the range of the current MCLG of 4 mg/L. This question cannot be an-

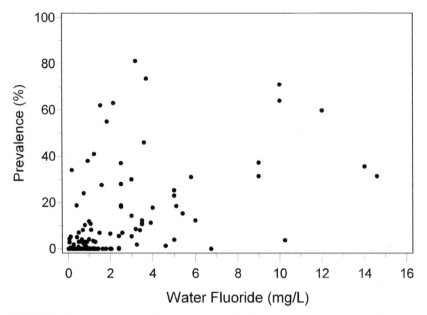

FIGURE 4-3 Prevalence of severe enamel fluorosis at the person level by water fluoride concentration, permanent teeth, age < 20 years, communities outside the United States.

swered by most studies, which use Dean's 1942 modification of his index combining "moderately severe" and "severe" classifications of his original system (Dean 1934) into a single category (Dean 1942; Rozier 1994). Three studies, however, in U.S. communities with water fluoride concentrations of approximately 4 mg/L have used enamel fluorosis indexes that draw severity distinctions within the "severe" category.

In Lowell, Indiana, with a water fluoride concentration of approximately 4 mg/L, 7% of a 1992 sample and 2% of a 1994 sample of children 7-14 years of age had at least one tooth surface assigned the highest possible TSIF score of 7 (Table 4-4). Expressed as a percentage of all tooth surfaces examined (mean, 32.3 per child), the prevalence of TSIF score 7 in the 1992 sample was substantially lower at 0.5% (Jackson et al. 1995). The lower prevalence using this metric is not surprising, as it includes surfaces on anterior teeth, which are not generally as susceptible to fluorosis as molars and other teeth located farther back in the mouth.

In Bushnell, Illinois, with a mean water fluoride concentration of 3.8 mg/L, samples of children age 8-10 years and 13-15 years were examined in 1980 and 1985 (Heifetz et al. 1988). As shown in Table 4-5, the TSIF score

TABLE 4-4 Maximum TSIF Scores in Two Samples of Children Age 7-14 Years in a U.S. Community with a Water Fluoride Concentration of 4.0 mg/L

Maximum TSIF Score	1992 study		1994 study	
	Number of Children	Percent	Number of Children	Percent
0	8	7.9	1	1.0
1	23	22.8	34	32.4
2	17	16.8	18	17.1
3	26	25.7	31	29.5
4	7	6.9	12	11.4
5	10	9.9	7	6.7
6	3	3.0	0	0.0
7	7	6.9	2	1.9
Total	101	100.0	105	100.0

SOURCE: Jackson et al. 1995, 1999; R.D. Jackson (Indiana University-Purdue University Indianapolis, personal commun., December 21, 2005).

TABLE 4-5 Percentage of Tooth Surfaces Assigned TSIF Scores in Four Samples of Children Age 8-10 Years and 13-15 Years in a U.S. Community with a Water Fluoride Concentration of 3.8 mg/L[a]

TSIF Score	1980 study		1985 study	
	Age 8-10 (n = 59)	Age 13-15 (n = 34)	Age 8-10 (n = 62)	Age 13-15 (n = 29)
0	30.3	36.9	24.2	22.5
1	28.5	25.6	32.2	30.8
2	17.1	16.7	18.7	18.8
3	19.7	18.6	19.7	22.1
4	0.3	0.3	0.6	0.5
5	2.8	1.3	3.1	3.9
6	0.1	0.1	0.1	0.0
7	1.2	0.5	1.4	1.5

[a]The numbers of children (n) are given in parentheses. The numbers of tooth surfaces examined were not reported.

SOURCE: Heifetz et al. 1988.

of 7 was assigned in all four samples. Detailed TSIF scores from this study are available only on as a percentage of all tooth surfaces examined. These results are consistent with those from the 1992 sample in Lowell, Indiana (Jackson et al. 1995) using the same fluorosis metric.

Confluent enamel pitting must be present for a tooth surface to be assigned a score of 7 on the TSIF scale (Table 4-3). In addition to the usual presence of dark brown staining, large areas of enamel may be missing and gross tooth structure may be altered as well. Thus, it has been sufficiently well documented that the most severe forms of enamel fluorosis for which classifications exist occur in children who reside in communities with water fluoride concentrations at or near the MCLG of 4 mg/L.

A third study, confined to the age range of 27-65 years, included a sample of 192 adults from Lordsburg, New Mexico, with a water fluoride concentration of 3.5 mg/L (Eklund et al. 1987). All members of this sample were native to Lordsburg and long-term residents of that community. The prevalence of severe fluorosis on Dean's 1942 scale was extremely high in this sample, 76% overall. The investigators modified Dean's scale specifically to split the "severe" category into 'severe' (discrete pitting) and 'very severe' (confluent pitting)" (Eklund et al. 1987). About half of those with more than moderate fluorosis were classified in the "very severe" category. These results for New Mexico adults are consistent with the results for children in Indiana and Illinois.

A reduction of all water fluoride concentrations to below 2 mg/L would be expected to make severe enamel fluorosis an extreme rarity in the United States, but would not be expected to eliminate it entirely. Isolated cases could still occur from excessive fluoride exposure from other sources, such as toothpaste swallowing and use of fluoride supplements and rinses. One can never rule out the possible existence of hypersusceptible individuals. Finally, though the ascertainment of severe enamel fluorosis is usually quite accurate in the United States, especially among children, it might be possible for dark yellow or brown staining and enamel pitting from other causes to be misdiagnosed as fluorosis. Such false positives might be particularly common among adults who are long-term users of smoked and smokeless tobacco products, heavy consumers of beverages such as coffee and tea, and perhaps some with special occupational exposures.

Aesthetic and Psychological Consequences of Enamel Fluorosis

Studies show that facial attractiveness is important and that attractive people are judged to be more socially desirable than less attractive people (Berscheid and Walster 1974; Adams and Huston 1975; Adams 1977; Jenny and Proshek 1986). Newton et al. (2003) assessed the impact of modified images of untreated cavities on front teeth on the appraisal of personal characteristics in the United Kingdom. Study participants associated decayed and discolored teeth with lower intelligence and social competence and with poor psychological adjustment. Interestingly, the ratings depended on the facial appearance studied, an indication that the impact of enamel fluorosis is less noticeable in a more attractive face. Although studies of the attractiveness of teeth are sparse, the orthodontic literature has shown that more than 80% of patients seek care out of concern for aesthetics, rather than health or function (Albino et al. 1981).

The potential for psychological and behavioral problems to develop from the aesthetically displeasing consequences of enamel fluorosis has been a long-standing concern. In 1984, an ad hoc panel of behavioral scientists convened by the U.S. Environmental Protection Agency (EPA) and the National Institute of Mental Health to evaluate the issue concluded that "individuals who have suffered impaired dental appearance as a result of moderate and severe fluorosis are probably at increased risk for psychological and behavioral problems or difficulties" (R.E. Kleck, unpublished report, Nov. 17, 1984, as cited in 50 Fed. Reg. 20164 [1985]). The panel recommended research on the social, emotional, and behavioral effects of enamel fluorosis.

Few studies have assessed the association between the public's perceived aesthetic problems and degree of enamel fluorosis. Only one of those studies was conducted in the United States. Lalumandier and Rozier (1998) found that parental satisfaction with the color of their children's teeth decreased as the severity of fluorosis increased. Although 73.9% of parents were satisfied with the color of teeth in the absence of enamel fluorosis, only 24.2% of parents were satisfied with the color of their children's teeth when the TSIF score was 4 or greater (moderate to severe forms). In a study of dental students' perceptions, Levy et al. (2002b) observed that fluorosis and nonfluorosis images were consistently rated more favorably by fourth-year students than by the same students in their first year. According to the authors, the results suggested that dentists might regard fluorosis with less concern given that they are exposed to a wide range of oral conditions, whereas those outside the dental profession might view fluorosis with more concern. Griffin et al. (2002) reviewed five published studies of aesthetic perception and enamel fluorosis and estimated that approximately 2% of U.S. schoolchildren might experience perceived aesthetic problems from exposure to fluoride at 0.7-1.2 mg/L. It should be noted that perceived aesthetic problems have also been reported even in the absence of enamel fluorosis because of nonfluorotic enamel opacities and hypoplasia, natural yellowish appearance of teeth, and discoloration due to dental caries. For example, Griffin et al. (2002) also noted that the percentage of respondents with no fluorosis who were not satisfied with the appearance of their teeth ranged from 18% to 41%.

In general, studies conducted in other parts of the world show that the level of satisfaction expressed by parents, children, and dentists with the appearance of enamel fluorosis decreases with increasing severity of enamel fluorosis (Clark et al. 1993; Riordan 1993; Clark 1995; Hawley et al. 1996;

Lalumandier and Rozier 1998; Griffin et al. 2002). In contrast with those studies, Ismail et al. (1993) did not find enamel fluorosis to be an aesthetic problem in Truro, Nova Scotia. The primary reason for disliking the color of front teeth was perceived yellowness unrelated to enamel fluorosis. Similarly, a study conducted in Brazil found that enamel fluorosis had no impact on children's self-perception of appearance (Peres et al. 2003).

A systematic review of water fluoridation estimated the proportion of the population likely to have aesthetic concerns about enamel fluorosis on the basis of a review of 88 studies (McDonagh et al. 2000a). The authors pointed out that the differences in the proportion of the population having enamel fluorosis of aesthetic concern with low concentrations of fluoride in drinking water and with fluoride at 1.2 mg/L were not statistically significant. However, the estimation of aesthetic concerns was based solely on a study conducted in Great Britain (Hawley et al. 1996) in which 14-year-old children from Manchester were asked to rate the appearance of life-sized pictures of two front teeth with enamel fluorosis (lips cropped off) classified by the TFI. According to the authors, the percentage of subjects who considered the appearance of the teeth unacceptable decreased from 29% for TF scores of 0 to 15% for TF scores of 2 and increased to 85% for TF scores of 4. Using those data, McDonagh et al. (2000a) defined enamel fluorosis of aesthetic concern as a case with a TF score of 3 or more, Dean's score of "mild" or worse, and a TSIF score of 2 or more. With this definition, McDonagh et al. (2000a) estimated the prevalence of fluorosis of aesthetic concern in the United Kingdom to be 63% at 4 mg/L and 25% at 2 mg/L. For lower water fluoride concentrations, the estimated prevalence ranged from 15% at 1.2 mg/L down to a baseline of 6% at 0.1 mg/L.

The committee judges that this analysis produced an overestimation of the prevalence of fluorosis of actual aesthetic concern for two main reasons. First, McDonagh et al. (2000a) applied the aesthetic concerns expressed by study participants about fluorosis on front teeth to fluorosis prevalence studies that included posterior teeth, which have much less potential to pose aesthetic problems. Second, the analysis did not take into account the observation by Hawley et al. (1996) that a higher percentage of children found teeth with milder forms of enamel fluorosis (TF scores lower than 3) aesthetically preferable to normal teeth; almost one-third of the children rated the photograph of teeth with no fluorosis as unacceptable.

There have been no new studies of the prevalence of moderate enamel fluorosis in U.S. populations since the early 1990s. Previous estimates ranged from 4% to 15% (50 Fed. Reg. 20164 [1985]). These estimates are based on studies that used classification indexes for scoring enamel fluorosis, and are not based on an assessment of aesthetics. None of the available indexes allow for making distinctions between fluorosis on the anterior and posterior teeth, so the percentage of children with moderate enamel fluorosis of aesthetic concern could not be determined, but the percentage would be lower than 15%.

The committee found only one study (Morgan et al. 1998) that specifically evaluated the psychological and behavioral impacts of enamel fluorosis on children with the condition. A group of 197 pediatric patients of a dental practice between the ages of 7 and 11 were examined for enamel fluorosis. Their parents completed the Child Behavior Checklist (CBCL), a widely used measure of behavioral problems in studies of children. The study found no substantial differences between groups classified by degree of fluorosis in overall CBCL scores or in scores on two subscales: externalizing (aggressive, hyperactive and antisocial behaviors typical of undercontrol or "acting out") and internalizing (behaviors of social withdrawal, depression and anxiety typical of overcontrol or inhibition). The study was limited by the fact that an aggregate measure of fluoride exposure was unrelated to enamel fluorosis and few if any of the children had severe enamel fluorosis.

Several methodologic issues have hindered the assessment of the aesthetic importance of unattractive teeth in general and enamel fluorosis in particular. First, assessing the perception of aesthetics is by its very nature subjective. Second, it is not clear who should make judgments about the aesthetic appearance of teeth. The perceptions of the affected individual, as a child and in subsequent life, as well as those of parents, friends, teachers, and other acquaintances can all be important. A sizeable proportion of parents and children have expressed dissatisfaction with the color of teeth even in the absence of enamel fluorosis. On the other hand, judgments made by professionals might not reflect the perception of the public. Third, it is difficult to place the condition of enamel fluorosis into the context of an overall aesthetic assessment of a person's appearance or facial attractiveness. Cultural influences can play a role in how the condition is perceived. It also appears that perceptions of the appearance of teeth can be modified by the attractiveness of other facial features. Fourth, when the public or dental professionals are asked to assess aesthetic acceptability, their perceptions might change during the evaluation session.

From the standpoint of this committee's charge to consider effects of relatively high levels of water fluoride, the main points to note are that the emphasis of research and discussion on psychological, behavioral, and social effects of enamel fluorosis has been almost entirely on children and on the mild and moderate forms of the condition that are more typical of lower fluoride exposure levels. Research needs to focus specifically on severe enamel fluorosis in those areas in which it occurs with appreciable frequency. In addition, research needs to include not only affected children while they are still children, but after they move into adulthood. Finally, parents might experience psychological and behavioral effects when their children develop enamel fluorosis, especially in its moderate and severe forms. Unfortunately, research on parental effects is completely lacking.

Dental Caries in Relation to Water Fluoride Concentrations of 2 mg/L and Higher

Many reports have discussed the inverse relationship between dental caries and water fluoride at concentrations considerably lower than the current MCLG of 4 mg/L and SMCL of 2 mg/L (Dean 1942; PHS 1991; McDonagh et al. 2000a; CDC 2001). Fewer studies have been conducted in the United States of overall caries experience in communities with naturally occurring fluoride concentrations higher than those produced by fluoridation. The studies of children are shown in Table 4-6. One study suggested that the overall frequency of caries is reduced at approximately 4 mg/L compared with approximately 1 mg/L (Englander and DePaola 1979). A study of New Mexico adults gave similar results (Eklund et al. 1987). Another study suggested little or no difference (Jackson et al. 1995) and another gave mixed results (Selwitz et al. 1995). The evidence from these studies is not persuasive that caries frequency is appreciably lower at approximately 4 mg/L than at approximately 2 mg/L or 3 mg/L. The evidence from studies conducted in other countries is no more consistent (Binder 1973; Olsson 1979; Kunzel 1980; Chen 1989; Lewis et al. 1992; Warnakulasuriya et al. 1992; Yoder et al. 1998; Angelillo et al. 1999; Grobler et al. 2001).

Dental Caries in Relation to Severe Enamel Fluorosis

As previously noted, it is suspected within the dental research community that the enamel pitting that occurs in severe fluorosis might increase caries risk by reducing the thickness of the protective enamel layer and by allowing food and plaque to become entrapped in enamel defects. The possibility is thus raised that in a community with a water fluoride concentration high enough to produce an appreciable prevalence of severe fluorosis, the specific subset of children who develop this condition might be placed at increased caries risk, independent of the effect of the fluoride itself on the remainder of the population. The population of interest consists of those children who develop severe enamel fluorosis at 4 mg/L. If the water fluoride concentration were reduced to below 2 mg/L, few if any of these children would still develop severe enamel fluorosis. Many of them would develop mild to moderate fluorosis, however, while others might develop no fluorosis. It would be unreasonable, however, to assume that some children would skip all the way down from severe fluorosis to no fluorosis when the water concentration is reduced, while others would have mild to moderate fluorosis at either concentration. As the desired fluorosis severity

TABLE 4-6 Mean Number of Decayed, Missing and Filled Surfaces (DMFS) in Permanent Teeth by Water Fluoride Concentration in Studies of Children in U.S. Communities with Water Fluoride Concentrations at or Near the MCLG of 4 mg/L

Reference	Age (years)	Year	Community	Number of Children	Approximate Water Fluoride Concentration (mg/L)	Mean DMFS
Englander and DePaola (1979)	12-15	NA	Kalamazoo, MI	315	1	5.1
			Stickney, IL	312	1	4.5
			Charlotte, NC	213	1	4.4
			Midland, TX	311	5-7	2.4
Driscoll et al. (1983)	8-11	1980	Kewanee, IL	157	1	2.0
			Monmouth, IL	80	2	1.4
			Abindgon and Elmwood, IL	110	3	1.0
			Bushnell, Ipava and Table Grove, IL	77	4	1.6
Driscoll et al. (1983)	12-16	1980	Kewanee, IL	179	1	4.1
			Monmouth, IL	63	2	2.7
			Abindgon and Elmwood, IL	82	3	2.0
			Bushnell, Ipava and Table Grove, IL	59	4	2.6
Heifetz et al. (1988)	8-10	1985	Kewanee, IL	156	1	1.5
			Monmouth, IL	102	2	1.1
			Abindgon and Elmwood, IL	112	3	0.8
			Bushnell, Ipava and Table Grove, IL	62	4	0.8
Heifetz et al. (1988)	13-15	1985	Kewanee, IL	94	1	5.1
			Monmouth, IL	23	2	2.9
			Abindgon and Elmwood, IL	47	3	2.5
			Bushnell, Ipava and Table Grove, IL	29	4	3.9
Selwitz et al. (1995)	8-10, 14-16	1990	Kewanee, IL	258	1	1.8
			Monmouth, IL	105	2	1.4
			Abindgon and Elmwood, IL	117	3	1.4
			Bushnell, Ipava and Table Grove, IL	77	4	1.8
Jackson et al. (1995)	7-14	1992	Brownsburg, IN	117	1	4.4
			Lowell, IN	101	4	4.3

NA: Not available.

distribution is inherently unknown, a conservative approach is to compare the children with severe fluorosis at 4 mg/L with children from their own communities with mild to moderate fluorosis.

Results for such comparisons are summarized in Table 4-7 for studies reporting the mean number of decayed, missing and filled tooth surfaces (DMFS), in Table 4-8 for studies reporting the number of decayed, missing and filled teeth (DMFT), and in Table 4-9 for studies reporting the per-

TABLE 4-7 Mean Number of Decayed, Missing, and Filled Permanent Tooth Surfaces (DMFS) among Children with Severe and Mild to Moderate Enamel Fluorosis

Country (reference)	Age (years)	Number of Children	Fluorosis Index and Range	Mean DMFS
United States (Driscoll et al. 1986)	8-16	218 54	Dean very mild to moderate Dean severe	1.6 3.0
Israel (Mann et al. 1987)	15-16	83 46	Dean very mild to moderate Dean severe	4.4 10.4
Israel (Mann et al. 1990)	8-10	55 6	Dean very mild to moderate Dean severe	1.2 1.8
Turkey (Ermiş et al. 2003)	12-14	24 105	TSIF 1-3 TSIF 4-7	1.7 1.9

TABLE 4-8 Mean Numbers of Decayed, Missing, and Filled Permanent Teeth (DMFT) among Children with Severe and Mild to Moderate Enamel Fluorosis

Country (reference)	Age (years)	Number of Children	Fluorosis Index and Range	Mean DMFT
Taiwan (Chen 1989)	6-16	1,290 10	Dean very mild to moderate Dean severe	1.7 2.5
Sri Lanka (Warnakulasuriya et al. 1992)	14	44 48	Dean mild Dean moderate to severe	3.4 3.3
Brazil (Cortes et al. 1996)	6-12	42 18	TFI 3-4 TFI ≥5	1.1 1.3
Turkey (Ermiş et al. 2003)	12-14	24 105	TSIF 1-3 TSIF 4-7	1.2 1.3
Ethiopia (Wondwossen et al. 2004)	12-15	87 89	TFI 3-4 TFI 5-7	1.5 2.4

TABLE 4-9 Percentage of Teeth Scored as Decayed, Missing, Filled, or with Caries among Children and Adults with Severe and Mild-to-Moderate Enamel Fluorosis

Country (reference)	Age (years)	Teeth	Number of Persons	Range of Dean's Fluorosis Index	Measure (%)
Ethiopia (Olsson 1979)	6-7, 13-14	All		Mild to moderate Severe	Cavities 25 9
United States (Driscoll et al. 1986)	8-16	All	218 54	Very mild to moderate Severe	Decayed or filled 4 20
United States (Eklund et al. 1987)	27-65	Molars	38 125	Mild to moderate Severe	Decayed, missing or filled 43 40
		Premolars	38 125	Mild to moderate Severe	11 19
		Anterior	38 125	Mild to moderate Severe	3 6

centage of decayed, missing and filled teeth. Not all researchers reported P-values for the specific contrasts in these tables. Moreover, the results are not independent, as some researchers studied more than one age group or reported results for more than one caries frequency measure or for more than one type of teeth. Nevertheless, in 11 of the 14 available contrasts, the measure of caries frequency was higher among those with severe fluorosis than among those with mild to moderate forms. In some comparisons, the differences were slight. Descriptively, the most pronounced differences were for all teeth among children age 15-16 years in Israel (Mann et al. 1987, Table 4-7), for all teeth among children age 8-16 years in Illinois (Driscoll et al. 1986, Table 4-9), for premolars among adults age 27-65 in New Mexico (Eklund et al. 1987, Table 4-9), and for all teeth among children ages 6-7 and 13-14 in Ethiopia (Olsson 1979, Table 4-9).

Mixed evidence comes from correlation or regression analyses. In studies in Uganda (Rwenyonyi et al. 2001) and Tanzania (Awadia et al. 2002), statistically significant correlations were not observed ($P > 0.05$) between severe fluorosis and caries frequency. A study of children in a South African community with a water fluoride concentration of 3 mg/L and a 30% prevalence of severe fluorosis reported a positive correlation ($P < 0.05$) between fluorosis scores on the Dean index and caries experience (DMFT) (Grobler et al. 2001). In the same study, no correlation between fluorosis and caries frequency was found in two other communities with water fluoride concentrations of 0.5 and 0.2 mg/L, in which the prevalence of severe fluorosis was 1% and 0%, respectively.

The studies on severe enamel fluorosis and caries are limited by being cross-sectional in design and conducted in a wide range locales. In most of the studies, there was no adjustment for oral hygiene, dental care, or other determinants of caries risk. Moreover, as previously noted, measures of the role of chance (i.e., confidence intervals or P-values) are not available for the specific contrasts of interest to the present report. Nevertheless, the hypothesis of a causal link between severe enamel fluorosis and increased caries risk is plausible and the evidence is mixed but supportive.

OTHER DENTAL EFFECTS

Fluoride may affect tooth dentin as well as enamel. The patterns of change observed in bone with age also occur in dentin, a collagen-based mineralized tissue underlying tooth enamel. Dentin continues to grow in terms of overall mass and mineral density as pulp cells deposit more matrix overall and more mineral in the dentin tubules. Several investigators have observed that, like older bone, older dentin is less resistant to fracture and tends to crack more easily (Arola and Reprogel 2005; Imbeni et al. 2005; Wang 2005). Aged dentin tends to be hypermineralized and sclerotic, where the dentin tubules have been filled with mineral and the apatite crystals are slightly smaller (Kinney et al. 2005), which could be significant because, as dentin ages in the presence of high amounts of fluoride, the highly packed fluoride-rich crystals might alter the mechanical properties of dentin as they do in bone (see Chapter 5). Unlike bone, however, dentin does not undergo turnover. Some preliminary studies show that fluoride in dentin can even exceed concentrations in bone and enamel (Mukai et al. 1994; Cutress et al. 1996; Kato et al. 1997; Sapov et al. 1999; Vieira et al. 2004). Enamel fluorosis, which accompanies elevated intakes of fluoride during periods of tooth development, results not only in enamel changes as discussed above but also in dentin changes. It has now been well established that fluoride is elevated in fluorotic dentin (Mukai et al. 1994; Cutress et al. 1996; Kato et al. 1997; Sapov et al. 1999; Vieira et al. 2004). Whether excess fluoride incorporation in fluorotic teeth increases the risk for dentin fracture remains to be determined, but the possibility cannot be ruled out.

Questions have also been raised about the possibility that fluoride may delay eruption of permanent teeth (Kunzel 1976; Virtanen et al. 1994; Leroy et al. 2003). The hypothesized mechanisms for this effect include prolonged retention of primary teeth due to caries prevention and thickening of the bone around the emerging teeth (Kunzel 1976). However, no systematic studies of tooth eruption have been carried out in communities exposed

EFFECTS OF FLUORIDE ON TEETH

to fluoride at 2 to 4 mg/L in drinking water. Delayed tooth eruption could affect caries scoring for different age groups.

FINDINGS

One of the functions of tooth enamel is to protect the dentin and, ultimately, the pulp from decay and infection. Severe enamel fluorosis compromises this health-protective function by causing structural damage to the tooth. The damage to teeth caused by severe enamel fluorosis is a toxic effect that the majority of the committee judged to be consistent with prevailing risk assessment definitions of adverse health effects. This view is consistent with the clinical practice of filling enamel pits in patients with severe enamel fluorosis and restoring the affected teeth.

In previous reports, all forms of enamel fluorosis, including the severest form, have been judged to be aesthetically displeasing but not adverse to health (EPA 1986; PHS 1991; IOM 1997; ADA 2005). This view has been based largely on the absence of direct evidence that severe enamel fluorosis results in tooth loss, loss of tooth function, or psychological, behavioral, or social problems. The majority of the present committee finds the rationale for considering severe enamel fluorosis only a cosmetic effect much weaker for discrete and confluent pitting, which constitutes enamel loss, than it is for the dark yellow to brown staining that is the other criterion symptom of severe fluorosis. Moreover, the plausible hypothesis of elevated caries frequency in persons with severe enamel fluorosis has been accepted by some authorities and has a degree of support that, though not overwhelmingly compelling, is sufficient to warrant concern. The literature on psychological, behavioral, and social effects of enamel fluorosis remains quite meager. None of it focuses specifically on the severe form of the condition or on parents of affected children or on affected persons beyond childhood.

Two of the 12 members of the committee did not agree that severe enamel fluorosis should now be considered an adverse health effect. They agreed that it is an adverse dental effect but found that no new evidence has emerged to suggest a link between severe enamel fluorosis, as experienced in the United States, and a person's ability to function. They judged that demonstration of enamel defects alone from fluorosis is not sufficient to change the prevailing opinion that severe enamel fluorosis is an adverse cosmetic effect. Despite their disagreement on characterization of the condition, these two members concurred with the committee's conclusion that the MCLG should prevent the occurrence of this unwanted condition.

Severe enamel fluorosis occurs at an appreciable frequency, approximately 10% on average, among children in U.S. communities with water fluoride concentrations at or near the current MCLG of 4 mg/L. Strong evidence exists of an approximate population threshold in the United States, such that the prevalence of severe enamel fluorosis would be reduced to nearly zero by bringing the water fluoride levels in these communities down to below 2 mg/L. There is no strong and consistent evidence that an appreciable increase in caries frequency would occur by reducing water fluoride concentrations from 4 mg/L to 2 mg/L or lower. At a fluoride concentration of 2 mg/L, severe enamel fluorosis would be expected to become exceedingly rare, but not be completely eradicated. Occasional cases would still arise for reasons such as excessive fluoride ingestion (e.g., toothpaste swallowing), inadvisable use of fluoride supplements, and misdiagnosis.

Despite the characterization of all forms of enamel fluorosis as cosmetic effects by previous groups, there has been general agreement among them, as well as in the scientific literature, that severe and even moderate enamel fluorosis should be prevented. The present committee's consensus finding that the MCLG should be set to protect against severe enamel fluorosis is in close agreement with conclusions by the Institute of Medicine (IOM 1997), endorsed recently by the American Dental Association (ADA 2005). As shown in Table 4-10, between 25% and 50% of U.S. children in communities with drinking water containing fluoride at 4 mg/L would be expected to consume more than the age-specific tolerable upper limits of fluoride intake set by IOM. Results from the Iowa Fluoride Study (Levy 2003) indicate that even at water fluoride levels of 2 mg/L and lower, some children's fluoride intake from water exceeds the IOM's age-specific tolerable upper limits (Table 4-11).

For all age groups listed in Table 4-10, the IOM's tolerable upper intake values correspond to a fluoride intake of 0.10 mg/kg/day (based on default body weights for each age group; see Appendix B). Thus, the exposure estimates in Chapter 2 also showed that the IOM limits would be exceeded at 2 mg/L for nonnursing infants at the average water intake level (Table 2-14). Specifically, as described in Chapter 2 (Tables 2-14 and 2-15), nonnursing

TABLE 4-10 Tolerable Upper Fluoride Intakes and Percentiles of the U.S. Water Intake Distribution, by Age Group

Age Group	Tolerable Upper Intake (IOM 1997)		Water Intake, mL/day (EPA 2004)	
	Fluoride, mg/day	Water, mL/day (at 4 mg/L)	50th Percentile	75th Percentile
0-6 months	0.7	175	42	585
7-12 months	0.9	225	218	628
1-3 years	1.3	325	236	458
4-8 years	2.2	550	316[a]	574[a]

[a]Ages 4-6 years. For ages 7-10 years, the 50th percentile is 355 mL/day and the 75th percentile is 669 mL/day.

TABLE 4-11 Comparison of Intakes from Drinking Water[a] from the Iowa Fluoride Study and IOM's Upper Tolerable Intakes

Age, months	IOM Tolerable Upper Intake (mg/day)	Percentiles of Iowa Fluoride Study Distribution (mg/day)		
		75th	90th	Maximum
3	0.7	0.7	1.1	6.7
12	0.9	0.4	0.7	6.0
24	1.3	0.4	0.6	2.1
36	1.3	0.5	0.7	1.7

[a]Fluoride concentrations in drinking water ranged from <0.3 to 2 mg/L.

SOURCE: Levy 2003.

infants have an average total fluoride intake (all sources except fluoride supplements) of 0.144 and 0.258 mg/kg/day at 2 and 4 mg/L fluoride in drinking water, respectively. Corresponding values are 0.090 and 0.137 mg/kg/day for children 1-2 years old and 0.082 and 0.126 mg/kg/day for children 3-5 years old. Furthermore, at EPA's current default drinking water intake rate, the exposure of infants (nursing and non-nursing) and children 1-2 years old would be at or above the IOM limits at a fluoride concentration of 1 mg/L (Table 2-13). For children with certain medical conditions associated with high water intake, estimated fluoride intakes from all sources (excluding fluoride supplements) range from 0.13-0.18 mg/kg/day at 1 mg/L to 0.23-0.33 mg/kg/day at 2 mg/L and 0.43-0.63 mg/kg/day at 4 mg/L.

IOM's tolerable upper limits were established to reduce the prevalence not only of severe fluorosis, but of moderate fluorosis as well, both of which ADA (2005) describes as unwanted effects. The present committee, in contrast, focuses specifically on severe enamel fluorosis and finds that it would be almost eliminated by a reduction of water fluoride concentrations in the United States to below 2 mg/L. Despite this difference in focus, the committee's conclusions and recommendations with regard to protecting children from enamel fluorosis are squarely in line with those of IOM and ADA.

The current SMCL of 2 mg/L is based on a determination by EPA that objectionable enamel fluorosis in a significant portion of the population is an adverse cosmetic effect. EPA defined objectionable enamel fluorosis as discoloration and/or pitting of teeth. As noted above, the majority of the committee concludes it is no longer appropriate to characterize enamel pitting as a cosmetic effect. Thus, the basis of the SMCL should be discoloration of tooth surfaces only.

The prevalence of severe enamel fluorosis is very low (near zero) at fluoride concentrations below 2 mg/L. However, from a cosmetic standpoint, the SMCL does not completely prevent the occurrence of moderate enamel fluorosis. EPA has indicated that the SMCL was intended to reduce the severity and occurrence of the condition to 15% or less of the exposed population. No new studies of the prevalence of moderate enamel fluorosis in U.S. populations are available. Past evidence indicated an incidence range of 4% to 15% (50 Fed. Reg. 20164 [1985]). The prevalence of moderate cases that would be classified as being of aesthetic concern (discoloration of the front teeth) is not known but would be lower than 15%. The degree to which moderate enamel fluorosis might go beyond a cosmetic effect to create an adverse psychological effect or an adverse effect on social functioning is also not known.

RECOMMENDATIONS

- Additional studies, including longitudinal studies, of the prevalence and severity of enamel fluorosis should be done in U.S. communities with fluoride concentrations higher than 1 mg/L. These studies should focus on moderate and severe enamel fluorosis in relation to caries and in relation to psychological, behavioral, and social effects among affected children, their parents, and affected children after they become adults.
- Methods should be developed and validated to objectively assess enamel fluorosis. Consideration should be given to distinguishing between staining or mottling of the anterior teeth and of the posterior teeth so that aesthetic consequences can be more easily assessed.
- More research is needed on the relation between fluoride exposure and dentin fluorosis and delayed tooth eruption patterns.

Turner model predicts that (1) fluoride uptake is positively associated with the bone remodeling rate and (2) fluoride clearance from the skeleton takes at least four times longer than fluoride uptake. A key correlate to the first prediction is that the concentration of fluoride in bone does not decrease with reduced remodeling rates. Thus, it appears that fluoride enters the bone compartment easily, correlating with bone cell activity, but that it leaves the bone compartment slowly. The model assumes that efflux occurs by bone remodeling and that resorption is reduced at high concentrations of fluoride because of hydroxyapatite solubility. Hence, it is reasonable that 99% of the fluoride in humans resides in bone and the whole body half-life, once in bone, is approximately 20 years (see Chapter 3 for more discussion of pharmacokinetic models).

The effects of fluoride on bone quality are evident but are less well characterized than its effects on bone cells. Bone quality is an encompassing term that may mean different things to different investigators. However, in general it is a description of the material properties of the skeleton that are unrelated to skeletal density. In other words, bone quality is a measure of the strength of the tissue regardless of the mass of the specimen being tested. It includes parameters such as extent of mineralization, microarchitecture, protein composition, collagen cross linking, crystal size, crystal composition, sound transmission properties, ash content, and remodeling rate. It has been known for many years that fluoride exposure can change bone quality. Franke et al. (1975) published a study indicating that industrial fluoride exposure altered hydroxyapatite crystal size and shape. Although the measurements in their report were made with relatively crude x-ray diffraction analyses, they showed a shorter and more slender crystal in subjects who were aluminum workers and known to be exposed to high concentrations of fluoride. Other reports documenting the effects of fluoride on ultrasound velocities in bone, vertebral body strength, ash content, and stiffness have shown variable results (Lees and Hanson 1992; Antich et al. 1993; Richards et al. 1994; Zerwekh et al. 1997a; Søgaard et al. 1994, 1995, 1997); however, the general conclusion is that, although there may be an increase in skeletal density, there is no consistent increase in bone strength. A carefully performed comparison study between the effects of fluoride (2 mg/kg/day) and alendronate in minipigs likely points to the true effect: "in bone with higher volume, there was less strength per unit volume, that is, . . . there was a deterioration in bone quality" (Lafage et al. 1995).

EFFECT OF FLUORIDE ON CELL FUNCTION

Two key cell types are responsible for bone formation and bone resorption, the osteoblast and osteoclast, respectively. Osteoprogenitor cells give rise to osteoblasts. Osteoprogenitor cells are a self-renewing population of cells that are committed to the osteoblast lineage. They originate from mesenchymal stem cells. Osteoblasts contain a single nucleus, line bone surfaces, possess active secretory machinery for matrix proteins, and produce very large amounts of type I collagen. Because they also produce and respond to factors that control bone formation as well as bone resorption, they play a critical role in the regulating skeletal mass. Osteoclasts are giant, multinucleated phagocytic cells that have the capability to erode mineralized bone matrix. They are derived from cells in the monocyte/macrophage lineage. Their characteristic ultrastructural features allow them to resorb bone efficiently by creating an extracellular lysosome where proteolytic enzymes, reactive oxygen species, and large numbers of protons are secreted. Osteoclastogenesis is controlled by local as well as systemic regulators.

Effect of Fluoride on Osteoblasts

Perhaps the single clearest effect of fluoride on the skeleton is its stimulation of osteoblast proliferation. The effect on osteoblasts was surmised from clinical trials in the early 1980s documenting an increase in vertebral bone mineral density that could not be ascribed to any effect of fluoride on bone resorption. Biopsy specimens confirmed the effect of fluoride on increasing osteoblast number in humans (Briancon and Meunier 1981; Harrison et al. 1981). Because fluoride stimulates osteoblast proliferation, there is a theoretical risk that it might induce a malignant change in the expanding cell population. This has raised concerns that fluoride exposure might be an independent risk factor for new osteosarcomas (see Chapter 10 for the committee's assessment).

The demonstration of an effect of fluoride on osteoblast growth in vitro was first reported in 1983 in avian osteoblasts (Farley et al. 1983). This study showed that fluoride stimulated osteoblast proliferation in a biphasic fashion with the optimal mitogenic concentration being 10 µM. The finding that fluoride displayed a biphasic pattern of stimulation (achieving a maximal effect at a specific concentration and declining from there) suggests that multiple pathways might be activated. It is possible that low, subtoxic doses do stimulate proliferation, but at higher doses other pathways responsible for decreasing proliferation or increasing apoptosis might become activated. This thinking suggested that fluoride might have multiple effects on osteoblasts and that might be the reason for some paradoxical findings in the clinical literature (see below). Nevertheless, the characteristics of the fluoride effect point clearly to a direct skeletal effect. Some of these characteristics are as follows: (1) the effects of fluoride on osteoblasts occur at low concentrations in vivo and in vitro (Lau and Baylink 1998); (2) fluoride effects are, for the most part, skeletal specific (Farley et al. 1983; Wergedal et al. 1988); (3) fluoride effects may require the presence of a bone-active growth factor (such as insulin-like-growth factor I or transforming growth factor β) for its action (Farley et al. 1988; Reed et al. 1993); and (4) fluoride affects predominantly osteoprogenitor cells as opposed to mature functioning osteoblasts (Bellows et al. 1990; Kassem et al. 1994).

Understanding the subcellular signaling mechanisms by which fluoride affects osteoblasts is of paramount importance. Information in this area has the potential to determine whether the fluoride effects are specific, whether toxicity is an issue, and what concentration may influence bone cell function. Moreover, as the pathways become more clearly defined, other targets might emerge. Two hypotheses in the literature describe the effect of fluoride. Both state that the concentration of tyrosine phosphorylated signal pathway intermediates is elevated after fluoride exposure. However, the means by which this occurs differs in the hypotheses. One view is that fluoride blocks or inhibits the activity of a phosphotyrosine phosphatase, thereby increasing the pool of tyrosine-phosphorylated proteins. The other view supports an action of fluoride (along with aluminum) on the stimulation of tyrosine phosphorylation that would also increase the pool of tyrosine-phosphorylated proteins. In the first hypothesis, growth factor activation of the Ras-Raf-MAP kinase pathway would involve stimulation of phosphotyrosine kinase activity. This is mediated by a family of cytosolic G proteins with guanosine triphosphate acting as the energy source. In the presence of fluoride, a sustained high concentration of tyrosine-phosphorylated proteins would be maintained because of the inability of the cell to dephosphorylate the proteins. This theory implicates the existence of a fluoride-sensitive tyrosine phosphatase in osteoblasts. Such an enzyme has been identified and purified. It appears to be a unique osteoblastic acid phosphatase-like enzyme that is inhibited by clinically relevant concentrations of fluoride (Lau et al. 1985, 1987, 1989; Wergedal and Lau 1992). The second hypothesis supports the belief that an AlF_x complex activates tyrosine phosphorylation directly. Data from this viewpoint indicate that fluoride alone does not stimulate tyrosine phosphorylation but rather that it requires the presence of aluminum (Caverzasio et al. 1996). The purported mechanism is that the MAP kinase pathway is activated by AlF_x, which triggers the proliferation response. A novel tyrosine kinase, Pyk2, has been identified that is known to be activated by AlF_x through a G-protein-coupled response and might be responsible for this effect (Jeschke et al. 1998). Two key pieces of evidence that support a G-protein-regulated tyrosine kinase activation step in the fluoride effect are that the mitogenic effect of fluoride can be blocked by genistein (a protein tyrosine kinase inhibitor) and pertussis toxin (a specific inhibitor of heterotrimeric G proteins) (Caverzasio et al. 1997; Susa et al. 1997).

At least two other potential mechanisms deserve mention. Kawase and Suzuki (1989) suggested that fluoride activates protein kinase C (PKC), and Farley et al. (1993) and Zerwekh et al. (1990) presented evidence that calcium influx into the cells might be a signal for the fluoride-mediated stimulation of proliferation.

In summary, the in vitro effects of fluoride on osteoblast proliferation appear to involve, at the least, a regulation of tyrosine-phosphorylated proteins. Whether this occurs through activation of MAP kinases, G proteins, phosphatases, PKC, or calcium (or a combination) remains to be determined. Whatever the mechanism, however, it is evident that fluoride has an anabolic activity on osteoblasts and their progenitors.

The effects of fluoride on osteoblast number and activity in in vivo studies and clinical trials essentially parallel the in vitro findings. Most reports document increased osteoblast number; however, some investigators have documented a complex and paradoxical effect of fluoride in patients with skeletal fluorosis. Boivin et al. (1989, 1990) reported that, in biopsy bone cores taken from 29 patients with skeletal fluorosis of various etiologies (0.79% ± 0.36% or 7,900 ± 3,600 milligrams per kilogram [mg/kg] of bone ash), there is an apparent increase in the production of osteoblasts with a concomitant increase in a toxic effect of fluoride at the cell level. They provided data to indicate that chronic exposure to fluoride in both endemic and industrially exposed subjects led to an increase in bone volume, an increase in cortical width, and an increase in porosity. However, there was no reduction in cortical bone mass. Osteoid parameters (unmineralized type I collagen) were also significantly increased in fluorotic patients. Interestingly, the fluorotic group had more osteoblasts than the control group, with a very high proportion of quiescent, flattened osteoblasts, but the mineral apposition rate was significantly decreased. It appeared as though the increased numbers of quiescent cells were in a prolonged inactive period. Thus, the conclusion drawn by these investigators was that fluoride exposure increased the birth rate of new osteoblasts, but at high concentrations there was an independent toxic effect on the cells that blocked the full manifestation for the increase in skeletal mass. Boivin et al. used a fluoride-specific electrode for measurements in acidified specimens of human bone. As a point of reference to the above findings, they found that normal control subjects (likely not to have lived in areas with water fluoridation) have mean fluoride content in bone ash (from iliac crest samples) ranging from 0.06% to 0.10% (600 to 1,000 mg/kg); untreated osteoporotic patients range from 0.05% to 0.08% (500 to 800 mg/kg); NaF-treated osteoporotic patients range from 0.24% to 0.67% (2,400 to 6,700 mg/kg) depending on duration of therapy; and skeletal fluorosis patients range from 0.56% to 1.33% (5,600 to 13,300 mg/kg) depending on the source and level of exposure (Boivin et al. 1988). All these ranges are of mean concentrations of fluoride and not individual measurements.

Effect of Fluoride on Osteoclasts

The effects of fluoride on osteoclast activity, and by extension the rate of bone resorption, are less well defined than its effects on osteoblasts. In general, there appears to be good evidence that fluoride decreases osteoclastogenesis and osteoclast activity in in vitro systems; however, its effect in in vivo systems is equivocal. This may be due, in part, to the systemic effects of fluoride in whole animals or humans. A further discussion on this point appears below.

Most reports in the literature studying the effect of fluoride on osteoclast function indicate an inhibition. In fact, the effect might be mediated through G-protein-coupled pathways as in the osteoblast. Moonga et al. (1993) showed that fluoride, in the form of AlF_{4-} resulted in a marked concentration-dependent inhibition of bone resorption. In association with this inhibition, they found a marked increase in the secretion of tartrate-resistant acid phosphatase (TRAP). TRAP presumably originated from the osteoclast; however, its function as a secreted enzyme is not known. The fluoride effect was reproduced with cholera toxin, another Gs stimulator. This effect does not appear to be mediated solely by an AlF_x complex because studies using NaF have reported similar findings (Taylor et al. 1989, 1990; Okuda et al. 1990).

Further evidence that fluoride might blunt osteoclastic bone resorption was reported in a study that investigated acid production as a critical feature of osteoclastic function. The pH within osteoclasts can be measured with the proton-sensitive dye acridine orange. Studies in which osteoclasts were observed found that parathyroid hormone induced osteoclast acidity but that calcitonin, cortisol, and NaF all blocked the effect. As acidification of the matrix is required for normal osteoclast function, fluoride, in this case, would act as an inhibitor to bone resorption (Anderson et al. 1986).

The effects of fluoride on bone resorption and osteoclast function in vivo present a complex picture. Some well-controlled animal studies document a decrease in osteoclast (as well as odontoclast) activity. In these studies, rodents and rabbits were exposed to doses of fluoride ranging from clinically relevant to high. Time courses ranged from days to weeks, and the findings indicated a statistically significant decrease in the number and activity of resorbing cells (Faccini 1967; Lindskog et al. 1989; Kameyama et al. 1994). Other studies documented little or no statistically significant effect of fluoride on osteoclast activity (Marie and Hott 1986; Huang 1987). Yet other work that utilized skeletal turnover and remodeling showed an increase in resorption after fluoride therapy (Kragstrup et al. 1984; Snow and Anderson 1986). These studies based their conclusions on the initiation of basic multicellular units (BMUs) and extent of remodeling surface. In the field of skeletal research, it has been accepted that adult bone remodels itself through the generation of BMUs. This unit is a temporal description of remodeling starting with osteoclastic bone resorption and progressing through a coupled stimulation of bone formation. All BMU activity, thus, is initiated with the action of an osteoclast. An increase in remodeling surface also implies an increase in BMUs. Snow and Anderson (1986) and Kragstrup et al. (1984) demonstrated an increase in resorption under the influence of fluoride by measuring BMU numbers and remodeling surface, respectively. Because these data were derived from intact in vivo animal models, the investigators could not conclude that the effects of fluoride on osteoclastic bone resorption were direct.

It is interesting that only a single report has appeared that links fluoride exposure to the receptor activator of NF kappaB (RANK) ligand, RANK receptor, or osteoprotegerin (OPG) concentrations. These molecules have recently been characterized as end-stage regulators of osteoclast formation and activity (Lee and Kim 2003). RANK ligand is produced by a variety of cells, with osteoblasts being the most prominent. In its usual form, it is a membrane-associated factor that binds to the RANK receptor on pre-osteoclasts and induces their further differentiation. OPG is a decoy RANK receptor that is an endogenous inhibitor of bone resorption by virtue of its ability to bind RANK ligand. A clinical trial by von Tirpitz et al. (2003) showed that both fluoride and bisphosphonate therapy decreased OPG concentrations. If this were a direct effect of fluoride, one would expect to see an increase in bone resorption. Conversely, if fluoride blocked bone resorption, the decrease in OPG concentrations could be due to a compensatory feedback pathway. Unfortunately, there were not enough histologic or biochemical marker data in this report to determine whether the fluoride effect was direct or indirect.

EFFECTS OF FLUORIDE ON HUMAN SKELETAL METABOLISM

Bone Strength and Fracture

Cellular and Molecular Aspects

Inducing a permanent alteration of skeletal mass in an adult human (or experimental animal) is quite difficult, because bone, as an organ system, possesses an innate mechanism for self-correction. That is, rates of bone formation are controlled, for the most part, by rates of bone resorption. As osteoclastic bone resorption increases or decreases, there is a compensatory increase or decrease in the rate of osteoblastic bone formation. This coupling between the two cell activities was first described by Hattner et al. (1965), and is responsible for the maintenance of a steady-state skeletal mass in adults. These early results indicate that effective management of skeletal mass would require controlling both cell processes. However, until recently, the only therapies approved by the U.S. Food and Drug Admnistration for treating osteoporosis in the United States targeted only osteoclastic bone resorption. They included molecules such as the bisphosphonates, estrogen and its analogs, and calcitonin derivatives. Currently, teraparitide is available as the only approved treatment that acts to stimulate osteoblastic bone formation. Fluoride falls into this category and that is the reason why there was such great interest in this ion as a potential therapy for osteoporosis. Unfortunately, fluoride did not prove to be an effective treatment for two major reasons. First, although it showed robust stimulation of bone mineral density (see below), its effects as an agent to reduce fractures have never been unequivocally documented. Second, because this naturally occurring element cannot be protected with a patent, the pharmaceutical industry has not been interested in investigating all its potential.

The first clinical trials of fluoride in humans were performed by Rich and Ensinck (1961). Since then many hundreds of reports have appeared in the medical literature. The overwhelming weight of evidence in these reports documents the effect of fluoride, at therapeutic doses, to be that of an increase in bone mineral density. The lowest dose of NaF to show a clear increase in bone mineral density was 30 mg/day, although there may be effects at lower doses (Hansson and Roos 1987; Kleerekoper and Balena 1991). Response was linear with time for at least 4 to 6 years (Riggs et al. 1990). This linear relationship was confirmed in another study lasting more than 10 years (Kleerekoper and Balena 1991). The observation that bone mineral density continues to increase with time is not surprising in and of itself; however, it differs from the action of the antiresorptive bisphosphonates. Whereas agents that depress bone resorption are most effective when the rate of bone remodeling is high, there appears to be no relationship between the rate of remodeling and the response to fluoride. Also, in contrast to the recent data demonstrating a persistence of bone density with the discontinuance of bisphosphonate therapy, discontinuance of fluoride therapy leads to immediate resumption of bone density loss (Talbot et al. 1996).

The dose and duration of fluoride exposure are critical components in determining the effects of the ingested ion on bone. In addition, approximately 30% of patients do not respond to fluoride at any dose (Kleerekoper and Mendlovic 1993). Moreover, there are wide variations in bioavailability among patients and fluoride preparations, and individual responses to the ion also vary widely (Boivin et al. 1993; Erlacher et al. 1995). Whereas the daily dose of fluoride in randomized therapeutic trials (20 to 34 mg/day) exceeds that for people drinking water with fluoride at 4 mg/L (4 to 8 mg/day for 1 to 2 L/day), the latter may be exposed much longer, leading to comparable or higher cumulative doses and bone fluoride concentrations (see discussion later in this chapter.)

Allolio and Lehmann (1999) noted that the peak blood concentrations of fluoride after swallowing 8 oz of water (at 1.0 µg/L) all at once will reach 8.75 µg/L. If peak blood concentrations are proportional to water concentration, then consumption of 8 oz of water containing fluoride at 4 mg/L would produce peak concentrations below the threshold for effects on osteoblasts examined in vitro (95 ng/mL) (Ekstrand and Spak 1990). Assuming that the blood fluoride concentrations decline between each episode of water consumption of 8 oz or less, such exposures may not achieve a concentration of fluoride in the extracellular fluids sufficient to affect bone cells. A caveat to this analysis is that bone cells may be exposed to potentially higher (but unknown) concentrations because of their proximity to the mineralized bone compartment. There have been no direct measurements of the local fluoride concentration around a site of bone resorption. However, a calculation based on estimated rates of resorption, diffusion kinetics, and starting concentration indicates that bone cells and other cells in the immediate vicinity may experience high concentrations of fluoride.

The conditions for an estimate of the fluoride concentration as a function of distance from the osteoclast are as follows:

1. The bone being resorbed has a fluoride content of 3,000 mg per kg of bone ash.
2. Bone ash is assumed to include 65% of the volume of viable bone and the density of viable bone is 1.2 g/cm^3. Thus, the concentration of fluoride in the bone compartment is approximately 5,500 µg/cm^3.
3. An osteoclast resorbs bone at an average rate of about 30,000 µm^3 in 2.5 weeks.
4. The osteoclast is delivering fluoride to the extracellular fluid space from a point source with a radius of 20 µm.
5. Diffusion occurs into a three-dimensional spherical space around the osteoclast.
6. The diffusion coefficient of fluoride in extracellular fluid is approximately 1.5 × 10^{-5} cm^2/s.

Under these conditions, the following equation describes the concentration of fluoride as a function of time and distance from the site of bone resorption (Saltzman 2004):

$$C_{(r,t)} = \frac{SA}{2Dr}\sqrt{\frac{4Dt}{\pi}}$$

where C is the concentration of fluoride as a function of distance and time, S is the delivery rate of fluoride from the resorption site, A is the radius of the point source from which the fluoride is delivered, D is the diffusion coefficient of the fluoride, r is the distance from the resorption site, and t is the time after commencement of the resorption. A graphical representation of this function is presented in Figure 5-1.

An examination of the curves in Figure 5-1 indicates that the fluoride concentration around a site of bone resorption can be quite high immediately adjacent to the osteoclast. The theoretical maximum concentration at 20 µm from the site (at the surface of the osteoclast) would be about 5,500 µg/cm^3. The concentration rapidly decays to zero in very short times at distances greater than 100 µm from the site. However, it appears that a sustained fluoride concentration is achieved in the range of hours and persists for the entire resorption process. Thus, by 2.5 weeks, the concentration of fluoride will be about 500 µg/cm^3 at a distance of 250 µm from the resorption site.

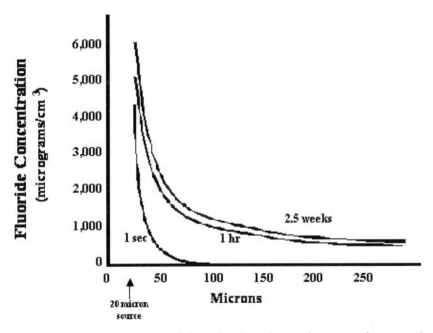

FIGURE 5-1 Concentration of fluoride plotted as a function of time and distance from the site of bone resorption. Release of fluoride from a site of bone resorption can achieve a near steady state concentration in a matter of hours. Twenty microns was defined as the radius of the point source from which fluoride was delivered to the extracellular fluid. Acknowledgement: Hani Awad, University of Rochester, Rochester, New York, assisted in this analysis.

The concentration of fluoride tends toward zero at longer distances. This modeling does not take into account any dissipation of fluoride due to flow of extracellular fluid through the bone marrow compartment. A more complete picture of the local concentration of fluoride around a resorption site should include this factor; however, there are no data on which to base this estimate. Thus, considering that within approximately 1 hour, the fluoride concentration achieves an equilibrium in the surrounding volume, it is likely that the actual fluoride concentration is less, but not substantially so.

Within 250 µm of a site of resorption, it is possible to encounter progenitor cells that give rise to bone, blood, and fat. Thus, one must assume that these cells would be exposed to high concentrations of fluoride. At this time, it is not possible to predict what effect this exposure would have on the functioning of skeletal elements, hematopoiesis, and adipose formation. It should also be pointed out that the number of resorbing sites in an adult skeleton at any point in time is quite small, on the order of 1×10^6 sites. That is, of the vast surface area of trabecular bone in a human skeleton, only about 1 million sites of bone resorption are occurring at any given moment. Whether these elevated concentrations of fluoride have a meaningful effect on bone metabolism can only be speculated at this time.

Some studies have measured the fluoride content of bone, but its effect on a direct measurement of bone strength in humans is not easy to determine. Animal studies have provided some clues. Some studies have reported a biphasic effect of fluoride on bone strength (Beary 1969; Rich and Feist 1970; Turner et al. 1992). For example, Turner et al. (1992) reported an increase in bone strength in rats with bone fluoride concentrations up to 1,200 mg/kg, but they found a decrease in strength back to that of untreated animals with concentrations around 6,000 to 7,000 mg/kg. Skeletal specimens with fluoride concentrations greater than this appeared to have less strength than control treated bone. A variable that may affect the analysis of bone strength is the age of the animal (see Chapter 3). Turner et al. (1995) performed another study in which they found little effect of fluoride on bone strength at any concentration in young rats but a significant effect in old rats. The predominant effect in the older animals clustered around bone fluoride concentrations of 6,000 to 8,000 mg/kg (Turner et al. 1995). Thus, whether fluoride has a biphasic effect on bone strength has not been firmly established.

Other reports in the literature suggesting that fluoride might diminish bone strength in animal models have appeared. Studies of rabbits by Turner et al. (1997) and Chachra et al. (1999) have put forward the point of view that fluoride exposure might decrease strength by altering the structural integrity of the bone microarchitecture. Turner et al. (1997) found no effects of fluoride on a number of bone serum markers, but an increase in bone formation and bone mass. However, this was accompanied by a decrease in bone strength at multiple sites. In a subsequent paper, these authors suggest that the decrease in strength might be due to alterations in mineral crystal structure (Chachra et al. 1999). Whether these results occur in humans remains to be shown. A decrease in bone strength in a human population will definitely increase the risk of fracture and there have been case reports to document this, especially in subjects who may be highly susceptible to accumulating fluoride, such as those with renal failure (Gerster et al. 1983). A more complete discussion of the effects of fluoride in larger population studies follows.

The applicability of rat studies to quantitatively assess risk of bone fracture in humans is uncertain because of the physiological differences between the skeletons of the species. For example, fluoride uptake into bone occurs more readily in humans than in rats (see Chapter 3 and Appendix D). Rats do not undergo Haversian remodeling in their cortical bones as humans do. On the other hand, if fluoride affects bone properties through crystal structure and the mineral-collagen interface, changes in rat bone strength may provide a model for human bone strength (Turner et al. 1992). In addition, whereas the relationship between bone strength and fracture has been studied in rodents, no comparable data are available for humans. The committee therefore judges that the rat experiments provide qualitative support for an effect of fluoride on fractures in humans but cannot yet be used to make quantitative risk estimates for this end point.

The qualifications noted above for rats do not apply as strongly to the rabbit model. Rabbits undergo Haversian remodeling (i.e., osteoclast bone resorption within cortical bone) as do humans (T. Hirano et al. 1999), and the rabbit growth plate behaves more like a human than does a rat or mouse (Zaleske et al. 1982; Irie et al. 2005). Thus, the rabbit is a better model for studying bone effects than rats or mice.

Epidemiology Data

The committee reviewed epidemiologic data on the relationship between fluoride exposure and fractures from two sources: observational studies of exposure to fluoride in water and randomized clinical trials of the use of fluoride in treating osteoporosis. Table 5-1 summarizes studies of bone fracture in populations exposed to fluoride in drinking water. Most of these studies have compared fluoridated (1 mg/L) and nonfluoridated areas. A meta-analysis by McDonagh et al. (2000a, b) evaluated bone fractures in relation to water fluoridation. Consequently, they excluded data from areas with drinking water fluoridated above 1 mg/L, if data at 1 mg/L were available. Results for fractures were reported as evenly distributed around the null—no effect—but statistical testing showed significant heterogeneity among studies. Because the exposures evaluated in this paper did not spe-

TABLE 5-1 Studies on Bone Fracture in Populations Exposed to Fluoride in Drinking Water

Study Design	Country	Subjects	Exposure
Ecologic	USA (national)	Residents of fluoridated and nonfluoridated communities (age ≥ 65; n (fluoridated communities) = 40 million; n (nonfluoridated communities) = 30 million; n (cases) = 218,951)	Fluoridated Nonfluoridated (concentrations not specified)
Ecologic	USA (national)	Patients discharged with hip fracture in counties throughout the USA (n = 541,985)	Fluoridated Nonfluoridated (concentrations not specified)
Ecologic	USA (national)	5% of Medicare population (ages 65 to 89; n [cases] = 59,383)	≤0.3 mg/L (natural) ≥0.7 mg/L (natural and artificial)
Ecologic	USA (national)	Data from National Health Interview Surveys (ages ≥ 45; n = 44,031)	≥0.7 mg/L (natural); groups assessed in terms of <20% or ≥80% of the population exposed to fluoridated water
Prospective cohort	USA (Oregon, Minnesota, Maryland, Pennsylvania)	Women (ages ≥ 65; n = 5,781)	Exposed to fluoridated or nonfluoridated (concentrations not specified) water for 20 years
Ecologic	USA (Minnesota)	Participants in another epidemiology project (ages ≥ 50)	10 years before and 10 years after fluoridation (1.1 mg/L) was implemented
Prospective cohort	USA (Pennsylvania)	Women participating in osteoporotic fracture study (ages ≥ 65; n = 2,076)	1.0 mg/L (artificial) 0.15 mg/L (natural) Number of years of exposure: 0, 1 to 10, 11 to 20, > 20 years
Ecologic	USA (Utah)	Hip fracture patients (ages ≥ 65; n = 246)	1 mg/L (artificial) <0.3 mg/L (natural)

Observations	Reference
Relative risk (RR) of hip fracture in fluoridated communities was 1.08 (95% confidence interval [CI] 1.06 to 1.10) for women and 1.17 (95% CI 1.13 to 1.22) for men. Lack of dose-response relationship between hip fracture risk and duration of water fluoridation. Analyses of annual age-adjusted incidence rates by duration of county water fluoridation showed a pattern of lowest risk in nonfluoridated counties and highest risk in counties fluoridated for up to 5 years, but rates gradually declined for longer durations.	Jacobsen et al. 1992
Weak positive association (before and after adjustment) between hip fracture incidence and percent of county residents who live in counties with fluoridated water.	Jacobsen et al. 1990
RR of hip fracture in the fluoridated group was 1.00 (95% CI 0.92 to 1.09) for men and 1.01 (95% CI 0.96 to 1.06) for women. For ankle fracture, it was 1.01 (95% CI 0.87 to 1.16) for men and 1.00 (95% CI 0.92 to 1.08) for women. For fractures of the distal forearm and proximal humerus, a gender difference in risk was found. For women, there was no association between fluoridation and the two types of fractures. Men in fluoridated areas had a 23% higher risk of proximal humerus fracture (RR 1.23; 95% CI 1.06 to 1.43) and a 16% higher risk of distal forearm fracture (RR 1.16; 95% CI .02 to 1.33).	Karagas et al. 1996
Rate of hip fracture hospitalization per 1,000 in the population with <20% exposed was 2.4 for women and 1.0 for men. For the group with ≥80% exposed, the rates were 2.2 for women and 1.1 for men.	Madans et al. 1983
RR after multivariate adjustment was 0.96 (95% CI 0.83 to 1.10; $P = 0.536$) for nonvertebral fractures, 0.73 (95% CI 0.55 to 0.97; $P = 0.033$) for vertebral fractures, 0.69 (95% CI 0.50 to 0.96; $P = 0.028$) for hip fractures, 0.85 (95% CI 0.58 to 1.23; $P = 0.378$) for humerus fractures, and 1.32 (95% CI 1.00 to 1.71; $P = 0.051$) for wrist fractures.	Phipps et al. 2000
Incidence of hip fracture was 484 per 100,000 residents before fluoridation and 450 per 100,000 residents after fluoridation. RR associated with fluoridation was 0.63 (95% CI 0.46 to 0.86).	Jacobsen et al. 1993
Axial and appendicular bone mass was similar between women exposed to fluoride for >20 years and those exposed for ≤20 years. No significant association was found between fluoride exposure and wrist, spinal, nonspinal, osteoporotic, or hip fractures.	Cauley et al. 1995
RR of hip fracture in the fluoridated population was 1.27 (90% CI 1.08 to 1.46) for women and 1.41 (95% CI 1.00 to 1.81) in men.	Danielson et al. 1992

continued

TABLE 5-1 Continued

Study Design	Country	Subjects	Exposure
Prospective cohort	USA (Iowa)	Women from three communities with different concentrations of fluoride in water (ages 20-92, n = 1,300)	1 mg/L (w/Ca at 60 mg/L) 1 mg/L (w/Ca at 375 mg/L) 4 mg/L (w/Ca at 15 mg/L)
Prospective cohort	USA (Iowa)	Women from 3 communities with different concentrations of fluoride in water (ages 20-35 and 55-80; n = 158 [referents], n = 230 [high fluoride])	1 mg/L (w/Ca at 67 mg/L) 1 mg/L (w/Ca at 375 mg/L) 4 mg/L (w/Ca at 15 mg/L)
Retrospective cohort	USA (Iowa)	Women from 3 communities with different concentrations of fluoride in water	1 mg/L (w/Ca at 60 mg/L) 1 mg/L (w/Ca at 375 mg/L) 4 mg/L (w/Ca at 15 mg/L)
Ecologic	USA (Michigan)	Female Medicaid recipients (ages ≥ 65)	≥89% of the population receives fluoridated water (2 groups) <15% of the population receives fluoridated water
Ecologic	Canada	Patients (ages 45 to 64, 65+) with hip fracture in two cities	0.3 mg/L 1 mg/L
Case-control	United Kingdom	Patients with hip fractures (ages ≥ 50; n [cases]) = 514; n [controls]= 527)	<0.9 mg/L (artificial) ≥0.9 mg/L (natural)
Ecologic	England, Wales	Patients discharged from hospital after hip fracture (ages ≥ 45; n = 20,393)	0.005 to 0.93 mg/L (natural and artificial)
Prospective cohort	France	Subjects enrolled in another study (ages ≥ 65; n = 3.216)	0.05 to 0.11 mg/L 0.11 to 0.25 mg/L >0.25 mg/L
Ecologic	France	Subjects enrolled in another study on aging (ages ≥ 65; n = 3,777)	0.05 to 0.11 mg/L 0.11 to 1.83 mg/L
Ecologic	Germany	Residents of fluoridated and nonfluoridated communities	0.08 to 0.36 mg/L (natural) 0.77 to 1.20 mg/L (artificial)

Observations	Reference
RR for osteoporotic fractures was 2.55 ($P = 0.07$) in the 4 mg/L group. Serum fluoride concentrations were not related to osteoporotic fractures or bone mineral density.	Sowers et al. 2005
In the 4-mg/L group, RR of any fracture was 1.81 (95% CI 0.45 to 8.22) in premenopausal women and 2.11 (95% CI 1.01 to 4.43) in postmenopausal women. RR for fractures of the hip, wrist, or spine was 2.70 (95% CI 0.16 to 8.28) in premenopausal women and 2.20 (95% CI 1.07 to 4.69) in postmenopausal women.	Sowers et al. 1991
Postmenopausal women in the 4 mg/L group reported significantly more fractures than the other two groups.	Sowers et al. 1986
Long-bone fracture rates were 94.3 per 1,000 and 81.1 per 1,000 in the two populations that are ≥ 89% fluoridated. The rate was 78.8 per 1,000 in the population that was < 15% fluoridated.	Avorn and Niessen 1986
For men, ages 45 to 64, standardized hospital admission rates were 0.59 and 0.55, respectively; for men over 65, rates were 5.09 and 4.52. For women, ages 45 to 64, corresponding rates were 0.60 and 0.71; and for ages over 65, they were 9.54 and 9.91.	Suarez-Almazor et al. 1993
Estimated average lifetime exposure to fluoride in drinking water ranged from 0.15 to 1.79 mg/L. Odds ratio associated with an average lifetime exposure to ≥ 0.9 mg/L was 1.0 (94% CI 0.7 to 1.5).	Hillier et al. 2000
Discharge rates ranged from 0.88 to 2.30. No correlation was found between discharge rates for patients with proximal femur fractures and water fluoride concentrations ($r = 0.16$, $P = 0.34$). Subsequent reanalysis of the data using a weighted least-squares technique showed a positive correlation between fluoride concentrations and hip fracture ($r = 0.41$, $P = 0.009$).	Cooper et al. 1990, 1991
Odds ratio for hip fractures was 1, 3.25 (95% CI 1.66 to 6.38), and 2.43 (95% CI 1.11 to 5.33), respectively. Odds ratio for non-hip fractures was 1, 0.88 (95% CI 0.63 to 1.22), and 1.05 (95% CI 0.74 to 1.51).	Jacqmin-Gadda et al. 1998
Odds ratio for hip fractures were 1 and 1.86 (90% CI 1.02 to 3.36), respectively. Odds ratio for non-hip fractures were 1 and 0.98 (95% CI 0.80 to 1.21), respectively.	Jacqmin-Gadda et al. 1995
Mean annual incidence of hip fracture in the fluoridated community was 173.36 per 100,000 for women and 56.79 per 100,000 men. In the nonfluoridated group, it was 189.35 per 100,000 in women and 56.60 per 100,000 in men.	Lehmann et al. 1998

continued

TABLE 5-1 Continued

Study Design	Country	Subjects	Exposure
Ecologic	Italy	Residents of two counties	1.45 mg/L (natural) 0.05 mg/L (natural)
Retrospective cohort	Finland	Residents of a rural location (n = 144,627)	≤0.1 mg/L 0.11 to 0.30 mg/L (natural) 0.31 to 0.50 mg/L (natural) 0.51 to 1.00 mg/L (natural) 1.10 to 1.50 mg/L (natural) >1.50 mg/L (natural)
Retrospective cohort	Finland	Premenopausal women in a province (ages 47 to 56; n = 3,222)	<0.3 mg/L (natural) 1 to 1.2 mg/L (artificial)
Ecologic	Finland	Patients with hip fracture (ages ≥ 50)	<0.3 mg/L (natural) 1.0 to 1.2 mg/L (artificial) >1.5 mg/L (natural)
Ecologic	Finland	Residents in two towns (n = 71,811 and n = 61,587)	<0.1 mg/L 1 mg/L
Retrospective cohort	China	Residents of rural communities exposed to various concentrations of fluoride in drinking water (ages ≥ 50; n = 8,266)	0.25 to 0.34 mg/L (natural) 0.58 to 0.73 mg/L (natural) 1.00 to 1.06 mg/L (natural) 1.45 to 2.19 mg/L (natural) 2.62 to 3.56 mg/L (natural) 4.32 to 7.97 mg/L (natural
Ecologic	Mexico	Children (ages 6-12 years) and adults (ages 13-60 years)	ND to 1.5 mg/L (natural) 1.51 to 4.99 mg/L (natural) 5.0 to 8.49 mg/L (natural) 8.5 to 11.9 mg/L (natural) >12 mg/L (natural)
Case-control	USA	Women participating in the Nurses' Health Study (ages 30-55; n [hip fracture] = 53; n [forearm fracture] = 188; n [controls] = 241)	Concentrations in toenails <2.00 ppm 2.00 to 3.35 ppm 3.36 to 5.50 ppm >5.50 ppm

Observations	Reference
Significantly greater rate of fracture incidence, particularly femur fractures (RR for males 4.28 and for females 2.64), in the low-exposure community.	Fabiani et al. 1999
Age-and area-adjusted RRs for men were 1.0, 1.05 (95% CI 0.90 to 1.22), 0.72 (95% CI 0.51 to 1.02), 1.03 (95% CI 0.81 to 1.32), 0.67 (95% CI 0.46 to 0.97), and 0.98 (95% CI 0.61 to 1.60). Corresponding values for women were 1.0, 0.93 (95% CI 0.84 to 1.02), 1.12 (95% CI 0.93 to 1.34), 1.12 (95% CI 0.96 to 1.31), 1.08 (95% CI 0.88 to 1.32), and 1.08 (95% CI 0.80 to 1.46). Among women aged 50 to 64 years, fluoride was associated with increased risk of hip fracture. Age- and area-adjusted rate ratio for this age group was 2.09 (95% CI 1.16 to 3.76) in the highest-exposure group (>1.5 mg/L) compared with the lowest-exposure group (≤0.1 mg/L).	Kurttio et al. 1999
No significant difference in fracture incidence among the fluoridated (15.4%) and nonfluoridated group (13.4%) ($P = 0.220$).	Kroger et al. 1994
No difference in incidence of hip fracture among exposure groups. Osteofluorosis was found in 22% of the high exposure group. Fluoride content of the bone was correlated with volumetric density of trabecular bone and osteoid-covered trabecular bone surface.	Arnala et al. 1986
In the <0.1-mg/L exposure group, RR was 2.5 (95% CI 1.6 to 3.9) for men and 1.5 (95% CI 1.2 to 1.8) for women. In the group exposed to 1 mg/L, RR was 1.0 for men and women.	Simonen and Laitinen 1985
Lowest prevalence of overall bone fracture was found in the 1.00 to 1.06 mg/L group and was significantly lower ($P < 0.05$) than that of the groups exposed to concentrations ≥4.32 and ≤0.34 mg/L. Prevalence of hip fracture was greatest in the in the 4.32 to 7.97 mg/L group and was significantly higher than the 1.0- to 1.06-mg/L group.	Li et al. 2001
Increased bone fracture (bone types not specified) incidence was observed at concentrations ranging from 1.5 to 4.99 mg/L. A plot of the incidence of fractures in adults versus the average corresponding fluoride concentration by zone indicated a third-order polynomial correlation ($R^2 = 0.9995$). Incidence in children was similar, except in one zone. Linear correlation between Dean index for dental fluorosis and the frequency of bone fracture in children ($R^2 = 0.94$) and adults ($R^2 = 0.98$).	Alarcón-Herrera et al. 2001
Women with higher concentrations of toenail fluoride appeared to be at greater risk of forearm fracture but to have a lesser risk of hip fracture than women with toenail concentrations <2 ppm. Odds ratio of hip fracture in women with >5.50 ppm compared with those with <2.00 ppm was 0.8 (95% CI 0.2 to 4.0). Corresponding adjusted odds ratio for forearm fracture was 1.6 (95% CI 0.8 to 3.1).	Feskanich et al. 1998

cifically address the committee's charge, this meta-analysis and most of the studies on which it was based were not critically evaluated. The committee restricted its attention to the observational studies that most directly address the study charge: studies that examined long-term exposure to fluoride in the range of 2 to 4 mg/L or above in drinking water. Randomized clinical trials that exposed subjects to higher doses over shorter periods of time were also considered.

The committee considered a number of factors as it evaluated the available data, including the following:

- The committee assumed that fluoride concentrations in bone are the most appropriate measure of exposure. Although difficult to measure in epidemiology studies, bone fluoride concentrations are positively associated with the amount of fluoride exposure, length of exposure, age, and certain diseases such as chronic renal insufficiency (see Chapter 3 for discussion of pharmacokinetic factors that affect fluoride uptake by bone). Use of other fluoride exposure measures is likely to cause measurement error. While exposure measurement error often biases results toward the null, there are many exceptions.
- U.S. exposure estimates presented in Chapter 2 indicate that water will be the major route of exposure for Americans drinking or cooking with water containing fluoride at 4 mg/L but that other sources become more important at concentrations closer to 1 mg/L.
- The incidence of fractures increases dramatically in old age. Minor or moderate traumas cause more fractures in the elderly than in healthy young adults. Other known or suspected risk factors include being female, being postmenopausal, diet (e.g., low calcium), physical inactivity, low body mass index, and use of certain drugs (e.g., corticosteroids) (Ross 1996; Woolf and Åkesson 2003). As a result, age is a very important covariate both as a potential confounder and as an effect modifier; control for age may need to be fairly detailed above age 50.
- Self-reports of fractures are reasonably accurate, although vertebral fractures are typically underreported. Elderly women may overreport total fractures, but the percent of false positives may be lower for fractures of the wrist and hip (Nevitt et al. 1992; Honkanen et al. 1999). Thus, although epidemiological studies would be better if they confirm the presence or absence of fractures, self-reports may be adequate. For example, relative risk measures (risk and rate ratios, but not odds ratios) are unbiased if the outcome is nondifferentially underreported but false positives are negligible (Poole 1985). We might expect the degree of false-positive reporting and underreporting not to differ by fluoride water concentrations, thus tending to attenuate associations.
- Fluoride may have different effects on fractures of different bones (as

suggested by Riggs et al. 1990). Consequently, epidemiologists need to be careful about the degree of aggregation of outcomes. If some bone sites are included that are not susceptible, then relative risk estimates will be biased toward the null; risk or rate differences would not.
• Studies that measure outcome and covariates individually but exposure by group (e.g., by water concentration) use a partially ecologic or group-level design. This design greatly improves the ability to measure and control for covariates relative to pure ecologic studies; control of covariates is one of the major problems in purely ecologic studies. See Appendix C for a description of these design differences.

Below is a review of the available epidemiologic data for evaluating the adequacy of EPA's maximum-contaminant-level goal (MCLG) for fluoride of 4 mg/L and secondary maximum contaminant level (SMCL) of 2 mg/L for protecting the public from bone fractures.

Studies Relevant to Assessing Risks at 4 mg/L

Observational Studies. The committee is aware of five published observational studies of fractures in subjects exposed to drinking water containing fluoride at 4 mg/L or higher (Sowers et al. 1986, 1991, 2005; Alarcón-Herrera et al. 2001; Li et al. 2001) and another (Kurttio et al. 1999) involving somewhat lower exposures that has some relevance. The first two Sowers papers examine the same cohort, one retrospectively (Sowers et al. 1986) and one prospectively (Sowers et al. 1991). Because the analysis in the 1986 paper is less detailed for fractures (particularly the discussion of potential confounders), it has been given less attention. Features of the key papers are highlighted in Table 5-2.

Sowers et al. (1991) directly assessed the risk of fracture from fluoride at 4 mg/L, reporting adjusted odds ratios (ORs) of 2.1 (95% CI = 1.0 to 4.4) for any fracture, and 2.2 (95% CI = 1.0 to 4.7) for fracture of the hip/wrist/spine in women 55 to 80 years of age at baseline (ORs were also elevated in younger women). The reference group was exposed to fluoride at 1 mg/L. This is a strong study, particularly because of its prospective cohort design. Although the 1993 National Research Council (NRC) report labeled it as ecologic, it is actually an individual-level study with an ecologic exposure measure (such designs are also called semi-individual; see Appendix C). Outcome and important covariates, including age, are measured at the individual level (control of covariates is particularly problematic in fully ecologic studies). This study has some weaknesses: confounding was assessed by using stepwise logistic regression (a common but less than optimal method for assessing confounding) and fractures were self-reported. Self-reports of fractures are often quite reliable (except for the spine, where

TABLE 5-2 Observational Studies of Bone Fractures in Populations Exposed to Fluoride Near 4 mg/L in Drinking Water

	Li et al. (2001)	Sowers et al. (1991)	Alarcón-Herrera et al (2001)
Design	Retrospective cohort with ecologic exposure measure	Prospective cohort with ecologic exposure measure	Ecologic
Location	China, 6 areas with fluoride ranging from 0.25 to 7.97 mg/L	3 areas in Iowa (USA) with fluoride at 1 or 4 mg/L	Guadiana Valley, Mexico, with fluoride ranging from <1.5 to 16 mg/L
No. subjects	8,262	827 at baseline, good follow-up	1,437 (333 less than 13 years old)
Exposure assessment	Ecologic; negligible sources other than water; very-long-term residents; very strong for this type of study	Ecologic; other sources likely in low-exposure groups; long residence time	Ecologic; inconsistent documentation (e.g., use of bottled water mentioned for only one area); permanent residents not defined
Outcomes	Self-reported fractures validated via x-ray, but lack of fracture not confirmed; recall bias seems unlikely; report all fractures since age 20 or 50, also hip since age 20; count number of subjects with fractures	Self-reported fractures (spine fractures likely underreported) for 5 year follow-up; report all fractures, plus fractures of hip/wrist/spine; count number of subjects with fractures	Self-reported fracture; any fracture "ever occurred without apparent cause, where a bone fracture would not normally be expected to occur"—highly subjective; counts multiple fractures per person?
Confounding	Very similar communities; many individual-level risk factors; imperfect method for covariate control (relying on significance tests)	Similar communities; many individual-level risk factors; imperfect method for covariate control (relying on significance tests)	No variables analyzed other than crude stratification by age (<13, ≥13); major weakness
Results	U-shaped results for all fractures, increasing trend for hip (age > 20): adjusted ORs (P values) versus 1 mg/L: Fluoride, mg/L All sites Hip 2.62 to 3.56 1.18 (0.35) 1.73 (0.34) 4.32 to 7.97 1.47 (0.02) 3.26 (0.02) Total fractures since age 50 also provided	Increased risk at 4 mg/L versus 1 mg/L Women 55 to 80 at baseline, adjusted ORs and (95% CI) versus 1 mg/L: 2.1 (1.0, 4.4) for any fracture 2.2 (1.0, 4.7) hip/wrist/spine	Effect measures not presented; percent of fractures increases in adults from 3.1% (<1.5 mg/L) to 7.9% (1.51 to 4.99 mg/L), 8.9% (5 to 8.99 mg/L), but then decreases. P values for the two intermediate levels were 0.046 and 0.041.

	Li et al. (2001)	Sowers et al. (1991)	Alarcón-Herrera et al (2001)
Overall	Strong study	Strong study	Weak study
Additional comments			Suggestive analysis of fracture versus dental fluorosis but insufficient detail

	Kurttio et al. 1999	Sowers et al. 2005
Design	Historical cohort	Prospective cohort with both ecologic and individual-level exposure measures
Location	Finland, rural communities nationwide	Same three areas of Iowa as earlier study
No. subjects	144,000+	1,300 women aged 20 to 92 (average, 55)
Exposure assessment	Groundwater measurements of almost 9,000 wells Fluoride concentrations estimated for each residence by using weighted medians, smoothed interpolations. Categories: <0.1, 0.1 to 0.3, 0.3 to 0.5, 0.5 to 1.0, 1 to 1.5, and >1.5 mg/L. Highest category corresponded to sampled concentrations of less than detection level to approximately 6 mg/L.	Ecologic (area of study) Individual (serum fluoride concentration)
Outcomes	First recorded hip fracture	Self-reported fracture, confirmed by medical records or x-ray copies, if available. Lack of fractures apparently not confirmed. Fractures separated into likely osteoporotic (hip, spine, wrist, ribs) and other.
Confounding	Analyzed controlling for age and geographic sector. Age adjustment was conducted within broad strata of 50 to 64 and 65 to 80. No information on nutrition, alcohol use, or physical activity.	Similar communities; many individual-level risk factors; imperfect method for covariate; control (relying on significance tests). Unclear if some covariates were included.

153

continued

TABLE 5-2 Continued

	Kurttio et al. 1999	Sowers et al. 2005
Results	For comparisons between the >1.5 mg/L group and the <0.1 mg/L group (ages 50 to 65), adjusted RR = 2.09 (95% CI, 1.16 to 3.76) in women, RR = 0.87 (95% CI, 0.35 to 2.16) in men. For all ages combined, no associations apparent. For fluoride as a continuous variable, RR = 1.44 (95% CI, 1.12 to 1.86) for women below age 65 at start of follow-up, and RR = 0.75 (95% CI, 0.51 to 1.12) for men in same age stratum (age and region adjusted). Women ages 55 to 69 had the most elevated RR in the continuous-variable analysis. Among separate 5-year components of follow-up period, the results were inconsistent.	Ecologic: 2.55-fold increased risk ($P = 0.07$) osteoporotic fracture at 4 mg/L versus 1 mg/L for all women (age breakdown not provided) after adjustment (including bone mineral density of femoral neck). Individual: RR = 1.16 ($P = 0.66$) for osteoporotic fracture versus log of serum fluoride for all women, after adjustment (including bone mineral density of femoral neck).
Overall	Strong study	Strong study
Additional comments	Suggestive of hip fracture risk, with continuous gradient from lowest to highest exposures	Weak association between bone density and serum fluoride (e.g., adjusted $\beta = 0.011 \pm 0.0073$ (SE), $P = 0.13$ for femoral neck). Use of serum fluoride concentration may bias results toward null if there is nondifferential error relative to bone fluoride concentrations. Bone mineral density may be, in part, an intermediate variable.

underreporting is typical). Details about the interviewers (training or blinding to exposure) were not provided. The paper also examined fractures in a community with high calcium concentrations in water: the adjusted OR for fracture of the hip/wrist/spine was 1.6 (95% CI = 0.71 to 3.4) for the older women and 0.30 (95% CI = 0.04 to 3.4) for younger women (the ORs for all fractures were similar). The regression analysis comparing the high fluoride and the reference communities was adjusted for calcium intake, but it produced no change.

The newest study by Sowers et al. (2005) investigated bone fracture in relation to fluoride concentration in drinking water. The authors measured serum fluoride, providing a potentially improved exposure assessment. In this cohort study, fractures were assessed prospectively for 4 years. Fractures were self-reported and then confirmed with medical records or x-ray copies, if available; lack of fractures was apparently not confirmed. Sowers et al. (2005) collected individual-based information on people from the same regions as the 1986 and 1991 studies. They collected serum fluoride concentrations and bone mineral density of the hip, radius, and spine. The number of subjects was considerably expanded (n = 1,300) from the earlier studies. Although there may be overlap in specific subjects, all the fracture events were recent. The authors reported risk ratios of fractures in the high fluoride area that were similar to those in the previous studies (risk ratio = 2.55, P = 0.07, even when adjusting for bone mineral density, which could function as an intervening variable between water ingestion and fracture outcome). Use of ecologic exposure measures need not cause bias due to exposure measurement error (see Appendix C).

Serum fluoride concentration was higher in the community with fluoride at 4 mg/L in drinking water. Bone and serum concentrations are related but the latter have more noise—potentially much more, depending on how samples were collected. Serum fluoride concentrations can vary within individuals, returning to baseline within hours of exposure.

Fasting serum fluoride concentrations are considered a good (although not necessarily perfect) measure of long-term exposure and of bone fluoride concentrations (Ericsson et al. 1973; Parkins et al. 1974; Taves and Guy 1979; Waterhouse et al. 1980; Whitford 1994; Clarkson et al. 2000; see also Chapter 2 for a discussion of biomarkers and Chapter 3 on pharmacokinetics). Although methods for serum collection are not described in the paper, Sowers stated that fasting serum concentrations were taken "whenever possible" (M.F. Sowers, University of Michigan, personal commun., July 1, 2005). Measured serum fluoride concentration was not statistically associated with fracture incidence in the adjusted model, including bone density, a potential intermediate variable (measured serum fluoride was only weakly associated with bone mineral density). However, it is unclear whether serum fluoride was a useful surrogate for concentrations in bone or chronic exposure here; random error would tend to bias results toward the null. Table 2 in the Sowers et al. (2005) paper indicated that long-term residency in the high-fluoride region was not associated with appreciably higher serum fluoride than short-term residency.

Besides differences in osteoporotic, but not other, fracture rates, these populations differed markedly with respect to smoking rates and hormone replacement (both lowest in the reference group) and physical activity (lowest in the high-fluoride group). It is unclear whether these factors were examined as potential confounders for fractures. Age subgroups were not presented in the new Sowers et al. study, so differences within age groups cannot be assessed and comparisons with the other observational studies on fractures cannot be made.

For all the Sowers studies, there is an unresolved question about whether the referent group (area with low fluoride and low calcium) might have a low fracture rate because of risk factors that are not controlled for in the studies, particularly as the high-calcium/low-fluoride region also showed increased fracture rates compared with the referent region. Potential bias due to such differences might be exacerbated by the use of an ecologic exposure measure (see Appendix C).

The study by Li et al. (2001) complements the Sowers studies in several ways, having a larger size and relatively strong exposure assessment for a partially ecologic study. It has a retrospective cohort design, increasing the potential for outcome and exposure misclassification, but these problems were addressed by the authors. Although exposure was assessed on the group level, exposure was finely categorized and other sources of fluoride exposure were estimated to be negligible. (Nonwater exposures to fluoride were presumably more important in the Sowers studies.) Communities were quite similar and individual-level risk factors were assessed. Fractures were self-reported; confirmation with x-rays showed very high validity (526 fractures confirmed among the 531 subjects reporting fractures). This study also has weaknesses. Confounding was assessed by statistical testing; the authors included a covariate in the logistic regression if they first found a statistically significant ($P < 0.05$) relationship between the variable and outcome analyzed bivariately. (Confounding should be judged by examining the effect measure, not statistical testing; see Rothman and Greenland 1998.) Absence of fractures was not confirmed, potentially biasing outcomes if false-positive reporting of fractures is expected to be more than an isolated occurrence. However, a limited number of sensitivity analyses of confounding performed by the committee did not explain the effect; recall bias seems an unlikely explanation for the U-shaped exposure-response curve (for all fractures since age 20), with the minimum fractures in the reference group of 1 mg/L. The dose-response curve for all fractures is plausible: some, but not all, animal studies suggest a biphasic relationship between bone fluoride concentrations and bone strength (see discussion earlier in this section on cellular and molecular aspects).

The Li et al. study did not directly assess fluoride at 4 mg/L. However the exposure group just above 4 mg/L (4.32 to 7.97 mg/L) showed an increase in all fractures since age 20 (OR = 1.47, P = 0.01, estimated 95% CI = 1.10 to 1.97), all fractures since age 50 (OR = 1.59, P = 0.02, estimated 95% CI = 1.08 to 2.35), and hip fractures since age 20 (OR = 3.26, P = 0.02, estimated 95% CI = 1.21 to 9.81). The exposure group just below 4 mg/L (2.62 to 3.56 mg/L) showed the following: all fractures since age 20 (OR = 1.18, P = 0.35, estimated 95% CI = 0.83 to 1.67), all fractures since age 50 (OR = 1.04, P = 0.87, estimated 95% CI = 0.65 to 1.66), and hip fractures since age 20 (OR = 1.73, P = 0.34, estimated 95% CI = 0.56 to 5.33). CI values were estimated by the committee using the approach of Greenland (1998). Although the latter results are not statistically significant at the 0.05 level, they are consistent with an upward trend (increasing dose-response relationship), particularly the result for hip fracture. The inclusion of all fractures is likely to bias ORs toward the null.

Although the authors did not estimate trend, Figures 2 and 3 presented in the paper by Li et al. (2001) suggest that linear trends in proportions from the 1.00 to 1.06 mg/L category up would provide a reasonable fit in that range. Using a generalized linear model with the binomial distribution and the identity link, and midranges for the exposure categories, the committee estimated absolute increases in fractures of 1.3% (95% CI = 0.3 to 2.2, P = 0.01) for the increment from 1.00 to 4.00 mg/L for overall fractures since age 20, 0.4% (95% CI = 0.0 to 0.8, P = 0.04) for hip fractures since age 20, and 0.9% (95% = CI 0.2 to 1.7, P = 0.02) for overall fractures since age 50.

The U-shaped exposure-response curve for all fractures combined (but not hip fractures) for this population of individuals provides an interesting and potentially important finding. Whereas the trend for fractures appears to increase from 1.00 to 4.00 mg/L, it must be appreciated that the fracture rate in the 1.00 to 1.06 mg/L category was lower than the rate in the category with the lowest intake.

Estimated fluoride exposure in the Li study is higher than for the Sowers studies (see Table 5-4 later in this chapter). Assuming that exposure was predominantly due to water, the committee estimated that participants in the Li study consumed on average about 2.5 L per day for the 2.62- to 3.56-mg/L group and 2.3 L per day for 4.32- to 7.97-mg/L group (versus 0.9 to 1.2 L per day for the Sowers studies). These water consumption levels are in the 90th to 95th percentile for the United States (see Appendix B).

Alarcón-Herrera et al. (2001) is a much weaker ecologic study with little attention to covariates other than a rough stratification by age (see

Table 5-2 for a brief discussion). The results are qualitatively similar to the stronger studies.

In addition, a retrospective cohort study in Finland by Kurttio et al. (1999) is pertinent to the issue of fracture risk at 4 mg/L, even though relatively few wells in that study had drinking water with fluoride concentrations that high. Residents were grouped into exposure categories based on modeled fluoride concentrations in well water closest to their residence: ≤0.1, 0.11 to 0.30, 0.31 to 0.50, 0.51 to 1.00, 1.10 to 1.50, and >1.5 mg/L (ranging up to 2.4 mg/L). Fluoride monitoring results among water samples for the highest modeled group varied from below detection to about 6 mg/L. Hospital discharge registers were tracked between 1981 and 1994 for reports of hip fracture among the cohort. For all ages combined, no associations were found between fluoride content in drinking water and hip fracture. However, analysis of age strata (50 to 64 and 65 to 80) indicated an increased risk of hip fracture in women aged 50 to 64 exposed to fluoride at >1.5 mg/L (adjusted rate ratio of 2.09%; 95% CI, 1.16 to 3.76; based on 13 cases [3,908 person years] compared with those in the least exposed group (≤0.1 mg/L). Some covariates were adjusted by using ecologic measures, an imperfect technique.

Clinical Trials of Osteoporosis Treatment. Using the Cochrane Handbook methodology, Haguenauer et al. (2000) performed a meta-analysis of randomized clinical trials of fluoride in postmenopausal women with primary osteoporosis. Eleven studies met the inclusion criteria; analyses of specific end points included only a subset. The summary relative risk estimate for new vertebral fractures was slightly decreased: 0.87 (95% CI = 0.51 to 1.46) for 2 years of treatment (four trials) and 0.90 (95% CI = 0.71 to 1.14) for 4 years (five trials). The summary relative risk estimate for new nonvertebral fractures was increased: 1.20 (95% CI = 0.68 to 2.10) after 2 years (one trial) and 1.85 (95% CI = 1.36 to 2.50) after 4 years (four trials); the latter association was strongest in trials using high-dose, non-slow-release forms of fluoride. See Table 5-3 for the four studies included in the analysis of nonvertebral fractures after 4 years. All four studies were prospective, double-blinded, and placebo-controlled; all subjects received supplemental calcium. There was loss to follow-up, particularly in the study of Kleerekoper et al. (1991), but it was similar in frequency in treated and placebo groups.

Table 5-3 reports relative risks of nonvertebral fractures at 4 years. Rate ratios are also provided for several studies. Hip fracture results are reported only for Riggs et al. (1990); the number of hip fractures in the other trials was at most one per group. Riggs et al. reported both complete fractures and total fractures. Total fractures equal complete plus incomplete "stress" fractures; the latter were observed by roentgenography in participants re-

TABLE 5-3 Four Randomized Clinical Trials Examining Nonvertebral Fractures

	Exposure	Enrollment: Exposed and Placebo	Participation[a] Exposed and Placebo	Relative Risk (95% CI) Nonvertebral Fractures[b]	Rate Ratio (95% CI) Nonvertebral Fracture[c]
Reginster et al. 1998	Fluoride at 20 mg/day as sodium monofluorophosphate, 4 years	100, 100	84%, 81%	1.1 (0.5, 2.4)[d]	1.1 (0.5, 2.3)
Pak et al. 1995	NaF at 50 mg/day slow-release, 4 cycles: 12 months on, 2 months off	54, 56	77%, 72%	0.6 (0.2, 2.5)[d]	
Kleerekoper et al. 1991	NaF at 75 mg/day, 4 years	46, 38	60%, 61%	1.5 (0.7, 3.5)[d]	3.0 (2.0, 4.6) "hot spots"[e]
Riggs et al. 1990	NaF at 75 mg/day, 4 years	101, 101	77%, 80%	1.6 (1.0, 2.5) complete 2.5 (1.7, 3.7) total[d,f] 2.3 (0.6, 8.8) complete, hip	1.9 (1.1, 3.2) complete 3.1 (2.0, 5.0) total[f]

[a]Participating person-time divided by total possible person-time.
[b]Risks were computed by dividing the number of first incident fractures (at most one per subject) by the number of enrolled subjects.
[c]Rates were computed by dividing the number of incident fractures (possibly more than one per subject) by participating person-time.
[d]The numbers that appear to have been used in the meta-analysis of Haguenauer et al. (2000); see their Figure 5.
[e]Areas of increased isotope uptake detected via radionuclide bone scan.
[f]In this study, total fractures = complete + incomplete "stress" fractures, the latter observed by roentgenography in participants reporting acute lower extremity pain syndrome.

porting acute lower extremity pain syndrome (see Kleerekoper et al. 1991 for a different interpretation).

Comparison of Exposure in Randomized Clinical Trials and Observational Studies. Despite the methodological strengths of the randomized clinical trials, their use in this review has limitations. In particular, fluoride exposures in the trials were higher in magnitude (20 to 34 mg/day) than in observational studies (5 to 10 mg/day for 4 mg/L) but shorter in time (years versus decades). One possibility is to compare studies using total fluoride exposure in absolute mass units. Because some biological effects (e.g., stimulation of osteoblasts) may occur only at high doses, results from clinical trials may not be directly comparable to risks due to long-term exposure to fluoride in water. On the other hand, the committee assumes that bone fluoride concentration is the most appropriate measure of exposure for examining fracture risk. Data permitting, it could be possible to compare the bone fluoride concentrations reached in the randomized clinical trials with those in the observational studies.

Of the four randomized clinical trails in the fracture meta-analysis, the committee was able to locate bone fluoride measurements for only one. Of the 202 postmenopausal women in the Riggs study, bone fluoride was measured before treatment and at 4 years in 43 treated and 35 placebo subjects (Lundy et al. 1995). Unfortunately, the data are presented only in a figure and in units of µmol of fluoride per mmol of calcium. The latter, however, can be approximately converted to mg/kg ash by using the following factors: 1 g of calcium per 7 g wet weight of bone (Mernagh et al. 1977) and 0.56 g of ash per g wet weight of bone (Rao et al. 1995). Using this conversion, the before-treatment bone ash fluoride concentrations were about 1,700 mg/kg in both the treated and the placebo groups. Taking the imprecision of the conversion factors into account, this value is consistent with reported concentrations for healthy, untreated persons living in areas without particularly high water fluoride concentrations and no other exceptional sources of fluoride intake (see Chapter 3). Four years later, bone ash concentrations were slightly higher in the placebo group and about 12,000 mg/kg in the treated group. The latter value should overestimate concentrations in the exposed group of the trial, because the average exposed subject in the Riggs study participated 3.1 years (Table 5-3).

Ideally, one would estimate bone concentrations in the other trials by using a pharmacokinetic model. Because the committee did not have an operational pharmacokinetic model, a regression model was used to estimate bone concentrations based on total fluoride exposure during clinical trials (see Chapter 3). Total exposures (Table 5-4) were estimated with the nominal daily dose and the average length of participation of the exposed group. The bone concentration for Riggs et al. estimated by this technique (7,400 mg/kg) is less than the value measured by Lundy et al. (roughly 12,000 mg/kg), but the latter examined a subset of subjects who had completed the full 4 years of the study. The regression model estimates 9,100 mg/kg in subjects completing 4 years. Although that estimate is still less than the measured concentration, Chapter 3 noted that the regression model may underestimate bone concentrations in fluoridated areas. Of the four clinical trials in Table 5-4, three were American. Fluoride exposure and concentra-

MUSCULOSKELETAL EFFECTS   161

tions in bone may be overestimated for the Pak study because of the use of a slow-release, less bioavailable form of fluoride. In sum, average fluoride bone concentrations among treated trial participants appear to range from about 5,400 to 12,000 mg/kg.

Comparison of Results of Randomized Clinical Trials and Observational Studies. Table 5-4 also includes estimates of total exposure and average bone fluoride for two observational studies. The committee estimated average fluoride concentrations in bone in the study by Sowers et al. (1991) using the regression model developed for chronic water exposure in Chapter 3. This model predicts bone concentrations based on concentration of fluoride in water, length of exposure, and sex. The result is in the same range as the clinical trials. Since the regression model does not take water consumption rates into account, it should underpredict bone fluoride concentrations for people with high water consumption. The bone fluoride estimates for Li et al. (2001) are, therefore, probably underestimates. Estimates of bone fluoride concentrations could be improved through the use of pharmacokinetic models (see Chapter 3).

Table 5-4 summarizes fracture outcomes for the four clinical trials (nonvertebral) and observational studies. There are a number of differences in the way the outcome data were collected and analyzed. For example, Li et al. counted fractures occurring since age 20 (or age 50, not shown), a longer observation period than the other studies; Li et al. and Sowers et al. measured fractures in different bones than those studied in the clinical trials; if trials use subjects from fluoridated areas, the subjects receiving placebos are from areas with fluoride close to 1 mg/L. Although the comparison involves several assumptions and uncertainties, the estimated concentrations of fluoride in bone and results of the randomized clinical trials generally appear consistent with those of the observational studies.

Interpretation of Weight of Evidence of the Fracture Data on Fluoride at 4 mg/L. For making causal inferences, many epidemiologists prefer to formulate and test specific competing hypotheses (e.g., Rothman and Greenland 1998). Other epidemiologists find it useful to weigh the evidence in light of some traditional "criteria" (more properly, guidelines) for examining whether observed associations are likely to be causal (Hill 1965). The discussion below provides a perspective on how the committee evaluated and viewed the strength of the collective evidence on possible causal associations.

• Consistency: Despite some design or data weaknesses, there is consistency among the results of all the observational studies using ecologic exposure measures. That is, none of the studies that included population ex-

TABLE 5-4 Estimated Bone Fluoride Concentrations and Bone Fracture Risks in Randomized Clinical Trials and Observational Studies

Reference	Fluoride Exposure (mg/day)	Average Length Exposure (years)	Estimated Total Exposure (g)	Estimated Bone Fluoride (mg/kg ash)	Relative Risks (RR) or Odds Ratios (OR)[a] and (95% CI)
Randomized clinical trials					
Reginster et al. 1998 (Belgium)	20	3.4	24	5,400[b]	1.1 (0.5, 2.4) RR nonvertebral, 4 years
Pak et al. 1995 (USA)	23 (slow-release)	3.1	25	5,500[b,c]	0.6 (0.2, 2.5) RR nonvertebral, 4.7 years
Kleerekoper et al. 1991 (USA)	34	2.4	30	6,200[b]	1.5 (0.7, 3.5) RR nonvertebral, 4 years
Riggs et al. 1990 (USA)	34	3.1	38	7,400[b] (12,000)[d]	1.6 (1.0, 2.5) RR complete nonvertebral, 4 years 2.5 (1.7, 3.7) RR total nonvertebral, 4 years 2.3 (0.6, 8.8) RR complete hip, 4 years
Observational studies					
Sowers et al. 1991 Baseline age 55 to 80 (4 mg/L area)	4.88[e]	35.9[f]	64	7,200[g]	2.1 (1.0, 4.4) OR any fracture, 5 years 2.2 (1.0, 4.7) OR hip/wrist/spine, 5 years
Sowers et al. 2005 Age 20 to 92 (4 mg/L area)	3.66	NA[h]	NA[h]	NA[h]	2.55 (P = 0.07) OR osteoporotic, ecologic 1.16 (P = 0.66) OR osteoporotic, log serum concentration
Li et al. 2001					
2.62 to 3.56 mg/L	7.85[i]	64[i]	180	>6,200[g]	1.18 (P = 0.35) OR, any site since age 20 1.73 (P = 0.34) OR, hip since age 20
4.32 to 7.97 mg/L	14.1[i]	61[i]	320	>11,000[g]	1.47 (P = 0.02) OR, any site since age 20 3.26 (P = 0.02) OR, hip since age 20

[a]When applied to cohort data, ORs overestimate RRs; the bias is small when odds are small as they are here.

[b]Estimated using regression model for clinical trials (Chapter 3) based on total exposure. Bone concentrations for U.S. studies may be underestimated because of background exposure.

[c]Possibly overestimated because of the use of a less bioavailable form of fluoride.

[d]Approximate bone concentration measured in a subset exposed for 4 years.

[e]Average estimated fluoride intake for ages 55 to 80 in 4-mg/L area from Sowers et al. (1986).

[f]Average residence time from Sowers et al. 1986 (baseline) plus 5 years.

[g]Estimated using regression model for water exposure (Chapter 3). Because of limitations in the regression model, these estimates do not take into account differences in water consumption. This may cause underestimation of bone fluoride concentrations for people with high water consumption rates, as estimated for participants in Li et al. (2001).

[h]Average length of exposure not available. Based on water fluoride concentrations alone, the average estimated bone concentrations are about 6,700 mg/kg ash (Chapter 3).

[i]Estimated exposures for these groups are from Li et al. (2001).

[j]Average exposure length equals average age, based on lifetime exposure.

posures above 4 mg/L found null or negative (inverse) associations between fluoride and bone fractures. There is probably minimal publishing bias here because of the intense interest on both sides of the fluoride controversy. Further, all the studies with exposure categories of approximately 2 mg/L and above in water showed elevated relative risks of fractures for these exposure estimates. However, the one study using an individual exposure measure found no association between fracture risk and serum fluoride. Because serum fluoride concentrations may not be a good measure of bone fluoride concentrations or long-term exposure, the ability to show an association might have been diminished.

• Strength of association: Although weak associations (e.g., small relative risks) can be causal, it is harder to rule out undetected biases. As indicated in Table 5-2, relative risk estimates generally varied from about 1.5 to 2.2 for studies with ecologic measures of exposure.

• Biologic plausibility/coherence: The weight of evidence of observational studies is increased when qualitative as well as quantitative; biochemical, physiological, and animal data suggest a biologically plausible mechanism by which a potential risk factor such as fluoride could cause adverse effects. In this instance, the type of physiological effect of fluoride on bone "quality" and the fractures observed in animal studies are consistent with the effects found in the observational studies. Furthermore, the results of the randomized clinical trials are consistent with an increased risk of nonvertebral fractures at fluoride concentrations in bone that may be reached by lifetime exposure to water at 4 mg/L.

• Dose-response (biological gradient): For the most part, the observational studies discussed above observed higher fracture risk with higher exposure compared with 1 mg/L. The combined findings of Kurttio et al. (1999), Alarcón-Herrera et al. (2001), and Li et al. (2001) lend support to gradients of exposure and fracture risk between 1 and 4 mg/L.

The remaining traditional guidelines of Hill and others are not major issues here: time sequence of effect after exposure is fulfilled in all the observational studies and the clinical trials; none of those designs was cross-sectional and all were able to assess sequence. Specificity of effect or exposure is rarely germane in environmental epidemiology. Experiment (that is, effect of removal of exposure) does not apply in this instance.

When papers using different designs or studying disparate populations are evaluated, findings of consistency among these studies do not require that the doses, exposures, or relative risks be the same. (Such quantitative reconciliation is pertinent for efforts to establish unit risks for quantitative risk assessment, pooling studies, or meta-analyses, and assignment of specific potencies goes far beyond the charge or assessment by the committee.) Further, it is not necessary that there be exact quantitative correspondence between animal and human data and physiologic, and epidemiologic exposures.

The weight of evidence supports the conclusion that lifetime exposure to fluoride at drinking water concentrations of 4 mg/L and higher is likely to increase fracture rates in the population, compared with exposure to fluoride at 1 mg/L, particularly in some susceptible demographic groups that are prone to accumulating fluoride into their bones.

Studies Relevant to Assessing Risks at 2 mg/L

The committee found four observational studies that involved exposures to fluoride around 2 mg/L (see Table 5-5). By far the strongest of those studies was by Kurttio et al. (1999). As described above, residents were grouped into exposure categories based on modeled fluoride concentrations in well water closest to their residence (≤0.1, 0.11 to 0.30, 0.31 to 0.50, 0.51 to 1.00, 1.10 to 1.50, and >1.5 mg/L [ranged up to 2.4 mg/L]) and hospital discharge registers were tracked for reports of hip fracture. Whereas no associations between fluoride content in drinking water and hip fracture were found for all ages combined, analysis of age strata (50 to 64 and 65 to 80 years) indicated an adjusted rate ratio of 2.09 (95% CI, 1.16 to 3.76) for hip fracture in women aged 50 to 64 exposed to fluoride at >1.5 mg/L.

Another study, performed in Finland, found no evidence of increased risk when hip fracture rates were compared in populations exposed to fluoride at ≤0.3, 1.0 to 1.2, and >1.5 mg/L (Arnala et al. 1986). However, this study had many weaknesses, including incomplete reporting methods, insufficient control of confounding, inability to assess cumulative exposure, and the possibility of nonsystematic or biased case ascertainment. It focused primarily on evaluating fluoride content and the histomorphometry of bone samples taken from the iliac crest of hip fracture patients and had the advantage of providing data on bone fluoride concentrations. Mean fluoride concentrations (± standard deviation) in bone were found to be 450 ± 190 mg/kg, 1,590 ± 690 mg/kg, and 3,720 ± 2,390 mg/kg in the low-, middle-, and high-exposure groups, respectively.

A study in France investigated fracture rates in relation to fluoride-using subjects enrolled in a different study on aging (Jacqmin-Gadda et al. 1995). Two fluoride exposure groups were compared: 0.05 to 0.11 mg/L and 0.11 to 1.83 mg/L. The odds ratio for hip fractures for the higher exposure group was 1.86 (95% CI, 1.02 to 3.36). The odds ratio for any fractures was 0.98 (95% CI, 0.80 to 1.21). These odds ratios were adjusted for age, gender, and Quetelet index for hip fractures and by age and gender for total fractures. (The authors selected confounders to include in their model on the basis of "statistical significance," although a more appropriate approach would have been to select covariates based on how much they change the odds

TABLE 5-5 Studies Relevant to Assessing Bone Fracture Risks from Exposure to Fluoride at 2 mg/L in Drinking Water

	Arnala et al. (1986)	Jacqmin-Gadda et al. (1995)
Design	Semiecologic; individual outcome data and ecologic exposure measure	Nested case control analysis drawn from cross-section study that was the first phase of a prospective cohort study.
Location	Finland, communities	France
No. subjects	462 fractures among a population of unspecified size	3,777 subjects age 65 and older from 75 civil parishes (mean residence time 41 years)
Exposure assessment and categories	Ecologic; exposure assignments drawn from a 1974 report by the National Board on Health on the fluoride content of drinking water in different communities. Communities with fluoride at <0.3 mg/L, 1.0 to 1.2 mg/L, and >1.5 mg/L	Two measurements were taken in 1991 and routinely thereafter (frequency not specified). Two exposure categories: 0.05 to 0.11 mg/L and 0.11 to 1.83 mg/L
Outcomes	Hip fractures among men and women combined, for age 50+. Factures due to severe trauma excluded.	Hip fractures
Effect measure	Comparison of age-adjusted 10-year incidence of hip fracture for ages 50+ and component age decades. Binomial t test used to compare age-adjusted hip fracture rates.	OR using multiple logistic regressions, controlling for confounders based on interview data.
Chance	No confidence intervals or P levels were provided.	95% CI and P values given
Confounding	Age-adjustment only. No information on whether women were postmenopausal. No distinction between rates for males and females.	Age, gender, Quetelet index (kg/height^2 in m), smoking, and sports activity

Fabiani et al. (1999)	Kurttio et al. (1999)
Semiecologic; individual outcome data and ecologic exposure measures	Historical cohort.
Two regions of central Italy Avezzano (lower fluoride in water) and Bracciano(higher fluoride in water)	Finland: rural communities nationwide
935 in Avezanno 190 in Bracciano; subjects treated in a public hospital from each region	144,000+
Drinking water sampled twice a year (years not specified), and one summary concentration was assigned to each region as a weighted mean. Avezanno (0.05 mg/L; range 0.040 to 0.058 mg/L; population of about 126,000) Bracciano (1.45 mg/L; range 0.15 to 3.40 mg/L; population of about 73,000)	Groundwater measurements of almost 9,000 wells. Fluoride concentrations estimated for each residence by using weighted medians, smoothed interpolations. Categories: <0.1, 0.1 to 0.3, 0.3 to 0.5, 0.5 to 1.0, 1 to 1.5, and > 1.5 mg/L. Highest category corresponded to sampled concentrations of less than detection level to approximately 6 mg/L.
Fractures at specific anatomical sites, reported by gender	First recorded hip fracture
Rates and 95% CI based on age-adjusted rates per 1,000 person years.	Crude and adjusted rate ratios using Cox regression based on person years, compared with lowest exposure group. Age stratification based on age at start of follow-up period. Fluoride analyzed as categorical and continuous variable.
95% CIs	95% CI around the rate ratio.
Authors relied on similarity of region to control for confounding. Analysis did not stratify or adjust for age, although mean ages of cases are shown (including whether the probabilities of their differences are $P < 0.05$).	Analyzed controlling for age and geographic sector. Age adjustment was conducted within broad strata of 50 to 64 and 65 to 80 years. No information on nutrition, alcohol use, or physical activity.

continued

TABLE 5-5 Continued

	Arnala et al. (1986)	Jacqmin-Gadda et al. (1995)
Results	Age-combined totals similar: 12.4/10,000 in low-fluoride, 11.9/10,000 in fluoridated, and 12.4/10,000 in high-fluoride areas. Component age groups generally similar to each other across exposure groups, except that age 80+ had lower incidence in the high-fluoride area.	For higher versus lower fluoride exposures: OR = 1.86 (1.02 to 3.36), $P = 0.04$ for hip fractures; OR = 0.98 (0.80 to 1.21) for all fractures. ORs adjusted for variables associated with hip fractures (age, gender, Quetelet) or total fractures (age, gender). Calcium in water did not appear to be included in the model.
Overall value of study regarding evaluation fracture risk at 2 mg/L	Weak	Weak
Comments	The paper was primarily devoted to histomorphology and bone fluoride concentrations in iliac crest. The results of that portion of the study are summarized in the accompanying text insofar as they bear on the incidence part of the paper. Incomplete reporting methods; insufficient control of confounding; inability to assess cumulative exposure; possibility of nonsystematic or biased case ascertainment/assignment; adjustment of group level covariate (region) rather than individual-level covariates.	Paper was short (a letter to the editor) and did not have sufficient detail to assess the distribution of fluoride exposure with the higher category; lacked information on age subgroups and on genders; inability to assess cumulative exposure; referent group has very low exposure (<0.11 mg/L).

Fabiani et al. (1999)	Kurttio et al. (1999)
Rates for low-fluoride area were statistically greater compared with Bracciano in the following categories: Females: femoral neck (hip), femur NOS (not otherwise specified), proximal humerus, nose, wrist Males: femoral neck (hip), femur NOS, nose, wrist Specifically for hip fracture (Avezanno/ Bracciano, rate per 1,000 person-years): males, 0.28/0.06, RR = 4.28 (95% CI, 4.16 to 4.40), average ages 70 and 52, respectively; females, 0.75/0.28, RR = 2.64 (95% CI 2.54 to 2.75), average ages 75 and 78, respectively.	For comparisons between the >1.5-mg/L group and the <0.1-mg/L group (ages 50 to 65): Adjusted RR = 2.09 (95 CI, 1.16 to 3.76) in women, RR = 0.87 (95% CI, 0.35 to 2.16) in men. For all ages combined, no associations apparent. For fluoride as a continuous variable: RR = 1.44 (95% CI, 1.12 to 1.86) for women below age 65 at start of follow-up, and RR = 0.75 (95% CI, 0.51 to 1.12) for men in same age stratum (age and region adjusted). Women ages 55 to 69 had the most elevated RR in the continuous-variable analysis. Among separate 5-year components of follow-up period, the results were inconsistent.
Weak	Strong
Serious design and analysis limitations. No data that would inform an assessment of a gradient. The dimension of the reported protective effect is not credible.	Suggestive of hip fracture risk, with continuous gradient from lowest to highest exposures.

ratio.) The committee found that because no data were presented on the distribution of fluoride exposure within the different groups, because data on gender and age were not reported separately, and because no parameters for assessing cumulative exposure were provided, reliable conclusions could not be drawn from this study.

Fabiani et al. (1999) conducted a study in two sociodemographically similar regions in central Italy. One region had fluoride concentrations in drinking water of 0.05 mg/L and the second region had fluoride at 1.45 mg/L. A significantly greater rate of fracture incidence, particularly femur fractures, were found in the low-exposure community. The relative risk was 4.28 (95% CI, 4.16 to 4.40) for males and 2.64 (95% CI, 2.54 to 2.75) for females. These risks were based on age-adjusted rates per 1,000 person-years. However, the number of cases was not provided and the mean age of cases in the two towns varied greatly in some instances. The investigators relied on similarity of regions to control for confounding, but it should be noted that the high-fluoride area included seven towns near Rome, whereas the lower-fluoride area included 35 towns further from Rome. Because of the serious design and analysis limitations of the study, the committee placed little weight on this study.

Overall, the committee finds that the available epidemiologic data for assessing bone fracture risk in relation to fluoride exposure around 2 mg/L is suggestive but inadequate for drawing firm conclusions about the risk or safety of exposures at that concentration. There is only one strong report to inform the evaluation, and, although that study (Kurttio et al. 1999) indicated an increased risk of fractures, it is not sufficient alone to base judgment of fracture risk for people exposed at 2 mg/L. It should be considered, however, that the Li et al. (2001) and Alarcón-Herrera et al. (2001) studies reported fracture increases (although imprecise with wide confidence intervals) between 1 and 4 mg/L, giving support to a continuous exposure-effect gradient in this range.

Skeletal Fluorosis

Excessive intake of fluoride will manifest itself in a musculoskeletal disease with a high morbidity. This pathology has generally been termed skeletal fluorosis. Four stages of this affliction have been defined, including a preclinical stage and three clinical stages that characterize the severity. The preclinical stage and clinical stage I are composed of two grades of increased skeletal density as judged by radiography, neither of which presents with significant clinical symptoms. Clinical stage II is associated with chronic joint pain, arthritic symptoms, calcification of ligaments, and osteosclerosis of cancellous bones. Stage III has been termed "crippling" skeletal fluorosis because mobility is significantly affected as a result of excessive calcifications in joints, ligaments, and vertebral bodies. This stage may also be associated with muscle wasting and neurological deficits due to spinal cord compression. The current MCLG is based on induction of crippling skeletal fluorosis (50 Fed. Reg. 20164 [1985]). Because the symptoms associated with stage II skeletal fluorosis could affect mobility and are precursors to more serious mobility problems, the committee judges that stage II is more appropriately characterized as the first stage at which the condition is adverse to health. Thus, this stage of the affliction should also be considered in evaluating any proposed changes in drinking-water standards for fluoride.

Descriptions of skeletal fluorosis date back to the 1930s, when the pathology was first recognized in India in areas of endemic fluoride exposure (Shortt et al. 1937) and in occupationally exposed individuals in Denmark (Roholm 1937). From an epidemiological standpoint, few cases of clinical skeletal fluorosis have been documented in the United States. Stevenson and Watson (1957) performed a large retrospective study involving 170,000 radiologic examinations[1] in people from Texas and Oklahoma, where many communities have fluoride water concentrations above 4 mg/L. They radiographically diagnosed only 23 cases of fluoride osteosclerosis in people consuming fluoride at 4 to 8 mg/L and no cases in people exposed to less (the number of people exposed in these categories was not provided). The cases (age 44 to 85) did not have unusual amounts of arthritis or back stiffness given their age (details not provided). Eleven had bone density of an extreme degree, and nine had more than minimal calcification of pelvic ligaments. The authors found no relationship between radiographic findings and clinical diagnosis or symptoms (details not provided). Cases were not classified as to the stage of the fluorosis (using the scheme discussed earlier). Based on the information in the paper, the committee could not determine whether stage II fluorosis was present. In a study of 253 subjects, Leone et al. (1955a) reported increased bone density and coarsened trabeculation in residents of a town with fluoride at 8 mg/L relative to another town with fluoride at 0.4 mg/L. Radiographic evidence of bone changes occurred in 10% to 15% of the exposed residents and was described as being slight and not associated with other physical findings except enamel mottling. The high-fluoride town was partially defluoridated in March 1952[2] (Maier 1953; Leone et al. 1954a,b; 1955b), a detail not mentioned in the radiographic study (Leone

---

[1]The number of patients represented by the 170,000 radiological examinations is not given.

[2]Maier (1953) indicates that "regular operation" of the defluoridation plant began March 11, 1952. At least one small pilot plant was operated for an unspecified period prior to that date (Maier 1953). Leone et al. (1954a,b) indicated initial defluoridation to 1.2 mg/L. Likins et al. (1956) reported a mean daily fluoride content of treated water in Bartlett of 1.32 mg/L over the first 113 weeks (27 months), with average monthly fluoride concentrations of 0.98-2.13 mg/L over the 18-month period referred to by Leone et al. (1954a,b; 1955b).

et al. 1955a) but which could have affected its results and interpretation. Leone et al. (1954a,b; 1955b) state that "any significant physiological manifestations of prolonged exposure would not be expected to have regressed materially in the 18 months of partial defluoridation." However, Likins et al. (1956) reported that urinary fluoride concentrations in males fell from means of 6.5 (children) and 7.7 (adults) mg/L before defluoridation to 4.9 and 5.1 mg/L, respectively, after 1 week, 3.5 and 3.4 mg/L, respectively, after 39 weeks, and 2.2 and 2.5 mg/L, respectively, after 113 weeks. These results indicate that, following defluoridation of the water supply, substantial changes in fluoride balance were occurring in the residents, including the apparent remobilization of fluoride from bone.

In patients with reduced renal function, the potential for fluoride accumulation in the skeleton is increased (see Chapter 3). It has been known for many years that people with renal insufficiency have elevated plasma fluoride concentrations compared with normal healthy persons (Hanhijärvi et al. 1972) and are at a higher risk of developing skeletal fluorosis (Juncos and Donadio 1972; Johnson et al. 1979). In cases in which renal disease and skeletal fluorosis were simultaneously present, it still took high concentrations of fluoride, such as from daily ingestion of 4 to 8 L of water containing fluoride at 2 to 3 mg/L (Sauerbrunn et al. 1965; Juncos and Donadio 1972), at least 3 L/day at 2 to 3 mg/L (Johnson et al. 1979), or 2 to 4 L/day at 8.5 mg/L (Lantz et al. 1987) to become symptomatic.

Most recently, the Institute of Medicine evaluated fluoride intake and skeletal fluorosis and was able to find only five reported cases of individuals with stage III skeletal fluorosis in the United States from approximately 1960 to 1997 (IOM 1997). Interestingly, however, a recent report has documented an advanced stage of skeletal fluorosis in a 52-year-old woman consuming 1 to 2 gal of double-strength instant tea per day throughout her adult life (Whyte et al. 2005). Her total fluoride intake was estimated at 37 to 74 mg/day from exposure to fluoride from well water (up to 2.8 mg/L) and instant tea. The report also documented the fluoride content of commercial instant teas and found substantial amounts in most brands. This illustrates the possibility that a combination of exposures can lead to higher than expected fluoride intake with associated musculoskeletal problems. Another case, documented by Felsenfeld and Roberts (1991), indicates the development of skeletal fluorosis from consumption of well water containing fluoride at 7 to 8 mg/L for 7 years. Renal insufficiency was not a factor in this case, but water consumption was considered likely to have been "increased" because of hot weather. Both cases mention joint stiffness or pain, suggesting at least stage II skeletal fluorosis.

From reports from the 1950s through the 1980s, it appears that preclinical bone changes and symptoms of clinical stages I and II may occur with bone concentrations between 3,500 and 12,900 mg/kg (Franke et al.

1975; Dominok et al. 1984; Krishnamachari 1986). The Public Health Service (PHS 1991) has reported that patients with preclinical skeletal fluorosis have fluoride concentrations between 3,500 and 5,500 mg/kg by ash weight. Clinical stage I patients have concentrations in the range of 6,000 to 7,000 mg/kg, stage II patients range from 7,500 to 9,000 mg/kg, and stage III patients have fluoride concentrations of 8,400 mg/kg and greater.[3]

However, a broader review of the literature on bone fluoride concentrations in patients with skeletal fluorosis revealed wider and overlapping ranges associated with different stages of the condition. Tables 5-6 and 5-7 show the reported concentrations of fluoride in bone ash and in bone (dry fat-free material) in cases of skeletal fluorosis. Most authors reported ash concentrations; others reported the dry weight concentrations or both types of results. Because ash contents (fraction of bone remaining in the ash) range widely,[4] the committee did not convert dry weight concentrations to ash concentrations. As reported ranges for various bones in individuals can differ, the tables list the type of bone sampled, distinguishing between measurements of iliac crest or pelvis and other bones.

On the basis of data on fluoride in the iliac crest or pelvis, fluoride concentrations of 4,300 to 9,200 mg/kg in bone ash have been found in cases of stage II skeletal fluorosis, and concentrations of 4,200 to 12,700 mg/kg in bone ash have been reported in cases of stage III fluorosis. The overall ranges for other bones are similar. These ranges are much broader than those indicated by PHS (1991). Baud et al. (1978) showed an overlap in the fluoride content in iliac crest samples between their controls (mean 1,036 mg/kg, range <500 to >2,500) and their cases (mean 5,617 mg/kg, range <2,500 to >10,000). The above ranges overlap the measurements reported by Zipkin et al. (1958), for which no evidence of fluorosis was reported (4,496 ± 2015 and 6,870 ± 1629 mg/kg ash in iliac crest at 2.6 and 4 mg/L, respectively). The expected degree of skeletal fluorosis was not found in two small groups of patients dialyzed with fluoride-containing water, who accumulated average bone-ash fluoride concentrations of 5,000 mg/kg and 7,200 mg/kg (Erben et al. 1984). Some of the cases with the lowest values (e.g., Teotia and Teotia 1973; Pettifor et al. 1989) were known to have hypocalcemia or secondary hyperparathyroidism; many of the industrial case reports described no hypocalcemia. Thus, it appears that fluoride content in bone may be a marker of the risk of skeletal fluorosis. In other words, the likelihood and severity of clinical skeletal fluorosis increase with the

---

[3]According to the sources cited by PHS (1991), these concentrations are based on measurements in iliac crest samples.

[4]From 38% to 60%, calculated from 100% minus the reported fraction lost during ashing (Franke and Auerman 1972); (41.8% standard error 1.94%) for the affected group and 49.9% (standard error 5.34%) for the control group (Krishnamachari 1982); and 32.7% to 68.4% (Zipkin et al. 1958).

TABLE 5-6 Reported Concentrations of Fluoride in Bone Ash in Cases of Skeletal Fluorosis

Stage of Skeletal Fluorosis	Fluoride Concentration in Bone Ash, mg/kg in Bone Ash		Number of Individuals	Reference
	Iliac Crest or Pelvis	Other Bones		
Preclinical stage				
Vague symptoms	4,100 4,300		2	Franke and Auermann 1972
Vague symptoms	3,500 to 4,500		Authors' summary	Franke et al. 1975
Stage 0 to 1				
Stage 0 to I	5,000		1	Franke and Auermann 1972
Stage 0 to I	6,900 (mean)		2	Schlegel 1974
Stage 0 to I	5,000 to 5,500		Authors' summary	Franke et al. 1975
Stage 1				
Stage I	6,000 6,400		2	Franke and Auermann 1972
Stage I	5,200 (mean)		8	Schlegel 1974
Stage I	6,000 to 7,000		Authors' summary	Franke et al. 1975
Stage 2				
Second phase	9,200	3,100 to 9,900	1	Roholm 1937
Stage I to II	8,700		1	Franke and Auermann 1972
Stage II	7,700 7,800		2	Franke and Auermann 1972
Stage II	7,500 (mean)		9	Schlegel 1974
Stage II	7,500 to 9,000		Authors' summary	Franke et al. 1975
Stage II	4,300 4,700[a]	2,500 to 5,000	1	Dominok et al. 1984
Stage II	8,800 8,900[a]	4,900 to 11,100	1	Dominok et al. 1984
Stage II		2,900 to 4,400	1	Dominok et al. 1984
Stage 3				
Third phase		7,600 to 13,100	1	Roholm 1937
Stage 3		6,300	1	Singh and Jolly 1961
Stage III		11,500	1	Franke and Auermann 1972
Crippling fluorosis	4,200		1	Teotia and Teotia 1973
Stage III	8,400		1	Schlegel 1974
Stage III	>10,000		Authors' summary	Franke et al. 1975

TABLE 5-6 Continued

Stage of Skeletal Fluorosis	Fluoride Concentration in Bone Ash, mg/kg in Bone Ash		Number of Individuals	Reference
	Iliac Crest or Pelvis	Other Bones		
Stage III	10,000	9,000 to 11,700	1	Dominok et al. 1984
Stage III	9,100	4,200 to 11,000	1	Dominok et al. 1984
Stage III	12,700	7,600 to 12,900	1	Dominok et al. 1984
Stage III	8,600 8,700^a	8,500 to 12,400	1	Dominok et al. 1984
Stage not given, or range of stages				
Skeletal fluorosis		700 to 6,800^b (mean, 3,430)	10	Singh and Jolly 1961; see also Singh et al. 1961
Old fluorosis, 7 years without fluoride exposure	3,000		1	Franke and Auermann 1972
Skeletal fluorosis	2,650 3,780 4,750 5,850		4	Teotia and Teotia 1973
Industrial fluorosis	5,617 (2,143)^c		43 (54 samples)	Baud et al. 1978; Boillat et al. 1980
Endemic genu valgum		7,283 (416)^d	20 (37 samples)	Krishnamachari 1982
Skeletal fluorosis	4,200 to 10,100		9	Boivin et al. 1986
Skeletal fluorosis	13,300 (2,700)^c 8,900 (3,400)^c 6,900 (1,900)^c 5,600 (2,100)^c 6,600 (2,700)^c 7,600 (4,800)^c		6 5 13 54 4 14	Boivin et al. 1988 (summary of studies^e)
Skeletal fluorosis	7,900 (3,600)^c (range: 4,200 to 22,000)		29	Boivin et al. 1989; 1990^f
Admitted to hospital for skeletal pain or skeletal deformities	5,580 (980)^c (range: 4,430 to 6,790)		7	Pettifor et al. 1989

^{a}Samples from right and left sides in same individual.
^{b}Tibia or iliac crest; includes 1 case of stage III fluorosis listed separately above.
^{c}Indicates mean and standard deviation.
^{d}Indicates mean and standard error.
^{e}Includes some studies (or individuals from studies) listed separately above.
^{f}Probably includes individuals from other studies listed above.

TABLE 5-7 Reported Concentrations of Fluoride in Bone (Dry Fat-Free Material) in Cases of Skeletal Fluorosis

Stage of Skeletal Fluorosis	Fluoride Concentration in Bone, mg/kg in Dry Fat-Free Material		Number of Individuals	Reference
	Iliac Crest or Pelvis	Other Bones		
Preclinical stage				
Vague symptoms	1,700 and 2,100		2	Franke and Auermann 1972
Stage 0 to 1				
Stage 0 to I	1,900		1	Franke and Auermann 1972
Stage 0 to I	3,000 (mean)		5	Schlegel 1974
Stage 1				
Early		5,000 to 7,000	1	Wolff and Kerr 1938 (cited in Jackson and Weidmann 1958)
Early		6,260 and 7,200	2	Sankaran and Gadekar 1964
Stage I	2,300 and 2,900		2	Franke and Auermann 1972
Stage I	3,200 (mean)		15	Schlegel 1974
Stage 2				
Moderate		7,680	1	Sankaran and Gadekar 1964
Stage I to II	4,300		1	Franke and Auermann 1972
Stage II	4,100 and 4,600		2	Franke and Auermann 1972
Stage II	3,000 (mean)		18	Schlegel 1974
Stage 3				
Skeletal fluorosis		8,600	1	Sankaran and Gadekar 1964
Advanced		8,800 and 9,680	2	Sankaran and Gadekar 1964
Stage III	3,600 (mean)		4	Schlegel 1974
Stage not given				
Old fluorosis, 7 years without fluoride exposure	1,700		1	Franke and Auermann 1972

bone fluoride content, but a given concentration of bone fluoride does not necessarily correspond to a certain stage of skeletal fluorosis in all cases. Other factors (e.g., calcium intake) appear to influence fluorosis severity at different concentrations of bone fluoride.

Overall, the committee finds that the predicted bone fluoride concentrations that can be achieved from lifetime exposure to fluoride at 4 mg/L (10,000 to 12,000 mg/kg bone ash) fall within or exceed the ranges of concentrations that have been associated with stage II and stage III skeletal fluorosis. Based on the existing epidemiologic literature, stage III skeletal fluorosis appears to be a rare condition in the United States. As discussed above, the committee judges that stage II skeletal fluorosis is also an adverse health effect. However, the data are insufficient to provide a quantitative estimate of the risk of this stage of the affliction. The committee could not determine from the existing epidemiologic literature whether stage II skeletal fluorosis is occurring in U.S. residents who drink water with fluoride at 4 mg/L. The condition does not appear to have been systematically investigated in recent years in U.S. populations that have had long-term exposures to high concentrations of fluoride in drinking water. Thus, research is needed on clinical stage II and stage III skeletal fluorosis to clarify the relationship of fluoride ingestion, fluoride concentration in bone, and clinical symptoms.

EFFECT OF FLUORIDE ON CHONDROCYTE METABOLISM AND ARTHRITIS

The two key chondrocyte cell types that are susceptible to pathological changes are articular chondrocytes in the joint and growth plate chondrocytes in the developing physis. The medical literature on fluoride effects in these cells is sparse and in some cases conflicting.

From physical chemical considerations, it might be expected that mineral precipitates containing fluoride would occur in a joint if concentrations of fluoride and other cations (such as Ca^{2+}) achieved a high enough concentration. A single case report by Bang et al. (1985) noted that a 74-year-old female who was on fluoride therapy for osteoporosis for 30 months had a layer of calcified cartilage containing 0.39% fluoride (or 3,900 mg/kg) by ash weight in her femoral head. The calcification was also visible on x-ray. Unfortunately, the limitation of this observation in a single patient is the lack of information on the preexistence of any calcified osteophytes. Nevertheless, it does indicate that at high therapeutic doses fluoride can be found in mineralizing nodules in articular cartilage.

Studies evaluating patient groups with a greater number of subjects found that the use of fluoride at therapeutic doses in rheumatoid patients showed a conflicting result. In one report (Duell and Chesnut 1991), fluoride exacerbated symptoms of rheumatoid arthritis, but, in another case

(Adachi et al. 1997), it was "well tolerated" with no evidence of worsening of the arthritis. No indications from either study implied that fluoride had a causal relationship with the rheumatoid arthritis. Perhaps the only study in the literature that attempts to link fluoride exposure to the induction of arthritis (osteoarthritis) is from Savas et al. (2001), who indicated that Turkish patients with demonstrated endemic fluorosis had a greater severity of osteoarthritic symptoms and osteophyte formation than age- and sex-matched controls.

The veterinary literature also contains a report indicating that, in 21 dairy herds consuming fluoride-containing feed and water, of the 100 cows examined and determined to have arthritic changes, the bone fluoride concentrations ranged from 2,000 to 8,000 mg/kg (Griffith-Jones 1977).

There are no data from which a dose-response relationship can be drawn regarding fluoride intake and arthritis in humans. However, in a rat study, Harbrow et al. (1992) showed articular changes with fluoride at 100 mg/L in drinking water but no effect at 10 mg/L. The changes with fluoride at 100 mg/L were a thickening of the articular surface (rather than a thinning as would be expected in arthritis) and there were no effects on patterns of collagen and proteoglycan staining. There are no comprehensive reports on the mechanism of fluoride effects in articular chondrocytes in vitro.

The effect of fluoride on growth plate chondrocytes is even less well studied than the effect on articular chondrocytes. It has been demonstrated that chronic renal insufficiency in a rat model can increase the fluoride content in the growth plate and other regions of bone (Mathias et al. 2000); however, this has not been known to occur in humans. Fluoride has also been shown to negatively influence the formation of mineral in matrix vesicles at high concentrations. Matrix vesicles are the ultrastructural particles responsible for initiating mineralization in the developing physis (Sauer et al. 1997). This effect could possibly account, in part, for the observation that fluoride may reduce the thickness of the developing growth plate (Mohr 1990).

In summary, the small number of studies and the conflicting results regarding the effects of fluoride on cartilage cells of the articular surface and growth plate indicate that there is likely to be only a small effect of fluoride at therapeutic doses and no effect at environmental doses.

FINDINGS

Fluoride is a biologically active ion with demonstrable effects on bone cells, both osteoblasts and osteoclasts. Its most profound effect is on osteoblast precursor cells where it stimulates proliferation both in vitro and in vivo. In some cases, this is manifested by increases in bone mass in vivo.

MUSCULOSKELETAL EFFECTS 179

The signaling pathways by which this agent works are slowly becoming elucidated.

Life-long exposure to fluoride at the MCLG of 4 mg/L may have the potential to induce stage II or stage III skeletal fluorosis and may increase the risk of fracture. These adverse effects are discussed separately below.

The current MCLG was designed to protect against stage III skeletal fluorosis. As discussed above, the committee judges that stage II is also an adverse health effect, as it is associated with chronic joint pain, arthritic symptoms, slight calcification of ligaments, and osteosclerosis of cancellous bones. The committee found that bone fluoride concentrations estimated to be achieved from lifetime exposure to fluoride at 2 mg/L (4,000 to 5,000 mg/kg ash) or 4 mg/L (10,000 to 12,000 mg/kg ash) fall within or exceed the ranges historically associated with stage II and stage III skeletal fluorosis (4,300 to 9,200 mg/kg ash and 4,200 to 12,700 mg/kg ash, respectively). This suggests that fluoride at 2 or 4 mg/L might not protect all individuals from the adverse stages of the condition. However, this comparison alone is not sufficient evidence to conclude that individuals exposed to fluoride at those concentrations are at risk of stage II skeletal fluorosis. There is little information in the epidemiologic literature on the occurrence of stage II skeletal fluorosis in U.S. residents, and stage III skeletal fluorosis appears to be a rare condition in the United States. Therefore, more research is needed to clarify the relationship between fluoride ingestion, fluoride concentrations in bone, and stage of skeletal fluorosis before any firm conclusions can be drawn.

Although a small set of epidemiologic studies were useful for evaluating bone fracture risks from exposure to fluoride at 4 mg/L in drinking water, there was consistency among studies using ecologic exposure measures to suggest the potential for an increased risk. The one study using serum fluoride concentrations found no appreciable relationship to fractures. Because serum fluoride concentrations may not be a good measure of bone fluoride concentrations or long-term exposure, the ability to shown an association might have been diminished. Biochemical and physiological data indicate a biologically plausible mechanism by which fluoride could weaken bone. In this case, the physiological effect of fluoride on bone quality and risk of fracture observed in animal studies is consistent with the observational evidence. Furthermore, the results of the randomized clinical trials were consistent with the observational studies. In addition, a dose-response relationship is indicated. On the basis of this information, all members of the committee agreed that there is scientific evidence that under certain conditions fluoride can weaken bone and increase the risk of fractures. The majority of the committee concluded that lifetime exposure to fluoride at drinking-water concentrations of 4 mg/L or higher is likely to increase fracture rates in the population, compared with exposure at 1 mg/L, particularly in some susceptible demographic groups that are more prone to accumulate fluoride in their bones. However, three of the 12 members judged that the evidence only supported a conclusion that the MCLG *might not* be protective against bone fracture. They judge that more evidence that bone fractures occur at an appreciable frequency in human populations exposed to fluoride at 4 mg/L is needed before drawing a conclusion that the MCLG is *likely* to be not protective.

Few studies have assessed fracture risk in populations exposed to fluoride at 2 mg/L in drinking water. The best available study was from Finland, which provided data that suggested an increased rate of hip fracture in populations exposed to fluoride at >1.5 mg/L. However, this study alone is not sufficient to determine the fracture risk for people exposed to fluoride at 2 mg/L in drinking water. Thus, the committee finds that the available epidemiologic data for assessing bone fracture risk in relation to fluoride exposure around 2 mg/L are inadequate for drawing firm conclusions about the risk or safety of exposures at that concentration.

RECOMMENDATIONS

- A more complete analysis of communities consuming water with fluoride at 2 and 4 mg/L is necessary to assess the potential for fracture risk at those concentrations. These studies should use a quantitative measure of fracture such as radiological assessment of vertebral body collapse rather than self-reported fractures or hospital records. Moreover, if possible, bone fluoride concentrations should be measured in long-term residents.

- The effects of fluoride exposure in bone cells in vivo depend on the local concentrations surrounding the cells. More data are needed on concentration gradients during active remodeling. A series of experiments aimed at quantifying the graded exposure of bone and marrow cells to fluoride released by osteoclastic activity would go a long way in estimating the skeletal effects of this agent.

- A systematic study of stage II and stage III skeletal fluorosis should be conducted to clarify the relationship of fluoride ingestion, fluoride concentration in bone, and clinical symptoms. Such a study might be particularly valuable in populations in which predicted bone concentrations are high enough to suggest a risk of stage II skeletal fluorosis (e.g., areas with water concentrations of fluoride above 2 mg/L).

- More research is needed on bone concentrations of fluoride in people with altered renal function, as well as other potentially sensitive populations (e.g., the elderly, postmenopausal women, people with altered acid-balance), to better understand the risks of musculoskeletal effects in these populations.

6

Reproductive and Developmental Effects of Fluoride

This chapter provides an update on studies of the reproductive and developmental effects of fluoride published since the earlier NRC (1993) review. Studies on reproductive effects are summarized first, primarily covering structural and functional alterations of the reproductive tract. This is followed by a discussion of developmental toxicity in animal and human studies.

REPRODUCTIVE EFFECTS

More than 50 publications since 1990 have focused on the reproductive effects of fluoride. Most of the studies used animal models, primarily rodents, and evaluated structural or functional alterations in the male reproductive tract associated with fluoride. Fewer animal studies evaluated the effects of fluoride on female reproductive tract structure or function. In this section, reports of fluoride effects on reproduction in animal models are reviewed first, followed by a discussion of the available studies of humans.

Animal Studies

The large number of studies gleaned from a search of the literature since 1990 that evaluated reproductive tract structure or function in animal models are outlined in Table 6-1, listing the fluoride dosing regimens and main observations. Most of the studies were conducted for the purpose of hazard identification and involved high doses of fluoride to reveal potentially sensitive reproductive-tract targets and pathways. A few selected

TABLE 6-1 Reproductive Toxicity Studies

Species, Sex, Number	Exposure Route	Concentration/ Dose	Exposure Duration	Effects	Reference
Mice, F, 15/group	Gavage	10 mg/kg/day (NaF)	30 days	Decreased protein in liver, muscle, and small intestine were observed. Significant accumulation of glycogen in gastrocnemius muscle and liver. Decline in succinate dehydrogenase activity in pectoralis muscle of treated mice. Administration of ascorbic acid and calcium to NaF-treated mice caused significant recovery from fluoride toxicity.	Chinoy et al. 1994
Mice, F, 25/group	Orally, feeding tube attached to hypodermic syringe	5 mg/kg/day (NaF)	45 days	Fluoride concentrations were increased in the urine, serum, and ovary compared with controls. In the ovary, there was impaired production of glutathione and impaired function of the protective enzymes—namely, glutathione peroxidase, superoxide dismutase, and catalase. There was increased ovarian lipid peroxidation. Enhanced concentrations of potassium and sodium were observed in the serum. The concentrations of serum calcium showed significant depletion. Withdrawal of NaF for 45 days showed partial recovery. Recovery was enhanced by treatment with ascorbic acid, calcium, vitamin E, and vitamin D.	Chinoy and Patel 1998
Mice, F, 20/group	Gavage	10 mg/kg/day (NaF)	30 days	Significant decline of ovarian protein and 3β- and 17β-hydroxysteroid dehydrogenase activities. Hypocholesterolemic effect in serum detected. Accumulation of glycogen in uterus.	Chinoy and Patel 2001

Mice, M, 40/group	Drinking water	10, 20 mg/kg/day (NaF)	30 days	Epithelial-cell pyknosis and absence of luminal sperm were observed. Disorganization of germinal epithelial cells of seminiferous tubules with absence of sperm in the lumina. Reduction in denudation of cells, epithelial cell height, nuclear pyknosis, and absence of sperm observed in the cauda epididymis. The vas deferens epithelium showed clumped sterocilia, nuclear pyknosis, and cell debris but no sperm in the lumen and an increase in the lamina propria. Marked recovery was observed with withdrawal of treatment. No effects observed in the prostate gland or seminal vesicles.	Chinoy and Sequeira 1989
Mice, M, 20/group	Gavage	10, 20 mg/kg/day (NaF)	30 days	NaF caused lessened fertility rate when normal cycling female mice were mated with treated mice. Large numbers of deflagellated spermatozoa with acrosomal, midpiece, and tail abnormalities were observed. Significant recovery in sperm count, sperm motility, and fertility rate was observed after withdrawal of treatment for 2 months.	Chinoy and Sequeira 1992
Mice, M, 20/group	Gavage	10 mg/kg/day (NaF)	30 days	Alterations in epididymal milieu as elucidated by the significant decrease in concentrations of sialic acid and protein as well as activity of ATPase in epididymides. Significant decrease in body and epididymis weight. Weight of vas deferens and seminal vesicle were not affected. Sperm maturation process was affected, leading to decline in cauda epididymal sperm motility and viability. Significant reduction in fertility rate and cauda epididymal sperm count. Treatment induced substantial metabolic alterations in the epididymides, vas deferens, and seminal vesicles of mice. Supplements of vitamin D and E during the withdrawal period enhanced recovery of all NaF-induced effects.	Chinoy and Sharma 1998

continued

TABLE 6-1 Continued

Species, Sex, Number	Exposure Route	Concentration/ Dose	Exposure Duration	Effects	Reference
Mice, M, 20/group	Gavage	10 mg/kg/day (NaF)	30 days	Significant decline in sperm acrosomal acrosin and hyaluronidase. Acrosomal damage and deflagellation observed. Sperm nuclear integrity not affected. Structural and metabolic alterations and reduced activity of the enzymes in sperm resulted in a significant decrease in sperm count and poor fertility rate. Cessation of NaF treatment for 30 days did not bring about complete recovery. Administration of ascorbic acid or calcium enhanced recovery and was more pronounced in groups treated with both ascorbic acid and calcium.	Chinoy and Sharma 2000
Mice, M, 10/group	Drinking water	100, 200, 300 mg/L (NaF) Mean doses during 4-week treatment: 12.53, 21.80, 39.19 mg/kg/day Mean doses during 10-week treatment: 8.85, 15.64, and 27.25 mg/kg/day	4 and 10 weeks	Fertility reduced significantly at 100, 200, and 300 mg/L after 10 weeks but not after 4 weeks. Implantation sites and viable fetuses were significantly reduced in females mated with males that had ingested NaF at a concentration of 200 mg/L for10 weeks. Relative weights of seminal vesicles and preputial glands were significantly increased in animals exposed to NaF 200 and 300 mg/L for 4 weeks but not in animals exposed for 10 weeks.	Elbetieha et al. 2000

Subject	Route	Dose	Duration	Effects	Reference
Rat, F, 25 (treated), 18 (control)	Drinking water	150 mg/L (NaF)	From 60 days before mating and through pregnancy and lactation	There was inhibition of lactation in rats with chronic fluorosis, as measured by slower rates of body weight gain in pups and lower amount of milk suckled in 30 minutes compared with control pups. Prolactin concentration was decreased in serum but increased in the pituitary gland. Microscopic examination showed accumulation of large mature secretory granules and appearance of extremely large abnormal secretory granules in lactotroph cytoplasma.	Yuan et al. 1994
Rat, F, 33-35/group	Drinking water	10, 25, 100, 175, 250 mg/L (NaF) Mean doses: 1.4, 3.9, 15.6, 24.7, and 25.1 mg/kg/day (NaF)	From day of sperm detection to gestation day 20.	Significant reductions in maternal water consumption in the two highest dose groups and a significant reduction in maternal feed consumption in the high-dose group. Body weights of dams were reduced in the higher-dose groups. No significant effect on any reproductive end points. Developmental effects of fluoride were minimal, with 250 mg/L (25.1 mg/kg/day being the lowest observed effect level due to skeletal variations).	Collins et al. 1995
Rat, F, 10/group	Drinking water	200, 400, and 600 mg/L (NaF) Mean doses: 22.58, 18.35, and 28.03 mg/kg/day (NaF)	30 days, before mating	None of the rats in the 28.03 mg/kg/day group survived the study period, and only three survived from the 18.35 mg/kg/day group. Clinical signs of toxicity (dehydration, lethargy, hunched posture) were observed in these groups. All the rats exposed to 22.58 mg/kg/day survived, and showed no signs of toxicity. Fetotoxicity observed at 22.58 mg/kg/day. Reduced number of viable fetuses, increased number of pregnant rats with resorptions, and increased total number of resorptions.	Al-Hiyasat et al. 2000
Rat, F, 10/group	Gavage	40 mg/kg/day (NaF)	Days 6 to 19 of gestation	Significant reductions in body weight, feed consumption, absolute uterine weight, and number of implantations. Significantly higher incidence of skeletal and visceral abnormalities. When NaF was administered with vitamin C, the total percentage of skeletal and visceral abnormalities was significantly lower compared with the group treated with NaF only. Vitamin E also had that effect but was not as great as vitamin C.	Verma and Guna Sherlin 2001

continued

TABLE 6-1 Continued

Species, Sex, Number	Exposure Route	Concentration/ Dose	Exposure Duration	Effects	Reference
Rat, M, 15-20/group	Single microdose injection into the vasa deferentia	50 µg/50 µL (NaF)	Single dose injection	Arrest of spermatogenesis and absence of spermatozoa in the lumina of the seminiferous tubules of the testes. This resulted in a decline in sperm count in caudae epididymides. Deflagellation and tail abnormalities were observed.	Chinoy et al. 1991a
Rat, M, 12/group	Drinking water	5 and 10 mg/kg/day (NaF)	30 days	Succinate dehydrogenase activity in the testes, adenosine triphosphatase activity, and sialic acid concentrations in epididymides in testes were inhibited. A more pronounced effect was observed on the cauda epididymis. Testicular cholesterol and serum testosterone concentrations were not affected. Significant decline in fertility attributed to decreased sperm motility and count.	Chinoy et al. 1992
Rat, M, 14/group	Drinking water	100 and 200 mg/L (NaF)	6 and 16 weeks	Severalfold increase in fluoride concentrations in the testes and bone at both test concentrations compared with controls. Fifty percent of the rats in both treatment groups exhibited histopathologic changes in the germinal epithelium of the testes after 16 weeks. Concentrations of copper and manganese in the testes, liver, and kidneys were not changed. Iron concentrations in the testes and plasma were not affected by fluoride but were increased in the liver, kidneys, and bone. Concentrations of zinc in the testes, plasma, liver, and kidneys decreased significantly, particularly in the 16-week groups. Zinc tended to increase in the bone.	Krasowska and Wlostowski 1992

Species	Route	Dose	Duration	Effects	Reference
Rat, M, 25-30/group	Gavage	10 mg/kg/day (NaF)	50 days	After 50 days of treatment, sperm acrosomal hyaluronidase and acrosin were reduced. Other observations included acrosomal damage and deflagellation of sperm, decline in sperm motility, decreased cauda epididymal sperm count, and reduced fertility. Incomplete recovery observed at withdrawal of NaF treatment for 70 days. Ascorbic acid and calcium produced significant recovery of NaF-induced effects.	Narayana and Chinoy 1994a
Rat, M, 10/group	Drinking water, administered before feeding	10 mg/kg/day (NaF)	50 days	No significant change in testicular cholesterol concentrations. Testicular 3β-HSD and 17β-HSD activities were modestly decreased by NaF ingestion. Histomorphometric analyses indicated a significant change in the Leydig cell diameter in correlation with androgen concentrations.	Narayana and Chinoy 1994b
Rat, M, 10-30/group	Gavage	10 mg/kg/day (NaF)	30 and 50 days	Significant elevation in serum fluoride concentrations (3.6 ± 0.11 ppm) with a simultaneous rise in sperm calcium. Treatment resulted in structural and metabolic alterations in sperm, leading to low sperm motility, low sperm mitochondrial activity index, reduced viability, and changes in sperm membrane phospholipids. A significant reduction in electrolyte concentrations of sperm was observed. Protein concentrations in cauda epididymal sperm suspension, vas deferens, seminal vesicle, and prostate significantly decreased after treatment. Glycogen accumulated in vas deferens and fructose decreased in seminal vesicles and vas deferens.	Chinoy et al. 1995
Rat, M, 18/group	Drinking water	100, 200 mg/L (NaF)	2, 4, 6 weeks	Serum testosterone concentration decreased with time in exposed rats. Testis cholesterol concentration was significantly decreased in the liver of rats exposed 4 and 6 weeks.	Zhao et al. 1995
Rat, M, 24/group	Injection, left testis	50, 175, 250 ppm (NaF)	Single injection	Seminiferous tubule damage observed in vehicle-injected control and exposed testes; no damage was observed in noninjected testes. Polymorphonuclear leukocyte infiltration was observed at injection site in both vehicle- and fluoride-injected groups after 24 hours. No effect on Leydig cells.	Sprando et al. 1996

continued

TABLE 6-1 Continued

Species, Sex, Number	Exposure Route	Concentration/ Dose	Exposure Duration	Effects	Reference
Rat, M, 12/group	Drinking water	25, 100, 175, 200 mg/L (NaF)	14 weeks (10 weeks pretreatment, 3 weeks mating, 1 week postmating)	No effects were observed within the P generation males and the F_1 generation groups in testis weights, prostate/seminal vesicle weights, nonreproductive organ weights, testicular spermatid counts, sperm production per gram of testis per day, sperm production per gram of testis, lutenizing hormone, follicle-stimulating hormone, or serum testosterone concentrations. No histological changes were observed in testicular tissues from either the P or the F_1 generation.	Sprando et al. 1997
Rat, M, 25	Drinking water	25, 100, 175, 250 mg/L (NaF)	In utero, during lactation, 14-weeks post-weaning	No significant effect on absolute volume of the seminiferous tubules, interstitial space, Leydig cells, blood vessel boundary layer, lymphatic space, macrophages, tubular lumen or absolute tubular length and absolute tubular surface area, mean Sertoli cell nucleoli number per tubular cross-section, mean seminiferous tubule diameter, and mean height of the seminiferous epithelium. Statistically significant decrease in the absolute volume and volume percent of the lymphatic endothelium was observed in NaF-treated groups (175 and 250 mg/L) and in the testicular capsule in the NaF-treated group (100 mg/L).	Sprando et al. 1998
Rat, M, F, 36-48/group 3 generations	Drinking water	0, 25, 100, 175, 250 mg/L (NaF)	10 weeks	Decreased fluid consumption observed at 175 and 250 mg/L attributed to decreased palatability; no effect on reproduction. No cumulative effects were observed in any generation. Mating, fertility, and survival, organ-to-body weight ratios, and organ-to-brain ratios were not affected. Treatment up to 250 mg/L did not affect reproduction.	Collins et al. 2001a

Species	Route	Dose	Duration	Effects	Reference
Rat, M, 6/group	Gavage	20 mg/kg/day (NaF)	29 days	Testicular 3β-HSD and 17β-HSD activities were decreased significantly. Substantial reduction in plasma concentrations of testosterone in the exposed group. Decreased epididymal sperm count and fewer mature luminal spermatozoa in the exposed group. NaF treatment was associated with oxidative stress, as indicated by an increased concentration of conjugated dienes in the testis, epididymis, and epididymal sperm pellet. Significant reduction in peroxidase and catalase activities in the sperm pellet in exposed group as compared with controls.	Ghosh et al. 2002
Rat, M, F, 10/group	Gavage	40 mg/kg/day (NaF)	Day 6 of gestation to day 21 of lactation	NaF treatment associated with significant reductions in body weight, feed consumption, concentration of glucose, and protein in the serum. Administration of vitamins C, D, and E helped to restore body weight loss as well as glucose, protein, sodium, and potassium concentrations in the serum of exposed rats. Withdrawal of NaF treatment during lactation caused significant amelioration in feed consumption and in serum sodium, potassium, glucose, and protein concentrations. Additional treatment with vitamin E caused substantial improvements in body weight reductions and in serum concentration of sodium, potassium, glucose, and protein.	Verma and Guna Sherlin 2002a
Rabbit, F, 10/group	Subcutaneous injection	5, 10, 20, 50 mg/kg/day (NaF)	100 days	Abnormal accumulation of lipids in testes observed in treated rabbits. Hyperphospholipidemia, hypertriglyceridemia, and hypercholesterolemia indicated enhanced lipid biosynthesis was observed in response to fluoride toxicosis. Significant ($P < 0.001$) increase in amount of free fatty acids observed in testes of treated animals.	Shashi 1992a
Rabbit, M, 5/group	Feed	20, 40 mg/kg/day (NaF)	30 days	Decline in fertility related to reduced sperm motility and count and changes in morphology and metabolism. No recovery after withdrawal for 30 days from treatment. With administration of ascorbic acid and calcium, marked recovery occurred.	Chinoy et al. 1991b

continued

TABLE 6-1 Continued

Species, Sex, Number	Exposure Route	Concentration/ Dose	Exposure Duration	Effects	Reference
Rabbit, M, 10/group	Drinking water	10 mg/kg/day (NaF)	18 or 29 months	Loss of cilia on the epithelial cells lining the lumen of the ductuli efferentes of the caput epididymidis and of stereocilia on the epithelial cells lining the lumen of the vas deferens was observed. The boundaries of cells peeled off and were not clear in some regions of the epithelial lining of the lumen of the ductuli efferentes and vas deferens. Cessation of spermatogenesis was noted only in rabbits treated for 29 months.	Susheela and Kumar 1991
Rabbit, M, 8/group	Drinking water	10 mg/kg/day (NaF)	18 months	Structural defects in the flagellum, the acrosome, and the nucleus of the spermatids and epididymal spermatozoa were observed in the treated rabbits. Absence of outer microtubules, complete absence of axonemes, structural and numeric aberrations of outer dense fibers, breakdown of the fibrous sheath, structural defects in the mitochondria of the middle piece of the flagellum, and detachment and peeling of the acrosome from the flat surfaces of the nucleus was observed.	Kumar and Susheela 1994
Rabbit, M, 12/group	Drinking water	10 mg/kg/day (NaF)	20 and 23 months	Fluoride concentrations in the sera of treated animals were significantly increased. Loss of stereocilia, significant decrease in the height of the pseudostratified columnar epithelium, and significant increase in the diameter of the caput and cauda ductus epididymis observed in the 23-month treatment group. Weights of the cauda epididymis and caput were significantly reduced in the 23-month-treated animals; the number of secretory granules in those organs was reduced.	Kumar and Susheela 1995

Species	Route	Dose	Duration	Effects	Reference
Rabbit, M, 12/group	Drinking water	10 mg/kg/day (NaF)	18 and 23 months	Fluoride concentrations in the sera were significantly increased in treated rabbits ($P < 0.001$). There was dilation of the smooth endoplasmic reticulum and mitochondrial cristae of the Leydig cells. Leydig cells had lower numbers of lipid droplets and smooth endoplasmic reticulum compared with Leydig cells of unexposed rabbits. Intranuclear filamentous inclusions observed in treated rabbits. Interstitial tissue of the testis was degenerated.	Susheela and Kumar 1997
Guinea pig, M, 10/group	Gavage	30 mg/kg/day (NaF)	30 days	Structural and metabolic alterations of the cauda epididymal spermatozoa led to substantial decreases in sperm mitochondrial activity index, motility, live/dead ratio. Increases in sperm membrane phospholipids were observed. ATPase, succinate dehydrogenase, and glutathione concentrations were decreased in testis of treated animals. Administration of ascorbic acid led to recovery in these parameters.	Chinoy et al. 1997
Sheepdog, F, M, 5/group	Feed	460 ppm (fluoride)	2 years	No adverse effect on reproduction attributable to treatment. Bony exostoses was observed in 4 of 10 treated dogs.	Schellenberg et al. 1990

ABBREVIATIONS: F, female; HSD, hydroxysteroid dehydrogenase; M, male.

examples illustrate the results of the many hazard identification studies: (1) cessation of spermatogenesis and alterations in the epididymis and vas deferens were observed in rabbits administered sodium fluoride (NaF) at 10 milligrams per kilogram (mg/kg) of body weight for 29 months (Susheela and Kumar 1991); (2) effects on Leydig cells and decreased serum testosterone were observed in rats exposed to NaF at 10 mg/kg for 50 days (Narayana and Chinoy 1994b); and (3) decreased protein in the ovary and uterus and decreased activity of steroidogenic enzymes (3β-hydroxysteriod dehydrogenase [HSD] and 17β-HSD) was found in mice treated with NaF at 10 mg/kg for 30 days (Chinoy and Patel 2001). In general, the hazard identification studies show that the reproductive tract is susceptible to disruption by fluoride at a concentration sufficiently high to produce other manifestations of toxicity.

For risk evaluation, a comprehensive multigenerational study of fluoride effects on reproduction using standard guidelines and adequate numbers of animals has been conducted in rats (Collins et al. 2001a). Rats were administered drinking water with NaF at 0, 25, 100, 175, and 250 mg/L over three generations. No compound-related effects were found on mating or fertility; gestation or lactation; or F_1 survival, development, and organ weights. No alterations in the teeth were seen except for mild whitening observed in rats exposed to fluoride at 100 mg/L or greater. That well-conducted study concluded that NaF at concentrations up to 250 mg/L in the drinking water did not alter reproduction in rats (Collins et al. 2001a).

Human Studies

The few studies gleaned from a search of the literature since 1990 that evaluated reproductive effects of fluoride ingestion in humans are outlined in Table 6-2, listing the estimated fluoride exposure and main observations. In highly exposed men with and without skeletal fluorosis (fluoride at 1.5-14.5 mg/L in the drinking water), serum testosterone concentrations were significantly lower than in a control cohort exposed to fluoride at less than 1.0 mg/L in drinking water (Susheela and Jethanandani 1996). Although there was a 10-year difference in the mean ages between the skeletal fluorosis patients (39.6 years) and control subjects (28.7 years), this study suggests that high concentrations of fluoride can alter the reproductive hormonal environment.

In an ecological study of U.S. counties with drinking-water systems reporting fluoride concentrations of at least 3 mg/L (Freni 1994), a decreased fertility rate was associated with increasing fluoride concentrations. Because methods for analyzing the potential amounts and direction of bias in ecological studies are limited, it is possible only to discuss some of the strengths and weaknesses of this complicated study (see Chapter 10 and

Appendix C for a more in-depth discussion of ecologic bias). Freni's study is actually partially ecologic; the outcome (fertility) is age-standardized at the individual level, while exposure to fluoride and covariates are measured at the group level. Controlling for age of the mother is a strength of the study, but to avoid bias all ecological variables should be standardized in the same fashion (Greenland 1992). The model adjusted for a number of important socioeconomic and demographic variables at the group level, but these might not adequately control for individual-level determinants of fertility such as family income and use of contraceptives. For example, median income (a group-level variable) and family income (an individual-level variable) may have independent and interactive effects on outcome. One of the two ecologic exposure measures examined the percentage of the population served by water systems with fluoride concentrations of at least 3 mg/L. That has the potential advantage of not assuming an effect at lower fluoride concentrations (as does the mean fluoride concentration, the other exposure measure), but it has the disadvantage that, unlike individual-level studies, nondifferential misclassification of dichotomous exposures within groups tend to bias ecologic results away from the null (Brenner et al. 1992). While the results of the Freni study are suggestive, the relationship between fertility and fluoride requires additional study.

A study of workers in Mexico, who were occupationally exposed to fluoride (estimated to range from 3 to 27 mg/day) producing hydrofluoric acid and aluminum fluoride, found alterations in serum hormone concentrations with normal semen parameters (Ortiz-Perez et al. 2003). However, that study involved a comparison of a high-fluoride-exposed group and a low-fluoride-exposed group with poorly defined exposures and overlapping exposure characteristics.

Overall, the available studies of fluoride effects on human reproduction are few and have significant shortcomings in design and power, limiting inferences.

DEVELOPMENTAL EFFECTS

There is wide variation with some correlation between fluoride concentration in maternal serum and cord blood, indicating that fluoride readily crosses the placenta. In general, average cord blood concentrations are approximately 60% of maternal serum concentrations, with proportionally lesser amounts present as higher maternal serum concentrations (Gupta et al. 1993; Malhotra et al. 1993; Shimonovitz et al. 1995). Therefore, potential toxicity to the developing embryo and fetus in the setting of high maternal ingestion of fluoride has been a concern evaluated in both animal and humans.

TABLE 6-2 Human Reproductive Studies

Subjects	Exposure Route, Duration	Concentration/Dose
Pregnant women (n = 25)	Drinking water	Maternal blood fluoride concentrations ranging from 0.1 to 2.4 ppm
Pregnant women (n = 25)	Drinking water	Maternal plasma fluoride concentrations ranging from 0.12 to 0.42 µg/mL
Pregnant women undergoing amniocentesis (n = 121, divided into 6 exposure groups)	Oral doses, 24 hours and 3 hours before amniocentesis	0.56, 1.12, 1.68, 2.30, or 2.80 mg of NaF corresponding to 0.25, 0.50, 0.75, 1.00, or 1.25 mg of F-
Men (ages 28-30; n = 8)	In vitro with spermatozoa, intervals of 5, 10, and 20 minutes	25, 50, 250 mM (NaF)
30 regions spread over nine states	Drinking water	≥ 3 mg/L (fluoride)
Pregnant women (n = 22)	Drinking water	Maternal serum fluoride concentrations ranging from 0.003-0.041µg/ml
Men with skeletal fluorosis (n = 30)	Drinking water	1.5-14.5 mg/L (fluoride)
Male workers in Mexico (ages 20-50; n = 126), who produce fluorohydric acid and aluminum fluoride	Drinking water	3-27.4 mg/day (fluoride)

ABBREVIATIONS: FSH, follicle-stimulating hormone.

Results	Reference
Fairly positive correlation ($r = 0.736$) between cord blood values and maternal blood fluoride concentrations. On average, the cord blood fluoride concentration was about 60% that in maternal blood. At a maternal fluoride concentration greater than 0.4 ppm, the cord blood fluoride concentration increased by only about 12%. The placenta was found to serve as an effective barrier within this range.	Gupta et al. 1993
Cord plasma fluoride concentrations ranged from 0.11-0.39 µg/ml. In 8% of the cases, cord plasma concentrations were higher than maternal plamsa concentrations. Positive correlation ($r = 0.97$) in fluoride concentrations between maternal and cord plasma indicates that the placenta allowed passive diffusion of fluoride from mother to fetus.	Malhotra et al. 1993
F-concentration in amniotic fluid was significantly higher than controls in the 1.25 mg/day F-group but not in any of the other exposure groups. No significant correlation between F-concentration in maternal plasma and in aminotic fluid.	Brambilla et al. 1994
Substantial enhancement of acid phosphatase and hyaluronidase activities after 5 and 10 minutes ($P < 0.001$). Decrease in lysosomal enzyme activity after 20 minutes. Analysis of sperm revealed elongated heads, deflagellation, splitting, loss of the acrosome, and coiling of the tail. Glutathione concentrations exhibited time-dependent decrease with complete depletion after 20 minutes ($P < 0.001$). Suppressed sperm motility after 20 minutes at a dose of 250 mM ($P < 0.001$).	Chinoy and Narayana 1994
In this ecological study, there was an association between decreasing total fertility rate and increasing fluoride concentrations in most regions. Combined result was a negative total fertility rate/fluoride association with a consensus combined P value of 0.0002-0.0004. Association was based on population means rather than individual women.	Freni 1994
Cord serum fluoride concentrations ranged from 0.003-0.078 µg/ml, and neonatal serum concentrations ranged from 0.017-0.078 µg/ml. No correlation in fluoride concentrations found between maternal and cord sera, maternal and neonatal sera, or maternal and neonatal sera.	Shimonovitz et al. 1995
Serum testosterone concentrations in patients were significantly lower than controls ($P < 0.01$).	Susheela and Jethanandani 1996
In the high-fluoride exposure group, a significant increase in FSH ($P < 0.05$) and a reduction of inhibin-B, free testosterone, and prolactin in serum ($P < 0.05$) were observed. Decreased sensitivity was found in the FSH response to inhibin-B ($P < 0.05$) when the high-exposure group was compared with the low-exposure group. Significant partial correlation was observed between urinary fluoride and serum concentrations of inhibin-B ($P < 0.028$). No abnormalities were found in the semen parameters in either the high- or low-fluoride exposure groups.	Ortiz-Perez et al. 2003

Animal Studies

Studies gleaned from a search of the literature since 1990 that evaluated developmental toxicity in animal models are outlined in Table 6-3, listing the fluoride dosing regimens and main observations. High-dose hazard identification studies, such as a recently reported *Xenopus* embryo development study using the FETAX assay (Goh and Neff 2003), suggest that developmental events are susceptible to disruption by fluoride.

For risk evaluation, several comprehensive studies of fluoride effects on development using standard guidelines and adequate numbers of animals have been conducted in rats and rabbits (Collins et al. 1995; Heindel et al. 1996; Collins et al. 2001b). Those high-quality studies evaluated fluoride concentrations in drinking water of 0-300 mg/L in rats and 0-400 mg/L in rabbits. Across the studies, there was a trend toward lower maternal body weights and lower maternal intake of food and water at the higher concentrations in both rats and rabbits (250-400 mg/L). Overall, developmental effects of fluoride were minimal, with 250 mg/L in rats being the lowest-observed-adverse-effect level due to skeletal variations (Collins et al. 1995, 2001b). For rabbits, the no-observed-adverse-effect level was >400 mg/L for administration during gestation days 6-19, the period of organogenesis (Heindel et al. 1996).

Human Studies

The few studies gleaned from a search of the literature since 1990 that evaluated developmental effects of fluoride ingestion in humans are outlined in Table 6-4, listing the type of study, estimated fluoride exposure, and main observations. These studies have focused on examining an association between fluoride and three different human developmental outcomes—spina bifida occulta, sudden infant death syndrome, and Down's syndrome. Two small studies have raised the possibility of an increased incidence of spina bifida occulta in fluorosis-prone areas in India (Gupta et al. 1994, 1995); larger, well-controlled studies are needed to evaluate that possibility further. Studies from New Zealand (Mitchell et al. 1991; Dick et al. 1999) found no association between fluoride and sudden infant death syndrome. In one of those studies (Dick et al. 1999), a nationwide case-control database of sudden infant death syndrome was evaluated for fluoride exposure status and controlled for the method of infant feeding (breast or reconstituted formula) with the conclusion that exposure to fluoridated water prenatally or postnatally at the time of death did not affect the relative risk of sudden infant death syndrome.

A small number of ecologic studies have examined Down's syndrome (trisomy 21) prevalence among populations in municipalities with differences in water fluoride concentrations. The possible association of cytogenetic effects with fluoride exposure (see Chapter 10) suggests that Down's syndrome is a biologically plausible outcome of exposure. There are other indications in the literature that environmental exposures could contribute to an increased incidence of Down's syndrome births among younger mothers (Read 1982; Yang et al. 1999; Hassold and Sherman 2000; Peterson and Mikkelsen 2000).[1] There are many difficulties with analyzing the available data on Down's syndrome and fluoride. First, the source of the data on Down's syndrome births must be considered. Sources have included birth certificates, hospital records, and reports from parents. Birth certificates are not an ideal source of data because signs of Down's syndrome are not always readily apparent at birth and the condition, even when diagnosed early, is not always recorded on the birth certificate. Thus, considerable differences can be expected in the data collected when different sources are used to determine the incidence of the disorder. At the present time, the only firm diagnosis of Down's syndrome comes from examination of chromosomes or DNA. Second, the mother's history of exposure to fluoride is difficult to determine. The fact that a woman has a baby in one city does not mean she is from that city or indicate how long she has been in the region. Third, the age of the mother is an important risk factor in the occurrence of children with Down's syndrome; the rates rise exponentially with age.

---

[1]Some fraction of maternal recombination events, prior to the first meiotic division, apparently result in a chromosome 21 tetrad (paired chromosomes each with two chromatids) that is more susceptible to nondisjunction, due to lack of a cross-over or to very proximal or very distal location of the cross-over (Lamb et al. 1996; 1997; Brown et al. 2000; Hassold and Sherman 2000; Petersen and Mikkelsen 2000; Pellestor et al. 2003). Production of the susceptible tetrad occurs during the mother's own fetal development and appears to be age-independent (Lamb et al. 1996; 1997; Brown et al. 2000; Hassold and Sherman 2000; Hassold et al. 2000; Petersen and Mikkelsen 2000). However, the likelihood that the susceptible tetrad will be processed abnormally—i.e., will give rise to nondisjunction rather than segregating normally—appears to be age-dependent, with an increased likelihood of nondisjunction with increased maternal age (Lamb et al. 1996; 1997; Brown et al. 2000; Hassold and Sherman 2000; Hassold et al. 2000; Wolstenholme and Angell 2000; Petersen and Mikkelsen 2000). This age-related effect involves a disturbance of the meiotic process (e.g., failure of the spindle apparatus or degradation of a meiotic protein), inhibition of a DNA repair enzyme, or an environmental exposure (Lamb et al. 1997; Brown et al. 2000; Hassold and Sherman 2000; Petersen and Mikkelsen 2000; Wolstenholme and Angell 2000; Pellestor et al. 2003), and is probably multifactorial (Pellestor et al. 2003). Environmental factors that disrupt the meiotic process could increase the likelihood of Down syndrome births in younger mothers, essentially increasing the likelihood of incorrect segregation of susceptible tetrads to that generally seen in older women. According to Petersen and Mikkelsen (2000), "the findings suggest that aging alone is sufficient to disrupt the meiotic process, whereas in younger women there is a higher requirement for a genetic or environmental factor for nondisjunction to occur." For example, Yang et al. (1999) reported that for a specific type of maternal meiotic error, for younger mothers, there was a significant association with environmental exposures (in this case, maternal smoking, especially in combination with the use of oral contraceptives) around the time of conception.

TABLE 6-3 Developmental Toxicity Studies

Species, Sex, Number	Exposure Route	Concentration/ Dose	Exposure Duration
Rat, F, 33-35/group	Drinking water	0, 10, 25, 100, 175, 250 mg/L (NaF) Mean doses: 0, 1.4, 3.9, 15.6, 24.7, and 25.1 mg/kg/day (NaF)	From day of sperm detection to gestation day 20
Rat, F, 10/group	Drinking water	40 mg/kg/day (NaF)	From day 6 to 19 of gestation
Rat, M, F, 40-50 animals/group from 4 or 5 litters at each age	Intraperitoneal injection	0, 30 and 48 mg/kg (NaF)	Single injection on postnatal day 1, 8, 15, or 29
Rat, M, F, 26/group Rabbit, M, F, 26/group	Drinking water	Rat: 0, 50, 150, 300 mg/L (NaF) (mean doses 6.6, 18.3, and 27.1 mg/kg/day) Rabbit: 0, 100, 200, 400 mg/L (NaF) (mean doses 10.3, 18.1, and 29.2 mg/kg/day)	Rat: from gestational day 6 to 15 Rabbit: from gestational day 6 to 19
Rat, M, F, 3 generations (F0, F_1, F_2), F_0: 48 M, 48 F/group; F_1: 36 M, 36 F/group; F_2: 238 fetuses	Drinking water	0, 25, 100, 175, 250 mg/L (NaF) Mean doses: (F0): 3.4, 12.4, 18.8, 28.0 mg/kg/day (NaF) (F_1): 3.4, 13.2, 19.3, 25.8 mg/kg/day (NaF)	F0: 10 weeks
Frog *(Xenopus)* embryo, 20/group	Incubated with NaF solution	100-1,000 ppm (NaF)	2, 3, 4, 5, 9, 14.75 hours after fertilization

ABBREVIATIONS: EC_{50}, median effective concentration; F, female; LC_{50}, median lethal concentration; M, male; NOAEL, no-observed-adverse-effect level.

Effects	Reference
Significant reductions in maternal water consumption in the two highest-dose groups and a significant reduction in maternal feed consumption in the high-dose group. Body weights of dams were reduced in the higher-dose groups. The only significant developmental effect was an increase in the average number of fetuses with three or more skeletal variations in the 25.1-mg/kg/day group.	Collins et al. 1995
NaF caused significantly lowered body weight, feed consumption, absolute uterine weight, and number of implantations. Higher incidence of skeletal (14th rib, dumbbell-shaped 5th sternebrae, incomplete ossification of skull, wavy ribs) and visceral abnormalities (subcutaneous hemorrhage) in fetuses. Vitamin D treatment improved reductions in body weight, feed consumption, and uterine weight.	Guna Sherlin and Verma 2001
Changes in renal function included decreased body weight after NaF treatment at 30 or 48 mg/kg; increased kidney/body weight ratio in the 48-mg/kg group; decreased urinary pH; decreased chloride excretion in the 48 mg/kg group, and increased urinary volume 120 hours after treatment. Renal toxicity was observed in postweaning day 29 rats. NaF exposure resulted in increased kidney/body weight ratio and kidney weight, profound diuresis, decreased urinary osmolality, and decreased ability to concentrate urine during water deprivation. Decrease in urinary chloride excretion was observed for the first 2 days after exposure; it was increased in water-deprived rats 120 hours after treatment. Hematuria and glucosuria were observed for 2 days after treatment with 48 mg/kg. Renal sensitivity noted after weaning in day 29 rats. Histological lesions noted in proximal tubules of treated day 29 rats.	Datson et al. 1985
In high-dose group, initial decreased body weight gain (recovered over time) and decreased water consumption. No clinical signs of toxicity were observed. In both the rabbit and rat, maternal exposure to NaF during organogenesis did not substantially affect frequency of postimplantation loss, mean fetal body weight/litter, and visceral or skeletal malformations. The NOAEL for maternal toxicity was 18 mg/kg/day (NaF) in drinking water for rats and rabbits. The NOAEL for developmental toxicity was greater than 27 mg/kg/day (NaF) for rats and greater than 29 mg/kg/day for rabbits.	Heindel et al. 1996
No dose-related feed consumption or mean body weight gain in either F_0 or F_1 females. Statistically significant decreases in fluid consumption for F_0 at 250 mg/L and F_1 at 175 and 250 mg/L. Corpora lutea, implants, fetal morphological development, and viable fetuses were similar in all groups. No dose-related anomalies in internal organs were observed in F_2 fetuses. Ossification of the hyoid bone was significantly decreased among F_2 fetuses at 250 mg/L.	Collins et al. 2001b
Reduction in head-tail lengths and dysfunction of the neuromuscular system of the tadpoles. EC_{50} for malformation in growth after exposure to NaF 5 hours after fertilization is 184 ppm. Calculated LC50 is 632 ppm. Values for EC_{50} and LC_{50} met the limits established for a teratogen in frog embryos.	Goh and Neff 2003

TABLE 6-4 Human Developmental Studies

Subjects	Exposure Route, Duration	Concentration/ Dose	Results	Reference
Pregnant women (mean age 29; n = 91), routine examination at 6th month of pregnancy, 4 groups	Oral doses, taken during final trimester of pregnancy	0, 1.5 mg of F (CaF$_2$) per day; 1.5 mg of F (NaF) per day; 0.75 mg of F (NaF) twice per day	Significant difference between cord plasma fluoride concentrations of newborns in untreated group (mean 27.8 µg/L) and of combined supplemented groups (mean 58.3 µg/L).	Caldera et al. 1988
Pregnant women (n = 25)	Drinking water	1.2 mg/L, continuous fluoride concentration in drinking water	Fluoride in maternal plasma varied from 12.00 µg/100 mL to 41.8 µg/100 mL. In cord blood, it ranged from 11.20 µg/100 mL to 38.8 µg/100 mL; 8% of cases showed cord plasma fluoride concentrations higher than that of maternal concentrations. A highly significant correlation was found between the plasma fluoride concentration of maternal and fetal blood ($P < 0.001$).	Malhotra et al. 1993
Children (ages 4-12; n = 30)	Drinking water	4.5-8.5 mg/L (fluoride)	Blood fluoride concentrations of children were 0.9 ppm and 1.1 ppm. Serum fluoride concentrations ranged from 1.6 to 1.9 ppm. Of 30 skiagrams of the lumbosacral region, 14 (47%) showed spina bifida occulta.	Gupta et al. 1994
Pregnant women (n = 22)	Drinking water	0.22-0.49 µg/L (fluoride in drinking water)	Serum fluoride concentrations were 0.018 ± 0.012 µg/mL in mothers, 0.030 ± 0.015 µg/mL for umbilical cord samples, and 0.038 ± 0.016 µg/mL for neonates. Statistically significant differences were found between maternal and cord serum fluoride ($P \leq 0.05$) and between neonatal and cord serum fluoride ($P \leq 0.05$). No statistical difference between maternal and neonatal serum fluoride. No correlation in fluoride concentrations between maternal and neonatal sera, between maternal and cord sera, or between neonatal and cord sera.	Shimonovitz et al. 1995

Fetuses (14-36 weeks of intrauterine life; n = 64)	Drinking water	0.2 mg/L (fluoride concentration in drinking water)	Higher contents of Ca, Mg, and P were disclosed in the diaphyseal part of the bones. Higher concentrations of fluoride were recorded in the metaphysis than in the shaft. Statistically significant correlations between fetal age and content of calcium and phosphorus in the bones; the fluoride contents in the shaft and in the metaphyseal part. No influence of fluoride on calcification of fetal bony tissue.	Mokrzynski and Machoy 1994
Children from India (ages 5-12; n = 50) with dental and/or skeletal fluorosis	Drinking water	≤1.5 (control), 4.5, and 8.5 mg/L (fluoride)	A total of 22 (44%) of the 50 children in the study group, and 6 (12%) of the children in the control group revealed spina bifida occulta in the lumbosacral region. Proportion of children with spina bifida occulta in fluoride-rich areas was 44%.	Gupta et al. 1995
Data for mothers under age 30, Down's syndrome birth rates in five counties of metropolitan Atlanta, Georgia (reanalysis of Erickson 1976)	Drinking water	Not specified; comparison of fluoridated and nonfluoridated communities; authors selected 0.1-0.3 mg/L as a reasonable range assumption for nonfluoridated areas	Highly significant association between fluoridated water and Down's syndrome births ($P < 0.005$) in a selected subset of previously published data.	Takahashi 1998

continued

TABLE 6-4 Continued

Subjects	Exposure Route, Duration	Concentration/ Dose	Results	Reference
Data from literature search on studies of Down's syndrome and exposure to fluoride	Drinking water	Range from all studies was 0-2.8 mg/L	Six ecological studies were included in the evaluation. Crude relative risk ranged from 0.84 to 3.0. Four studies found no significant association between Down's syndrome and water fluoride concentration. Two studies showed increased incidence of Down's syndrome with increased water fluoride concentrations ($P < 0.05$). All the studies scored poorly on the validity assessment. Only two studies controlled for confounding factors, only one of which presented summary outcome measures.	Whiting et al. 2001
Data from literature search on SIDS mortality rate for 1980-1984 in New Zealand	Drinking water	Median fluoridation was ≤1 g/m³	Strong negative correlation between SIDS and mean daily temperature of -0.83 ($P = 0.0001$). Nonsignificant correlation between SIDS and average fluoridation ($P = 0.24$). Mean daily temperature was significant while average fluoridation was not. Daily temperature was a significant predictor of SIDS after removing average fluoridation from the model.	Mitchell et al. 1991
485 postneonatal deaths attributed to SIDS; 1,800 control infants	Drinking water	0.7-1.0 mg/L (artificial) 0.1-0.3 mg/L (natural)	Exposed infants to fluoridated water in utero were not at increased risk for SIDS, adjusted odds ratio 1.19. Fluoridated water was not associated with increased risk for SIDS among breastfed infants. Fluoridated formula feeding, compared with unfluoridated formula, showed no increase of SIDS. No interaction between fluoridation and infant feeding.	Dick et al. 1999

ABBREVIATIONS: SIDS, sudden infant death syndrome.

Two early papers (Rapaport 1956, 1963) reported an association between elevated rates of Down's syndrome and high water fluoride concentrations. Rapaport also was the first to suggest that maternal age might be an important consideration, with the association between drinking water fluoride concentrations and elevated rates of Down's syndrome particularly pronounced among young mothers. However, the impact of Rapaport's observations is limited by some significant methodological concerns, including the use of crude rates as opposed to maternal age-specific rates, limited case ascertainment, and the presentation of crude rates per 100,000 population as opposed to per live births. Several subsequent reports (Berry 1958; Needleman et al. 1974; Erickson et al. 1976; Erickson 1980) studied the association of Down's syndrome with fluoride or water fluoridation. Berry (1958) found little difference in rates of Down's syndrome between communities with relatively high and low water fluoride concentrations; however, the populations evaluated were small, and maternal age was not considered in the analysis. Needleman et al. (1974) found a positive association between water fluoride concentration and Down's syndrome incidence when crude incidence rates were compared; however, this apparent association was largely lost when the comparison was limited to before and after fluoridation for a subset of towns that introduced water fluoridation, an attempt to partially control for maternal age. Erickson et al. (1976) used data from two sources, the Metropolitan Atlanta Congenital Malformations Surveillance Program and the National Cleft Lip and Palate Intelligence Service. The metropolitan Atlanta database is particularly robust, with detailed retrospective ascertainment. Erickson et al. (1976) found no overall association between the crude incidence rates of Down's syndrome and water fluoridation; however, their data suggested a possible increased rate of Down's syndrome among births to mothers below age 30. Takahashi (1998) grouped Erickson's metropolitan Atlanta data for mothers under 30 and calculated a highly significant association ($P < 0.005$) between fluoridated water and Down's syndrome births to young mothers. A recent review (Whiting et al. 2001) has evaluated the quality of the literature and concluded that an association between water fluoride concentration and Down's syndrome incidence is inconclusive. While the committee agrees with this overall characterization, the review by Whiting et al. was problematic. For example, it described all six studies as ecological and all but one (Rapaport 1956) as having found the majority of cases. However, some studies were partially ecologic, assigning exposure at the group level but categorizing case status and limited covariates (age, race) at the individual level. Erickson (1980) ascertained cases via birth certificates and explicitly acknowledged problems with this approach.

Overall, the available studies of fluoride effects on human development are few and have some significant shortcomings in design and power, limiting their impact.

FINDINGS

A large number of reproductive and developmental studies in animals have been conducted and published since 1990, and the overall quality of the database has improved significantly. High-quality studies in laboratory animals over a range of fluoride concentrations (0-250 mg/L in drinking water) indicate that adverse reproductive and developmental outcomes occur only at very high concentrations. A few studies of human populations have suggested that fluoride might be associated with alterations in reproductive hormones, fertility, and Down's syndrome, but their design limitations make them of little value for risk evaluation.

RECOMMENDATIONS

- Studies in occupational settings are often useful in identifying target organs that might be susceptible to disruption and in need of further evaluation at the lower concentrations of exposure experienced by the general population. Therefore, carefully controlled studies of occupational exposure to fluoride and reproductive parameters are needed to further evaluate the possible association between fluoride and alterations in reproductive hormones reported by Ortiz-Perez et al. (2003).
- Freni (1994) found an association between high fluoride concentrations (3 mg/L or more) in drinking water and decreased total fertility rate. The overall study approach used by Freni has merit and could yield valuable new information if more attention is given to controlling for reproductive variables at the individual and group levels. Because that study had design limitations, additional research is needed to substantiate whether an association exists.
- A reanalysis of data on Down's syndrome and fluoride by Takahashi (1998) suggested a possible association in children born to young mothers. A case-control study of the incidence of Down's syndrome in young women and fluoride exposure would be useful for addressing that issue. However, it may be particularly difficult to study the incidence of Down's syndrome today given increased fetal genetic testing and concerns with confidentiality.

7

Neurotoxicity and Neurobehavioral Effects

This chapter evaluates the effects of fluoride on the nervous system and behavior, with particular emphasis on studies conducted since the earlier NRC (1993) review. The human data include epidemiologic studies of populations exposed to different concentrations of fluoride and individual case studies. In addition, laboratory studies of behavioral, biochemical, and neuroanatomical changes induced by fluoride have been reviewed and summarized. At the end of the chapter, conclusions and recommendations for future research are presented.

HUMAN STUDIES

Cognitive Effects

Several studies from China have reported the effects of fluoride in drinking water on cognitive capacities (X. Li et al. 1995; Zhao et al. 1996; Lu et al. 2000; Xiang et al. 2003a,b). Among the studies, the one by Xiang et al. (2003a) had the strongest design. This study compared the intelligence of 512 children (ages 8-13) living in two villages with different fluoride concentrations in the water. The IQ test was administered in a double-blind manner. The high-fluoride area (Wamiao) had a mean water concentration of 2.47 ± 0.79 mg/L (range 0.57-4.50 milligrams per liter [mg/L]), and the low-fluoride area (Xinhuai) had a mean water concentration of 0.36 ± 0.15 mg/L (range 0.18-0.76 mg/L). The populations studied had comparable iodine and creatinine concentrations, family incomes, family educational levels, and other factors. The populations were not exposed to other significant sources of fluoride, such as smoke from coal fires, industrial pollution, or consumption of brick tea. Thus, the difference in fluoride exposure was attributed to the amount in the drinking water. Mean urinary fluoride[1] concentrations were found to be 3.47 ± 1.95 mg/L in Wamiao and 1.11 ± 0.39 mg/L in Xinhuai. Using the combined Raven's Test for Rural China, the average intelligence quotient (IQ) of the children in Wamiao was found to be significantly lower (92.2 ± 13.00; range, 54-126) than that in Xinhuai (100.41 ± 13.21; range, 60-128).

The IQ scores in both males and females declined with increasing fluoride exposure. The distribution of IQ scores from the females in the two villages is shown in Figure 7-1. A comparable illustration of the IQ scores of males is shown in Figure 7-2. The number of children in Wamiao with scores in the higher IQ ranges was less than that in Xinhuai. There were corresponding increases in the number of children in the lower IQ range. Modal scores of the IQ distributions in the two villages were approximately the same. A follow-up study to determine whether the lower IQ scores of the children in Wamiao might be related to differences in lead exposure disclosed no significant difference in blood lead concentrations in the two groups of children (Xiang et al. 2003b).

A study conducted by Lu et al. (2000) in a different area of China also compared the IQs of 118 children (ages 10-12) living in two areas with different fluoride concentrations in the water (3.15 ± 0.61 mg/L in one area and 0.37 ± 0.04 mg/L in the other). The children were lifelong residents of the villages and had similar social and educational levels. Urinary fluoride concentrations were measured at 4.99 ± 2.57 mg/L in the high-fluoride area and 1.43 ± 0.64 mg/L in the low-fluoride area. IQ measurements using the Chinese Combined Raven's Test, Copyright 2 (see Wang and Qian 1989), showed significantly lower mean IQ scores among children in the high-fluoride area (92.27 ± 20.45) than in children in the low-fluoride area (103.05 ± 13.86). Of special importance, 21.6% of the children in the high-fluoride village scored 70 or below on the IQ scale. For the children in the low-fluoride village, only 3.4% had such low scores. Urinary fluoride concentrations were inversely correlated with mental performance in the IQ test. Qin and Cui (1990) observed similar negative correlation between IQ and fluoride intake through drinking water.

Zhao et al. (1996) also compared the IQs of 160 children (ages 7-14)

---

[1] In the following sections of the chapter, the word "fluoride" is used frequently to indicate what is being measured in blood or urine of people or animals after some treatment with a fluoride. According to medical dictionaries, the word fluoride refers to any binary compound containing fluorine. In many studies, the amount of fluoride reported in urine, blood, or tissue of subjects is the amount of fluorine in the specimen(s). The measurements are frequently referred to as the amount of fluoride present. Furthermore, it is virtually impossible to distinguish between the species of fluoride measured.

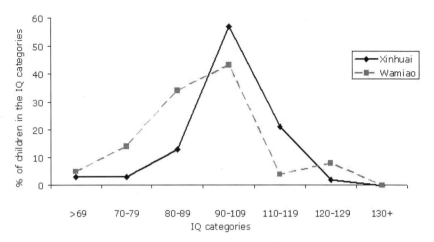

FIGURE 7-1 Distribution of IQ scores from females in Wamiao and Xinuai. SOURCE: data from Xiang et al. 2003a.

living in a high-fluoride area (average concentration of 4.12 mg/L) with those of children living in a low-fluoride area (average concentration 0.91 mg/L). Using the Rui Wen Test, the investigators found that the average IQ of children in the high-fluoride area (97.69) was significantly lower than that of children in the low-fluoride area (105.21). No sex differences were found, but, not surprisingly, IQ scores were found to be related to parents'

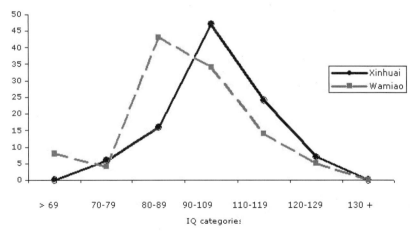

FIGURE 7-2 Distribution of IQ scores from males in Wiamiao and Xinuai. SOURCE: data from Xiang et al. 2003a.

education. The investigators also reported that enamel fluorosis was present in 86% of the children in the high-exposure group and in 14% of the children in the low-exposure group and that skeletal fluorosis was found only in the high-exposure group at 9%.

Another Chinese study evaluated fluoride exposure due to inhalation of soot and smoke from domestic coal fires used for cooking, heating, and drying grain (Li et al. 1995). Many of the children exhibited moderate to severe enamel fluorosis. The average IQ of 900 children (ages 8-13) from an area with severe enamel fluorosis was 9-15 points lower than the average IQ of children from an area with low or no enamel fluorosis. Urinary fluoride concentrations were found to be inversely correlated with IQ, as measured by the China Rui Wen Scale for Rural Areas, and were monotonically related to the degree of enamel fluorosis. Studies based on fluoride exposure from the inhalation of smoke from coal fires are difficult to interpret because of exposure to many other contaminants in smoke.

The significance of these Chinese studies is uncertain. Most of the papers were brief reports and omitted important procedural details. For example, some studies used a modification of the Raven Progressive Matrix test but did not specify what the modifications were or describe how the test was administered. Most of the studies did not indicate whether the IQ tests were administered in a blinded manner. Some of the effects noted in the studies could have been due to stress induced by the testing conditions. Without detailed information about the testing conditions and the tests themselves, the committee was unable to assess the strength of the studies. Despite this, the consistency of the collective results warrants additional research on the effects of fluoride on intelligence in populations that share similar languages, backgrounds, socioeconomic levels, and other commonalities.

It should be noted that many factors outside of native intelligence influence performance on IQ tests. One factor that might be of relevance to fluoride is impairment of thyroid gland function (see Chapter 8). For example, hypothyroidism produces tiredness, depression, difficulties in concentration, memory impairments, and impaired hearing. In addition, there is some evidence that impaired thyroid function in pregnant women can lead to children with lower IQ scores (Klein et al. 2001).

Mental and Physiological Changes

There are numerous reports of mental and physiological changes after exposure to fluoride from various routes (air, food, and water) and for various time periods (Waldbott et al. 1978). A number of the reports are, in fact, experimental studies of one or more individuals who underwent withdrawal from their source of fluoride exposure and subsequent re-exposures under "blind" conditions. In most cases, the symptoms disappeared with the elimination of exposure to fluoride and returned when exposure was reinstated. In some instances, when the fluoride was given in water, this procedure was repeated several times under conditions in which neither the patient nor the provider of the fluoride knew whether the water contained fluoride. Also reported are instances when fluoride-produced symptoms occurred when people moved into a community with fluoridated water but disappeared when the individuals moved to a nonfluoridated community.

Spittle (1994) reviewed surveys and case reports of individuals exposed occupationally or therapeutically to fluoride and concluded there was suggestive evidence that fluoride could be associated with cerebral impairment. A synopsis of 12 case reports of fluoride-exposed people of all ages showed common sequelae of lethargy, weakness, and impaired ability to concentrate regardless of the route of exposure. In half the cases, memory problems were also reported. Spittle (1994) described several of the biochemical changes in enzymatic systems that could account for some of the psychological changes found in patients. He suggested that behavioral alterations found after excessive exposure could be due to the disruption of the N-H bonds in amines, and subsequently in proteins, by the production of N-F bonds (Emsley et al. 1981). This unnatural bond would distort the structure of a number of proteins with the collective potential to cause important biological effects. Fluorides also distort the structure of cytochrome-c peroxidase (Edwards et al. 1984). Spittle also noted the likelihood of fluoride interfering with the basic cellular energy sources used by the brain through the formation of aluminum fluorides (Jope 1988) and subsequent effects on G proteins.

Effects of Silicofluorides

It has been suggested that the silicofluorides used to fluoridate drinking water behave differently in water than other fluoride salts (see Chapter 2 for further discussion) and produce different biological effects. For example, adding sodium silicofluoride (Na_2SiF_6) or fluorosilicic acid (H_2SiF_6) to drinking water has been reported to increase the accumulation of the neurotoxicant lead in the body (Masters and Coplan 1999; Masters et al. 2000). This association was first attributed to increased uptake of lead (from whatever source) caused by fluoride. However, enhanced lead concentrations were found only when the water treatments were made with a fluorosilicate and in children already in a high-lead exposure group.

Urbansky and Schock (undated, 2000) took exception to almost all aspects of the studies by Masters and Coplan on the fluorosilicates. They argued that, under the conditions prevailing at the time of the addition of silicofluorides to drinking water, silicofluorides would be completely hydrolyzed before they reached the consumer's tap (Urbansky and Schock 2000). Measurement techniques and statistical methods were also questioned. They concluded that there is no "credible evidence" that water fluoridation has any quantifiable effect on the solubility, bioavailability, or bioaccumulation of any form of lead.

Another issue that has been raised about differential effects of silicofluorides comes from the dissertation of Westendorf (1975). In that study, silicofluorides were found to have greater power to inhibit the synthesis of cholinesterases, including acetylcholinesterase, than sodium fluoride (NaF). For example, under physiological conditions, one molar equivalent of silicofluoride is more potent in inhibiting acetylcholinesterase than six molar equivalents of NaF (Knappwost and Westendorf 1974). This could produce a situation in which acetylcholine (ACh) accumulates in the vicinity of ACh terminals and leads to excessive activation of cholinergic receptors in the central and peripheral nervous system. At high concentrations, agents with this capability are frequently used in insecticides and nerve gases. At intermediate concentrations, choking sensations and blurred vision are often encountered. Modifications of the effectiveness of the acetylcholinergic systems of the nervous system could account for the fact that, even though native intelligence per se may not be altered by chronic ingestion of water with fluoride ranging from 1.2 to 3 mg/L, reaction times and visuospatial abilities can be impaired. These changes would act to reduce the tested IQ scores. Such noncognitive impairments in children were reported in a meeting abstract (Calderon et al. 2000), but a full publication has not been issued. Extended reaction times have been associated with impaired function of the prefrontal lobes, a behavioral change not directly tied to alterations in IQ (Winterer and Goldman 2003). Because almost all IQ tests are "time-restricted," slow reaction times would impair measured performance.

An interesting set of calculations was made by Urbansky and Schock (undated)—namely, compilation of the binding strengths of various elements with fluorine. They studied eight different complexes. Aluminum and fluorine have the highest binding affinity. Fluorine also forms complexes with other elements including sodium, iron, calcium, magnesium, copper, and hydrogen. Associations with some of these other elements may have implications for some of the neurotoxic effects noted after fluoride or SiF exposure.

Dementia

For more than 30 years it has been known that Alzheimer's disease is associated with a substantial decline in cerebral metabolism (Sokoloff 1966). This original observation has been replicated many times since then. The decrease is reflected in the brain's metabolic rate for glucose, cerebral rate for oxygen, and cerebral blood flow. In terms of reduced cerebral blood flow, the reduction found in Alzheimer's patients is about three times greater than in patients with multi-infarct dementia. As early as 1983, Foster et al. (1983) demonstrated a general decline in the rate of utilization of glucose with the marker F-2-fluorodeoxyglucose with a positron-emission tomography scan. Recently, over and above the general decline in aerobic metabolism, several patterns of enhanced decreases in energy utilization have been demonstrated. The temporal, parietal, and frontal regions are areas with some of the greatest reductions (Weiner et al. 1993; Starkstein et al. 1995). It is possible that the decline in glucose utilization is an early sign of the onset of dementia (Johnson et al. 1988; Silverman and Small 2002). In addition there is evidence from a number of sources that alterations induced by Alzheimer's disease can be observed in many body regions and in blood. This indicates that the disease has system-wide effects in the body. One system particularly sensitive to carbohydrate utilization is the collection of areas involved with the synthesis of ACh. The release of this transmitter is also negatively affected by the interruption of aerobic metabolism and the effect can be noticed in the projection fields of the cholinergic systems. Fluoride produces additional effects on the ACh systems of the brain by its interference with acetylcholinesterase.

Most of the drugs used today to treat Alzheimer's disease are agents that enhance the effects of the remaining ACh system. Nevertheless, it must be remembered that one certain characteristic of Alzheimer's disease is a general reduction of aerobic metabolism in the brain. This results in a reduction in energy available for neuronal and muscular activity.

Because of the great affinity between fluorine and aluminum, it is possible that the greatest impairments of structure and function come about through the actions of charged and uncharged AlF complexes (AlF_x). In the late 1970s and through the early 1990s there was considerable interest in the possibility that elemental aluminum was a major contributing factor to the development of dementia of the Alzheimer's variety as well as to other neurological disorders. In a study of more than 3,500 French men and women above the age of 65 (Jacqmin et al. 1994), a significant decrease in cognitive abilities was found when their drinking water contained calcium, aluminum, and fluorine. Only aluminum showed any relation to cognitive impairment and that depended on the pH of the drinking water being below 7.3. Curiously, at higher pH values, a favorable effect on cognitive actions was found. In recent work with animals, aluminum-induced behavioral changes similar to those found in human dementia, as well as correlated histological changes in animals' brains, were found (Miu et al. 2003). Active research continues at the cellular level on the neural mechanisms disturbed by aluminum (Becaria et al. 2003; Millan-Plano et al. 2003). On the epidemiological side there are inconsistencies in the results of different studies. For example, a recent review concludes that "the toxic effects of aluminum cannot be ruled out either, and thus exposure to aluminum should be monitored and limited as far as possible" (Suay and Ballester 2002). In addition to a depletion of acetylcholinesterase, fluoride produces alterations in phospholipid metabolism and/or reductions in the biological energy available for normal brain functions (see section later in this chapter on neurochemical effects). In addition, the possibility exists that chronic exposure to AlF_x can produce aluminum inclusions with blood vessels as well as in their intima and adventitia. The aluminum deposits inside the vessels and those attached to the intima could cause turbulence in the blood flow and reduced transfer of glucose and O_2 to the intercellular fluids. Finally histopathological changes similar to those traditionally associated with Alzheimer's disease in people have been seen in rats chronically exposed to AlF (Varner et al. 1998).

ANIMAL STUDIES

Behavioral Changes

Studies of NaF

One of the most frequently cited and much discussed studies reporting a link between fluoride and behavior is by Mullenix et al. (1995). The study involved administering NaF to rats at different ages. Two groups of rats were exposed to NaF during gestation by subcutaneous injections given to pregnant dams. Other groups of rats received NaF in water beginning at weaning. Another set of rats was exposed to NaF in water in adulthood. Because of differences in the treatment regimes, procedures involved with the transport of animals at different ages, and other alterations in methods between the age groups, the data from the study are meaningful only if they are considered separately.

In "experiment 1," pregnant dams were subcutaneously injected with NaF at 0.13 mg/kg either on gestational days 14-18 (one or two injections per day, for a total of nine injections) or on days 17-19 (three injections per day). In "experiment 2," NaF at 75, 100, 125, or 175 mg/L was administered in the drinking water to rats at 21 days of age for 6-20 weeks. In "experiment 3," 12-week-old rats were given NaF at 100 mg/L in drinking water for 5-6 weeks. Behavioral tests were performed on prenatally treated and weanling rats at 9 weeks of age, and adult-treated rats were tested at the end of their exposure period. Concentrations of fluoride in plasma in seven brain regions were measured at the time of sacrifice.

To appreciate the data generated by the testing procedures, some details of the testing methods and data analysis used in the Mullenix et al. study must be considered. The methods used were ones developed earlier to quantify animal behavior by using computer-based methods (Kernan et al. 1987, 1988: Kernan and Mullenix 1991). The basic procedures involved the following: The animals were tested in pairs consisting of a treated and a control rat. They were placed in a Plexiglas chamber divided in the middle by a Plexiglas wall to make two adjacent testing chambers. This wall had several holes in it. Thus, each rat could see, hear, and smell its pair-mate. The actual floor space available to each animal was approximately 10 in by 10 in. The chamber was an unusual trapezoidal design with the walls slanting outward from the floor. This shape was created to enhance the clarity of images of the rats recorded by two video cameras. One camera was placed above the testing chambers and another was off to one side. Both were aligned so as to encompass the testing areas of both animals. Sprague-Dawley albino rats were used in the experiments and, to further enhance the pictures, the side away from the horizontally placed camera was black. The floor was also black.

The two video cameras recorded the behavior of both animals simultaneously. The cameras were programmed to take still photos of the animals every second for the 15-minute testing period. Thus, the cameras sent 900 pictures of each animal during a single test period. The computer was programmed to detect five bodily positions, eight "modifiers" (apparently this term means an action with a presumptive goal), and several combinations of postures and modifiers. In all, the computer could record more than 100 combinations of positions, modifiers, and combinations of one or more of the measures indicating the "presumed intentions" of the animals (e.g., groom/attention). For each of these postures or actions or combinations, the number of times it was initiated, the total time spent doing it, and the distribution of the act throughout the 15-minute period were calculated separately for each rat.

In experiment 1, none of the rats treated on gestational days 14-18 showed any behavioral differences from controls. However, among rats treated on gestation days 17-19, male rats were reported to be more active than controls. The increase in activity was attributed to increased instances of grooming and head turning and not enhanced locomotor movement. Plasma concentrations of fluoride were comparable to those of the controls. Fluoride concentrations in the brain were not measured in this group.

In experiment 2, high mortality was observed in the highest treatment group (175 mg/L), and testing was discontinued at that concentration. Female rats exposed to NaF at 125 mg/L had fewer instances of sitting, spent less time sitting, had fewer head turns, and had fewer clusters of grooming bouts than controls. They also showed a reduction in the groom/attention composite index. Females exposed to fluoride in drinking water at 100 mg/L for 6 weeks showed behavioral changes related to grooming, including reduced grooming bouts, reductions in persistent grooming periods, and the grooming/attention cluster. However, these effects were not seen among the females treated for longer periods (20 weeks). Among male rats, changes in behavior were observed only in the 125 mg/L group evaluated after 16 weeks of treatment. Changes included less sitting, less head turning, more standing, and reductions in grooming behavior. Standing and seeming attention postures were increased in these weanling-exposed rats. Measurements of fluoride in plasma showed an increase in concentration after 6 weeks of exposure to NaF at 100 mg/L in male and female rats. All seven areas of the brain analyzed showed increased concentrations of fluoride. As noted in Chapter 3, the accuracy of these measurements has been questioned (Whitford 1996), because other studies have shown that brain fluoride concentrations are considerably lower than, but proportionate to, those in plasma (Carlson et al. 1960; Whitford et al. 1979).

The computer program used in the behavior analyses also generated a statistic named "RS" that combines all the detected alterations in every recognized mode or modified mode of behavior. This overall index of change was reported as significant in females 6 weeks after the start of NaF treatment at concentrations of 100 and 125 mg/L. The statistic was not changed in males treated with NaF at a concentration of 125 mg/L for 11 weeks.

In experiment 3, only female rats showed behavioral changes compared with controls. Changes included reductions in sitting and grooming. Plasma fluoride concentrations were increased in males and females. Testing of fluoride concentrations in the brain found increased concentrations in the medulla of both sexes and in the hippocampal region of females. As noted above, the accuracy of these measurements has been questioned.

The results from these three experiments are difficult to interpret. One difficulty is interpreting the computer-derived categorization of activity patterns compared with behavioral descriptions commonly used by most animal researchers. For example, increased activity usually refers to increased locomotor activity measured in relatively large open fields or mazes. In the Mullenix et al. study, increased activity is characterized by head turning, grooming behaviors, and sniffing and exploration of the corners of the box, which traditionally are not characterized as part of locomotor activity. The small chambers in which the animals were tested would have prevented much locomotor movement at all.

Another aspect of the study that is a modifying issue is the stress-related experience of the rats before the experiments began. The transportation and associated handling of animals over long distances are known stressors to rats and mice. For experiment 1, the pregnant rats were shipped on day 6 of gestation and were housed singly thereafter. The rats used in experiment 2 were shipped to the laboratory at 17 days of age, along with their dams. The adult rats of experiment 3 were shipped at 10 weeks of age. Because the animals were from the Charles River Laboratories in Kingston, New York, the means of transportation to the laboratory in Boston was likely by truck. The transportation of animals by land or air has been shown to produce lasting effects on rodents (Isaacson et al. 2003). The histological effects of transportation and relocation include neuronal losses and substantial instances of shrunken or bloated cells, including some with condensed cytoplasmic inclusions. Other signs of stress and neural insult can be seen, including the presence of reactive microglia throughout the brain. These changes might well interact with later fluoride treatments. In essence, this means comparisons between groups can be legitimately made within the several experiments but not between them. Mullenix et al. (1995) interpreted their behavioral results to imply the interruption of hippocampal dysfunction. Another plausible interpretation is that the behavioral change might have involved alterations in the adrenal-pituitary axis (Gispen and Isaacson 1986).

The results of the Mullenix studies are difficult to compare with studies from other laboratories. The apparatus used has a unique configuration, the chambers were small, and the paired animals were in visual, olfactory, and auditory contact with each other. The data generated are largely derived in idiosyncratic ways by the hardware and software of a relatively complex computer program. From a practical standpoint, it would be extremely difficult for other investigators to replicate the study. The committee is aware there has been debate about the interpretation and significance of the findings of this study. For example, Ross and Daston (1995) note that decreased grooming can be an indication of illness. Because of the high concentrations of fluoride used in the study, it is possible that the animals had gastrointestinal or renal disturbances (Whitford and Taves 1973; Pashley et al. 1984; also see Chapter 9). As discussed above, the committee agrees there are difficulties with interpreting the results of the study, but those difficulties do not warrant dismissal of the results. The study provided some evidence that exposure to fluoride (prenatal, weaning, or in adulthood) might have affected the behavior of rats, albeit almost always in a gender-specific fashion.

In a different type of study, Swiss albino mice were treated with NaF at 30, 60, and 120-mg/L in water for 30 days and behavioral tests were performed daily 1 hour after treatment. The testing included akinesia, catalepsy, swim endurance, and simple maze tests. Animals in the 120 mg/L group scored more poorly in all the tests. Histological changes observed in the brains of these animals are discussed later in this chapter (Bhatnagar et al. 2002).

Paul et al. (1998) investigated the effects of NaF on the motor activity and coordination of female Wistar rats. The rats were treated with NaF at 20 or 40 mg/kg/day by gastric intubation for 2 months and were tested in an activity chamber and on a rota-rod apparatus. Only female rats were used because of the high mortality rates among males in preliminary studies. In both treatment groups, food intake and body weight gain were reduced in a dose-dependent manner. A reduction in spontaneous motor activity was based on results from an apparatus that recorded every type of movement, bodily adjustment, or twitch. This should not be confused with increased activity as measured by locomotor movements in a large arena. In the rotarod motor coordination test, no significant changes were observed between the treated and control rats. There was a dose-related decrease in cholinesterase in the blood but not in the brain. Similar effects on motor activity have been observed in other studies in which rats were treated with NaF at 500 mg/L in drinking water. Alterations of acetylcholinesterase concentrations were found in the brain at this concentration (Ekambaram and Paul 2001, 2002).

Studies of AlF$_3$

Varner et al. (1994) studied the effects of chronic administration of aluminum fluoride (AlF$_3$), on the behavior of Long-Evans rats. AlF$_3$ was administered in drinking water at concentrations of 0.5, 5.0, or 50 mg/L. In terms of fluorine, these values translate into the equivalent of 0.34, 3.4, and 34 mg/L. The animals were between 130 and 154 days old at the beginning of the experiment and were maintained on this program for 45 weeks. In the animals treated with AlF$_3$ at 5 and 50 mg/L, no differences in behavior were found in activity in an open field, in patterns of stride when walking, in spontaneous alternation of arms in a T-maze, in a motor coordination test, or in two tests of learning and memory in the Morris water maze. (Rats in the 0.5-mg/L group were too few to provide meaningful results.) The only behavioral change noted was a lack of preference of the location of a banana odor over the location of a lemon odor. Control animals generally prefer the banana odor. This overall lack of behavioral effects occurred in spite of extensive histological changes associated with neuronal damage and cell death in the hippocampus and other parts of the forebrain.

Anatomy

The complete analyses of the changes found in the brains of rats given one of the three doses of AlF$_3$ used by Varner et al. (1994) were reported in a separate paper (Varner et al. 1993). All groups of the AlF$_3$-exposed rats had significant losses of cells in the CA1 and CA3 areas of the hippocampus, but the losses were not dose dependent. Two types of cellular anomalies were found in the treated animals: (1) argentophilic cells throughout the hippocampus and dentate gyrus with considerable sparing of cells in the CA2 region; and (2) increased aluminum fluorescence in most of the brain, especially in the inner and outer linings of a large number of blood vessels, both large and small. Intravascular inclusions of aluminum particles were sometimes noted within blood vessels. Cells containing aluminum inclusions were not uncommon. This enhancement of aluminum deposits is not surprising because the amount of aluminum found in the brain was almost double that found in control animals.

Varner et al (1998) undertook a second study to determine the relative contribution of fluoride to the high mortality found in the 0.5-mg/L group of the earlier study, to extend the histological procedures used to evaluate the brains, and to determine whether the high death rates after this low dose would be found on replication. Three groups of nine adult rats were administered AlF_3 at 0.5 mg/L, NaF at 2.1 mg/L (containing the same amount of fluoride as the AlF_3 group), or double-distilled deionized water for 1 year. During that time six of nine animals drinking the AlF_3 water died, three of the nine animals drinking the NaF died, and one animal from the control group died. Aluminum content in brain, kidney, and liver was measured by a direct current plasma technique modified for use with tissues containing substantial fat. Brains from both the NaF and the AlF_3 groups had more than twice as much aluminum as the brains of the control animals. This supports the work of Strunecka et al. (2002) indicating that fluoride enhances the uptake of aluminum. But, the uptake was organ specific. There was no increase of aluminum found in the kidneys or liver. Sections from the brains of all animals were processed in a manner that allowed their staining with hematoxylin and eosin, the Morin stain for aluminum (and counterstained with cresyl violet), and a modified Bielschowsky silver stain as well as with antisera specific for IgM, β-amyloid, or amyloid A.

There was a progressive decline in the appearance of the AlF_3 treated rats compared with the NaF or control animals before their demise. Their hair was sparse and their skin had a copper color. Toenails and teeth indicated a condition reflecting a hypermelanosis. Body weights, however, did not vary among the groups. Hemispheric differences in the brain were found in the distribution of aluminum using the Morin staining ultraviolet microscopic procedure. A greater amount of aluminum fluorescence was seen in layers 5 and 6 of the parietal neocortex and hippocampus of the left relative to the right hemisphere in the AlF_3-treated rats. Areas CA3 and CA4 were the most affected regions of the hippocampus.

The occurrence of abnormal cells was also determined for all brains. Signs of neuronal anomalies included chromatin clumping, enhanced protein staining, pyknosis, vacuolation, ghost-like swollen appearances of cells, and enhanced silver staining in cell bodies and their processes. Both NaF and AlF_3 treatments produced cellular distortions in cortical layers 2 and 3 of both hemispheres, but enhanced cellular abnormalities in layers 5 and 6 were found only in the left hemisphere. Both treatments also produced a diminished number of cells in the left CA3 region of the hippocampus but only the AlF_3 treatment reduced cell numbers in this region of the left hemisphere. These observations are similar to previous findings reported in the brains of cats after intracerebroventricular administration of aluminum chloride (Crapper and Dalton 1973).

Both the AlF_3 and the NaF treatments increased staining of neurons for IgM in the right hemisphere. No differences were found among the groups in the presence of IgM on the left side of the brain. Minor amounts of IgM were found in the hippocampus and dentate gyrus but without any group differences. The control group had few instances of β-amyloid but the brains of the AlF_3-treated animals demonstrated a bimodal distribution of deposits in the vasculature of the dorsal thalamus. Staining was either very high or nonexistent. The NaF-treated group showed a similar bimodality of accumulation of β-amyloid in the right lateral posterior thalamic region.

The pattern of neuronal degeneration found by Varner et al. (1998) was also found in two other studies (Bhatnager et al. 2002; Shivarajashankara et al. 2002). In the study by Bhatnagar et al. (2002) described earlier in this chapter, the investigators observed a significant number of degenerated nerve cell bodies in hippocampal subregions CA3 and CA4 and in the dentate gyrus. Shivarajashankara et al. (2002) exposed Wistar rats to NaF in utero during the last week of gestation and for 10 weeks after birth. Animals received either 30 or 100 mg/L in their drinking water. At the end of the 10 weeks the animals were sacrificed and their brains were sectioned and stained with cresyl violet. Little change was seen in the 30-mg/L treated animals but the brains of the 100-mg/L treated animals showed large amounts of neurodegeneration. There were only a few normal appearing pyramidal cells in regions CA1 and CA3 of the hippocampus. Almost all the cells in these areas were pyknotic and showed intensely stained protein in their shrunken cytoplasm. Neuronal degeneration, but to a lesser degree, was found in the upper layers of neocortex, the amygdala, and the cerebellum. These areas were not extensively studied by Varner et al. (1998).

The interactions between fluoride and aluminum have been studied in laboratories and in the environment. There is evidence that fluoride enhances the uptake of aluminum and that aluminum reduces the uptake of fluoride (Spencer et al. 1980, Ahn et al. 1995). This complicates predicting the effect of exposure to aluminum- or fluorine-containing complexes in natural situations.

NEUROCHEMICAL EFFECTS AND MECHANISMS

A number of studies have examined biochemical changes in the brain associated with fluoride. For example, Guan et al. (1998) reported alterations in the phospholipid content of the brain of rats exposed to NaF at 30 or 100 mg/L for 3-7 months. The most prominent changes were found in phosphatidylethanolamine, phosphatidylcholine, and phosphatidylserine.

After 7 months of treatment, ubiquinone was clearly elevated, likely due as a compensatory reaction to the increase in free radicals in the brain. Fluoride has been shown to decrease the activities of superoxide dismutase (Guan et al. 1989) and glutathione peroxidase (Rice-Evans and Hoschstein 1981), the consequences being increased free radicals.

NaF injected subcutaneously into rabbits altered brain lipid metabolism (Shashi 1992b) and concentrations of protein, free amino acid, and RNA in the brain (Shashi et al. 1994).

Using slices of rat neocortex, Jope (1988) found that NaF stimulated the hydrolysis of phosphoinositide by activation of a G protein, Gp. This protein acts as a transducer between receptors and phospholipase C. He also found that a metal chelator added to the preparation eliminated this effect. This information and other observations led to the conclusion that the effective agent in the hydrolysis was an AlF_x complex. Under his experimental conditions, the AlF_4 was most likely formed from trace amounts of aluminum derived from the glass or from a fluorine-containing contaminant in a reagent. The addition of increasing amounts of aluminum did not increase the hydrolysis effect. In fact, adding substantial amounts of aluminum inhibited it. As in several types of experiment, it is the low aluminum fluoride concentrations that produce the greatest biochemical or physiological effects. In this regard, it is important to note that, even if aluminum bioavailability is low in rats and in other laboratory species, only a small amount is needed to produce untoward effects (Yokel et al. 2001).

Many of the untoward effects of fluoride are due to the formation of AlF_x complexes. AlF_x and BeF_x complexes are small inorganic molecules that mimic the chemical structure of a phosphate. As such they influence the activity of phosphohydrolases and phospholipase D. Only micromolar concentrations of aluminum are needed to form AlF_x (Sternweis and Gilman 1982). The G protein effects produced by AlF_x are not limited to enzymes that bind phosphates or nucleoside-polyphosphate (Chabre 1990). AlF_x also impairs the polymerization-depolarization cycle of tubulin. This could account for some of the intensely stained neurofilaments in cells in the brains of animals exposed to chronic NaF (Varner et al. 1993, 1998). AlF_x appears to bind to enzyme-bound GDP or ADP, thus imitating GTP or ATP and, in a sense, generating "false messages" within the brain. This binding ability is probably due to the molecular similarities between $AlF_3(OH)$ and a phosphate group in the molecular structure, in particular, a tetrahedral arrangement (Strunecka and Patocka 2002).

G protein-coupled receptors mediate the release of many neural transmitters including the catecholamines, serotonin, ACh, and the excitatory amino acids. They also are involved in regulating glucagons, vasopressin, neuropeptides, endogenous opioids, prostaglandins, and other important systemic influences on brain and behavior. AlF_x is also involved in regulating the pineal melatonin system as well as the thyroid-stimulating hormone-growth hormone connection. It has been said in this regard "every molecule of AlF_x is the messenger of false information" (Strunecka and Patocka 2002, p. 275). This may be an accurate synopsis of the AlF_x effect at a single synapse, but the brain is a highly redundant and dispersed communication system containing millions of synapses. Because of this, observable alterations in mental or motor actions might require the formation of a multitude of false messages in a number of brain circuits acting over a prolonged period of time. Thus, the number of false messages required to disrupt an "action pattern" in the brain probably will vary according to the nature of the ongoing activities.

An especially important neurochemical transmitter that reaches almost all areas of the brain is ACh. As discussed above, some studies show that NaF and SiF inhibit cholinesterases, including acetylcholinesterase. The progressive accumulation of ACh at synaptic locations produced by the diminished esterase activity leads to a number of complex effects that can be summarized as an initial increase in stimulation of the target cells but ultimately leads to diminished stimulation—even a blockade of all activity. This earlier dialogue properly emphasized the behavioral importance of cholinergic activity in the brain and body more generally.

Long et al. (2002) reported changes in the number of acetylcholine receptors (nAChRs) in the rat brain due to fluoride. Rats were administered NaF in drinking water at 30 or 100-mg/L for 7 months. Decreased numbers of nAChRα7 subunits were found in the brains of rats from both treatment groups, but only the brains of the 100-mg/L group had diminished nAChRα4 subunits of this receptor. These results are of interest because changes in the nicotinic receptors have been related to the development of Alzheimer's disease (Lindstrom 1997; Newhouse et al. 1997) and, in frontal brain areas, to schizophrenia (Guan et al. 1999).

FINDINGS

Human Cognitive Abilities

In assessing the potential health effects of fluoride at 2-4 mg/L, the committee found three studies of human populations exposed at those concentrations in drinking water that were useful for informing its assessment of potential neurologic effects. These studies were conducted in different areas of China, where fluoride concentrations ranged from 2.5 to 4 mg/L. Comparisons were made between the IQs of children from those populations with children exposed to lower concentration of fluoride ranging from 0.4 to 1 mg/L. The studies reported that while modal IQ scores were unchanged, the average IQ scores were lower in the more highly exposed children. This was due to fewer children in the high IQ range. While the studies lacked sufficient detail for the committee to fully assess their quality and their relevance to U.S. populations, the consistency of the collective results warrant additional research on the effects of fluoride on intelligence. Investigation of other mental and physiological alterations reported in the case study literature, including mental confusion and lethargy, should also be investigated.

Behavioral Effects on Animals

A few animal studies have reported alternations in the behavior of rodents after treatment with fluoride. However, the observed changes were not striking in magnitude and could have been due to alterations in hormonal or peptide activity. Animal studies to date have used conventional methodologies to measure learning and memory abilities or species-typical behaviors in novel locations. The tasks used to measure learning and memory did not require any significant mental effort. No studies were available on higher order mental functions, altered reactions to stress, responses to disease states, or supplemental reactions to known neurotoxins. Procedures are available that could test for cognitive functions, but they are labor intensive and have seldom been used in the past 60 years. One example is the reasoning test designed by Maier (1929), who found that even a small lesion of the neocortex impaired performance on the reasoning test (Maier 1932). A more recent example is the delayed matching to position test with different outcomes (Savage 2001), which have shown that damage to the hippocampus can affect learning.

Fluorosilicates

As noted in Chapter 2, exposure to fluorosilicates could occur under some conditions. There are reports that such chemicals enhance the uptake of lead into the body and brain, whereas NaF does not. Further research is needed to elucidate how fluorosilicates might have different biological effects from fluoride salts.

Neurochemical and Biochemical Changes

Lipids and phospholipids, phosphohydrolases and phospholipase D, and protein content have been shown to be reduced in the brains of laboratory animals subsequent to fluoride exposure. The greatest changes were found in phosphatidylethanolamine, phosphotidylcholine, and phosphotidylserine. Fluorides also inhibit the activity of cholinesterases, including acetylcholinesterase. Recently, the number of receptors for acetylcholine has been found to be reduced in regions of the brain thought to be most important for mental stability and for adequate retrieval of memories.

It appears that many of fluoride's effects, and those of the aluminofluoride complexes are mediated by activation of Gp, a protein of the G family. G proteins mediate the release of many of the best known transmitters of the central nervous system. Not only do fluorides affect transmitter concentrations and functions but also are involved in the regulation of glucagons, prostaglandins, and a number of central nervous system peptides, including vasopressin, endogenous opioids, and other hypothalamic peptides. The AlF_x binds to GDP and ADP altering their ability to form the triphosphate molecule essential for providing energies to cells in the brain. Thus, AlF_x not only provides false messages throughout the nervous system but, at the same time, diminishes the energy essential to brain function.

Fluorides also increase the production of free radicals in the brain through several different biological pathways. These changes have a bearing on the possibility that fluorides act to increase the risk of developing Alzheimer's disease. Today, the disruption of aerobic metabolism in the brain, a reduction of effectiveness of acetylcholine as a transmitter, and an increase in free radicals are thought to be causative factors for this disease. More research is needed to clarify fluoride's biochemical effects on the brain.

Anatomical Changes in the Brain

Studies of rats exposed to NaF or AlF_3 have reported distortion in cells in the outer and inner layers of the neocortex. Neuronal deformations were also found in the hippocampus and to a smaller extent in the amygdala and the cerebellum. Aluminum was detected in neurons and glia, as well as in the lining and in the lumen of blood vessels in the brain and kidney. The substantial enhancement of reactive microglia, the presence of stained intracellular neurofilaments, and the presence of IgM observed in rodents are related to signs of dementia in humans. The magnitude of the changes was large and consistent among the studies. Given this, the committee concludes further research is warranted in this area, similar to that discussed at a February 2-3,1999, EPA workshop on aluminum complexes and neurotoxicity and that recommended for study by NTP (2002).

RECOMMENDATIONS

On the basis of information largely derived from histological, chemical, and molecular studies, it is apparent that fluorides have the ability to interfere with the functions of the brain and the body by direct and indirect means. To determine the possible adverse effects of fluoride, additional data from both the experimental and the clinical sciences are needed.

- The possibility has been raised by the studies conducted in China that fluoride can lower intellectual abilities. Thus, studies of populations exposed to different concentrations of fluoride in drinking water should include measurements of reasoning ability, problem solving, IQ, and short- and long-term memory. Care should be taken to ensure that proper testing methods are used, that all sources of exposure to fluoride are assessed, and that comparison populations have similar cultures and socioeconomic status.
- Studies of populations exposed to different concentrations of fluoride should be undertaken to evaluate neurochemical changes that may be associated with dementia. Consideration should be given to assessing effects from chronic exposure, effects that might be delayed or occur late-in-life, and individual susceptibility (see Chapters 2 and 3 for discussion of subpopulations that might be more susceptible to the effects of fluoride from exposure and physiologic standpoints, respectively).
- Additional animal studies designed to evaluate reasoning are needed. These studies must be carefully designed to measure cognitive skills beyond rote learning or the acquisition of simple associations, and test environmentally relevant doses of fluoride.
- At the present time, questions about the effects of the many histological, biochemical, and molecular changes caused by fluorides cannot be related to specific alterations in behavior or to known diseases. Additional studies of the relationship of the changes in the brain as they affect the hormonal and neuropeptide status of the body are needed. Such relationships should be studied in greater detail and under different environmental conditions.
- Most of the studies dealing with neural and behavioral responses have tested NaF. It is important to determine whether other forms of fluoride (e.g., silicofluorides) produce the same effects in animal models.

8

Effects on the Endocrine System

The endocrine system, apart from reproductive aspects, was not considered in detail in recent major reviews of the health effects of fluoride (PHS 1991; NRC 1993; Locker 1999; McDonagh et al. 2000a; WHO 2002; ATSDR 2003). Both the Public Health Service (PHS 1991) and the World Health Organization (WHO 2002) mentioned secondary hyperparathyroidism in connection with discussions of skeletal fluorosis, but neither report examined endocrine effects any further. The Agency for Toxic Substances and Disease Registry (ATSDR 2003) discussed four papers on thyroid effects and two papers on parathyroid effects and concluded that "there are some data to suggest that fluoride does adversely affect some endocrine glands." McDonagh et al. (2000a) reviewed a number of human studies of fluoride effects, including three that dealt with goiter and one that dealt with age at menarche. The following section reviews material on the effects of fluoride on the endocrine system—in particular, the thyroid (both follicular cells and parafollicular cells), parathyroid, and pineal glands. Each of these sections has its own discussion section. Detailed information about study designs, exposure conditions, and results is provided in Appendix E.

THYROID FOLLICULAR CELLS

The follicular cells of the thyroid gland produce the classic thyroid hormones thyroxine (T4) and triiodothyronine (T3); these hormones modulate a variety of physiological processes, including but not limited to normal growth and development (Larsen et al. 2002; Larsen and Davies 2002; Goodman 2003). Between 4% and 5% of the U.S. population may be affected by deranged thyroid function (Goodman 2003), making it among the most prevalent of endocrine diseases (Larsen et al. 2002). The prevalence of subclinical thyroid dysfunction in various populations is 1.3-17.5% for subclinical hypothyroidism and 0.6-16% for subclinical hyperthyroidism; the reported rates depend on age, sex, iodine intake, sensitivity of measurements, and definition used (Biondi et al. 2002). Normal thyroid function requires sufficient intake of iodine (at least 100 micrograms/day [µg/d]), and areas of endemic iodine deficiency are associated with disorders such as endemic goiter and cretinism (Larsen et al. 2002; Larsen and Davies 2002; Goodman 2003). Iodine intake in the United States (where iodine is added to table salt) is decreasing (CDC 2002d; Larsen et al. 2002), and an estimated 12% of the population has low concentrations of urinary iodine (Larsen et al. 2002).

The principal regulator of thyroid function is the pituitary hormone thyroid-stimulating hormone (TSH), which in turn is controlled by positive input from the hypothalamic hormone thyrotropin-releasing hormone (TRH) and by negative input from T4 and T3. TSH binds to G-protein-coupled receptors in the surface membranes of thyroid follicular cells (Goodman 2003), which leads to increases in both the cyclic adenosine monophosphate (cAMP) and diacylglycerol/inositol trisphosphate second messenger pathways (Goodman 2003). T3, rather than T4, probably is responsible for the feedback response for TSH production (Schneider et al. 2001). Some T3, the active form of thyroid hormone, is secreted directly by the thyroid along with T4, but most T3 is produced from T4 by one of two deiodinases (Types I and II[1]) in the peripheral tissue (Schneider et al. 2001; Larsen et al. 2002; Goodman 2003). T3 enters the nucleus of the target cells and binds to specific receptors, which activate specific genes.

Background

An effect of fluoride exposure on the thyroid was first reported approximately 150 years ago (Maumené 1854, 1866; as cited in various reports). In 1923, the director of the Idaho Public Health Service, in a letter to the Surgeon General, reported enlarged thyroids in many children between the ages of 12 and 15 using city water in the village of Oakley, Idaho (Almond 1923); in addition, the children using city water had severe enamel deficiencies in their permanent teeth. The dental problems were eventually attributed to the presence in the city water of 6 mg/L fluoride, and children born after a change in water supply (to water with <0.5 mg/L fluoride) were not

---

[1]Type I deiodinase, along with Type III, is also responsible for deactivating T4 and T3 by removing the iodine atoms (Schneider et al. 2001; Larsen et al. 2002; Goodman 2003).

so affected (McKay 1933); however, there seems to have been no further report on thyroid conditions in the village.

More recently, Demole (1970) argued that a specific toxicity of fluoride for the thyroid gland does not exist, because (1) fluoride does not accumulate in the thyroid; (2) fluoride does not affect the uptake of iodine by thyroid tissue; (3) pathologic changes in the thyroid show no increased frequency in regions where water is fluoridated (naturally or artificially); (4) administration of fluoride does not interfere with the prophylactic action of iodine on endemic goiter; and (5) the beneficial effect of iodine in threshold dosage to experimental animals is not inhibited by administration of fluoride, even in excessive amounts. Bürgi et al. (1984) also stated that fluoride does not potentiate the consequences of iodine deficiency in populations with a borderline or low iodine intake and that published data fail to support the hypothesis that fluoride has adverse effects on the thyroid (at doses recommended for caries prevention). McLaren (1976), however, pointed out the complexity of the system, the difficulties in making adequate comparisons of the various studies of fluoride and the thyroid, and evidence for fluoride accumulation in the thyroid and morphological and functional changes (e.g., changes in activity of adenylyl cyclase), suggesting that analytical methods could have limited the definitiveness of the data to date. His review suggested that physiological or functional changes might occur at fluoride intakes of 5 mg/day.

Although fluoride does not accumulate significantly in most soft tissue (as compared to bones and teeth), several older studies found that fluoride concentrations in thyroid tissue generally exceed those in most other tissue except kidney (e.g., Chang et al. 1934; Hein et al. 1954, 1956); more recent information with improved analytic methods for fluoride was not located. Several studies have reported no effect of fluoride treatment on thyroid weight or morphology (Gedalia et al. 1960; Stolc and Podoba 1960; Saka et al. 1965; Bobek et al. 1976; Hara 1980), while others have reported such morphological changes as mild atrophy of the follicular epithelium (Ogilvie 1953), distended endoplasmic reticulum in follicular cells (Sundström 1971), and "morphological changes suggesting hormonal hypofunction" (Jonderko et al. 1983).

Fluoride was once thought to compete with iodide for transport into the thyroid, but several studies have demonstrated that this does not occur (Harris and Hayes 1955; Levi and Silberstein 1955; Anbar et al. 1959; Saka et al. 1965). The iodide transporter accepts other negatively charged ions besides iodide (e.g., perchlorate), but they are about the same size as iodide (Anbar et al. 1959); fluoride ion is considerably smaller and does not appear to displace iodide in the transporter.

Animal Studies

A number of studies have examined the effects of fluoride on thyroid function in experimental animals or livestock (for details, see Appendix E, Tables E-1, E-2, and E-3). Of these, the most informative are those that have considered both the fluoride and iodine intakes.

Guan et al. (1988) found that a fluoride intake of 10 mg/L in drinking water had little apparent effect on Wistar rats with sufficient iodine intake, but a fluoride intake of 30 mg/L in drinking water resulted in significant decreases in thyroid function (decreases in T4, T3, thyroid peroxidase, and 3H-leucine), as well as a decrease in thyroid weight and effects on thyroid morphology (Table E-2). In iodine-deficient rats, fluoride intake of 10 mg/L in drinking water produced abnormalities in thyroid function beyond that attributable to low iodine, including decreased thyroid peroxidase, and low T4 without compensatory transformation of T4 to T3.

Zhao et al. (1998), using male Kunmin mice, found that both iodine-deficient and iodine-excess conditions produced goiters, but, under iodine-deficient conditions, the goiter incidence at 100 days increased with increased intake of fluoride. At 100 days, the high-fluoride groups had elevated serum T4 at all concentrations of iodine intake and elevated T3 in iodine-deficient animals. High fluoride intake significantly inhibited the radioiodine uptake in the low- and normal-iodine groups.

Stolc and Podoba (1960) found a decrease in protein-bound iodine in blood in fluoride-treated female rats (3-4 mg/kg/day) fed a low-iodine diet but not in corresponding rats fed a larger amount of iodine. Both groups (low- and high-iodine) of fluoride-treated rats showed a reduced rate of biogenesis of T3 and T4 after administration of ^{131}I compared with controls (Stolc and Podoba 1960).

Bobek et al. (1976) found decreases in plasma T4 and T3 as well as a decrease in free T4 index and an increase in T3-resin uptake in male rats given 0.1 or 1 mg of fluoride per day (0.4-0.6 or 4-6 mg/kg/day) in drinking water for 60 days.[2] The authors suggested the possibility of decreased binding capabilities and altered thyroid hormone transport in blood.

Decreases in T4 and T3 concentrations have been reported in dairy cows at estimated fluoride doses up to 0.7 mg/kg/day with possible iodine deficiency (Hillman et al. 1979; Table E-3). Reduced T3 (Swarup et al. 1998) and reduced T3, T4, and protein-bound iodine (Cinar and Selcuk 2005) have also been reported in cows diagnosed with chronic fluorosis in India and Turkey, respectively.

---

[2]The decrease in T3 in the group receiving 0.1 mg/day was not statistically significant (Table E-1). Note that ATSDR (2003) stated that an intermediate-duration minimal risk level (MRL) derived from this study of thyroid effects in rats would have been lower (more protective) than the chronic-duration MRL derived from a human study of bone effects (0.05 mg/kg/day).

Hara (1980) found elevated T3 and T4 at the lowest dose (approximately 0.1 mg/kg/day), decreased T3 and normal T4 at intermediate doses (3-4 mg/kg/day), and decreased TSH and growth hormone (indicating possible effects on pituitary function) at the highest doses (10-20 mg/kg/day). This was the only animal study of fluoride effects on thyroid function to measure TSH concentrations; however, full details (e.g., iodine intake) are not available in English.

Other studies have shown no effect of fluoride on the end points examined (Gedalia et al. 1960; Siebenhüner et al. 1984; Clay and Suttie 1987; Choubisa 1999; Table E-1). Choubisa (1999) looked only for clinical evidence of goiter in domestic animals (cattle and buffaloes) showing signs of enamel or skeletal fluorosis; no hormone parameters (e.g., T4, T3, TSH) were measured. Gedalia et al. (1960) also did not measure T4, T3, or TSH; radioiodine uptake, protein-bound iodine, and total blood iodine were all normal in rats receiving fluoride doses up to approximately 1 milligram per kilogram of body weight per day (mg/kg/day). Clay and Suttie (1987) reported no significant differences from control values for T4 concentration and T3 uptake in heifers fed up to 1.4 mg/kg/day; iodine intake is not stated but probably was adequate, and TSH was not measured.

Siebenhüner et al. (1984) carried out a special experiment involving iodine depletion of the thyroid before 6 days of fluoride treatment. No effects were seen on the parameters measured, including T3 and T4 concentrations; however, TSH was not measured. In addition, propylthiouracil (PTU), the agent used to deplete the thyroid of iodine, also has an inhibitory effect on deiodinases (Larsen et al. 2002; Larsen and Davies 2002); Siebenhüner et al. (1984) did not mention this second action of PTU and its relevance to the interpretation of the experimental results, and there was no control group without the PTU treatment.

Human Studies

Several authors have reported an association between endemic goiter and fluoride exposure or enamel fluorosis in human populations in India (Wilson 1941; Siddiqui 1960; Desai et al. 1993), Nepal (Day and Powell-Jackson 1972), England (Wilson 1941; Murray et al. 1948), South Africa (Steyn 1948; Steyn et al. 1955; Jooste et al. 1999), and Kenya (Obel 1982). Although endemic goiter is now generally attributed to iodine deficiency (Murray et al. 1948; Obel 1982; Larsen et al. 2002; Belchetz and Hammond 2003), some of the goitrogenic areas associated with fluoride exposure were not considered to be iodine deficient (Steyn 1948; Steyn et al. 1955; Obel 1982; Jooste et al. 1999). Obel (1982) indicated that many cases of fluorosis in Kenya occur concurrently with goiter. Several authors raise the possibility that the goitrous effect, if not due to fluoride, is due to some other substance in the water (e.g., calcium or water hardness) that was associated with the fluoride concentration (Murray et al. 1948; Day and Powell-Jackson 1972) or that enhanced the effect of fluoride (Steyn 1948; Steyn et al. 1955). Dietary selenium deficiencies (e.g., endemic in parts of China and Africa or due to protein-restricted diets) can also affect normal thyroid function[3] (Larsen et al. 2002); no information on dietary selenium is available in any of the fluoride studies. Appendix E summarizes a number of studies of the effects of fluoride on thyroid function in humans (see Table E-4).

Three studies illustrated the range of results that have been reported: (1) Gedalia and Brand (1963) found an association between endemic goiter in Israeli girls and iodine concentrations in water but found no association with fluoride concentrations (<0.1-0.9 mg/L). (2) Siddiqui (1960) found goiters only in persons aged 14-17 years; the goiters, which became less visible or invisible after puberty, were associated with mean fluorine content of the water (5.4-10.7 mg/L) and were inversely associated with mean iodine content of the water. (3) Desai et al. (1993) found a positive correlation ($P < 0.001$) between prevalence of goiter (9.5-37.5%) and enamel fluorosis (6.0-59.0%), but no correlation between prevalence of goiter and water iodine concentration ($P > 0.05$).

Day and Powell Jackson (1972) surveyed 13 villages in Nepal where the water supply was uniformly low in iodine (≤ 1 µg/L; see Figure 8-1). Here the goiter prevalence (5-69%, all age groups) was directly associated with the fluoride concentration (<0.1 to 0.36 mg/L; $P < 0.01$) or with hardness, calcium concentration, or magnesium concentration of the water (all $P < 0.01$). Goiter prevalence of at least 20% was associated with all fluoride concentrations ≥ 0.19 mg/L, suggesting that fluoride might influence the prevalence of goiter in an area where goiter is endemic because of low iodine intake. The possibility of a nutritional component (undernutrition or protein deficiency) to the development of goiter was also suggested.

Jooste et al. (1999) examined children (ages 6, 12, and 15) who had spent their entire lives in one of six towns in South Africa where iodine concentrations in drinking water were considered adequate (median urinary iodine concentration exceeding 201 µg/L [1.58 µmol/L]; see Appendix E, Tables E-4 and E-5; Figure 8-2). For towns with low (0.3-0.5 mg/L) or near "optimal" (0.9-1.1 mg/L) fluoride concentrations in water, no relationship between fluoride and prevalence of mild goiter was found (5-18%); for the other two towns (1.7 and 2.6 mg/L fluoride), however, goiter prevalences were 28% and 29%, respectively, and most children had severe enamel mottling. These two towns (and one low-fluoride town) had very low proportions (0-2.2%) of children with iodine deficiency, defined as urinary

---

[3]All three deiodinases contain selenocysteine at the active sites and therefore have a minimum requirement for selenium for normal function (Larsen et al. 2002).

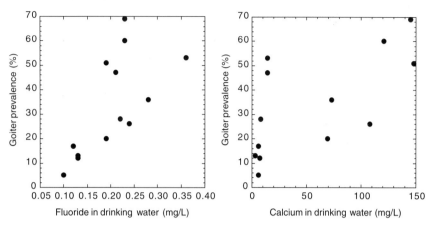

FIGURE 8-1 Goiter prevalence versus fluoride (left) and calcium (right) concentration in drinking water for 13 villages in Nepal with very low iodine concentrations. SOURCE: Day and Powell-Jackson 1972.

iodine concentrations <100 µg/L (<0.79 µmol/L). The town with the lowest prevalence of goiter also had the lowest prevalence of undernutrition; the two towns with the highest prevalence of goiter (and highest fluoride concentrations) did not differ greatly from the remaining three towns with

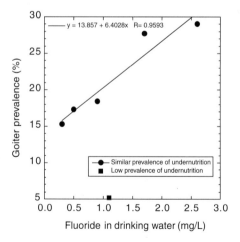

FIGURE 8-2 Goiter prevalence versus drinking water fluoride concentrations in six South African towns with adequate iodine concentrations. One town had a significantly lower prevalence of undernutrition than the other five towns and is not included in the line fitting. SOURCE: Jooste et al. 1999.

respect to prevalence of undernutrition. The authors suggested that fluoride or an associated goitrogen might be responsible for the goiters seen in the two towns with the highest fluoride concentrations but that some other factor(s) was involved in development of goiter in the other towns.

Several studies have compared various aspects of thyroid status in populations with different fluoride intakes (for details, see Appendix E, Table E-4). Leone et al. (1964) and Baum et al. (1981) reported no significant differences in thyroid status between populations with low (0.09-0.2 mg/L) and high (3-3.5 mg/L) fluoride concentrations in the drinking water. Leone et al. (1964) looked only at protein-bound iodine and physical examination of the thyroid in adults; Baum et al. (1981) measured a number of parameters in teenagers, including T4, T3, and TSH. Neither study reported iodine status of the groups. Baum et al. (1981) showed but did not explain a decrease in thyroglobulin in girls in the high-fluoride group.

Bachinskii et al. (1985) examined 47 healthy persons, 43 persons with hyperthyroidism, and 33 persons with hypothyroidism. Prolonged consumption of "high-fluoride" drinking water (2.3 mg/L, as opposed to "normal" concentrations of 1 mg/L) by healthy persons was associated with statistically significant changes in TSH concentrations (increased), T3 concentrations (decreased), and uptake of radioiodine (increased), although the mean values for TSH and T3 were still within normal ranges (see Appendix E, Table E-6). The mean value of TSH for the healthy group (4.3 ± 0.6 milliunits/L; Table E-6) is high enough that one expects a few individuals to have been above the normal range (typically 0.5-5 milliunits/L; Larsen et al. 2002). These results were interpreted as indicating disruption of iodine metabolism, stress in the pituitary-thyroid system, and increased risk of developing thyroidopathy (Bachinskii et al. 1985).

Lin et al. (1991) examined 769 children (7-14 years old) for mental retardation in three areas of China, including an area with "high" fluoride (0.88 mg/L) and low iodine, an area with "normal" fluoride (0.34 mg/L) and low iodine, and an area where iodine supplementation was routine (fluoride concentration not stated). Ten to twelve children in each area received detailed examinations, including measuring thyroid ^{131}I uptake and thyroid hormone concentrations. Children in the first area had higher TSH, slightly higher ^{131}I uptake, and lower mean IQ than children in the second area. Children in the first area also had reduced T3 and elevated reverse T3, compared with children in the second area. The authors suggested that high fluoride might exacerbate the effects of iodine deficiency. In addition, the authors reported a difference in T3/rT3 (T3/reverse-T3) ratios between high- and low-fluoride areas and suggested that excess fluoride ion affects normal deiodination.

A recent study by Susheela et al. (2005) compared thyroid hormone status (free T4, free T3, and TSH) of 90 children with enamel fluorosis (drinking water fluoride ranging from 1.1 to 14.3 mg/L) and 21 children without enamel fluorosis (0.14-0.81 mg/L fluoride in drinking water) in areas where iodine supplementation was considered adequate.[4] Forty-nine children (54.4%) in the sample group had "well-defined hormonal derangements"; findings were borderline in the remaining 41 children. The types of hormonal derangements included elevated TSH and normal T4 and T3 (subclinical hypothyroidism); low T3 and normal T4 and TSH ("low T3 syndrome"); elevated T3 and TSH and normal T4 (possible T3 toxicosis); elevated TSH, low T4, and normal T3 (usually indicative of primary hypothyroidism and iodine deficiency); and low T3, high TSH, and normal T4. All but the first category are considered to be associated with or potentially caused by abnormal activity of deiodinases. The authors concluded that fluoride in excess may be inducing diseases that have usually been attributed to iodine deficiency and that iodine supplementation may not be adequate when excess fluoride is being consumed.

Thyroid hormone disturbances were also noted in the control children, and urine and fluoride concentrations in the control children reflect higher fluoride intake than can be accounted for by the drinking water alone (Susheela et al. 2005). Thus, the authors recommend that end points such as hormone concentrations should be examined with respect to serum or urinary fluoride concentrations, not just drinking water fluoride concentrations. In addition, they note that all hormone endpoints (T3, T4, and TSH) should be examined, lest some of the abnormalities be missed.

Mikhailets et al. (1996) detected thyroid abnormalities (moderate reduction of iodine uptake, low T3, normal T4, and increased TSH) in 165 aluminum workers with signs of chronic fluorosis and an estimated average fluoride intake of 10 mg/working day. A tendency toward increased TSH was observed with increased exposure time and with more severe fluorosis. Workers with more than 10 years of service had a significant decrease in T3 concentration in comparison to controls. The frequency of individuals with low concentrations of T3 (corresponding to hypothyroidism) was 65% among workers with more than 10 years of service and 54% among workers with Stage 2 fluorosis. The highest frequency of occurrence of low T3 (76%) was observed in people with chronic fluoride intoxication including liver damage (moderate cytolysis), suggesting a disorder in peripheral conversion of T4 to T3 (deiodination). The possibility of indirect effects of fluorine on enzymatic deiodination was also suggested.

Tokar' et al. (1989) and Balabolkin et al. (1995) have also reported

---

[4]The lower range of fluoride in drinking water in the fluorosis group is not much different from the higher range for the controls; however, in India, fluoride concentrations below 1 mg/L in drinking water are considered "safe" (Trivedi et al. 1993; Susheela et al. 2005) so the demarcation is at least a logical one.

thyroid effects in fluoride- or fluorine-exposed workers; full details of these studies are not available in English. Balabolkin et al. (1995) found that 51% of the workers examined had subclinical hypothyroidism with reduced T3.

No changes in thyroid function were detected in two studies of osteoporosis patients treated with NaF for 6 months or several years (Eichner et al. 1981; Hasling et al. 1987; for details, see Appendix E, Table E 7). These study populations are not necessarily representative of the general population, especially with respect to age and the fact that they usually receive calcium supplements. In an earlier clinical study to examine the reported effects of fluoride on individuals with hyperthyroidism, Galletti and Joyet (1958) found that, in 6 of 15 patients, both basal metabolic rate and protein-bound iodine fell to normal concentrations, and the symptoms of hyperthyroidism were relieved after fluoride treatment. Fluoride was considered clinically ineffective in the other 9 patients, although improvement in basal metabolic rate or protein-bound iodine was observed in some of them. In the 6 patients for whom fluoride was effective, tachycardia and tremor disappeared within 4-8 weeks, and weight loss was stopped. The greatest clinical improvement was observed in women between 40 and 60 years old with a moderate degree of thyrotoxicosis; young patients with the classic symptoms of Graves' disease did not respond to fluoride therapy. Radioiodine uptake tests were performed on 10 of the patients, 7 of whom showed an inhibitory effect on initial ^{131}I uptake by the thyroid.

Discussion (Effects on Thyroid Function)

In studies of animals with dietary iodine sufficiency, effects on thyroid function were seen at fluoride doses of 3-6 mg/kg/day (Stolc and Podoba 1960; Bobek et al. 1976; Guan et al. 1988; Zhao et al. 1998); in one study, effects were seen at doses as low as 0.4-0.6 mg/kg/day (Bobek et al. 1976). In low-iodine situations, more severe effects on thyroid function were seen at these doses (Stolc and Podoba 1960; Guan et al. 1988; Zhao et al. 1998). Effects on thyroid function in low-iodine situations have also been noted at fluoride doses as low as 0.06 mg/kg/day (Zhao et al. 1998), ≤0.7 mg/kg/day (Hillman et al. 1979), and 1 mg/kg/day (Guan et al. 1988). Studies showing no effect of fluoride on thyroid function did not measure actual hormone concentrations (Gedalia et al. 1960; Choubisa 1999), did not report iodine intakes (Gedalia et al. 1960; Clay and Suttie 1987; Choubisa 1999), used fluoride doses (<1.5 mg/kg/day) below those (3-6 mg/kg/day) associated with effects in other studies (Gedalia et al. 1960; Clay and Suttie 1987), or did not discuss a possibly complicating factor of the experimental procedure used (Siebenhüner et al. 1984). Only one animal study (Hara 1980) measured TSH concentrations, although that is considered a "precise and specific barometer" of thyroid status in most situations (Larsen et al. 2002). Full details of Hara's report are not available in English.

Goiter prevalence of at least 20% has been reported in humans exposed to water fluoride concentrations ≥ 0.2 mg/L (low-iodine situation; Day and Powell-Jackson 1972) or 1.5-3 mg/L (undernutrition, but adequate iodine; Jooste et al. 1999); however, other causes of goiter have not been ruled out. Bachinskii et al. (1985) showed increased TSH concentrations and reduced T3 concentrations in a population with a fluoride concentration of 2.3 mg/L in their drinking water (in comparison to a group with 1.0 mg/L), and Lin et al. (1991) showed similar results for a population with 0.88 mg/L fluoride in the drinking water (in comparison to a group with 0.34 mg/L); another study showed no effect at 3 mg/L (Baum et al. 1981). Among children considered to have adequate iodine supplementation, Susheela et al. (2005) found derangements of thyroid hormones in 54% of children with enamel fluorosis (1.1-14.3 mg/L fluoride in drinking water), and in 45-50% of "control" children without enamel fluorosis but with elevated serum fluoride concentrations. Mikhailets et al. (1996) observed an increase in TSH in workers with increased exposure time and with more severe fluorosis; low T3 was found in 65% of workers with more than 10 years of service and in 54% of workers with Stage 2 fluorosis. Several studies do not include measurements of T4, T3, or TSH (Siddiqui 1960; Gedalia and Brand 1963; Leone et al. 1964; Day and Powell-Jackson 1972; Teotia et al. 1978; Desai et al. 1993; Jooste et al. 1999).

Nutritional information (especially the adequacy of iodine and selenium intake) is lacking for many (iodine) or all (selenium) of the available studies on humans. As with the animal studies, high fluoride intake appears to exacerbate the effects of low iodine concentrations (Day and Powell-Jackson 1972; Lin et al. 1991). Uncertainty about total fluoride exposures based on water fluoride concentrations, variability in exposures within population groups, and variability in response among individuals generally have not been addressed. Although no thyroid effects were reported in studies using controlled doses of fluoride for osteoporosis therapy, the study populations are not necessarily representative of the general population with respect to age, calcium intake, and the presence of metabolic bone disease.

Thus, several lines of information indicate an effect of fluoride exposure on thyroid function. However, because of the complexity of interpretation of various parameters of thyroid function (Larsen et al. 2002), the possibility of peripheral effects on thyroid function instead of or in addition to direct effects on the thyroid, the absence of TSH measurements in most of the animal studies, the difficulties of exposure estimation in human studies, and the lack of information in most studies on nutritional factors (iodine, selenium) that are known to affect thyroid function, it is difficult to predict exactly what effects on thyroid function are likely at what concentration of fluoride exposure and under what circumstances.

Suggested mechanisms of action for the results reported to date include decreased production of thyroid hormone, effects on thyroid hormone transport in blood, and effects on peripheral conversion of T4 to T3 or on normal deiodination processes, but details remain uncertain. Both peripheral conversion of T4 to T3 and normal deiodination (deactivation) processes require the deiodinases (Types I and II for converting T4 to T3 and Types I and III for deactivation; Schneider et al. 2001; Larsen et al. 2002; Goodman 2003). Several sets of reported results are consistent with an inhibiting effect of fluoride on deiodinase activity; these effects include decreased plasma T3 with normal or elevated T4 and TSH and normal T3 with elevated T4 (Bachinskii et al. 1985; Guan et al. 1988; Lin et al. 1991; Balabolkin et al. 1995; Michael et al. 1996; Mikhailets et al. 1996; Susheela et al. 2005). The antihyperthyroid effect that Galletti and Joyet (1958) observed in some patients is also consistent with an inhibition of deiodinase activity in those individuals.

The available studies have generally dealt with mean values of various parameters for the study groups, rather than with indications of the clinical significance, such as the fraction of individuals with a value (e.g., TSH concentration) outside the normal range or with clinical thyroid disease. For example, in the two populations of asymptomatic individuals compared by Bachinskii et al. (1985), the elevated mean TSH value in the higher-fluoride group is still within the normal range, but the number of individuals in that group with TSH values above the normal range is not given.

In the absence of specific information in the reports, it cannot be assumed that all individuals with elevated TSH or altered thyroid hormone concentrations were asymptomatic, although many might have been. For asymptomatic individuals, the significance of elevated TSH or altered thyroid hormone concentrations is not clear. Belchetz and Hammond (2003) point out that the population-derived reference standards (e.g., for T4 and TSH) reflect the mean plus or minus two standard deviations, meaning that 5% of normal people have results outside a given range. At the same time, healthy individuals might regulate plasma T4 within a "personal band" that could be much more narrow than the reference range; this brings up the question of whether a disorder shifting hormone values outside the personal band but within the population reference range requires treatment (Davies and Larsen 2002; Belchetz and Hammond 2003). For example, early hypothyroidism can present with symptoms and raised TSH but with T4 concentrations still within the reference range (Larsen et al. 2002; Belchetz and Hammond 2003).

Subclinical hypothyroidism is considered a strong risk factor for later development of overt hypothyroidism (Weetman 1997; Helfand 2004). Biondi et al. (2002) associate subclinical thyroid dysfunction (either hypo- or hyperthyroidism) with changes in cardiac function and corresponding increased risks of heart disease. Subclinical hyperthyroidism can cause bone demineralization, especially in postmenopausal women, while subclinical hypothyroidism is associated with increased cholesterol concentrations, increased incidence of depression, diminished response to standard psychiatric treatment, cognitive dysfunction, and, in pregnant women, decreased IQ of their offspring (Gold et al. 1981; Brucker-Davis et al. 2001). Klein et al. (2001) report an inverse correlation between severity of maternal hypothyroidism (subclinical or asymptomatic) and the IQ of the offspring (see also Chapter 7).

A number of authors have reported delayed eruption of teeth, enamel defects, or both, in cases of congenital or juvenile hypothyroidism (Hinrichs 1966; Silverman 1971; Biggerstaff and Rose 1979; Noren and Alm 1983; Loevy et al. 1987; Bhat and Nelson 1989; Mg'ang'a and Chindia 1990; Pirinen 1995; Larsen and Davies 2002; Hirayama et al. 2003; Ionescu et al. 2004). No information was located on enamel defects or effects on eruption of teeth in children with either mild or subclinical hypothyroidism. The possibility that either dental fluorosis (Chapter 4) or the delayed tooth eruption noted with high fluoride intake (Chapter 4; see also Short 1944) may be attributable at least in part to an effect of fluoride on thyroid function has not been studied.

THYROID PARAFOLLICULAR CELLS

The parafollicular cells (C cells) of the thyroid produce a 32-amino acid peptide hormone called calcitonin (Bringhurst et al. 2002; Goodman 2003). Calcitonin acts to lower blood calcium and phosphate concentrations, primarily or exclusively by inhibiting osteoclastic (bone resorption) activity. Calcitonin does not play a major role in calcium homeostasis in humans, and its primary importance seems to be to protect against excessive bone resorption (Bringhurst et al. 2002; Goodman 2003). At high concentrations, calcitonin can also increase urinary excretion of calcium and phosphate, but these effects in humans are small and not physiologically important for lowering blood calcium (Goodman 2003). Parafollicular cells express the same G-protein-coupled, calcium-sensing receptors in their surface membranes as do the chief cells of the parathyroid glands, receptors that respond directly to ionized calcium in blood; however, the secretory response of the parafollicular cells is opposite that of the parathyroid chief cells (Bringhurst et al. 2002; Goodman 2003).

Animal Studies

Very few animal studies have examined the effects of fluoride exposure on parafollicular cells or calcitonin secretion (see Appendix E, Table E-8). Sundström (1971) found no evidence for short-term release of calcitonin in response to fluoride treatment in rats, in line with the view that NaF administration to rats by lavage resulted in hyperparathyroidism, secondary to the calcitonin-like (blood calcium-lowering) action of fluoride on bone tissue. Rantanen et al. (1972) reported that fluoride exposure had a retarding effect on cortical bone remodeling in female pigs and that an intact thyroid gland was necessary for this effect. Replacing thyroid hormone (but not calcitonin) in thyroidectomized pigs eliminated the retarding effect of fluoride, suggesting that the effect involved the formation, release, or enhanced action of calcitonin.

Human Studies

Teotia et al. (1978) found elevated calcitonin concentrations in seven patients with skeletal fluorosis in a high-fluoride area and in one of two patients who had moved to low-fluoride areas and showed improvement in various parameters (see Appendix E, Tables E-9 and E-10). Elevated calcitonin was found in all patients with an estimated fluoride intake of at least 9 mg/day and in one patient with an estimated current fluoride intake of 3.8 mg/day and a previous (until 2 years before) intake of 30 mg/day. Four of the individuals also had elevated parathyroid hormone (PTH), and radiographs of two suggested secondary hyperparathyroidism. Plasma calcium in the fluorosis patients was generally in the normal range, but urinary calcium concentrations were lower than those of controls; dietary calcium intakes were considered to be adequate. Vitamin D deficiency was not found.

In a review of skeletal fluorosis, Krishnamachari (1986) mentioned, but did not elaborate on, "significant alterations" in the "parathyroid-thyrocalcitonin axis," also stating that the sequence of the hormonal changes was not clear and that the changes did not occur to the same degree in all patients, possibly reflecting the adequacy of calcium intake. Elevated calcitonin was found in some but not all cases of skeletal fluorosis in a series of epidemiologic studies reviewed by Teotia et al. (1998).

Tokar' et al. (1989) reported elevated concentrations of calcitonin in the blood of workers employed in fluorine production, indicating stimulation of thyroid gland parafollicular cells. Huang et al. (2002) reported significantly elevated concentrations of serum PTH and calcitonin in 50 male fluoride workers and concluded that an excess of fluoride might affect secretion of both calcium-adjusting hormones.

Discussion (Effects on Parafollicular Cell Function)

Calcitonin concentrations do not seem to have been routinely measured in cases of skeletal fluorosis, but elevated calcitonin does seem to be present when looked for. The effect has been noted at fluoride intakes as low as 3.8 mg/day in humans (approximately 0.06 mg/kg/day) and was found routinely at intakes of at least 9 mg/day (approximately 0.15 mg/kg/day). No animal studies have reported calcitonin concentrations after fluoride exposure. Teotia et al. (1978) proposed several possible mechanisms (direct and indirect) of fluoride action with respect to effects on calcitonin and PTH secretion, but currently the significance of the elevated calcitonin concentrations associated with skeletal fluorosis is not clear.[5]

PARATHYROID GLANDS

In humans, four small parathyroid glands are normally situated on the posterior surface of the thyroid. These glands produce PTH, a simple 84-peptide hormone, which is the principal regulator of extracellular calcium (Bringhurst et al. 2002; Goodman 2003).[6] The primary effect of PTH is to increase the calcium concentration and decrease the phosphate concentration in blood (Bringhurst et al. 2002; Goodman 2003). The major mechanisms by which this effect occurs include the mobilization of calcium phosphate from the bone matrix, primarily from increased osteoclastic activity; in the kidney, increased reabsorption of calcium, decreased reabsorption of phosphate, and increased activation of vitamin D; and increased intestinal absorption of calcium (Bringhurst et al. 2002; Goodman 2003). PTH is also important for skeletal homeostasis (bone remodeling). Regulation of PTH secretion is inversely related to the concentration of ionized calcium (Bringhurst et al. 2002; Goodman 2003).

Healthy individuals secrete PTH throughout the day (1-3 pulses per hour); blood concentrations of PTH also exhibit a diurnal pattern, with peak values after midnight and minimum values in late morning (el-Hajj

---

[5]Calcitonin inhibits bone resorption by acting directly on the osteoclast, but it appears to play only a small role in regulating bone turnover in adults (Raisz et al. 2002). Elevated calcitonin concentrations are often present in certain types of malignancy, especially medullary thyroid carcinoma (carcinoma arising from the thyroid parafollicular cells; Bringhurst et al. 2002; Schlumberger et al. 2002), but are considered a marker for the malignancy or for certain other severe illnesses, rather than an adverse consequence. One source suggests that subtle alterations in calcitonin production or response may play a role in metabolic bone disease (Raisz et al. 2002).

[6]It is important to note that assays of PTH have varied over the years (Bringhurst et al. 2002; Goodman 2003), making it difficult to compare reported PTH concentrations among different studies; in this report, PTH concentrations (when given) are compared with the controls or healthy individuals reported for the specific studies.

Fuleihan et al. 1997; Goodman 2003). Circadian patterns of PTH concentrations differ in men and women (Calvo et al. 1991) and between healthy and osteoporotic postmenopausal women (Eastell et al. 1992; Fraser et al. 1998). The diurnal fluctuations might be important for urinary calcium conservation (el-Hajj Fuleihan et al. 1997) and might be involved in anabolic responses of bone to PTH (Goodman 2003). Alterations in PTH rhythms might contribute to or be associated with osteoporosis (el-Hajj Fuleihan et al. 1997; Fraser et al. 1998).

In Vitro Studies

Fluoride ion has been shown to be a potent inhibitor of PTH secretion in bovine and human parathyroid cells in vitro (Chen et al. 1988; Shoback and McGhee 1988; Sugimoto et al. 1990; Ridefelt et al. 1992); PTH inhibition was observed at concentrations ranging from 0.5 to 20 mM (9.5-380 mg/L) with maximum effect at or above 5 mM (95 mg/L). This action by fluoride either requires or is potentiated by Al^{3+}, consistent with a mechanism of G-protein stimulation. Fluoride (or aluminum fluoride), via the G proteins, suppresses cAMP accumulation, increases cytosolic Ca^{2+} (probably by stimulating a calcium channel), increases inositol phosphate accumulation, and also might directly inhibit the PTH secretory process (Chen et al. 1988; Shoback and McGhee 1988; Sugimoto et al. 1990; Ridefelt et al. 1992). No single mechanism is clearly responsible for inhibiting PTH secretion, suggesting that several mechanisms might be involved in its regulation.

Animal Studies

A number of animal studies of the effects of fluoride on parathyroid function are summarized below (for more details, see Appendix E, Table E-11). Administration of NaF as a lavage was found to elicit hyperparathyroidism in rats (Yates et al. 1964, as cited by Sundström 1971); the hyperparathyroidism was thought to be secondary to a direct, calcitonin-like, action of fluoride on bone tissue (Rich and Feist 1970, as cited by Sundström 1971). Levy et al. (1970) demonstrated increased resistance (suppressed sensitivity) of alveolar bone to PTH (in pharmacologic doses) in marmosets fed fluoride in drinking water (50 mg/L) for 5 months. More recently, increased serum inorganic fluoride due to use of the anesthetic isoflurane was associated with decreased ionized calcium and increased PTH and osteocalcin in cynomolgus monkeys (Hotchkiss et al. 1998).

A fivefold increase in blood PTH was seen as early as 1 week in lambs given drinking water with fluoride at 90 mg/L (Faccini and Care 1965); by 1 month, ultrastructural changes considered to be indicative of increased activity were observed in the parathyroid glands. The overactivity of the parathyroid might be a response to a "more stable mineral system, i.e. fluoroapatite" that is "resistant to the normal processes of resorption," thus requiring an increase in PTH activity to maintain normal serum calcium concentrations (Faccini 1969).

Chavassieux et al. (1991) reported a significant decrease in serum calcium and phosphorus and increases in serum PTH in sheep fed 1 or 5 mg of NaF per kg per day for 45 days, without calcium supplementation. Because of wide variation, the increased serum PTH is not considered statistically significant, but mean serum PTH in both groups at 45 days was at least twice as high as at the beginning of the experiment. This study and those of Faccini and Care (1965) and Hotchkiss et al. (1998) suggest a hypocalcemic response to the fluoride, followed by increased PTH secretion in response to the hypocalcemia.

Two longer-term animal studies with "high" concentrations of calcium and vitamin D intake have reported no effect of fluoride exposure on calcium homeostasis or parathyroid function (Andersen et al. 1986; Turner et al. 1997). However, two other studies with low-calcium situations found an altered parathyroid response. In one of these studies, Li and Ren (1997) reported that rats fed fluoride (100 mg/L in drinking water) for 2 months along with a low-calcium diet exhibited osteomalacia, osteoporosis, accelerated bone turnover, increased serum alkaline phosphatase, increased osteocalcin,[7] and increased PTH. Fluoride-treated animals with adequate dietary calcium showed only slightly increased osteoblastic activity after 2 months but elevated serum alkaline phosphatase activity and increased average width of trabecular bone after 1 year.

In an earlier study, Rosenquist et al. (1983) fed drinking water containing fluoride at 50 mg/L to male Wistar rats from the age of 5 weeks until age 51 weeks; half the animals were given a calcium-deficient diet for the last 16 weeks. Control animals were fed drinking water containing fluoride at <0.5 mg/L. At 35 weeks, average serum immunoreactive PTH was reduced, but not significantly, in the fluoride-treated rats. At 51 weeks, calcium-deficient rats without fluoride showed elevated PTH (the normal response), whereas calcium-deficient rats with fluoride showed very slightly less PTH than calcium-sufficient, fluoride-treated rats. All groups had normal serum calcium concentrations. The authors concluded that fluoride in the amount used does not increase parathyroid activity and that fluoride supplementation "seems to prevent the profound changes in parathyroid activity that result from calcium deficiency" (Rosenquist et al. 1983). However, a better interpretation of the data is that the normal increase in PTH in response to a dietary calcium deficiency did not occur in the fluoride-treated animals

---

[7]Elevated osteocalcin and alkaline phosphatase are considered markers for bone turnover (Raisz et al. 2002).

(although some morphological changes occurred), suggesting that normal parathyroid function was inhibited. These animals were adults when the calcium deficiency was imposed, and the effect of fluoride treatment on animals with a preexisting calcium deficiency was not examined. Substantially wider standard deviations were observed for all fluoride-treated and calcium-deficient groups than in the controls (no fluoride, calcium sufficiency), suggesting variable responses in the animals.

Dunipace et al. (1995, 1998) examined the effects of fluoride (up to 50 mg/L in drinking water) on male Sprague-Dawley rats with a normal diet (Dunipace et al. 1995) or with either a calcium-deficient diet or a diet deficient in protein, energy, or total nutrients (Dunipace et al. 1998). Fluoride reportedly had no effect on various clinical parameters monitored in normal, calcium-deficient, or malnourished animals; however, the papers showed results only for combinations of fluoride treatment groups, and calcium-related parameters such as PTH and calcitonin concentrations were not measured. The combination of general malnutrition and calcium deficiency was not examined.

Verma and Guna Sherlin (2002b) reported hypocalcemia in female rats and their offspring when the mothers were treated with NaF (40 mg/kg/day) during gestation and lactation. PTH was not measured.

Tiwari et al. (2004) reported decreased serum calcium, increased serum alkaline phosphatase, increased concentrations of vitamin D metabolites (both 25(OH)D3 and 1,25(OH)2D3), and lower whole body bone mineral density (suggestive of deficient mineralization) in rats born to mothers given a calcium-deficient diet and high fluoride (50 mg/L in drinking water) from day 11 of gestation; after weaning the pups were given the same low-calcium, high-fluoride regimen. Although the authors did not measure PTH or examine bone histomorphometry, they did demonstrate specific changes in gene transcription in the duodenal mucosa, including decreased transcription of the genes for the vitamin D receptor and calbindin D 9 k (a vitamin-D regulated protein that enhances calcium uptake) and altered (decreased at 9 weeks) transcription of the gene for the calcium-sensing receptor (which senses changes in extracellular calcium concentrations and regulates serum calcium concentrations by influencing PTH secretion). Excess fluoride continued to produce alterations in gene expression even when calcium was restored to the diet. The changes in gene expression are thought to result in decreased absorption of calcium from the gut.

Human Studies (Clinical, Occupational, or Population)

Clinical, occupational, and population studies of the effects of fluoride on human parathyroid function are summarized below (for more detail, see Appendix E, Table E-12). In one study with healthy subjects, a single oral dose of 27 mg of fluoride was followed by decreases in serum calcium and phosphorus and an increase in serum immunoreactive PTH (Larsen et al. 1978), suggesting a rise in PTH in response to the decrease in serum calcium. The fall in serum calcium was attributed to increased mineralization of bone in response to the fluoride dose. Oral doses of fluoride at 27 mg/day for 3 weeks in healthy adults produced a significant increase in serum osteocalcin at the end of the 3-week period but not in total or ionized calcium, alkaline phosphatase, PTH, and several other parameters (Dandona et al. 1988). The mean PTH concentration at 3 weeks was elevated slightly over the initial (pretreatment) values, and the standard deviation was considerably larger, suggesting that a few individuals might have had significant increases. In a follow-up letter, Gill et al. (1989) suggested that the age of the subjects and the sensitivity of the PTH assay might influence the findings.

Stamp et al. (1988, 1990) reported increased concentrations of biologically active PTH in osteoporosis patients receiving both calcium and sodium fluoride during short- and long-term treatments. In the short-term (8-day) study, two groups of patients were identified with respect to stability of serum calcium and phosphorus concentrations (Stamp et al. 1988). In the group with more stable serum calcium, NaF inhibited intestinal calcium and phosphorus absorption and reduced calcium balance; this inhibition is not explainable by the formation of calcium-fluoride complexes and might be due to inhibition by fluoride of some step(s) in active transport (Stamp et al. 1988).

In patients treated for 15 ± 10 months, the treated group as a whole had statistically significant elevation of biologically active PTH and serum alkaline phosphatase (Stamp et al. 1990). In those patients (32% of the treated group) in whom biologically active PTH was above the upper limit of normal, serum alkaline phosphatase was not elevated above control concentrations; elevated PTH also was associated with relative hypophosphatemia and relative hypercalciuria. Thus, in some individuals, fluoride stimulated the synthesis or release of serum alkaline phosphatase, and PTH concentrations were in the normal range; in others, serum alkaline phosphatase was not increased, indicating failure of the osteoblastic response, and PTH concentrations were above the normal range.

Duursma et al. (1987) also found that individuals varied in their responses to fluoride treatment for osteoporosis. Those individuals who had a femoral neck fracture during the treatment period (6 of 91 patients) also appeared to have lower serum alkaline phosphatase concentrations and higher serum PTH concentrations than other patients.

In a comparison of 25 fluoride-treated osteoporosis patients with calcium supplementation and 38 controls with no fluoride treatment (but in most cases calcium supplementation), Jackson et al. (1994) reported no significant difference in mean calcium concentrations between groups, although 2 of 25 individuals were outside the normal range (versus 0 of 38 controls). A significant elevation in mean alkaline phosphatase concentration was observed in the treated group, with 8 of 25 individuals outside the normal range (versus 0 of 38 controls); for those 8 individuals, the significant elevation was largely due to an increased concentration of bone isoenzymes. For the 24 patients for whom baseline (pretreatment) information was available, mean calcium concentrations were significantly lower and alkaline phosphatase was significantly higher. PTH was not measured in these patients, and individuals with a history of thyroid, parathyroid, or gastrointestinal problems were not included in the study. The authors stated that "none of the mean differences between groups were considered to be clinically significant," but whether some individuals had clinically significant situations was not addressed.

Dure-Smith et al. (1996) reported that fluoride-treated osteoporosis patients who showed a rapid increase in spinal bone density also showed a general state of calcium deficiency and secondary hyperparathyroidism; similarly treated patients with a decrease or slow increase in spinal bone density were much less likely to be calcium deficient. The degree of calcium deficiency appeared to be related to the previous fluoride-dependent increase in spinal bone density, indicating that an osteogenic response to fluoride can increase the skeletal requirement for calcium, even in patients with a high calcium intake. Reasons for the differences in response to fluoride treatment (rapid increase versus decrease or slow increase in spinal bone density) were not identified.

Osteoporosis patients treated either with slow-release NaF or with a placebo (both with concurrent calcium supplementation) showed decreases in immunoreactive PTH from initial pretreatment values, presumably due to the calcium supplementation (Zerwekh et al. 1997b). PTH values in the fluoride-treated group stayed slightly (but not significantly) higher than those in the placebo group.

Li et al. (1995) described a population study in China that examined adults in regions with various fluoride concentrations in the drinking water and either "normal" or "inadequate" nutrition in terms of protein and calcium intake; people in the sample were "healthy" rather than randomly selected. A significant decrease in blood calcium concentration was associated with an increase in fluoride exposure in the populations with inadequate nutrition but was not detected in subjects with normal nutrition. Elevated alkaline phosphatase activity with increased fluoride exposure was observed in all populations, with higher values in subjects with inadequate nutrition. PTH concentrations were not measured. For calcium, alkaline phosphatase, and several other blood parameters, all values were stated to be within the normal range regardless of fluoride exposure and nutritional condition, but it is not clear whether "all values" refers to mean or individual values.

Jackson et al. (1997) examined adult volunteers in the United States who had lived at least 30 years in communities with natural fluoride concentrations in drinking water of 0.2, 1.0, or 4.0 mg/L. Mean values for plasma calcium, phosphate, and alkaline phosphatase for all groups were within the normal ranges, although there were statistically significant differences among groups for calcium and phosphate concentrations. On the basis of plasma fluoride concentrations, the group in the 0.2-mg/L community was thought to have higher fluoride intake than expected solely from their drinking water. Calcium intakes and general nutritional status were not discussed, and PTH concentrations were not measured.

Human Studies (Endemic Skeletal Fluorosis)

Six papers (five from India and one from South Africa) describe parathyroid function in cases of endemic skeletal fluorosis (see Appendix E, Table E-13). An additional paper describes a U.S. patient with renal insufficiency, systemic fluorosis attributed to the renal insufficiency (and resulting polydipsia), and serum immunoreactive PTH more than three times the normal value (Juncos and Donadio 1972). The patient's fluoride intake at the time of the study was about 20 mg/day, or 0.34 mg/kg/day. Johnson et al. (1979) refer to that patient and 5 others with renal disease in whom fluoride (approximately 1.7-3 mg/L in drinking water) "may have been the cause of detectable clinical and roentgenographic effects." They state that plasma PTH concentrations were elevated in all 6, albeit the concentrations were considered "relatively low" for the severity of the bone disease. Two other U.S. patients with skeletal fluorosis but no renal disease did not have elevated PTH concentrations (Felsenfeld and Roberts 1991; Whyte et al. 2005).

Singh et al. (1966) found significantly higher serum alkaline phosphatase values in individuals with fluorosis but no significant differences between patients and controls in serum calcium or inorganic phosphate. They did not measure PTH.

Teotia and Teotia (1973) reported that 5 of 20 patients with skeletal fluorosis had clear evidence of secondary hyperparathyroidism. The estimated mean fluoride intake was ≥25 mg/day; dietary calcium and vitamin D were considered adequate. Laboratory results showed increased plasma alkaline phosphatase, increased phosphate clearance, decreased tubular reabsorption of phosphate, increased urinary fluoride, and decreased urinary calcium. Plasma calcium and phosphate were normal in four of the patients. Elevated serum immunoreactive parathyroid hormone was observed in all five, especially in the person with elevated plasma calcium and decreased plasma phosphate. This person, who was thought to have been developing tertiary hyperparathyroidism, was later found to have a parathyroid

EFFECTS ON THE ENDOCRINE SYSTEM 245

adenoma. Radiological findings in all five people were consistent with hyperparathyroidism.

Teotia et al. (1978) reported increased PTH concentrations in four of seven patients with endemic skeletal fluorosis (including the patient with the lowest fluoride intake); increased alkaline phosphatase was seen in at least three, and increased calcitonin was seen in all seven (Figure 8-3; Table E-10). Radiographs of two persons were consistent with secondary hyper parathyroidism. Dietary intakes of fluoride were estimated to range from 8.7 to 52 mg/day. Plasma calcium concentrations in the fluorosis patients were generally in the normal range, but urinary calcium concentrations were lower than those of controls; dietary calcium intakes were considered to be adequate. Vitamin D deficiency was not found. The finding that not everyone had elevated PTH is consistent with other observations of variability in individual responses.

Srivastava et al. (1989) described four siblings in India with skeletal fluorosis, normal total and ionized calcium concentrations, and normal vitamin D concentrations. The mother of the four had subnormal total and ionized calcium and subnormal vitamin D. All five individuals had significantly elevated PTH, elevated osteocalcin, and elevated alkaline phosphatase (Figure 8-4). Fluoride intakes were estimated to be between 16 and 49

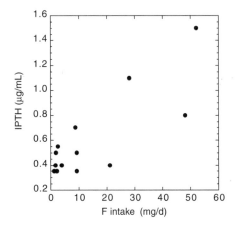

FIGURE 8-3 Plasma immunoreactive parathyroid hormone (IPTH) versus fluoride intake for nine skeletal fluorosis patients (two of whom had moved to a low-fluoride area) and five controls (data from Teotia et al. 1978; see Appendix E, Tables E-10 and E-13). Note that two of the control patients shown with IPTH values of 0.35 µg/mL were actually reported as "< 0.35" µg/mL. The four IPTH values of 0.7 µg/mL or greater were considered elevated above the values found in healthy controls.

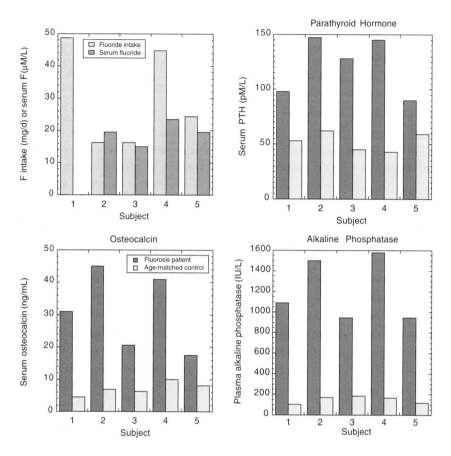

FIGURE 8-4 Fluoride intake and serum fluoride (upper left) in four Indian siblings (subjects 2-5) and their mother (subject 1). Serum PTH and osteocalcin and plasma alkaline phosphatase are shown for the same subjects and for normal age-matched Indian controls. SOURCE: Srivastava et al. 1989.

mg/day, primarily from a water source containing fluoride at 16.2 mg/L. The findings of elevated PTH in the presence of low or normal total and ionized calcium concentrations suggest secondary hyperparathyroidism in these individuals.

Pettifor et al. (1989) described a study of 260 children between 6 and 16 years old in an area of South Africa with endemic skeletal fluorosis (water fluoride concentrations of 8-12 mg/L). Hypocalcemia was present in 23% of these children and in six of nine children presenting with skeletal symptoms who were studied individually. In comparable areas with low fluoride concentrations, the prevalence of hypocalcemia was only 2% to 13%. Bone fluoride was elevated about 10-fold in the seven children measured. The children exhibited a reduced phosphaturic response during a PTH-stimulation test, suggestive of pseudohypoparathyroidism Type II; the response was directly related to the presence of hypocalcemia and could be corrected by correcting the hypocalcemia. Biopsies of iliac crest bone gave a picture of severe hyperosteoidosis associated with secondary hyperparathyroidism and a mineralization defect. The authors suggested that fluoride ingestion might increase calcium requirements and exacerbate the prevalence of hypocalcemia. The usual result of low calcium intake is classical rickets and generalized osteopenia; in this case, the combination of low calcium and high fluoride resulted in a different presentation at a later age. The degree of hypocalcemia appears to play a major role in determining the severity of osteomalacia present in endemic skeletal fluorosis and influences the renal response to hyperparathyroidism (in terms of variable serum phosphate values). The authors also pointed out the "striking male predominance" of skeletal fluorosis in their study and cited similar findings in previous studies.

Gupta et al. (2001) described a one-time study of children aged 6-12 in four regions of India with different fluoride intakes (for details, see Appendix E, Table E-14). Mean serum calcium concentrations were within the normal range for all groups. The serum PTH in all groups was correlated with the fluoride intake (Figure 8-5) and with the severity of clinical and skeletal fluorosis. The authors concluded that the increased serum PTH was related to high fluoride ingestion and could be responsible for maintaining serum calcium concentrations as well as playing a role in the toxic manifestations of fluorosis. Calcium intake is not stated in the paper, but the primary author has indicated that calcium intake in the study areas was normal (S. K. Gupta, Satellite Hospital, Banipark, Jaipur, personal communication, December 11, 2003).

In a review of skeletal fluorosis, Krishnamachari (1986) indicated that the nature (osteosclerotic, osteomalacic, osteoporotic) and severity of the fluorosis depend on factors such as age, sex, dietary calcium intake, dose and duration of fluoride intake, and renal efficiency in fluoride handling. In some cases, secondary hyperparathyroidism is observed with associated characteristic bone changes. He also noted the preponderance of males among fluorosis patients and discussed a possible protective effect of estrogens. In his review, Krishnamachari (1986) described a twofold model for the body's handling of fluoride.

1. In the presence of adequate calcium, absorbed fluoride is deposited in the bone as calcium fluorapatite. Bone density increases, urinary fluoride

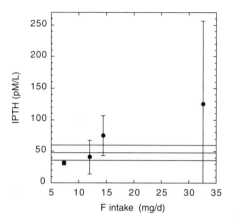

FIGURE 8-5 Parathyroid hormone (IPTH) versus fluoride intake for children in four villages with different mean fluoride intakes (Gupta et al. 2001; also see Appendix E, Tables E-13 and E-14). Vertical lines indicate standard deviations on the means. Horizontal lines indicate normal range of IPTH (48.1 ± 11.9 pM/L) for this method of measurement.

increases, but urinary calcium and phosphorus are not altered. Osteosclerosis and calcification of many tendons and ligaments occur. Serum alkaline phosphatase activity is elevated, but no specific changes occur in other constituents of serum. There are minimal hormonal changes and only mild secondary hyperparathyroidism. If the situation progresses, there will be osteophytosis (bony outgrowths), neurological complications,[8] and late crippling, producing an osteosclerotic form of fluorosis that primarily affects adults.

2. In the presence of inadequate calcium, fluoride directly or indirectly stimulates the parathyroid glands, causing secondary hyperparathyroidism leading to bone loss. Bone density is variably increased, with areas of sclerosis or porosis; there is evidence (radiological and densitometrical) of bone loss. There is renal conservation of calcium in spite of hyperparathyroidism, with no significant changes in serum biochemistry; urinary hydroxyproline excretion is significantly increased. In these conditions, an osteoporotic type of skeletal fluorosis occurs at a younger age, and growing children develop deformities due to bone softening.

---

[8]"Neurological complications" probably refers to the effects of compression of the spinal cord, e.g., those described by Singh et al. (1961).

Teotia et al. (1998) compared a number of epidemiologic studies of skeletal fluorosis from 1963 to 1997, including 45,725 children consuming water with fluoride at 1.5-25 mg/L. They observed that the combination of fluoride exposure and calcium deficiency led to more severe effects of fluoride, metabolic bone diseases, and bone deformities, resulting from excess fluoride, low calcium, high PTH, and high 1,25-dihydroxy vitamin D3. Fluoride exposure in the presence of calcium sufficiency led to an osteosclerotic form of fluorosis, with minimal secondary hyperparathyroidism. For comparable fluoride intake, metabolic bone disease occurs in 90% of children with calcium deficiency versus 25% of children with adequate calcium intake. The authors concluded that the toxic effects of fluoride occur at a lower fluoride intake (>2.5 mg/day) when there is a calcium deficiency and that fluoride appears to exaggerate the metabolic effects of calcium deficiency on bone.

Discussion (Parathyroid Function)

Of the animal studies that actually measured PTH, two studies have shown no effect of fluoride on PTH concentrations in blood (Liu and Baylink 1977; Andersen et al. 1986); animals in these studies were supplied with adequate or high dietary calcium. An additional three studies reported no effect of fluoride on serum or plasma calcium concentrations but did not measure PTH concentrations (Rosenquist and Boquist 1973; Dunipace et al. 1995, 1998). Rosenquist and Boquist (1973) gave no information on dietary calcium. One experiment by Dunipace et al. (1998) specifically used low dietary calcium for some treatment groups. Turner et al. (1997) found decreased serum calcium and elevated (but not significantly so) PTH in fluoride-treated animals with high dietary calcium. Both Verma and Guna Sherlin (2002b) and Tiwari et al. (2004) reported hypocalcemia due to combined calcium deficiency and fluoride exposure, but PTH was not measured. Tiwari et al. (2004) described changes in gene expression that would result in reduced calcium absorption from the gut. Elevated PTH concentrations were reported for fluoride-treated animals in three papers, including one with no information on dietary calcium (Faccini and Care 1965), one with normal dietary calcium and decreased serum calcium (Chavassieux et al. 1991), and one with low dietary calcium (Li and Ren 1997). In one other study, the normal response to a calcium deficiency (elevated PTH) did not occur in fluoride-exposed animals (Rosenquist et al. 1983).

Human studies show elevated PTH concentrations in at least some individuals at doses of 0.4-0.6 mg/kg/day (Teotia and Teotia 1973; Larsen et al. 1978; Duursma et al. 1987; Dandona et al. 1988; Stamp et al. 1988, 1990; Srivastava et al. 1989; Dure-Smith et al. 1996; Gupta et al. 2001) and in some cases at doses as low as 0.15 mg/kg/day (Teotia et al. 1978) and 0.34 mg/kg/day (Juncos and Donadio 1972). Li et al. (1995) found a significant decrease in mean plasma calcium concentrations with increased fluoride exposure in populations of apparently healthy adults with inadequate nutrition, but PTH was not measured. Jackson et al. (1994) found calcium concentrations outside the normal range in 2 of 25 persons treated with fluoride for osteoporosis, but the mean value for the group was within the normal range; these persons also received calcium supplementation. Calcium concentrations in 24 patients decreased from pretreatment concentrations; however, PTH concentrations were not measured. Jackson et al. (1997) also found no significant effect of fluoride on blood calcium concentrations in people who lived in communities with different fluoride concentrations but presumably had adequate nutrition; PTH concentrations were not measured.

The indirect action of fluoride on parathyroid function is relatively straightforward: fluoride induces a net increase in bone formation (Chavassieux et al. 1991) and also decreases calcium absorption from the gastrointestinal tract (beyond the degree expected by formation of calcium-fluoride complexes; Krishnamachari 1986; Stamp et al. 1988; Ekambaram and Paul 2001); both of these effects lead to an increase in the body's calcium requirement (Pettifor et al. 1989; Ekambaram and Paul 2001). If dietary calcium is inadequate to support the increased requirement, the response is an increase in PTH (secondary hyperparathyroidism). PTH acts to increase resorption of bone, but the effect is uneven; low-fluoride bone is resorbed first (Faccini 1969). As bone fluoride increases, the "solubility" of the bone, or the ease with which it is resorbed, is decreased (because of the greater stability of fluorapatite), giving an apparent resistance to the effects of PTH (Faccini 1969; Levy et al. 1970; Messer et al. 1973a,b). The indirect action of fluoride to cause an increased calcium requirement is consistent with reports of reduced milk production (due to inadequate mobilization of calcium from bone) in livestock with excessive fluoride consumption and of more severe fluorosis in lactating animals (due to the higher calcium utilization during lactation) (e.g., Eckerlin et al. 1986a,b; Jubb et al. 1993). The work of Tiwari et al. (2004) provides an initial description of a mechanism by which fluoride exposure in the presence of a calcium deficiency further increases the dietary requirement for calcium, namely by altering the expression of genes necessary for calcium absorption from the gastrointestinal tract.

Some studies also indicate direct effects of fluoride on the parathyroid gland. Elevated PTH in the presence of normal serum calcium might indicate a stimulatory effect of fluoride (Gill et al. 1989; Srivastava et al. 1989). The absence of the normal elevation of PTH in response to calcium deficiency suggests an inhibitory effect (Rosenquist et al. 1983), as do several in vitro studies (Chen et al. 1988; Shoback and McGhee 1988; Sugimoto et al. 1990; Ridefelt et al. 1992). The possibility also exists that a direct effect on either the parathyroid or the thyroid parafollicular cells leads to a compensatory response from the other, but this has not been examined.

Several studies have reported different responses among individuals or variability in group responses (Teotia and Teotia 1973; Teotia et al. 1978; Krishnamachari 1986; Duursma et al. 1987; Dandona et al. 1988; Stamp et al. 1988; 1990; Jackson et al. 1994; Dure-Smith et al. 1996; Gupta et al. 2001); the reasons for these differences are not clear but might include genetic differences in addition to variability in nutritional factors. The effects also might vary with age, sex, and the duration (as well as degree) of hypocalcemia.

Any cause of hypocalcemia or vitamin D deficiency can lead to secondary hyperparathyroidism (elevated PTH) in an attempt by the body to maintain calcium homeostasis (Ahmad and Hammond 2004).[9] Fluoride clearly has the effect of decreasing serum calcium and increasing the calcium requirement in some or many exposed persons. In those studies which have measured it, PTH is elevated in some persons in response to fluoride exposure, indicating secondary hyperparathyroidism. No information has been reported in those studies on the clinical effects, if any, in those persons. In general, secondary hyperparathyroidism in response to calcium deficiency may contribute to a number of diseases, including osteoporosis, hypertension, arteriosclerosis, degenerative neurological diseases, diabetes mellitus, some forms of muscular dystrophy, and colorectal carcinoma (Fujita and Palmieri 2000). McCarty and Thomas (2003) suggest that down-regulation of PTH (by calcium and/or vitamin D supplementation) could assist in control of weight and prevention of diabetes.

Calcium deficiency induced or exacerbated by fluoride exposure may contribute to other adverse health effects. For example, Goyer (1995) indicates that low dietary calcium increases the concentration of lead in critical organs and the consequent toxicity. A recent increase in the number of cases of nutritional rickets in the United States appears to reflect calcium-deficient diets rather than vitamin D deficiencies (DeLucia et al. 2003). These cases occur in children whose diet lacks dairy products;[10] circulating PTH concentrations are elevated, as are alkaline phosphatase concentrations. The authors "emphasize that nutritional calcium deficiency may occur in North American infants and is not limited to the setting of developing countries" and state that "factors that affect calcium absorption may be important in determining a susceptibility to the development of rickets."

---

[9]Renal failure is the most common cause of secondary hyperparathyroidism (Ahmad and Hammond 2004).

[10]A diet low in dairy products will have not only a lower calcium content but probably also a higher fluoride content, due to greater use of beverages such as juices that have been manufactured with fluoridated municipal water (see Chapter 2); absorption and retention of fluoride will be higher because of the calcium deficiency.

PINEAL GLAND

The pineal gland is a small organ (150 mg in humans) located near the center of the brain. One of the major components of the mammalian circadian system, it lies in the upper margins of the thalamus in the dorsal aspects of the third ventricle and has both physical and neuronal connections with the brain. Although the pineal gland lies outside the blood-brain barrier, it has access to the cerebrospinal fluid. The pineal gland's major neuronal connections with the brain are the sympathetic nerve fibers coming from the superior cervical ganglion; the activity of these sympathetic nerves controls synthesis and release of the pineal hormone melatonin (Cone et al. 2002).[11] Other substances (primarily peptides) are also secreted from the pineal gland and have been reported to have various physiological effects, including antigonadotropic, metabolic, and antitumor activity (Anisimov 2003).

Most melatonin production occurs during darkness (Reiter 1998; Salti et al. 2000; Cone et al. 2002; Murcia García et al. 2002). Peak serum concentrations of melatonin occur during childhood in humans, with decreasing concentrations during adolescence before stabilization at the low concentration characteristic of adults (García-Patterson et al. 1996; Murcia García et al. 2002); further decreases in melatonin occur at menopause in women and at a corresponding age in men (Reiter 1998).

Melatonin affects target tissues, such as the hypophyseal pars tuberalis, that have a high density of melatonin receptors. The primary effect seems to be temporally specific activation of cAMP-sensitive gene expression in the pars tuberalis by the sensitization of adenylyl cyclase, thus synchronizing the suprachiasmatic nucleus of the hypothalamus and clock-controlled genes in peripheral tissue (Stehle et al. 2003). In humans, changes in melatonin are associated with the status of the reproductive system—onset of puberty, stage of puberty, menstrual cyclicity, menopause (Reiter 1998; Salti et al. 2000)—but the functional relationships are not fully understood. The elevated melatonin concentrations characteristic of prepubertal age suggest an inhibitory effect on pubertal development (Aleandri et al. 1997; Salti et al. 2000); sexual maturation begins when serum melatonin starts to decrease (Aleandri et al. 1997; Reiter 1998). Melatonin also seems to be involved with anxiety reactions; for example, the beneficial effects of fluoxetine (Prozac) in mice during an anxiety test are not found if the pineal gland has been removed (Uz et al. 2004).

Melatonin and pineal peptides have been associated with a number of other physiological effects, including regulation of circadian rhythms and

---

[11]Melatonin is also found in cells lining the gut from stomach to colon. Its functions are mainly protective, including free radical scavenging. Some of melatonin's actions are receptor-mediated and involve the central and peripheral sympathetic nervous systems (Reiter et al. 2003a).

EFFECTS ON THE ENDOCRINE SYSTEM   253

sleep (Arendt 2003; Cajochen et al. 2003); regulation of reproductive physiology in seasonal breeders (Aleandri et al. 1997; Reiter 1998; Stehle et al. 2003); effects on calcium and phosphorus metabolism, parathyroid activity, bone growth, and development of postmenopausal osteoporosis (Chen et al. 1990, 1991; Sandyk et al. 1992; Shoumura et al. 1992; el-Hajj Fuleihan et al. 1997; Roth et al. 1999; Cardinali et al. 2003; Goodman 2003); oncostatic or anticarcinogenic effects (Cohen et al. 1978; García-Patterson et al. 1996; Panzer 1997; Anisimov 2003); antioxidant actions (Srinivasan 2002; Reiter et al. 2003b); and effects on the central nervous system, psychiatric disease, and sudden infant death syndrome (García-Patterson et al. 1996; Reiter 1998; Delagrange et al. 2003). Panzer (1997) suggested that the simultaneous decrease in melatonin concentrations and the exponential increase in bone growth during puberty could be a factor in the typical age distribution of osteosarcoma.

Pineal Gland Calcification

The pineal gland is a calcifying tissue; in humans, calcified concretions can be found at any age, although the likelihood increases with age (Vígh et al. 1998; Akano and Bickler 2003) and may be associated with menopause (Sandyk et al. 1992). The occurrence of pineal calcifications varies among different populations and nations (Vígh et al. 1998), possibly in association with the degree of industrialization (Akano and Bickler 2003), rates of breast cancer (Cohen et al. 1978), and high circannual light intensity near the equator (Vígh et al. 1998). Osteoporosis might be associated with fewer concretions (Vígh et al. 1998).

Melatonin secretion is well correlated with the amount of uncalcified pineal tissue (Kunz et al. 1999) but not with the size of pineal calcification (Vígh et al. 1998; Kunz et al. 1999). An increase in calcification of the pineal gland in humans probably represents a decrease in the number of functioning pinealocytes and a corresponding decrease in the individual's ability to produce melatonin (Kunz et al. 1999). The degree of calcification, relative to the size of an individual's pineal gland, has been suggested as a marker of the individual's decreased capability to produce melatonin (Kunz et al. 1999).

As with other calcifying tissues, the pineal gland can accumulate fluoride (Luke 1997, 2001). Fluoride has been shown to be present in the pineal glands of older people (14-875 mg of fluoride per kg of gland in persons aged 72-100 years), with the fluoride concentrations being positively related to the calcium concentrations in the pineal gland, but not to the bone fluoride, suggesting that pineal fluoride is not necessarily a function of cumulative fluoride exposure of the individual (Luke 1997, 2001). Fluoride has not been measured in the pineal glands of children or young adults, nor has there been any investigation of the relationship between pineal fluoride concentrations and either recent or cumulative fluoride intakes.

In Vitro Studies

Few studies have examined the effects of fluoride on pineal function. NaF (2.5-20 mM, or fluoride at 47.5-380 mg/L) produces markedly increased adenylyl cyclase activity (up to four times control activity) of rat pineal homogenates in vitro (Weiss 1969a,b), as it does in other tissues (Weiss 1969a); ATPase activity in the homogenates was inhibited by up to 50% (Weiss 1969a). Potassium fluoride (7-10 mM, or fluoride at 133-190 mg/L) has been used experimentally to increase adenylyl cyclase activity in rat pineal glands in vitro (Zatz 1977, 1979).

Animal Studies

Details of the effect of fluoride on pineal function are presented in Appendix E, Table E-15. Luke (1997) examined melatonin production as a function of age and time of day in Mongolian gerbils (*Meriones unguiculatus*). On an absolute basis, melatonin production by the low-fluoride group was constant at ages 7-28 weeks, with no difference between males and females. Relative to body weight, melatonin output declined progressively with age until adulthood (by 11.5 weeks in females and 16 weeks in males). In contrast, prepubescent gerbils fed the high-fluoride diet had significantly lower pineal melatonin production than prepubescent gerbils fed the low-fluoride diet. Relative to body weight, the normal higher rate of melatonin production in sexually immature gerbils did not occur.

Sexual maturation in females occurred earlier in the high-fluoride animals (Luke 1997); males had increases in melatonin production relative to body weight between 11.5 and 16 weeks (when a decrease normally would occur), and testicular weight at 16 weeks (but not at 9 or 28 weeks) was significantly lower in high-fluoride than in low-fluoride animals. The circadian rhythm of melatonin production was altered in the high-fluoride animals at 11.5 weeks but not at 16 weeks. In high-fluoride females at 11.5 weeks, the nocturnal peak (relative to body weight) occurred earlier than in the low-fluoride animals; also, the peak value was lower (but not significantly lower) in the high-fluoride animals. In males, a substantial reduction ($P < 0.00001$) in the nocturnal peak (relative to body weight) was observed in the high-fluoride animals.

Human Studies

Although no studies are available that specifically address the effect of fluoride exposure on pineal function or melatonin production in humans, two studies have examined the age of onset of menstruation (age of menarche) in girls in fluoridated areas (Schlesinger et al. 1956; Farkas et al. 1983; for details, see Appendix E, Table E-15);[12] the earlier study was discussed by Luke (1997) as part of the basis for her research. No comparable information on sexual maturation in boys is available.

In girls examined approximately 10 years after the onset of fluoridation (1.2 mg/L, in 1945) in Newburgh, New York, the average age[13] at menarche was 12 years, versus 12 years 5 months among girls in unfluoridated Kingston (Schlesinger et al. 1956).[14] The authors stated that this difference was not statistically significant. Note that those girls who reached menarche during the time period of the study had not been exposed to fluoride over their entire lives, and some had been exposed perhaps for only a few years before menarche (they would have been 8-9 years old at the time fluoridation was started). Those girls in Newburgh who had been exposed to fluoridated water since birth (or before birth) had not yet reached menarche by the time of the study.

A later study in Hungary (Farkas et al. 1983) reported no difference in the menarcheal age of girls in a town with "optimal" fluoride concentration (1.09 mg/L in Kunszentmárton, median menarcheal age 12.779 years) and a similar control town (0.17 mg/L in Kiskunmajsa; median menarcheal

---

[12]Both Schlesinger et al. (1956) and Farkas et al. (1983) referred to tables of the distribution of ages at the time of first menstruation, but, in fact, both studies provided only frequencies by age (presumably at the time of study, in either 1-year or 0.5-year increments) of girls having achieved menarche by the stated age. Farkas et al. (1983) specifically indicated use of the probit method for ascertainment of the median age at menarche; the data provided by Schlesinger et al. (1956) appear to correspond to that method, but they do not specifically mention it. The probit (or status quo) method appears to be routinely used to estimate the median (or other percentiles of) age at menarche, sometimes in conjunction with an estimated mean age at menarche based on recall data (e.g., Wu et al. 2002; Anderson et al. 2003; Chumlea et al. 2003; Padez and Rocha 2003). According to Grumbach and Styne (2002), "The method of ascertainment of the age of menarche is of importance. Contemporaneous recordings are performed with the probit method of asking, 'yes' or 'no,' are you menstruating? These may be incorrect because of social pressures of the culture and socioeconomic group considered. Recalled ages of menarche are used in other studies and considered to be accurate within 1 year (in 90% of cases) during the teenage years and in older women, too."

[13]Probably the median age, although the text simply says "average." Similar studies appear to use the term "average age at menarche" to refer to the "estimated median age at menarche" (Anderson et al. 2003).

[14]For comparison purposes, estimates of mean or median age at menarche for the white population in the United States include 12.80 years for 1963-1970 (Anderson et al. 2003) and 12.55-12.7 years for 1988-1994 (Wu et al. 2002; Anderson et al. 2003; Chumlea et al. 2003).

age 12.79 years). This study shows postmenarcheal girls present at younger ages in the higher fluoride town than in the low-fluoride town, although the reported median ages were the same (Farkas et al. 1983).

Discussion (Pineal Function)

Whether fluoride exposure causes decreased nocturnal melatonin production or altered circadian rhythm of melatonin production in humans has not been investigated. As described above, fluoride is likely to cause decreased melatonin production and to have other effects on normal pineal function, which in turn could contribute to a variety of effects in humans. Actual effects in any individual depend on age, sex, and probably other factors, although at present the mechanisms are not fully understood.

OTHER ENDOCRINE ORGANS

The effects of fluoride exposure have been examined for several other endocrine organs, including the adrenals, the pancreas, and the pituitary (for details, see Appendix E, Tables E-16 and E-17). Effects observed in animals include changes in organ weight, morphological changes in tissues, increased mitotic activity, decreased concentrations of pituitary hormones, depressed glucose utilization, elevated serum glucose, and elevated insulin-like growth factor-1 (IGF-1). Effects reported in humans include "endocrine disturbances," impaired glucose tolerance, and elevated concentrations of pituitary hormones. Studies of the effects of fluoride on glucose metabolism and in diabetic animals are discussed below; information on other effects is extremely limited.

Animal Studies (Diabetic Animals)

Two studies have examined the effects of fluoride exposure in diabetic rats. In the first study, Dunipace et al. (1996) compared male Zucker fatty diabetic rats and Zucker age-matched controls given drinking water with fluoride at 5, 15, or 50 mg/L.[15] For the physiological, biochemical, and genetic variables that were monitored, no "measurable adverse effects" were noted. Statistically significant differences with respect to fluoride intake (as opposed to differences between normal and diabetic animals) were observed only for diabetic rats with fluoride at 50 mg/L. No endocrinological parameters (e.g., PTH) were measured. Dunipace et al. (1996) reported that fluoride intake, excretion, and balance were generally similar in this study and

---

[15]These fluoride intakes were considered to be equivalent to intakes by humans of 1, 3, and 10 mg/L (Dunipace et al. 1996).

in a previous study with Sprague-Dawley rats but that there were "strain-specific differences in fluoride sensitivity"; these differences were not defined or explained. The Zucker fatty diabetic rat is considered to be an animal model for human Type II (noninsulin-dependent) diabetes mellitus, although the diabetic rats in this study did not experience renal insufficiency, and the study was terminated before an age that might be more comparable to ages associated with late-onset diabetes and diabetic complications in humans. The authors concluded that the diabetic rats "were not at increased risk of fluorosis," even though femoral fluoride concentrations (2,700-9,500 µg/g in ash for diabetic rats given fluoride at 15 or 50 mg/L versus 2,500-3,600 in normal rats given fluoride at 50 mg/L) were in the range associated with fluorosis in humans and exceeded concentrations of bone fluoride associated with decreased bone strength in rabbits (6,500-8,000 ppm in ash; Turner et al. 1997); no basis for their conclusion was given.

In the second study, Boros et al. (1998) compared the effects of fluoride at 10 mg/L in drinking water for 3 weeks on young female rats (Charles River, Wistar), either normal (nondiabetic) or with streptozotocin-induced, untreated diabetes. An additional group of normal rats was given an amount of fluoride in drinking water corresponding to the fluoride intake by the diabetic rats (up to about 3 mg/day per rat). Both feed and water consumption increased significantly in the diabetic rats (with and without fluoridated water); water consumption was significantly higher in the diabetic rats on fluoridated water than in those on nonfluoridated water. Fasting blood glucose concentrations were increased significantly in both diabetic groups, but more so in the group on fluoridated water. Fluoride treatment of nondiabetic animals did not cause any significant alteration in blood glucose concentrations. Plasma fluoride was higher, and bone fluoride was lower, in diabetic than in nondiabetic animals given the same amount of fluoride, indicating lower deposition of fluoride into bone and lower renal clearance of fluoride in the diabetic animals. The increased kidney weight found in diabetic animals on nonfluoridated water was not seen in the fluoride-treated diabetic animals. Additional biochemical and hormonal parameters were not measured.

In contrast to the Zucker fatty diabetic rats in the study by Dunipace et al. (1996), the streptozotocin-induced diabetic rats in this study (Boros et al., 1998) provide an animal model considered representative of Type I (insulin-dependent) diabetes mellitus in humans. In these rats, the general severity of the diabetes (blood glucose concentrations, kidney function, weight loss) was worse in animals given fluoride at 10 mg/L in their drinking water. In both types of diabetic rats, fluoride intake was very high because of the several-fold increase in water consumption, and corresponding plasma, soft tissue, and bone fluoride concentrations were elevated accordingly. Thus, any health effects related to plasma or bone fluoride concentrations, for example, would be expected to occur in animals or humans with uncontrolled (or inadequately controlled) diabetes at lower fluoride concentrations in drinking water than for nondiabetics, because of the elevated water intakes. In addition, the results reported by Boros et al. (1998) suggested that, for some situations (e.g., diabetes in which kidney function is compromised), the severity of the diabetes could be increased with increasing fluoride exposure.

Animal Studies (Normal Animals)

Turner et al. (1997) reported a 17% increase in serum glucose in female rabbits given fluoride in drinking water at 100 mg/L for 6 months. IGF-1 was also significantly increased (40%) in these rabbits, but other regulators of serum glucose, such as insulin, were not measured. The authors suggested that IGF-1 concentrations might have changed in response to changes in serum glucose concentrations. Dunipace et al. (1995, 1998) found no significant differences with chronic fluoride treatment in mean blood glucose concentrations in rats; specific data by treatment group were not reported, and parameters such as insulin and IGF-1 were not measured.

Suketa et al. (1985) and Grucka-Mamczar et al. (2005) have reported increases in blood glucose concentrations following intraperitoneal injections of NaF; Suketa et al. (1985) attributed these increases to fluoride stimulation of adrenal function. Rigalli et al. (1990, 1992, 1995), in experiments with rats, reported decreases in insulin, increases in plasma glucose, and disturbance of glucose tolerance associated with increased plasma fluoride concentrations. The effect of high plasma fluoride (0.1-0.3 mg/L) appeared to be transient, and the decreased response to a glucose challenge occurred only when fluoride was administered before (as opposed to together with or immediately after) the glucose administration (Rigalli et al. 1990). In chronic exposures, effects on glucose metabolism occurred when plasma fluoride concentrations exceeded 0.1 mg/L (5 µmol/L) (Rigalli et al. 1992, 1995). The in vivo effect appeared to be one of inhibition of insulin secretion rather than one of insulin-receptor interaction (Rigalli et al. 1990). Insulin secretion (both basal and glucose-stimulated) by isolated islets of Langerhans in vitro was also inhibited as a function of fluoride concentrations (Rigalli et al. 1990, 1995). Rigalli et al. (1990) pointed out that recommended plasma fluoride concentrations for treatment of osteoporosis are similar to those shown to affect insulin secretion.

Human Studies

Jackson et al. (1994) reported no differences in mean fasting blood glucose concentrations between osteoporosis patients treated with fluoride and untreated controls, although 3 of 25 treated individuals had values outside the normal range (versus 1 of 38 controls). No significant differences were found between groups of older adults with different fluoride concentrations in drinking water in studies in China (Li et al. 1995; subjects described as "healthy" adults) and the United States (Jackson et al. 1997), and all mean values were within normal ranges.[16] Glucose tolerance tests were not conducted in these studies.

Trivedi et al. (1993) reported impaired glucose tolerance in 40% of young adults with endemic fluorosis, with fasting serum glucose concentrations related to serum fluoride concentrations; the impaired glucose tolerance was reversed after 6 months of drinking water with "acceptable" fluoride concentrations (<1 mg/L). It is not clear whether individuals with elevated serum fluoride and impaired glucose tolerance had the highest fluoride intakes of the group with endemic fluorosis or a greater susceptibility than the others to the effects of fluoride. For all 25 endemic fluorosis patients examined, a significant positive correlation between serum fluoride and fasting serum immunoreactive insulin (IRI) was observed, along with a significant negative correlation between serum fluoride and fasting glucose/insulin ratio (Trivedi et al. 1993).

The finding of increased IRI contrasts with findings of decreased insulin in humans after exposure to fluoride (Rigalli et al. 1990; de la Sota et al. 1997) and inhibition of insulin secretion by rats, both in vivo and in vitro (Rigalli et al. 1990, 1995). However, the assay for IRI used by Trivedi et al. (1993) could not distinguish between insulin and proinsulin, and the authors suggested that the observed increases in both IRI and serum glucose indicate either biologically inactive insulin—perhaps elevated proinsulin—or insulin resistance. Inhibition of one of the prohormone convertases (the enzymes that convert proinsulin to insulin) would result in both elevated proinsulin secretion and increased blood glucose concentrations and would be consistent with the decreased insulin secretion reported by Rigalli et al. (1990, 1995) and de la Sota et al. (1997). Although Turner et al. (1997) suggested fluoride inhibition of insulin-receptor activity as a mechanism for increased blood glucose concentrations, Rigalli et al. (1990) found no difference in response to exogenous insulin in fluoride-treated versus control rats, consistent with no interference of fluoride with the insulin-receptor interaction.

Discussion (Other Endocrine Function)

More than one mechanism for diabetes or impaired glucose tolerance exists in humans, and a variety of responses to fluoride are in keeping with

---

[16]In the study by Jackson et al. (1997), samples were nonfasting; in the study by Li et al. (1995), it is not clear whether samples were fasting or nonfasting.

variability among strains of experimental animals and among the human population. The conclusion from the available studies is that sufficient fluoride exposure appears to bring about increases in blood glucose or impaired glucose tolerance in some individuals and to increase the severity of some types of diabetes. In general, impaired glucose metabolism appears to be associated with serum or plasma fluoride concentrations of about 0.1 mg/L or greater in both animals and humans (Rigalli et al. 1990, 1995; Trivedi et al. 1993; de al Sota et al. 1997). In addition, diabetic individuals will often have higher than normal water intake, and consequently, will have higher than normal fluoride intake for a given concentration of fluoride in drinking water. An estimated 16-20 million people in the United States have diabetes mellitus (Brownlee et al. 2002; Buse et al. 2002; American Diabetes Association 2004; Chapter 2); therefore, any role of fluoride exposure in the development of impaired glucose metabolism or diabetes is potentially significant.

SUMMARY

The major endocrine effects of fluoride exposures reported in humans include elevated TSH with altered concentrations of T3 and T4, increased calcitonin activity, increased PTH activity, secondary hyperparathyroidism, impaired glucose tolerance, and possible effects on timing of sexual maturity; similar effects have been reported in experimental animals. These effects are summarized in Tables 8-1 and 8-2, together with the approximate intakes or physiological fluoride concentrations that have been typically associated with them thus far. Table 8-2 shows that several of the effects are associated with average or typical fluoride intakes of 0.05-0.1 mg/kg/day (0.03 with iodine deficiency), others with intakes of 0.15 mg/kg/day or higher. A comparison with Chapter 2 (Tables 2-13, 2-14, and 2-15) will show that the 0.03-0.1 mg/kg/day range will be reached by persons with average exposures at fluoride concentrations of 1-4 mg/L in drinking water, especially the children. The highest intakes (>0.1 mg/kg/d) will be reached by some individuals with high water intakes at 1 mg/L and by many or most individuals with high water intakes at 4 mg/L, as well as by young children with average exposures at 2 or 4 mg/L.

Most of the studies cited in this chapter were designed to ascertain whether certain effects occurred (or in cases of skeletal fluorosis, to see what endocrine disturbances might be associated), not to determine the lowest exposures at which they do occur or could occur. Estimates of exposure listed in these tables and in Appendix E are, in most cases, estimates of average values for groups based on assumptions about body weight and water intake. Thus, individual responses could occur at lower or higher exposures than those listed. Although the comparisons are incomplete, similar effects

TABLE 8-1 Summary of Major Observed Endocrine Effects of Fluoride in Experimental Animals, with Typical Associated Intakes and Physiological Fluoride Concentrations

End Point	Fluoride Intake, mg/kg/day	Fluoride in Serum or Plasma, mg/L	Fluoride in Urine, mg/L	Fluoride in Bone, ppm in ash	Key References
Altered thyroid function (altered T4 and T3 concentrations)	3-6 (lower with iodine deficiency)	NA[a]	≥6 (possibly ≥2-3)	≥2,400	Stolc and Podoba 1960; Bobek et al. 1976; Hillman et al. 1979; Guan et al. 1988; Zhao et al. 1998; Cinar and Selcuk 2005
Altered calcitonin activity	2	NA	NA	3,200-3,500[b]	Rantanen et al. 1972
Altered melatonin production; altered timing of sexual maturity	3.7	NA	NA	2,800	Luke 1997
Inhibited parathyroid function	5.4	NA	NA	NA	Rosenquist et al. 1983
Increased serum glucose; increased severity of diabetes	7-10.5	0.1-0.7[c,d]	NA	>1,000	Rigalli et al. 1990, 1992, 1995; Turner et al. 1997; Boros et al. 1998
Increased parathyroid hormone concentrations, secondary hyperparathyroidism	9-10	≥ 0.2[c]	NA	2,700-3,200	Faccini and Care 1965; Chavassieux et al. 1991

[a]Not available.
[b]ppm.
[c]Serum.
[d]Plasma.

are seen in humans at much lower fluoride intakes (or lower water fluoride concentrations) than in rats or mice, but at similar fluoride concentrations in blood and urine. This is in keeping with the different pharmacokinetic behavior of fluoride in rodents and in humans (Chapter 3) and with the variability in intake, especially for humans.

TABLE 8-2 Summary of Major Observed Endocrine Effects of Fluoride in Humans, with Typical Associated Intakes and Physiological Fluoride Concentrations

End Point	Fluoride Intake, mg/kg/day[a]	Fluoride in Serum or Plasma, mg/L	Fluoride in Urine, mg/L	Key References
Altered thyroid function (altered T4 and/or T3 concentrations)	0.05-0.1 (0.03 with iodine deficiency)	≥0.25[a]	2.4	Bachinskii et al. 1985; Lin et al. 1991; Yang et al. 1994; Michael et al. 1996; Susheela et al. 2005
Elevated TSH concentrations	0.05-0.1 (0.03 with iodine deficiency)	≥0.25[a]	≥2	Bachinskii et al. 1985; Lin et al. 1991; Yang et al. 1994; Susheela et al. 2005
Elevated calcitonin concentrations	0.06-0.87	0.11-0.26[b]	2.2-18.5 mg/day	Teotia et al. 1978
Goiter prevalence ≥ 20%	0.07-0.13 (≥ 0.01 with iodine deficiency)	NA[c]	NA	Day and Powell-Jackson 1972; Desai et al. 1993; Jooste et al. 1999
Impaired glucose tolerance in some individuals	0.07-0.4	0.08[a] 0.1-0.3[b]	2-8	Rigalli et al. 1990, 1995; Trivedi et al. 1993; de la Sota 1997
Increased parathyroid hormone concentrations, secondary hyperparathyroidism, in some individuals	0.15-0.87	0.14-0.45[b]	3-18.5 mg/day	Juncos and Donadio 1972; Teotia and Teotia 1973; Larsen et al. 1978; Teotia et al. 1978; Duursma et al. 1987; Dandona et al. 1988; Stamp et al. 1988, 1990; Pettifor et al. 1989; Srivastava et al. 1989; Dure-Smith et al. 1996; Gupta et al. 2001

[a]Serum.
[b]Plasma.
[c]Not available.

Thyroid Function

Fluoride exposure in humans is associated with elevated TSH concentrations, increased goiter prevalence, and altered T4 and T3 concentrations; similar effects on T4 and T3 are reported in experimental animals, but TSH has not been measured in most studies. In animals, effects on thyroid function have been reported at fluoride doses of 3-6 mg/kg/day (some effects at

0.4-0.6 mg/kg/day) when iodine intake was adequate (Table 8-1); effects on thyroid function were more severe or occurred at lower doses when iodine intake was inadequate. In humans, effects on thyroid function were associated with fluoride exposures of 0.05-0.13 mg/kg/day when iodine intake was adequate and 0.01-0.03 mg/kg/day when iodine intake was inadequate (Table 8-2).

Several sets of results are consistent with inhibition of deiodinase activity, but other mechanisms of action are also possible, and more than one might be operative in a given situation. In many cases, mean hormone concentrations for groups are within normal limits, but individuals may have clinically important situations. In particular, the inverse correlation between asymptomatic hypothyroidism in pregnant mothers and the IQ of the offspring (Klein et al. 2001) is a cause for concern. The recent decline in iodine intake in the United States (CDC 2002d; Larsen et al. 2002) could contribute to increased toxicity of fluoride for some individuals.

Thyroid Parafollicular Cell Function

Only one study has reported calcitonin concentrations in fluoride-exposed individuals. This study found elevated calcitonin in all patients with fluoride exposures above about 0.15 mg/kg/day and in one patient with a current intake of approximately 0.06 mg/kg/day (Table 8-2); these exposures corresponded to plasma fluoride concentrations of 0.11-0.26 mg/L. Results attributed to altered calcitonin activity have also been found in experimental animals at a fluoride exposure of 2 mg/kg/day (Table 8-1). It is not clear whether elevated calcitonin is a direct or indirect result of fluoride exposure, nor is it clear what the clinical significance of elevated calcitonin concentrations might be in individuals.

Parathyroid Function

In humans, depending on the calcium intake, elevated concentrations of PTH are routinely found at fluoride exposures of 0.4-0.6 mg/kg/day and at exposures as low as 0.15 mg/kg/day in some individuals (Table 8-2). Similar effects and exposures have been found in a variety of human studies; these studies indicate that elevated PTH and secondary hyperparathyroidism occur at fluoride intakes higher than those associated with other endocrine effects. In the single study that measured both calcitonin and PTH, all individuals with elevated PTH also had elevated calcitonin, and several individuals had elevated calcitonin without elevated PTH (Teotia et al. 1978). Elevated concentrations of PTH and secondary hyperparathyroidism have also been reported at fluoride intakes of 9-10 mg/kg/day (and as low as 0.45-2.3 mg/kg/day in one study) in experimental animals (Table 8-1). One animal study found what appears to be inhibition of the normal parathyroid response to calcium deficiency at a fluoride intake of 5.4 mg/kg/day.

As with calcitonin, it is not clear whether altered parathyroid function is a direct or indirect result of fluoride exposure. An indirect effect of fluoride by causing an increased requirement for calcium is probable, but direct effects could occur as well. Also, although most individuals with skeletal fluorosis appear to have elevated PTH, it is not clear whether parathyroid function is affected before development of skeletal fluorosis or at lower concentrations of fluoride exposure than those associated with skeletal fluorosis. Recent U.S. reports of nutritional (calcium-deficiency) rickets associated with elevated PTH (DeLucia et al. 2003) suggest the possibility that fluoride exposure, together with increasingly calcium-deficient diets, could have an adverse impact on the health of some individuals.

Pineal Function

The single animal study of pineal function indicates that fluoride exposure results in altered melatonin production and altered timing of sexual maturity (Table 8-1). Whether fluoride affects pineal function in humans remains to be demonstrated. The two studies of menarcheal age in humans show the possibility of earlier menarche in some individuals exposed to fluoride, but no definitive statement can be made. Recent information on the role of the pineal organ in humans suggests that any agent that affects pineal function could affect human health in a variety of ways, including effects on sexual maturation, calcium metabolism, parathyroid function, postmenopausal osteoporosis, cancer, and psychiatric disease.

Glucose Metabolism

Increased serum glucose and increased severity of existing diabetes have been reported in animal studies at fluoride intakes of 7-10.5 mg/kg/day (Table 8-1). Impaired glucose tolerance in humans has been reported in separate studies at fluoride intakes of 0.07-0.4 mg/kg/day, corresponding to serum fluoride concentrations above about 0.1 mg/L. The primary mechanism appears to involve inhibition of insulin production.

General Considerations

The available studies of the effects of fluoride exposure on endocrine function have several limitations. In particular, many studies did not measure actual hormone concentrations, several studies did not report nutritional status (e.g., iodine or calcium intake), and, for thyroid function, other possible goitrogenic factors have not been ruled out. Most studies have too few exposure groups, with, for example, the "high"-fluoride group in one study having lower concentrations of fluoride in drinking water than the "normal"-fluoride group in another study. In general, the human exposures are not well characterized. Nevertheless, there is consistency among the available studies in the types of effects seen in humans and animals and in the concentrations or fluoride exposures associated with the effects in humans.

For all the endocrine effects reported to occur from fluoride exposure, the variability in exposure and response among populations (or strains of an experimental animal) or within a human population requires further attention. For example, correlations between the fluoride intake or the presence or degree of fluorosis and the presence (or prevalence) or severity of other effects generally have not been examined on an individual basis, which could permit identification of individual differences in susceptibility or response. Several reports have identified subgroups within an exposed population or group, in terms of the response observed, even when group means are not statistically different.

Variability in response to fluoride exposures could be due to differences in genetic background, age, sex, nutrient intake (e.g., calcium, iodine, selenium), general dietary status, or other factors. Intake of nutrients such as calcium and iodine often is not reported in studies of fluoride effects. The effects of fluoride on thyroid function, for instance, might depend on whether iodine intake is low, adequate, or high, or whether dietary selenium is adequate. Dietary calcium affects the absorption of fluoride (Chapter 3); in addition, fluoride causes an increase in the dietary requirements for calcium, and insufficient calcium intake increases fluoride toxicity. Available information now indicates a role for aluminum in the interaction of fluoride on the second messenger system; thus, differences in aluminum exposure might explain some of the differences in response to fluoride exposures among individuals and populations.

The clinical significance of fluoride-related endocrine effects requires further attention. For example, most studies have not mentioned the clinical significance for individuals of hormone values out of the normal range, and some studies have been limited to consideration of "healthy" individuals. As discussed in the various sections of this chapter, recent work on borderline hormonal imbalances and endocrine-disrupting chemicals indicates that significant adverse health effects, or an increased risk for development of clearly adverse health outcomes, could be associated with seemingly mild imbalances or perturbations in hormone concentrations (Brucker-Davis et al. 2001). In addition, the different endocrine organs do not function entirely separately: thyroid effects (especially elevated TSH) may be associated with parathyroid effects (Stoffer et al. 1982; Paloyan Walker et al. 1997), and glucose metabolism may be affected by thyroid or parathyroid status (e.g., McCarty and Thomas 2003; Procopio and Borretta 2003; Cettour-Rose et al. 2005). Adverse effects in individuals might occur when hormone concentrations are still in the normal ranges for a population but are low or high for that individual (Brucker-Davis et al. 2001; Belchetz and Hammond 2003). Some investigators suggest that endocrine-disrupting chemicals could be associated with nonmonotonic dose-response curves (e.g., U-shaped or inverted-U-shaped curves resulting from the superimposition of multiple dose-response curves) and that a threshold for effects cannot be assumed (Bigsby et al. 1999; Brucker-Davis et al. 2001).

In summary, evidence of several types indicates that fluoride affects normal endocrine function or response; the effects of the fluoride-induced changes vary in degree and kind in different individuals. Fluoride is therefore an endocrine disruptor in the broad sense of altering normal endocrine function or response, although probably not in the sense of mimicking a normal hormone. The mechanisms of action remain to be worked out and appear to include both direct and indirect mechanisms, for example, direct stimulation or inhibition of hormone secretion by interference with second messenger function, indirect stimulation or inhibition of hormone secretion by effects on things such as calcium balance, and inhibition of peripheral enzymes that are necessary for activation of the normal hormone.

RECOMMENDATIONS

- Further effort is necessary to characterize the direct and indirect mechanisms of fluoride's action on the endocrine system and the factors that determine the response, if any, in a given individual. Such studies would address the following:
  — the in vivo effects of fluoride on second messenger function
  — the in vivo effects of fluoride on various enzymes
  — the integration of the endocrine system (both internally and with other systems such as the neurological system)
  — identification of those factors, endogenous (e.g., age, sex, genetic factors, or preexisting disease) or exogenous (e.g., dietary calcium or iodine concentrations, malnutrition), associated with increased likelihood of effects of fluoride exposures in individuals
  — consideration of the impact of multiple contaminants (e.g., fluoride and perchlorate) that affect the same endocrine system or mechanism
  — examination of effects at several time points in the same individuals to identify any transient, reversible, or adaptive responses to fluoride exposure.

- Better characterization of exposure to fluoride is needed in epidemiology studies investigating potential endocrine effects of fluoride. Important exposure aspects of such studies would include the following:

EFFECTS ON THE ENDOCRINE SYSTEM

— collecting data on general dietary status and dietary factors that could influence the response, such as calcium, iodine, selenium, and aluminum intakes

— characterizing and grouping individuals by estimated (total) exposure, rather than by source of exposure, location of residence, fluoride concentration in drinking water, or other surrogates

— reporting intakes or exposures with and without normalization for body weight (e.g., mg/day and mg/kg/day), to reduce some of the uncertainty associated with comparisons of separate studies

— addressing uncertainties associated with exposure and response, including uncertainties in measurements of fluoride concentrations in bodily fluids and tissues and uncertainties in responses (e.g., hormone concentrations)

— reporting data in terms of individual correlations between intake and effect, differences in subgroups, and differences in percentages of individuals showing an effect and not just differences in group or population means.

— examining a range of exposures, with normal or control groups having very low fluoride exposures (below those associated with 1 mg/L in drinking water for humans).

- The effects of fluoride on various aspects of endocrine function should be examined further, particularly with respect to a possible role in the development of several diseases or mental states in the United States. Major areas for investigation include the following:

— thyroid disease (especially in light of decreasing iodine intake by the U.S. population);

— nutritional (calcium deficiency) rickets;

— calcium metabolism (including measurements of both calcitonin and PTH);

— pineal function (including, but not limited to, melatonin production); and

— development of glucose intolerance and diabetes.

9

Effects on the Gastrointestinal, Renal, Hepatic, and Immune Systems

This chapter evaluates the effects of fluoride on the gastrointestinal system (GI), the kidney, the liver, and the immune system, focusing primarily on new data that have been generated since the earlier NRC (1993) review. Studies that involved exposures to fluoride in the range of 2-4 milligrams per liter (mg/L) are emphasized, so that the safety of the maximum-contaminant-level goal (MCLG) can be evaluated.

GI SYSTEM

Fluoride occurs in drinking water primarily as free fluoride. When ingested some fluorides combine with hydrogen ions to form hydrogen fluoride (HF), depending on the pH of the contents of the stomach (2.4% HF at pH 5; 96% HF at pH 2). HF easily crosses the gastric epithelium, and is the major form in which fluoride is absorbed from the stomach (see Chapter 3). Upon entering the interstitial fluid in the mucosa where the pH approaches neutrality, HF dissociates to release fluoride and hydrogen ions which can cause tissue damage. Whether damage occurs depends on the concentrations of these ions in the tissue. It appears that an HF concentration somewhere between 1.0 and 5.0 mmol/L (20 and 100 mg/L), applied to the stomach mucosa for at least 15 minutes, is the threshold for effects on the function and structure of the tissue (Whitford et al. 1997). Reported GI symptoms, such as nausea, may not be accompanied by visible damage to the gastric mucosa. Thus, the threshold for adverse effects (discomfort) is likely to be lower than that proposed by Whitford et al. This review is concerned primarily with the chronic ingestion of fluoride in drinking water containing fluoride at 2-4 mg/L. Single high doses of ingested fluoride are known to elicit acute GI symptoms, such as nausea and vomiting, but whether chronic exposure to drinking water with fluoride at 4 mg/L can elicit the same symptoms has not been documented well.

The primary symptoms of GI injury are nausea, vomiting, and abdominal pain (see Table 9-1). Such symptoms have been reported in case studies (Waldbott 1956; Petraborg 1977) and in a clinical study involving double-blind tests on subjects drinking water artificially fluoridated at 1.0 mg/L (Grimbergen 1974). In the clinical study, subjects were selected whose GI symptoms appeared with the consumption of fluoridated water and disappeared when they switched to nonfluoridated water. A pharmacist prepared solutions of sodium fluoride (NaF) and sodium silicofluoride (Na_2SiF_6) so that the final fluoride ion concentrations were 1.0 mg/L. Eight bottles of water were prepared with either fluoridated water or distilled water. Patients were instructed to use one bottle at a time for 2 weeks. They were asked to record their symptoms throughout the study period. Neither patients nor the physician administering the water knew which water samples were fluoridated until after the experiments were completed. The fluoridation chemicals added to the water at the time of the experiments were likely the best candidates to produce these symptoms. Despite those well-documented case reports, the authors did not estimate what percentage of the population might have GI problems. The authors could have been examining a group of patients whose GI tracts were particularly hypersensitive. The possibility that a small percentage of the population reacts systemically to fluoride, perhaps through changes in the immune system, cannot be ruled out (see section on the immune system later in this chapter).

Perhaps it is safe to say that less than 1% of the population complains of GI symptoms after fluoridation is initiated (Feltman and Kosel 1961). The numerous fluoridation studies in the past failed to rigorously test for changes in GI symptoms and there are no studies on drinking water containing fluoride at 4 mg/L in which GI symptoms were carefully documented. Nevertheless, there are reports of areas in the United States where the drinking water contains fluoride at concentrations greater than 4 mg/L and as much as 8 mg/L (Leone et al. 1955b). Symptoms of GI distress or discomfort were not reported. In the United Kingdom, where tea drinking is more common, people can consume up to 9 mg of fluoride a day (Jenkins 1991). GI symptoms were not reported in the tea drinkers. The absence of symptoms might be related to the hardness of the water, which is high in some areas of the United Kingdom. Jenkins (1991) reported finding unexpectedly high concentrations of fluoride (as high as 14 mg/L) in soft water compared with hard water when boiled. In contrast, in India, where endemic fluorosis is well documented, severe GI symptoms are common (Gupta et al. 1992; Susheela et al. 1993; Dasarathy et al. 1996). One cannot rule out the

TABLE 9-1 Studies of Gastrointestinal Effects in Humans

Approximate Concentration of Fluoride in the Stomach[a]	Study Design	Findings	Application/ Proposed Mechanisms	Comments	Reference
Water Fluoridation					
1.0 mg/L	Case reports of patients (n = 52) drinking artificially fluoridated water.	Stomach cramps, abdominal pain, and nausea resolved when patients stopped drinking fluoridated water.	Possible gastrointestinal hypersensitivity.	Low daily dose of fluoride; cluster of subjects selected on the basis of symptoms.	Waldbott 1956
1.0 mg/L	Double-blinded test of patients (n = 60) drinking artificially fluoridated water in Haarlem, Netherlands.	50% of subjects had stomach and intestinal symptoms; 30% had stomatitis.	Possible gastrointestinal hypersensitivity.	Low daily dose of fluoride; self-reporting of symptoms.	Grimbergen 1974
1.0 mg/L	Case reports of symptoms in subjects (n = 20) drinking fluoridated water in Milwaukee.	Fatigue, pruritis, polydipsia, headaches, and gastrointestinal symptoms.	Possible gastrointestinal hypersensitivity.	Low daily dose; cluster of subjects selected on the basis of symptoms.	Petraborg 1977
Water Fluoridation Accidents					
75-300 mg/L[b] (range due to differences found in 2 fluoride feeders)	Symptoms reported in 34 children during accidental overfeed in school water supply.	Fluoride concentrations in water were 93.5 and 375 mg/L. 68% of the children had gastrointestinal upset.	Acute fluoride toxicity of the gastric epithelium.	Symptoms resolved after problem was corrected; doses of fluoride in mg/kg were not reported.	Hoffman et al. 1980
250 mg/L, (based on 50-mL ingestion)	Symptoms reported in 22 subjects during accidental overfeed in school water supply.	Fluoride concentration in water was 1,041 mg/L. 91% of the subjects had nausea and vomiting.	Acute fluoride toxicity of the gastric epithelium.	Only small amounts of the beverages made with the school's water were consumed.	Vogt et al. 1982

41 mg/L	Symptoms reported in 321 subjects during accidental overfeed in water supply.	Of the 160 persons who drank water; 52% had gastroenteritis. Only 2% of subjects who did not drink water reported gastroenteritis. Itching and skin rash also reported. Fluoride concentration in water peaked at 51 mg/L.	Acute fluoride toxicity of the gastric epithelium.	Petersen et al. 1988
150 mg/L (assuming no dilution with stomach fluid)	Symptoms reported in 47 residents of a town during accidental fluoride overfeed of the water supply.	90% had nausea, vomiting, diarrhea, abdominal pains, or numbness or tingling of the face or extremities. One person in the town died. Fluoride concentration in water was 150 mg/L.	Acute fluoride toxicity of the gastric epithelium.	Death occurred due to nausea, vomiting, diarrhea, and repeated ingestion of water (large acute dose).	Gessner et al. 1994
20-30 mg/L (based on 100-mL ingestion)	Symptoms reported in 39 patrons of a restaurant who consumed water or ice during an overfeed accident.	34 subjects had acute gastrointestinal illness in a 24-hour period after exposure. Fluoride concentration in water was 40 mg/L.	Acute fluoride toxicity of the gastric epithelium.	Symptoms resolved after problem was corrected; dose of fluoride was not reported but was estimated to be 3 mg/kg.	Penman et al. 1997
46-69 mg/L	Symptoms reported in 7 school children during accidental overfeed.	Nausea and vomiting. Fluoride concentration in water was 92 mg/L.	Acute fluoride toxicity of the gastric epithelium.	Dose in mg/kg was not reported.	Sidhu and Kimmer 2002

continued

TABLE 9-1 Continued

Approximate Concentration of Fluoride in the Stomach[a]	Study Design	Findings	Application/ Proposed Mechanisms	Comments	Reference
Other Exposures					
5 ppm	Symptoms reported in pregnant women and their children from birth to 9 years taking NaF (1.2 mg) supplements. 672 cases (461 controls)	1% of cases had dermatologic, gastrointestinal, and neurologic effects. Comparisons with controls treated with binder placebo tablets established the effects to be from fluoride and not the binder.	Chronic or acute toxicity.	Details of clinical trial (e.g., randomization, stratification) not reported; dose in mg/kg was not reported; gastrointestinal systems were probably worse in small children (due to higher dose per kilogram of body weight).	Feltman and Kosel 1961
20 ppm, (assuming 100 of mL stomach fluid)	Symptoms observed in 10 adult volunteers who ingested 3 g of gel containing fluoride at 0.42% (4,200 mg/L)	Petechiae and erosion found in 7 of 10 subjects. Surface epithelium was most affected portion of the mucosa.	Acute fluoride toxicity of the gastric epithelium.	Approximately 10% of a probably toxic dose.	Spak et al. 1990
136 ppm (calculated from 30 mg of NaF ingested in 100 mL of stomach fluid)	Symptoms observed in 10 patients with otosclerosis treated with NaF at 30 mg/day for 3-12 months.	7 subjects had abdominal pains, vomiting, and nausea. Endoscopy revealed petechiae, erosion, and erythema. Histological exams showed chronic atrophic gastritis in all patients and in only one of the controls.	Acute fluoride toxicity of the gastric epithelium.		Das et al. 1994

200 ppm (using the 0.05% NaF mouthwash example)	Evaluation of reports to the American Association of Poison Control Centers of suspected overingestion of fluoride to estimate toxic amounts of home-use fluoride products.	Authors estimate a "probably toxic dose" of fluoride to children less than 6 years of age to be 50 mg. That dose was based on examples of a 10-kg child ingesting 10.1 g of 1.1% NaF gel; 32.7 g of 0.63% SnF$_2$ gel; 33.3 g of toothpaste with 1,500 ppm of fluoride; 50 g of toothpaste with 1,000 ppm of fluoride; or 221 mL of 0.05% NaF rinse.	Acute fluoride toxicity of the gastric epithelium.	Similar total acute doses as the water fluoridation overfeed accidents.	Shulman and Wells 1997

[a]In most studies, the concentration of fluoride in the stomach was not determined, so estimates were made by the committee. The actual concentrations could vary widely depending on the volume in the stomach and the rate of gastric secretions. The latter could also vary depending on the effect of fluoride (or any other agent) on the secretory process.

[b]Estimated from ingesting 400 mL of fluoridated water (unless dose was reported) diluted 0.8 with 100 mL of stomach fluid with fluoride at 1 mg/: (empty stomach).

influence of poor nutrition (the absence of dietary calcium in the stomach) contributing to the GI upset from fluoride ingestion. Chronic ingestion of drinking water rich in fluoride on an empty stomach is more likely to elicit symptoms.

GI Symptoms Relating to the Concentration of Fluoride Intake

It is important to realize that GI effects depend more on the net concentration of the aqueous solution of fluoride in the stomach than on the total fluoride dose in the fluid or solid ingested. The presence of gastric fluids already in the stomach when the fluoride is ingested can affect the concentration of the fluoride to which the gut epithelium is exposed. The residual volume of stomach fluid ranges between 15 and 30 mL in people fasting overnight (Narchi et al. 1993; Naguib et al. 2001; Chang et al. 2004). Such volumes would decrease the fluoride concentration of a glass of drinking water by only about 10%. In Table 9-1, the concentrations of fluoride in the stomach were estimated from the mean reported fluoride exposures. A dilution factor was used when it was clear that the subjects already had fluid in their stomach. The results from the water fluoridation overfeed reports (concentrations of fluoride in the stomach between 20 and 250 mg/L) indicate that GI symptoms, such as nausea and vomiting, are common side effects from exposure to high concentrations of fluoride.

Fluoride supplements are still routinely used today in areas where natural fluoride in the drinking water falls below 0.7 mg/L. In an early clinical trial using fluoride supplements, Feltman and Kosel (1961) administered fluoride tablets containing 1.2 mg of fluoride or placebo tablets to pregnant mothers and children up to 9 years of age. They determined that about 1% of the subjects complained of GI symptoms from the fluoride ingredient in the test tablets. If it is assumed that the stomach fluid volume after taking the fluoride supplement was approximately 250 mL, the concentration to which the stomach mucosal lining was exposed was in the neighborhood of 5 mg/L. GI effects appear to have been rarely evaluated in the fluoride supplement studies that followed the early ones in the 1950s and 1960s. Table 9-1 suggests that, as the fluoride concentration increases in drinking water, the percentage of the population with GI symptoms also increases. The table suggests that fluoride at 4 mg/L in the drinking water results in approximately 1% of the population experiencing GI symptoms (see Feltman and Kosel 1961).

Chronic Moderate Dose Ingestion of Fluoride

It is clear from the fluoride and osteoporosis clinical trial literature (also see Chapter 5) that gastric side effects were common in these studies (e.g.,

Mamelle et al. 1988; Hodsman and Drost 1989; Kleerekoper and Mendlovic 1993). Slow-release fluorides and calcium supplementation helped to reduce GI side effects (Kleerekoper and Mendlovic 1993; Das et al. 1994; Haguenauer et al. 2000). In areas of endemic fluorosis, such as parts of India, most subjects suffer from GI damage and adverse GI symptoms (Gupta et al. 1992; Susheela et al. 1993; Dasarathy et al. 1996). In one study (Susheela et al. 1993), every fourth person exposed to fluoride in drinking water (<1 to 8 mg/L) reported adverse GI symptoms. The results from these studies cannot be compared with the water fluoridation studies summarized in Table 9-1, because in the osteoporosis trials fluoride was nearly always administered as enteric coated tablets along with calcium supplements and the nutrition status of populations in endemic fluorosis areas is different from that in the United States.

Fluoride Injury Mechanisms in the GI Tract

Because 1% of the population is likely to experience GI symptoms, and GI symptoms are common in areas of endemic fluorosis, especially where there is poor nutrition (Gupta et al. 1992; Susheela et al. 1993; Dasarathy et al. 1996), it is important to understand the biological and physiological pathways for the effects of fluoride on the GI system. Those mechanisms have been investigated in many animal studies. In those studies, the concentrations of fluoride used were generally 100- to 1,000-fold higher than what occurs in the serum of subjects drinking fluoridated water. Although some tissues encounter enormous elevations in fluoride concentrations relative to the serum (e.g., kidney, bone), it is unlikely that the gut epithelium would be exposed to millimolar concentrations of fluoride unless there has been ingestion of large doses of fluoride from acute fluoride poisoning. During the ingestion of a large acute dose of fluoride such as fluoride-rich oral care products, contaminated drinking water during fluoridation accidents, and fluoride drugs for the treatment of osteoporosis, the consumption of large amounts of drinking water containing fluoride at 4 mg/L would serve only to aggravate the GI symptoms. Animal studies (see Table 9-2) have provided some important information on the mechanisms involved in GI toxicity from fluoride. Fluoride can stimulate secretion of acid in the stomach (Assem and Wan 1982; Shayiq et al. 1984), reduce blood flow away from the stomach lining, dilate blood vessels, increase redness of the stomach lining (Fujii and Tamura 1989; Whitford et al. 1997), and cause cell death and desquamation of the GI tract epithelium (Easmann et al. 1984; Pashley et al. 1984; Susheela and Das 1988; Kertesz et al. 1989; NTP 1990; Shashi 2002).

Because fluoride is a known inhibitor of several metabolic intracellular enzymes, it is not surprising that, at very high exposures, there is cell

TABLE 9-2 Animal Studies of Gastrointestinal Effects and Mechanisms of Fluoride

Species	Study	Findings	Possible Mechanisms/Comments	Reference
In Vitro Studies				
Rat	Circular muscle strips from the colons of colitic rats were treated with 10 mM NaF. Colitis was experimentally induced by intracolonic instillation of acetic acid.	NaF-induced contractions were significantly reduced in tissues from colitic rats compared with controls on days 2 and 3 postenema but not 14 days after enema. Results suggest that colitis alters smooth muscle contractility by disturbing elements in the signal transduction pathway distal to receptor activation of the G proteins.	Purpose of the study was to investigate whether colitis-induced decreases in the contraction of colonic smooth muscle is due to alteration in the excitation-contraction-coupling process at a site distal to receptor occupancy. Decrease in contractility might be due to impaired utilization of intracellular calcium.	Myers et al. 1997
Mouse	Isolated distended stomachs were treated with NaF at various concentrations (1-10 mM NaF).	Dose-related stimulation of H^+ ion secretion. Stimulation of H^+ ion secretion might be due to histamine release and increased formation of cyclic AMP (cAMP) in the gastric mucosa.	Fluoride might contribute to excess acid production in gastrointestinal tract.	Assem and Wan 1982
Guinea pig	Isolated gastric chief cells treated with NaF (0-30 mM).	NaF increased intracellular diacylglycerol and Ca^{2+}; 0.1 mM $AlCl_3$ increased the effect of NaF.	Possible activation of a pertussis-toxin insensitive G protein coupled to a signal transducing mechanism. Action appears to be distinct from that activated by cholecystokinin.	Nakano et al. 1990
Rabbit	Electronic chloride secretion by the jejunum was assessed by measuring short-circuit current variations (ΔI_{sc}) due to alterations in ionic transport.	NaF induced a transient increase in I_{sc} at >5 mM; inhibited the antisecretory effect of peptide PYY and its analog P915 at 2 mM and decreased the stimulation of secretion by forskolin and dibutyryl cAMP by 50% at 2 mM. At 5 mM, inhibition of protein kinase C by bisindolylmaleimide caused a sustained increase in I_{sc}.	NaF might reduce PYY-induced inhibition via a G-protein-dependent and a G-protein-independent functional pathway.	Eto et al. 1996

Rabbit	Fluoride transport in intestinal brush border membrane vesicles examined.	Fluoride uptake by brush border membrane vesicles occurred rapidly and with an overshoot only in the presence of an inward-directed proton gradient.	Fluoride transport occurs via a carrier-mediated process that might involve cotransport of fluoride with H^+ or exchange of fluoride with OH^-	He et al. 1998
Rabbit	Effect of NaF on the transport of bovine serum albumin across the distal and proximal colonic epithelium.	Transport of bovine serum albumin was significantly reduced by NaF.	Fluoride affected transport mechanisms in the colon.	Hardin et al. 1999
In Vivo Studies				
Rat	25 mg/kg in drinking water for 60 days.	Increased gastric acidity and output.	Elevation of cAMP concentrations in the gastric mucosa can stimulate H^+ output, which might account for gastric symptoms reported in endemic fluorosis areas or from occupational exposure by inhalation.	Shayiq et al. 1984
Rat	1 or 10 mM NaF (in 0.1 M HCl) placed in rat stomach for 1 hour.	Concentration- and time-dependent histological damage to the surface mucous cells.	The higher concentration of NaF increased gastric permeability to small but not large molecules.	Pashley et al. 1984
Rat	1, 10, or 50 mM NaF (in 0.1 M HCl) placed in rat stomach.	At 10 mM, desquamation of the surface mucous epithelial cells. At 50 mM, substantial damage to cells around the gastric gland openings and interfoveolar cell loss.	Possible toxicity of the gut epithelium.	Easmann et al. 1984

continued

TABLE 9-2 Continued

Species	Study	Findings	Possible Mechanisms/Comments	Reference
Rat	100 mM NaF and 50 mM CaF_2 intragastrically.	NaF-treated rats had extensive desquamation and cell injury. CaF_2-treated animals showed some desquamation and decrease in secretory activity.	Injury to stomach lining might affect secretion.	Kertesz et al. 1989
Rat	Single oral dose of NaF at 300 mg/kg.	Reduced blood flow from the stomach, reduced blood calcium, dilated blood vessels in the stomach, and redness.	Redness in the pyloric region of the stomach and intestine is likely due to a relaxation of the small veins, resulting in an accumulation of circulating blood in the mucosa of the intestinal tract.	Fujii and Tamura 1989
Rat	300 mg/L in drinking water for 6 months.	Gross lesions of the stomach in male rats. Diffuse mucosal hyperplasia with cellular necrosis in female rats.	Chronic fluoride toxicity of the gut epithelium.	NTP 1990
Rat	Stomachs of rats were instilled with 5 and 20 mM NaF for 1 hour.	Increased output of fluid, fucose, and galactose; marked reduction of titratable acidity of the lumen was pH dependent; and reduced amount of Alcian blue was bound to adherent mucus in a pH-independent manner.	Authors suggest that NaF accumulates with acid and acts as a barrier-breaking agent, rather than as a mucus-secretion stimulating agent.	Gharzouli et al. 2000
Mouse	NaF at 100 mg/L in drinking water for 30 days.	Organosomatic index decreased. Histopathologic changes of the intestine included increased number of goblet cells in the villi and crypts, cytoplasmic degranulation and vacuolation, nuclear pyknosis, abnormal mitosis, and lymphatic infiltration of submucosa and lamina propria.		Sondhi et al. 1995

Rabbit	10 mg/kg/day by gavage for 2 years.	Morphologic abnormalities observed in all treated animals. "Cracked-clay" appearance of the microvilli surface of the duodenal epithelium and epithelial cell degeneration.		Susheela and Das 1988
Rabbit	Rabbits were given subcutaneous injections of 5, 10, 20, and 20 mg/kg/day for 15 weeks, and the duodenum was examined by histology.	Erosion and cell death of the surface mucosa, hemorrhage, cell death of Brunner's gland, clumped submucosa, and hypertrophy of muscles in muscularis mucosae. Loss of mucosal layer was in direct proportion to NaF exposure.	Injury to the intestine caused by cell death in the mucosal lining.	Shashi 2002
Dog	Stomach mucosa with vascular supply intact was exposed to 1, 5, 10, 50, and 100 mM fluoride in different experiments.	At 5 and 10 mM fluoride, marked increases in the fluxes of water and sodium potassium, and hydrogen ions, mucus secretion, and tissue swelling and redness observed. Histopathologic exams showed marked thinning of the surface cell layer, reduced uptake of periodic acid Schiff stain, localized exfoliation and necrosis of surface cells, acute gastritis, and edema.	Authors concluded that the threshold for effects on the structure and function of the gastric mucosa was approximately 1 mM fluoride.	Whitford et al. 1997

death and desquamation of the GI gut epithelium wall. The mechanisms involved in altering secretion remain unknown but are likely the result of fluoride's ability to activate guanine nucleotide regulatory proteins (G proteins) (Nakano et al. 1990; Eto et al. 1996; Myers et al. 1997). Whether fluoride activates G proteins in the gut epithelium at very low doses (e.g., from fluoridated water at 4.0 mg/L) and has significant effects on the gut cell chemistry must be examined in biochemical studies.

THE RENAL SYSTEM

The kidney is the organ responsible for excreting most of the fluoride. It is exposed to concentrations of fluoride about five times higher than in other organs, as the tissue/plasma ratio for the kidney is approximately 5 to 1, at least in the rat (Whitford 1996). Kidneys in humans may be exposed to lower fluoride concentrations than in rats. Human kidneys, nevertheless, have to concentrate fluoride as much as 50-fold from plasma to urine. Portions of the renal system may therefore be at higher risk of fluoride toxicity than most soft tissues. In this section, three aspects of kidney function are discussed in the context of fluoride toxicity. First, can long-term ingestion of fluoride in drinking water at 4 mg/L contribute to the formation of kidney stones? Second, what are the mechanisms of fluoride toxicity on renal tissues and function? And third, what special considerations have to be made in terms of residents who already have kidney failure and who are living in communities with fluoride at 4 mg/L in their drinking water?

Does Fluoride in Drinking Water Contribute to Kidney Stones?

Early water fluoridation studies did not carefully assess changes in renal function. It has long been suspected that fluoride, even at concentrations below 1.2 mg/L in drinking water, over the years can increase the risk for renal calculi (kidney stones). Research on this topic, on humans and animals, has been sparse, and the direction of the influence of fluoride (promotion or prevention of kidney stones) has been mixed (Table 9-3; Juuti and Heinonen 1980; Teotia et al. 1991; Li et al. 1992; Shashi et al. 2002). Singh et al. (2001) carried out an extensive examination of more than 18,700 people living in India where fluoride concentrations in the drinking water ranged from 3.5 to 4.9 mg/L. Patients were interviewed for a history of urolithiasis (kidney stone formation) and examined for symptoms of skeletal fluorosis, and various urine and blood tests were conducted. The patients with clear signs and symptoms of skeletal fluorosis were 4.6 times more likely to develop kidney stones. Because the subjects of this study were likely at greater risk of kidney stone formation because of malnutrition, similar research should be conducted in North America in areas with fluoride at 4 mg/L in the drinking water. It is possible that the high incidence of uroliths is related to the high incidence of skeletal fluorosis, a disorder that has not been studied extensively in North America. If fluoride in drinking water is a risk factor for kidney stones, future studies should be directed toward determining whether kidney stone formation is the most sensitive end point on which to base the MCLG.

Mechanisms of Fluoride Toxicity on Kidney Tissue and Function

Fluoride in acute and chronic doses can dramatically affect the kidney, but, again, it is the dose that is important. People living in fluoridated areas (at 1.0 mg/L) drinking 1.0 L of water a day will consume 1 mg of fluoride a day (less than 0.014 mg/kg for the average 70-kg person). There are no published studies that show that fluoride ingestion on a chronic basis at that concentration can affect the kidney. However, people living in an area where the drinking water contains fluoride at 4 mg/L who consume 2-3 L of water per day will ingest as much as 12 mg fluoride per day on a chronic basis (see Chapter 2). On the basis of studies carried out on people living in regions where there is endemic fluorosis, ingestion of fluoride at 12 mg per day would increase the risk for some people to develop adverse renal effects (Singh et al. 2001).

Humans can be exposed to even higher acute doses of fluoride either unintentionally (water fluoridation accidents, hemodialysis accidents, accidental poisoning) or intentionally, such as from fluorinated general anesthetics. Administration of certain halothane anesthetics, which are defluorinated by the liver, can result in serum fluoride concentrations that are 50-fold higher than normal, and those concentrations are maintained during surgery and well afterward (see Table 9-3 and Chapter 2). These concentrations of fluoride in the serum have been associated with nephrotoxicity, but most of the symptoms resolve after surgery when fluoride concentrations are allowed to decline. Although it is unlikely that consuming fluoridated drinking water could lead to such high serum fluoride concentrations, one has to consider that subjects who already have impaired kidney function and are unable to excrete fluoride efficiently will retain more fluoride. At this time, there are no studies to distinguish between adverse effects produced by fluoride and the defluorinated metabolites of fluorinated general anesthetics. Therefore, it is plausible that the defluorinated metabolites are responsible for some, most, or even all of the side effects on the kidneys.

Animal studies have helped in determining just how the kidney responds to high doses of fluoride. Borke and Whitford (1999) showed that ATP-dependent calcium uptake in rat kidneys was significantly affected by exposures equivalent to that of patients undergoing hemodialysis. Cittanova et al. (2002) showed that high concentrations of fluoride affected the ATPase

TABLE 9-3 Renal Effects of Fluoride

Species	Study	Findings
Renal Stone Formation		
Human	Incidence of renal stones in Finnish hospital districts with different concentrations of fluoride in drinking water, in a fluoridated community, and a nonfluoridated city.	At fluoride concentrations of 1.5 mg/L or greater, the standardized hospital admission rates for urolithiasis was increased about one-sixth. No differences were found with fluoride concentrations of ≤0.49 mg/L and 0.50-1.49 mg/L. A separate comparison of a fluoridated city (1 mg/L) and a referent city (<0.49 mg/L) found a 25% lower rate of urolithiasis in the fluoridated city.
Human	20 children with vesical stones were evaluated for fluoride intake and content of renal stones.	Mean fluoride intake was 2.5 ± 0.8 mg in 24 hours. Subjects had normal plasma and urinary excretion of fluoride. No statistically significant difference in fluoride content between the center and periphery of the stones. Fluoride content was higher in stones with calcium than in those with uric acid or ammonium urate. Authors conclude that fluoride does not cause initiation or growth of the nucleus of vesical stones.
Human	18,706 tribal people from fluoride endemic and nonendemic areas of India were evaluated for history of renal stones.	In endemic areas, fluoride in drinking water was 3.5-4.9 mg/L. Prevalence of urolithiasis was 4.6 times higher in the endemic area than in the nonendemic area. In the endemic area, subjects with fluorosis had nearly double the prevalence of urolithiasis compared with those without fluorosis.
Rat	Effect of NaF on ethylene glycol-induced renal stone formation in rats.	NaF reduced oxalate stone production.
Toxic Effects of Fluoride on Kidney Tissues and Function		
Human	Renal function evaluated in 50 patients exposed by inhalation to sevoflurane compared with 25 controls exposed to isoflurane.	Mean peak plasma fluoride was 29.3 ± 1.8 μmol/L 2 hours after anesthesia and 18 μmol/L after 8 hours. Five patients had peak concentrations of greater than 50 μmol/L. No lasting renal or hepatic functional changes found.

Proposed Mechanisms	Comments	Reference
		Juuti and Heinonen 1980
	Fluoride's role as a promoter of kidney stones was ruled out but this is based on a small sample size. The authors did not study nephrolithiasis and excessive chronic fluoride intake.	Teotia et al. 1991
Lack of nutrition in the population leads to increases in oxalate excretion. Oxalate increases oxidative load, which increases cellular damage where urinary crystals have an opportunity to grow. Fluoride contributes to the oxidative load and passively participates in renal crystal formation.	Water fluoride concentration was at EPA's current MCLG, but malnutrition among the study population probably made risk for renal stones higher.	Singh et al. 2001
NaF inhibition of induced renal stones appears to be due to its ability to decrease oxalate synthesis and urinary oxalate excretion.	Decreased urinary oxalate secretion might be a toxic effect on the kidneys.	Li et al. 1992
		Frink et al. 1992

continued

TABLE 9-3 Continued

Species	Study	Findings
Human	Renal damage evaluated in 23 patients exposed by inhalation to sevoflurane compared with 11 controls exposed to isoflurane.	8 patients had serum fluoride concentrations > 50 µmol/L. An inverse correlation was found between peak fluoride concentration and maximal urinary osmolality after the injection of vasopressin ($r = -0.42$, $P < 0.05$). Increased urinary N-acetyl-β-glucosaminidase excretion, but no lasting damage to the kidney.
Human (in vitro)	Immortalized ascending duct cells of kidneys were incubated with 0-100 mM fluoride.	Fluoride decreased cell number by 23% ($P < 0.05$), total protein content by 30% ($P < 0.05$), and hydrogen-leucine incorporation by 43% ($P < 0.05$). LDH release was increased by 236% ($P < 0.05$), with a threshold of 5 mM. There was also a 58% reduction in Na,K-ATPase activity at 5 mM ($P < 0.05$). Crystal formations found in mitochondria.
Human	Renal function evaluated in 50 patients exposed by inhalation to sevoflurane.	Mean peak plasma fluoride was 28.2 ± 14 µmol/L 1 hour after exposure. 2 patients had concentrations > 50 µmol/L 12-24 hours after anesthesia and raised blood urea nitrogen and creatinine concentrations.
Human	Health survey of residents of rural areas in China exposed to airborne fluoride from combustion of coal.	Glomerular filtration rate was affected, as shown by significantly lower urinary inorganic phosphate concentrations in exposed populations compared with control populations.
Human (in vitro)	Effects of fluoride on renal acid phophatases in the afferent arterioles and in glomeruli.	Alkaline fixation-resistant and lysosomal acid phosphatase activities were significantly inhibited at 75 µM. Tartrate-resistant activity was also significantly inhibited at 250 µM.
Human	Renal function in Chinese children (n = 210) exposed to different concentrations of fluoride in drinking water. Subjects stratified into 7 groups (n = 30), including controls. Comparisons made between subjects with "high fluoride load" and enamel fluorosis (details not provided) in areas with fluoride at <1.0, 1.0-2.0, 2.0-3.0, and >3.0 mg/L.	Significant increase in urine NAG and gamma-GT activities in children with enamel fluorosis exposed to fluoride at 2.58 mg/L and in children exposed at 4.51 mg/L. Dose-response relationship observed between fluoride concentration and these two measures of renal damage.

GASTROINTESTINAL, RENAL, HEPATIC, AND IMMUNE SYSTEMS   285

Proposed Mechanisms	Comments	Reference
		Higuchi et al. 1995
Mitochondrion appears to be the target of fluoride toxicity in collecting duct cells. Effects are partly responsible for the urinary concentrating defects in patients after administration of biotransformed inhaled anesthetics.		Cittanova et al. 1996
	Authors concluded that sevoflurane might induce nephrotoxicity.	Goldberg et al. 1996
		Ando et al. 2001
		Partanen 2002
	Subjects were similar with respect to age, gender, and nutritional status.	Liu et al. 2005

continued

TABLE 9-3 Continued

Species	Study	Findings
Rat	NaF 10, 50, 150 mg/L in drinking water for 6 weeks.	Plasma fluoride concentrations were <0.4, 2, 7, and 35 µmol/L, respectively. ATP-dependent 45Ca uptake was significantly lower in the high exposure group than in controls ($P < 0.05$). Thapsigargin treatment showed that the lower uptake was associated with significantly lower activities of both the plasma membrane Ca^{2+}-pump (in high-dose group compared with controls, $P < 0.05$) and endoplasmic reticulum Ca^{2+}-pump (in the mid- and high-dose groups compared with controls, $P < 0.05$).
Rat	30 and 100 mg/L in drinking water for 7 months.	Decreased phosphatidylethanolamine and phosphatidylcholine phospholipids and ubiquinon in the kidney. Increased lipid peroxidation. Electron microscopy revealed alterations in renal structures, including mitochondrial swelling in the proximal convoluted tubules and decreased numbers of microvilli and disintegrated brush border at the luminal surface.
Rat (in vitro)	Kidney epithelial cells (NRK-52E) were cultured with NaF.	Calcium accumulation was significantly increased.
Rabbit (in vitro)	Immortalized kidney cells of the thick ascending limb were cultured with 1, 5, or 10 mmol of NaF for 24 hours; or 5 mmol for 1, 5, and 10 hours.	At 5 mmol after 24 hours, fluoride decreased cell numbers by 14% ($P < 0.05$), protein content by 16%, leucine incorporation by 54%, and Na-K-2Cl activity by 84%. There was a 145% increase in LDH and a 190% increase in N-acetyl-β-glucosaminidase release. Na,K-ATPase activity was significantly impaired at 1 mmol for 24 hours and after 2 hours at 5 mmol.
Rabbit	NaF at 5, 10, 20, and 50 mg/kg/day injected subcutaneously for 15 weeks.	At 10 mg/kg/day and higher, increased cloudy swellings, degeneration of the tubular epithelium, cell death, vacuolization of the renal tubules, hypertrophy and atrophy of the glomeruli, exudation, interstitial edema, and interstitial nephritis.

Proposed Mechanisms	Comments	Reference
Ca^{2+} homeostasis appears to have been affected by an increase in turnover or breakdown or decreasing the expression of plasma membrane and endoplasmic reticulum Ca^{2+}-pump proteins.		Borke and Whitford 1999
The pathogenesis of chronic fluorosis might be due to oxidative stress and modification of cellular membrane lipids. Those alterations might explain observed systemic effects, especially in soft tissues and organs.		Guan et al. 2000
Elevation of ER-type Ca^{2+}ATPase activity appears to operate as a regulatory system to protect against large increases in cytosolic calcium concentrations due to increased influx of calcium into the ER.		Murao et al. 2000
Na,K-ATPase pump appears to be a major target of fluoride toxicity in the loop of Henle.		Cittanova et al. 2002
Mechanism for the damage not proposed		Shashi et al. 2002

continued

TABLE 9-3 Continued

Species	Study	Findings
Fluoride Toxicity in Hemodialysis Patients		
Human	Plasma and bone concentrations of fluoride and renal osteodystrophy in HD patients	Mean plasma concentration of fluoride was 10.8 µmol/L in 34 patients with residual glomerular filtration rates (RGFR) and 15.6 µmol/L in 25 patients with anuria. Mean bone ash concentration of fluoride was 5,000 mg/kg in 14 patients with RGFR and 7,200 mg/kg in 26 patients with anuria. Evidence of secondary hyperparathyroidism. Evidence of osteodystrophy reported, but did not appear to be of the advanced degree found with skeletal fluorosis.
Human	Comparison of serum fluoride concentrations in 17 HD patients and 17 CAPD patients.	Higher serum fluoride concentrations found in HD patients (4.0 ± 0.5 µmol/L) compared with CAPD patients (2.5 ± 0.3 µmol/L), $P < 0.005$.
Human	Renal osteodystrophy in 209 HD patients in Saudi Arabia.	Bone and joint pain reported in 25.8% of patients. The major radiological finding was osteosclerosis in 70% of patients. Mean serum concentration of aluminum was 25.4 ± 17.7 µg/L; of 1,25-dihydroxy vitamin D3 was 8.1 ± 4.2 ng/L; and of fluoride was 92.2 ± 31.4 µg/L.
Human	Effects on plasma potassium concentration of 25 HD patients from mineral water containing fluoride at 9 mg/L.	There was a significant correlation between plasma fluoride and potassium concentrations before dialysis ($P < 1 \times 10^{-7}$) but not after. Group-by-group comparisons indicated that the correlation was linked to the group consuming the mineral water ($P < 1 \times 10^{-7}$), which had higher plasma potassium concentrations before dialysis than the group that did not drink the mineral water ($P < 0.005$).
Human	Serum fluoride concentrations evaluated in 29 HD patients.	Serum fluoride was significantly higher in patients before and after HD than in healthy subjects. Despite net clearance of fluoride during HD, serum fluoride did not return to normal concentrations.

Proposed Mechanisms	Comments	Reference
	The bone concentrations of fluoride fall within the ranges historically associated with stage II and stage III skeletal fluorosis (see Chapter 5). The study reported no skeletal fluorosis, but it was unclear what criteria were used for assessment of the condition. Suggests bone concentrations alone do not adequately predict skeletal fluorosis. The patients were supplemented with calcium, and were given aluminum hydroxide if serum phosphate was too high.	Erben et al. 1984
	Authors noted that fluoride content of the HD fluids, which were prepared with fluoridated water, was significantly higher than in commercially prepared peritoneal dialysis fluid.	Bello and Gitelman 1990
	Osteodystrophy could be related to aluminum exposure. Water quality in Saudi Arabia is not the same as in the United States.	Huraib et al. 1993
		Nicolay et al. 1999
		Usuda et al. 1997

continued

TABLE 9-3 Continued

Species	Study	Findings
Human	Serum fluoride concentrations evaluated in 39 patients with end stage renal disease living in an area with fluoride at 47.4 ± 3.28 µM/L in drinking water. 30 patients treated with HD and 9 with CAPD.	Mean serum fluoride was significantly higher in dialysis patients (2.67 ± 1.09 µM/L) than in controls. CAPD patients had higher mean fluoride concentrations (3.1 ± 1.97 µM/L) than HD patients (2.5 ± 1.137 µM/L). 39% of dialysis patients had serum fluoride concentrations > 3.0 µM/L, a concentration believed to pose a risk of osteodystrophy.
Human	Plasma fluoride concentrations measured in 35 dialysis patients.	Highly significant correlation between fluoride concentrations before and after dialysis ($P < 0.00001$) and between the months of hemodialysis and average fluoride concentration before dialysis ($r = 0.624$; $P = 0.008$).
Human	Serum fluoride concentrations measured in 150 dialysis patients.	Serum fluoride concentrations were approximately 3.3 times higher in dialysis patients than in healthy subjects.
Human	153 iliac crest bone biopsies from renal osteodystrophy patients were analyzed.	Increase in bone fluoride was weakly associated with increased osteoid volume, surface, and thickness. Bone fluoride had a negative correlation with bone microhardness.

Hemodialysis Accidents

Human	Evaluation of 12 patients who became severely ill after HD treatment and 20 patients who did not become ill after treatment in the same unit.	12 of 15 patients treated in one room had severe pruritus, multiple nonspecific symptoms, and/or fatal ventricular fibrillation (3 patients). Serum fluoride concentration in ill patients was as high as 716 µmol/L. 20 patients treated in a different room did not become ill ($P < 0.0001$).

ABBREVIATIONS: CAPD, continuous ambulatory peritoneal dialysis; ER, endoplasmic reticulum; GT, glutamyltransferase; HD, hemodialysis; LDH, lactate dehydrogenase; NAG, N-acetyl-beta-D-glucosaminidase.

Proposed Mechanisms	Comments	Reference
		Al-Wakeel et al. 1997
		Nicolay et al. 1997
		Torra et al. 1998
Fluoride incorporation at the mineralizing front increases mineralization lag time.	The authors speculated that accumulated fluoride interacted with aluminum in dialysis patients, altering bone properties.	Ng et al. 2004
	Water used for dialysis in the ill patients was found to have excessive concentrations of fluoride because of errors in maintenance of the deionization system.	Arnow et al. 1994

pump in cultured rabbit ascending loop cells. Guan et al. (2000) showed that the same concentrations of fluoride that caused dental fluorosis in rats affected kidney phospholipids. Rat studies show that the animals that had most of their renal tissue surgically removed retained more fluoride in their bones, which became more susceptible to fracture (Turner et al. 1996). Turner's rat studies were also conducted to simulate the concentrations that humans would be exposed to in regions where the drinking water contained fluoride at 3-10 mg/L.

Patients with Renal Impairment

Several investigators have shown that patients with impaired renal function, or on hemodialysis, tend to accumulate fluoride much more quickly than normal. Patients with renal osteodystrophy can have higher fluoride concentrations in their serum (see Table 9-3). Whether some bone changes in renal osteodystrophy can be attributed to excess bone fluoride accumulation alone, or in combination with other elements such as magnesium and aluminum, has not been clearly established (Erben et al. 1984; Huraib et al. 1993; Ng et al. 2004). Extreme caution should be used in patients on hemodialysis because failures of the dialysis equipment have occurred in the past, resulting in fluoride intoxication (Arnow et al. 1994).

HEPATIC SYSTEM

Although some studies have observed histopathologic changes in the liver in response to high doses of fluoride (Kapoor et al. 1993; Grucka-Mamczar et al. 1997), the changes have not been carefully quantified. In a study to examine the histologic effects of NaF directly on the liver, rats fed 5-50 mg/kg/day showed vacuolization of the hepatic cells, cellular necrosis, and dilated and engorged liver tissue that was not seen in the control animals (Shashi and Thapar 2001).

In some of the studies in which effects of chronic or acute fluoride doses were observed in kidneys, the livers were also examined for signs of toxicity. Tormanen (2003) showed that fluoride caused substrate inhibition of rat liver arginase at substrate concentrations above 4 mM, and rat kidney arginase was more sensitive than liver arginase to inhibition by fluoride. de Camargo and Merzel (1980) first reported significant increases in fatty deposits in the livers of rats but not in their kidneys when they were given NaF at 1, 10, or 100 mg/L in tap water for 180 days. Twenty years later, Wang et al. (2000) used high-performance liquid chromatography to document the changes in liver lipids after rats were fed drinking water with fluoride at 30 or 100 mg/L for 7 months. The higher concentration of fluoride reduced total phospholipids. Within the phospholipids, the saturated fatty acid components increased and polyunsaturated fatty acids decreased. Liver cholesterol and dolichol were unchanged. The authors concluded that fluoride-induced alteration in liver membrane lipids could be an important factor in the pathogenesis of chronic fluorosis.

Whether any of these changes has relevance to the long-term daily ingestion of drinking water containing fluoride at 4 mg/L will require careful analysis of liver function tests in areas with high and low concentrations of fluoride in the drinking water. The clinical trials involving fluoride therapy for treating osteoporosis require that subjects be administered fluoride at concentrations approaching 1.0 mg/kg/day. Although such studies are rarely carried out for more than 5 years, this period of time should be sufficient to measure any changes in hepatic function. Jackson et al. (1994) reported that there was a significant increase in liver function enzymes in test subjects taking 23 mg of fluoride a day for 18 months, but the enzyme concentrations were still within the normal range. It is possible that a lifetime ingestion of 5-10 mg/day from drinking water containing fluoride at 4 mg/L might turn out to have long-term effects on the liver, and this should be investigated in future epidemiologic studies.

Finally, because the liver is the primary organ for defluorinating toxic organofluorides, there is a concern that added fluoride body burden that would be experienced in areas where the drinking water had fluoride at 4 mg/L might interfere with the activity of the cytochrome P450 complex (Baker and Ronnenberg 1992; Kharasch and Hankins 1996).

IMMUNE SYSTEM

Hypersensitivity

In the studies by physicians treating patients who reported problems after fluoridation was initiated, there were several reports of skin irritation (Waldbott 1956; Grimbergen 1974; Petraborg 1977). Although blinded experiments suggested that the symptoms were the result of chemicals in the water supply, various anecdotal reports from patients complaining, for example, of oral ulcers, colitis, urticaria, skin rashes, nasal congestion, and epigastric distress, do not represent type I (anaphylactic), II (cytotoxic), III (toxic complex), or IV (delayed type reactivity) hypersensitivity, according to the American Academy of Allergy (Austen et al. 1971). These patients might be sensitive to the effects of silicofluorides and not the fluoride ion itself. In a recent study, Machalinski et al. (2003) reported that the four different human leukemic cell lines were more susceptible to the effects of sodium hexafluorosilicate, the compound most often used in fluoridation, than to NaF.

Nevertheless, patients who live in either an artificially fluoridated community or a community where the drinking water naturally contains fluoride at 4 mg/L have all accumulated fluoride in their skeletal systems and potentially have very high fluoride concentrations in their bones (see Chapter 3). The bone marrow is where immune cells develop and that could affect humoral immunity and the production of antibodies to foreign chemicals. For example, Butler et al. (1990) showed that fluoride can be an adjuvant, causing an increase in the production of antibodies to an antigen and an increase in the size and cellularity of the Peyer's patches and mesenteric lymph nodes. The same group (Loftenius et al. 1999) then demonstrated that human lymphocytes were more responsive to the morbilli antigen. Jain and Susheela (1987), on the other hand, showed that rabbit lymphocytes exposed to NaF had reduced antibody production to transferrin.

At the very early stages of stem cell differentiation in bone, fluoride could affect which cell line is stimulated or inhibited. Kawase et al. (1996) suggested that NaF (0.5 mM for 0-4 days) stimulates the granulocytic pathway of the progenitor cells in vitro. This was confirmed by Oguro et al. (2003), who concluded that "NaF [<0.5 mM] induces early differentiation of bone marrow hemopoietic progenitor cells along the granulocytic pathway but not the monocytic pathway."

It has long been claimed that cells do not experience the concentrations of fluoride that are used in vitro to demonstrate the changes seen in cell culture. Usually millimolar concentrations are required to observe an effect in culture. Because serum fluoride normally is found in the micromolar range, it has been claimed that there is no relevance to the in vivo situation. However, studies by Okuda et al. (1990) on resorbing osteoclasts reported that: "NaF in concentrations of 0.5-1.0 mM decreased the number of resorption lacunae made by individual osteoclasts and decreased the resorbed area per osteoclast. We argue that the concentration of fluoride in these experiments may be within the range 'seen' by osteoclasts in mammals treated for prolonged periods with approximately 1 mg of NaF/kg body weight (bw) per day." Sodium fluoride intake at 1 mg/kg/day in humans could result in bone fluoride concentrations that might occur in an elderly person with impaired renal function drinking 2 L of water per day containing fluoride at 4 mg/L (see Chapters 3 and 5 for more information on bone fluoride concentrations).

Cellular Immunity

Macrophage function is a major first line of defense in immunity. When macrophage function is impaired, the body could fail to control the invasion of foreign cells or molecules and their destructive effects. The studies that have investigated the function of the cells involved in humoral immunity are summarized in Table 9-4.

Fluoride, usually in the millimolar range, has a number of effects on immune cells, including polymorphonuclear leukocytes, lymphocytes, and neutrophils. Fluoride interferes with adherence to substrate in vitro. The variety of biochemical effects on immune cells in culture are described in Table 9-4. Fluoride also augments the inflammatory response to irritants. Several mechanisms have been proposed, and the main route is thought to be by means of activation of the G-protein complex. It appears that aluminum combines with fluoride to form aluminum fluoride, a potent activator of G protein. In a study by O'Shea et al. (1987), for example, AlF_4 had a greater influence on lymphocyte lipid metabolism than did fluoride in the absence of aluminum. On the other hand, Goldman et al. (1995) showed that the aluminofluoride effect of activating various enzymes in macrophages is independent of the G-protein complex.

There is no question that fluoride can affect the cells involved in providing immune responses. The question is what proportion, if any, of the population consuming drinking water containing fluoride at 4.0 mg/L on a regular basis will have their immune systems compromised? Not a single epidemiologic study has investigated whether fluoride in the drinking water at 4 mg/L is associated with changes in immune function. Nor has any study examined whether a person with an immunodeficiency disease can tolerate fluoride ingestion from drinking water. Because most of the studies conducted to date have been carried out in vitro and with high fluoride concentrations, Challacombe (1996) did not believe they warranted attention. However, as mentioned previously in this chapter, bone concentrates fluoride and the blood-borne progenitors could be exposed to exceptionally high fluoride concentrations. Thus, more research needs to be carried out before one can state that drinking water containing fluoride at 4 mg/L has no effect on the immune system.

FINDINGS

The committee did not find any human studies on drinking water containing fluoride at 4 mg/L where GI, renal, hepatic, or immune effects were carefully documented. Most reports of GI effects involve exposures to high concentrations of fluoride from accidental overfeeds of fluoride into water supplies or from therapeutic uses. There are a few case reports of GI upset in subjects exposed to drinking water fluoridated at 1 mg/L. Those effects were observed in only a small number of cases, which suggest hypersensitivity. However, the available data are not robust enough to determine whether that is the case.

Studies of the effects of fluoride on the kidney, liver, and immune system indicate that exposure to concentrations much higher than 4 mg/L can affect renal tissues and function and cause hepatic and immunologic alterations

TABLE 9-4 Effects of Fluoride on Immune System Cells

Species	Study	Findings
In vitro		
Human	Metabolism factors measured in cultured PMNs incubated with mM concentrations of fluoride.	Significant inhibition of PMN metabolic activity at 0.1 mM fluoride for O_2 generation. Activity was also inhibited at 0.5 mM for $^{14}CO_2$ release from labeled glucose and at 1.0 mM for nitroblue tetrazolium-reduction.
Human	Leukocyte capillary migration inhibition assay.	8% inhibition with 0.5 ppm fluoride and 20% inhibition with 20 ppm fluoride.
Various	Evaluated signal transduction in cultured macrophages exposed to NaF with or without aluminum.	NaF reduced intracellular ATP concentrations, suppressed agonist-induced protein tyrosine phosphorylation and reactive oxygen species formation. There was in situ activation of nitrogen-activated protein kinase, phospholipase A2, and phosphatidylinositol-phospholipase C. Little or no effect on NaF-mediated enzyme action was observed when cells were treated with $AlCl_3$ or deferoxamine.
Human	Cell migration assay and micropore filter assay used to assess effect of NaF on locomotion and chemotaxis of human blood leukocytes.	Significant reduction in chemotaxis and locomotion observed with 1 mM fluoride.
Human	Cultured neutrophils treated with fluoride.	Fluoride activated diacylglycerol generation and phospholipase D activity. Increased diradylglycerol mass, with kinetics similar to superoxide generation.
Human	Electropermeabilized neutrophils treated with fluoride.	O_2 production was increased by electropermeabilization. That effect was antagonized by GDP[β-S], required Mg^{2+}, and was blocked by staurosporine and H-7.
Human	Adherence assay of PMNs cultured with 0.0625-4.0 μM with or without autologous serum.	No effect in the absence of serum. With serum, adherence significantly decreased at 0.5 μM. Decrease was 1.1% at 0.125 μM and 52.7% at 1.5 μM.

Application/Proposed Mechanisms	Comments	Reference
Inhibition was primarily due to suppression of nonoxidative glucose metabolism. Peak effect was at 20 mM, a lethal dose to the cells.		Gabler and Leong 1979
	Effect at 0.5 ppm fluoride likely not significant. 20 ppm fluoride is 100 times higher than serum fluoride concentrations expected if 1.5 L of 4 ppm fluoride in water is consumed.	Gibson 1992
Authors suggest that some of the pleiotropic effects of NaF in intact cells might be due to depletion of ATP and not by G-protein activation.		Goldman et al. 1995
	1 mM fluoride is a high concentration relative to blood fluoride, but such a concentration might be possible within the Haversian canal system of bone, restricting migration of leukocytes through bone.	Wilkinson 1983
Data are consistent with the activation of phosphatidic acid and diglyceride generation by both phopholipase D-dependent and independent mechanisms.		Olson et al. 1990
Supports the hypothesis that fluoride activates G protein, most likely Gp, by interacting with the nucleotide-binding site on the G α subunit.		Hartfield and Robinson 1990
Effect is not direct and is probably modulated by a seric factor.	Concentrations of fluoride tested are similar to those found in blood.	Gomez-Ubric et al. 1992

continued

TABLE 9-4 Continued

Species	Study	Findings
Human	Promyelocytic HL-60 cells treated with 0.5 mM NaF for 0-4 days.	Cell proliferation was inhibited by NaF and was augmented by the addition of 1,25-dihydroxyvitamin D3. Other observations were changes in cellular morphology, increased cellular adhesion to plastic, reduced nuclear/cytoplasmic ratio, and increased cellular expression of chloroacetate esterase. No effect on cellular nonspecific esterase activity.
Human	Blood lymphocytes incubated with NaF at 0.31, 0.62, or 1.2 mM.	NaF augmented lymphocyte response to a mitogen (PHA) or a specific antigen (morbilli antigen from infected cells). Simultaneous incubation of NaF at 0.62 mM with PHA significantly increased cytokine INF-γ release from activated T and/or NK cells compared with treatment with PHA alone ($P < 0.01$).
Human	CD34$^+$ cells isolated from umbilical cord blood were incubated with 1, 10, and 50 mM NaF for 30 and 120 minutes.	At 10 and 50 mM NaF, there was damage to CFU-GM and significantly decreased cloning potential of these cells. Growth of BFU-E was also inhibited.
Rat	Liver macrophages treated with fluoride.	Arachidonic acid and prostaglandins were released (required extracellular calcium), but there was no formation of inositol phosphates or superoxide. Those effects were inhibited by staurosporine and phorbol ester. Protein kinase C was translocated from the cytosol to membranes.
Mouse	Cultured lymphocytes treated with NaF and AlCl$_3$.	With NaF, there was a breakdown of polyphosphoinositides, decreased production of phosphoinositols, increased cytosolic Ca^{2+}, and start of phosphorylation of the T-cell receptor. Effects were potentiated by addition of AlCl$_3$.
Mouse	Bone marrow progenitor cells cultured with 0.1-0.5 mM NaF.	Upregulation in the activities of intracellular enzymes (LDH, β-glucuronidase, acid phosphatase), cellular reduction of nitroblue tetrazolium, and nitric oxide production.

Application/Proposed Mechanisms	Comments	Reference
NaF stimulates the early stages of HL-60 differentiation toward a granulocyte-like cell. 1,25-Dihydroxyvitamin D3 acts as a cofactor with NaF, primarily through interaction with an endogenous NaF-induced cyclooxygenase product(s), possibly PGE2.		Kawase et al. 1996
Authors concluded that NaF's effect on INF-γ release during an immune response might be one of the primary ways that fluoride ion influences the immune system.		Loftenius et al. 1999
		Machalinski et al. 2000
Calcium-dependent protein kinase C appears to be involved in fluoride's action on liver macrophages.		Schulze-Specking et al. 1991
The active moiety is AlF_4^-. AlF_4^-–induced effects were insensitive to cyclic adenosine monophosphate.		O'Shea et al. 1987
Authors suggest that NaF induces early differentiation of bone marrow hemopoietic progenitor cells along the granulocytic pathway but not the monocytic pathway linked to osteoclast formation.		Oguro et al. 2003

continued

TABLE 9-4 Continued

Species	Study	Findings
In vivo		
Rabbit	Rabbits immunized with transferrin before or after 9 months treatment with 10 mg/kg/day. Circulating anti-transferrin titers were measured during the 9 months. DNA and protein synthesis were determined by [^{3}H]thymidine and [^{14}C]leucine incorporation.	NaF inhibited antibody formation and had a threshold of 0.78 ppm in circulation. DNA and protein synthesis were also inhibited.
Rat	Sensitization assay performed with rats administered 5 mL of a 100-mmol solution of NaF twice a week for 2-3 weeks and given ovalbumin in drinking water.	Significant increase in surface immunoglobulin expression on lymphocytes from the Peyer's patches and mesenteric lymph nodes.
Rat	0.1, 0.2, and 0.4 mg of fluoride administered intratracheally.	Significant PMN-leukocyte infiltration in the lungs observed 24 hours after treatment with 0.2 and 0.4 mg. mRNA of chemokines and proinflammatory cytokines was increased. Increased adhesion of PMNs to plastic dish.
Mouse	Antibacterial defense mechanisms and lung damage were assessed in mice exposed to 2, 5, 10 mg/m^3 of a fluoride aerosol in an inhalation chamber for 4 hours per day for 14 days.	Suppression of pulmonary bactericidal activity against *Staphylococcus aureus* at 5 and 10 mg/m^3. Significant decrease in the number of alveolar macrophages in bronchoalveolar lavage fluid at 10 mg/m^3 in mice not bacterially challenged. Significant increase in PMNs and lymphocytes at 10 mg/m^3.

ABBREVIATIONS: BFU-E, burst forming unit of erythrocytes; CFU-GM, colony-forming unit of granulocyte-macrophages; GDP[β-S], guanosine 5'-[β-thio]diphosphate; INF-γ, interferon γ; LDH, lactate dehydrogenase; PGE2, prostaglandin E; PHA, phytohemaggultinin ; PMN, polymorphonuclear leukocyte.

Application/Proposed Mechanisms	Comments	Reference
Antibody formation appears to be inhibited because of the decrease in lymphocyte proliferation and inhibition of protein synthetic ability of immunocytes.	General inhibition of metabolic function.	Jain and Susheela 1987
Microulcerations of the gastric mucosa.	Authors note that the concentrations tested were within the range that could be inadvertently ingested by infants/children or adults from fluoride supplements or gels.	Butler et al. 1990
		S. Hirano et al. 1999
Authors concluded that inhalation of fluoride can cause cellular alterations in the lung that diminish the ability to respond to infectious bacteria.		Yamamoto et al. 2001

in test animals and in vitro test systems. For example, a few studies suggest that fluoride might be associated with kidney stone formation, while other studies suggest that it might inhibit stone formation. Some effects on liver enzymes have been observed in studies of osteoporosis patients treated with fluoride, but the available data are not sufficient to draw any conclusions about potential risks from low-level long-term exposures. Little data is available on immunologic parameters in human subjects exposed to fluoride from drinking water or osteoporosis therapy, but in vitro and animal data suggest the need for more research in this area.

As noted earlier in Chapters 2 and 3, several subpopulations are likely to be susceptible to the effects of fluoride from exposure and pharmacokinetic standpoints. With regard to the end points covered in this chapter, it is important to consider subpopulations that accumulate large concentrations of fluoride in their bones (e.g., renal patients). When bone turnover occurs, the potential exists for immune system cells and stem cells to be exposed to concentrations of fluoride in the interstitial fluids of bone that are higher than would be found in serum. From an immunologic standpoint, individuals who are immunocompromised (e.g., AIDS, transplant, and bone-marrow-replacement patients) could be at greater risk of the immunologic effects of fluoride.

RECOMMENDATIONS

Gastric Effects

- Studies are needed to evaluate gastric responses to fluoride from natural sources at concentrations up to 4 mg/L and from artificial sources. Data on both types of exposures would help to distinguish between the effects of water fluoridation chemicals and natural fluoride. Consideration should be given to identifying groups that might be more susceptible to the gastric effects of fluoride.
- The influence of fluoride and other minerals, such as calcium and magnesium, present in water sources containing natural concentrations of fluoride up to 4 mg/L on gastric responses should be carefully measured.

Renal and Hepatic Effects

- Rigorous epidemiologic studies should be carried out in North America to determine whether fluoride in drinking water at 4 mg/L is associated with an increased incidence of kidney stones. There is a particular need to study patients with renal impairments.
- Additional studies should be carried out to determine the incidence, prevalence, and severity of renal osteodystrophy in patients with renal im-

pairments in areas where there is fluoride at up to 4 mg/L in the drinking water.
- The effect of low doses of fluoride on kidney and liver enzyme functions in humans needs to be carefully documented in communities exposed to different concentrations of fluoride in drinking water.

Immune Response

- Epidemiologic studies should be carried out to determine whether there is a higher prevalence of hypersensitivity reactions in areas where there is elevated fluoride in the drinking water. If evidence is found, hypersensitive subjects could then be selected to test, by means of double-blinded randomized clinical trials, which fluoride chemicals can cause hypersensitivity. In addition, studies could be conducted to determine what percentage of immunocompromised subjects have adverse reactions when exposed to fluoride in the range of 1-4 mg/L in drinking water.
- More research is needed on the immunotoxic effects of fluoride in animals and humans to determine if fluoride accumulation can influence immune function.
- It is paramount that careful biochemical studies be conducted to determine what fluoride concentrations occur in the bone and surrounding interstitial fluids from exposure to fluoride in drinking water at up to 4 mg/L, because bone marrow is the source of the progenitors that produce the immune system cells.

10

Genotoxicity and Carcinogenicity

This chapter reviews research publications and relevant review articles published since the earlier NRC (1993) report and other relevant papers not included in that review, and also considers salient earlier papers. Evaluation of the plausibility and potential for carcinogenicity is based on human epidemiologic studies, laboratory animal lifetime bioassays, shorter-term genotoxicity tests, metabolism and pharmacokinetic data, and mechanistic information. Genotoxicity tests indicate the potential for fluoride to cause mutations, affect the structure of chromosomes and other genomic material; affect DNA replication, repair, and the cell cycle; and/or transform cultured cell lines to enable them to cause tumors when implanted into host animals. In interpreting the experimental studies and the consistency among disparate tests and systems, factors to be considered include the chemical form, concentrations, duration of exposure or application, vehicle or route of exposure, presence or absence of dose response, and information that each study provides about the potential stage of cancer development at which the chemical might operate. The degree of consistency of genotoxicity tests with the epidemiologic studies and whole animal bioassays on these points was evaluated.

GENOTOXICITY

Genotoxicity tests comprise in vitro and in vivo assays to assess the effects on DNA and chromosomal structure and/or function. The results of these assays serve as indicators of the potential interaction of chemicals with the genetic material. Changes in chromosomal or DNA structure or function may be a step in the pathway to carcinogenesis. More often, they indicate interference with the normal duplication, function, and control of cell division and genetic activity that also might result in precancer or early neoplastic processes. Genotoxicity also encompasses the ability to cause germ cell and somatic cell mutations that cause malformations, disease, and other adverse health outcomes.

Many cell systems derived from various organisms have been used to the assess genotoxicity of a large array of chemicals. In evaluating the applicability of the results of these tests to human risk from fluoride ingestion, some of the key parameters are the concentrations used in the assays compared with physiologic concentrations, the form and vehicle for fluoride exposure in the assay, and existing data on overall applicability of the various assays to risk in humans. Tennant (1987) and Tennant et al. (1987) concluded that the *Salmonella* reverse mutation assay was the best short-term genotoxicity assay available for predicting carcinogenicity in mammals. However, Parodi et al. (1991) reviewed the results of various genotoxicity tests in comparison with animal carcinogenicity studies, and found that in vitro cytogenetic tests, particularly sister-chromatid exchange tests (SCEs), were more predictive of carcinogenicity than the *Salmonella* reverse mutation assay. Tice et al. (1996) subsequently reviewed relative sensitivities of rodents and humans to genotoxic agents and concluded that humans are more than an order of magnitude more sensitive than rodents to most of the genotoxic agents they examined using the genetic activity profile database.

The available new genotoxicity studies of fluoride are detailed in Table 10-1. The most extensive and important additions to the genotoxicity literature on fluoride since 1993 are in vivo assays in human populations and, to a lesser extent, in vitro assays using human cell lines and in vivo experiments with rodents. These studies are discussed below.

Gene Mutation

Mutagenicity indicates direct action of a substance on DNA. Alterations in DNA suggest that the chemical has the potential to cause genetic effects as well as carcinogenic potential. In 1993, the existing literature did not indicate that fluoride posed a mutation hazard. The literature included assays with *Salmonella* (virtually all negative results), various mammalian cells lines (virtually all negative), and cultured human lymphocytes. Positive results in the human lymphocytes were seen at fluoride concentrations above 65 micrograms per milliliter (µg/mL) (parts per million [ppm]) and generally at more than 200 µg/mL, (much greater concentrations than those to which human cells in vivo typically would be exposed). No pertinent studies have been found since those reviewed in the 1993 NRC report. The committee interprets the weight of evidence from in vivo rodent studies to

TABLE 10-1 Summary of Recent Genotoxicity Studies of Fluoride

Population or System/Method and Assay	Findings	Remarks	References
In vivo human studies			
Subjects (n = 746) with normal or inadequate nutrition living in regions of China with water concentrations of fluoride at 0.2, 1.0, or 4.8 mg/L. Assay: SCE in blood lymphocytes.	Subjects in the 4.8-mg/L region had lower average SCEs per cell.	Plasma and urine fluoride concentrations also measured; these were proportional to water concentrations.	Y. Li et al. 1995
Comparison of 100 residents of North Gujarat exposed to drinking water with fluoride at 1.95 to 2.2 mg/L with 21 subjects in Ahmedabad exposed at 0.6 to 1.0 mg/L. Assay: SCE in blood lymphocytes and cell cycle proliferative index.	SCE rate was significantly greater in subjects from North Gujarat, but there was no difference in the cell cycle proliferative index.	Insufficient documentation of subject ascertainment or control for potential demographic confounding.	Sheth et al. 1994
Phosphate fertilizer workers with inhalation exposure. Assay: chromosome aberrations, micronucleus, SCE.	Exposed workers had elevation in all cytogenetic outcomes tested.		Meng et al. 1995; Meng and Zhang 1997
Peripheral blood lymphocytes from inhabitants of the Hohhot region in inner Mongolia (n = 53 with fluorosis; n = 20 with no fluorosis) exposed to fluoride in drinking water at 4 to 15 mg/L compared with controls (n = 30) exposed to fluoride at < 1 mg/L. Assay: SCE and micronucleus.	SCE: higher frequency in individuals with fluorosis (87% increase in SCEs), than no fluorosis (13% increase) compared with controls. Micronucleus: higher frequency in individuals with fluorosis (3.4-fold increase) than no fluorosis (1.8-fold increase) compared with controls.	Insufficient documentation of subject ascertainment or control for potential demographic confounding.	Wu and Wu 1995

Human populations with long-term residency in communities with water concentrations of fluoride at 0.2, 1.0, and 4.0 mg/L. Measured plasma and urine fluoride concentrations. Assay: SCE in blood lymphocyte.	SCEs higher in 4.0-mg/L community. Follow-up study in the 4.0-mg/L community comparing residents using well water ($\leq$0.3 mg/L) and city water (4.0 mg/L) found no difference in SCE frequency between these two groups in the 4-mg/L town.	Jackson et al. 1997
Cultured peripheral blood lymphocytes from 7 female osteoporosis patients treated with disodium monofluorophosphate and NaF for 15 to 49 months (22.6 to 33.9 mg of fluoride/day). Measured serum fluoride. Assay: chromosomal aberration, micronuclei, cell cycle progression.	No cytogenetic effects compared with the matched controls.	Van Asten et al. 1998
Comparison of residents of South Gujarat exposed fluoride at approximately 0.7 mg/L (control village) with residents exposed at 1.5 to 3.5 mg/L (3 villages). Assay: SCE in peripheral lymphocytes.	One of the high-fluoride villages had elevated SCEs. No difference was found between the other two and the control village.	Insufficient documentation of subject ascertainment and demographic characteristics. Joseph and Gadhia 2000
Case series in India of osteosarcoma (n = 20) compared with population distribution regarding bone tumor fluoride concentration and p53 mutations. Assay: p53 mutation and fluoride concentrations in tumor tissue.	Two (10%) cases had p53 mutants in osteosarcoma tissue, and those two had the highest bone tumor fluoride concentrations.	Only patients undergoing prosthesis fitting at one hospital were selected; selection bias was possible. If replicated with systematic ascertainment, this design could indicate a mechanism for carcinogenic activity by fluoride. Ramesh et al. 2001

continued

TABLE 10-1 Continued

Population or System/Method and Assay	Findings	Remarks	References
In vivo animal studies			
Mice (B6C3F$_1$) exposed via drinking water for 6 weeks. Measured fluoride concentrations in bone. Assay: micronuclei in peripheral red blood cells, chromosome aberrations in bone marrow.	No micronuclei increase in peripheral red blood cells, and no chromosome aberration increase in bone marrow. Bone concentrations of fluoride increased with dose to >7,000 ppm.	Method addresses some of the conflicts in previous in vitro and in vivo studies.	Zeiger et al. 1994
Four Zucker rats, diabetic and nondiabetic males. Fluoride in water at 5 to 50 mg/L for 6 months. Assay: SCE.	No SCE elevation in any exposed subgroup.		Dunipace et al. 1996
Wistar rats exposed to NaF at 0, 7, and 100 mg/L in drinking water. Assay: single cell gel electrophoresis (Comet assay)	No increase in single-strand DNA damage.		Ribeiro et al. 2004a
In vitro human studies			
Synchronized human diploid fibroblasts. Attempt to reconcile disparate methods of classifying aberrations (e.g., gaps). Assay: chromosome aberrations.	50 ppm NaF is lowest concentration inducing aberrations.	Proposes mechanism of inhibition of DNA synthesis and repair.	Aardema and Tsutsui 1995
Cultured human diploid cells. NaF treatment for 2.5 hours or continuous. Assay: clastogenicity.	Fluoride clastogenic at >5 ppm. No effect on ploidy.		Oguro et al. 1995

Human diploid fibroblasts at quiescent stage treated with NaF at 1 to 10 ppm (fluoride ion at 0.45 to 4.5 ppm), 1 to 3 weeks. Assay: clastogenicity.	No clastogenicity.	Fluoride concentrations in range of water supplies.	Tsutsui et al. 1995
Human lymphocytes from 50 individuals cultured in 10 to 30 ppm NaF. Assays: chromosomal aberration and SCE.	Chromosomal aberration: 23% and 8% increased frequency of total aberrations at 20 and 30 ppm, respectively, but not at 10 ppm. SCE: no effects reported.		Gadhia and Joseph 1997
Human embryo hepatocytes. Treated with NaF at 40, 80, and 160 mg/L for 24 hours. Assay: single cell gel electrophoresis (Comet assay) Lipid peroxidase and glutathione also assayed.	Dose-related increase in single-strand DNA damage.	Dose-related increase in lipid peroxidase, decrease in gluthathione, and increase in the percentage of apoptotic cells.	Wang et al. 2004
In vitro animal studies			
Cell cultures of rodents, prosimians, apes, and humans. Assay: chromosome aberration.	Clastogenicity of fluoride in great apes and human cells only at 42 to 252 ppm NaF.		Kishi and Ishida 1993
BALB/c-3T3 mouse cells. Examined numerous chemicals, including NaF. Assay: cell transformation.	1.2 to 4.6 mM (19 to 193 ppm) NaF negative for transformation.	Standard transformation assay modified to increase sensitivity.	Matthews et al. 1993
Rats (Sprague-Dawley) cultured bone marrow cells. NaF and KF at 0.1 to 100 µM for 12, 24, or 36 hours. Assay: cytotoxicity and SCE.	Dose-response observed for cytotoxicity. No inhibition of cell proliferation. No effect on SCE.		Khalil and Da'dara 1994

continued

TABLE 10-1 Continued

Population or System/Method and Assay	Findings	Remarks	References
Rat (Sprague-Dawley) cultured bone marrow cells. Treated with NaF and KF 0.1 to 100 µM for 12, 24, or 36 hours. Assay: chromosomal aberration and break.	Weak effects at 1.0 µM, NaF and KF. Effects slightly greater for KF than NaF.		Khalil 1995
Chinese hamster ovary cells. Treated with NaF at 7.28, 56, and 100 µg/mL for 3 hours. Assay: single cell gel electrophoresis (Comet assay)	No increase in single-strand DNA damage.		Ribeiro et al. 2004b
Rat (F344/N) vertebral cells. NaF treatments 1 to 3 days. Assay: chromosomal aberration.	Dose-related increases of chromosome aberrations at 0.5 and 1.0 mM for 24 and 48 hours.	Potential target organ of NTP carcinogenicity studies that yielded osteosarcomas. Provides possible mechanism for carcinogenesis of vertebrae.	Mihashi and Tsutsui 1996

ABBREVIATIONS: KF, potassium fluoride; NaF, sodium fluoride; NTP, National Toxicology Program.

indicate very low probability of a mutagenic risk for humans (NRC 1993; WHO 2002; ATSDR 2003).

Chromosomal Changes and DNA Damage

This section describes studies of fluoride's effects on chromosomes and chromatids, formation of micronuclei, and DNA damage. Chromosomal alterations can include changes in chromosome number (aneuploidy) and aberrations of the chromosomes (before DNA synthesis) or chromatids (after DNA synthesis). (Nondisjunction or translocation of chromosome 21, producing Down's syndrome, is discussed in Chapter 6 on Reproductive and Developmental Effects.) Classification of chromosome/chromatid aberrations has become standardized in recent years: some types of aberrations (e.g., chromatid gaps) are judged to be less important in evaluating effects on chromosomes than other major aberrations (e.g., breaks and translocations). SCE is not known to be on the causal pathway of any adverse health effects, but it is considered a generic indication of exposure to substances that can affect chromosomal structure, many of which are also carcinogens. The SCE assay is a helpful and widely used assay because of its greater sensitivity at lower concentrations than chromosome aberrations. Fewer cells need to be scored in order to establish with confidence whether an increase in SCEs has occurred in a specific test system.

Micronuclei are DNA-containing bodies derived from chromosomal material that is left behind during mitosis. Either a faulty mitotic process or chromosomal breaks can cause this phenomenon. Micronuclei can be visualized in nondividing cells. The relatively new "Comet assay" detects single-strand DNA damage in individual cells using microgel electrophoresis.

Effects on cell survival (cytotoxicity) and effects on cell division are commonly investigated and reported in the course of conducting in vitro cytogenetic studies, and they are included in the summary below.

Human Cells In Vitro

Interpreting the health significance of observed cytogenetic effects on human cells in culture depends on the dose, timing of application relative to the point in the cell cycle, and type of cultured cells, among other factors. As of the 1993 NRC report, the existing data of this type were inconsistent regarding the cytogenetic effects of fluorides. Since that time, Tsutsui et al. (1995) applied sodium fluoride (NaF) at or near concentrations found in water supplies (1 to 10 ppm, equivalent to 0.45 to 4.5 ppm fluoride ion) to diploid fibroblasts for up to 3 weeks and did not observe clastogenicity. Aardema and Tsutsui (1995) using a similar cell system found aberrations only above 50 ppm. The cell phases at which these effects were observed suggested that the underlying mechanism of the chromosomal aberrations might be interference by fluoride with DNA synthesis and repair. In human diploid IMR90 cells, Oguro et al. (1995) observed clastogenicity only above 5 ppm NaF after short- and long-term applications. Gadhia and Joseph (1997) noted that 20 and 30 ppm NaF, but not 10 ppm, caused aberrations. No effects on SCEs were seen in their study. Recently, Wang et al. (2004) used the Comet assay to study genotoxicity in human embryo hepatocytes after treatment with NaF. They observed a dose-related increase in single-strand DNA damage at concentrations of 40, 80, and 160 mg/L.

Other Mammalian Systems In Vitro

Previous studies with a wide variety of test systems found cytogenetic effects in some but not all systems used (NRC 1993; WHO 2002; ATSDR 2003).

Recent studies with in vitro rodent systems include those by Khalil and Da'dara (1994) and Khalil (1995). They evaluated effects on cultured bone marrow cells of Sprague-Dawley rats after exposure to NaF or potassium fluoride (KF) at concentrations ranging from 0.1 µM to 0.1 µM (up to 2 ppm fluoride) for 12 to 36 hours. They did not observe increased SCE levels at any concentration, although there was dose-dependent cytotoxicity. Both NaF and KF induced chromosomal aberrations in a dose-dependent manner between 0.1 and 100 µM. Mihashi and Tsutsui (1996) studied effects on cultured vertebral cells of F344/N rats after 1 to 3 days of 9 to 18 ppm NaF treatment and found dose-dependent increases in chromosomal aberrations based on time and concentrations. Kishi and Ishida (1993) compared activity of NaF on chromosome aberrations for a series of cell lines from rodents, prosimians, great apes, and humans. Clastogenicity by 42 to 252 ppm NaF was seen only in the great ape and human cell lines. Their work thus indicates a greater sensitivity to fluoride in human than in rodent cells. In an older study not included in the NRC (1993) report, Jagiello and Lin (1974) reported that in vitro exposure of oocytes to NaF disrupted meiotic anaphase of ewes and cows but not of mice. The effective doses were the same order of magnitude as those reported by NRC in 1993 to cause chromosome aberrations in human lymphocytes. In vivo tests performed only in mice indicated that fluoride was not genotoxic, even at high doses. Ribeiro et al. (2004b) used the Comet assay to assess effects of NaF on Chinese hamster ovary cells in vitro. No damage was observed at concentrations of up to 100 µg/mL.

Rodent Systems In Vivo

Zeiger et al. (1994) administered NaF in drinking water for 6 weeks to B6C3F$_1$ mice and assayed micronuclei and chromosome aberration occurrences. They observed no increases over unexposed controls. Similarly, Dunipace et al. (1996) exposed diabetic and nondiabetic Zucker male rats to fluoride concentrations up to 50 mg/L in water for up to 6 months. They found no increase in the rate of SCEs for any test group.

Ribeiro et al. (2004a) exposed Wistar rats to NaF at 7 and 100 mg/L in drinking water for 6 weeks. Comet assays of peripheral blood, oral mucosa, and brain cells in vivo showed no increase in single-strand DNA damage.

Nonmammalian Systems In Vivo

Previous work on nonmammalian systems was sparse but did not indicate consistent cytogenetic effects. No new relevant studies have been reported.

Human Cells In Vivo

The NRC 1993 report noted the absence of human in vivo genotoxicity studies. Since 1993, important contributions to the evaluation of genotoxicity of fluoride have been in the area of cytogenetic studies of human *populations* exposed via diverse routes to various fluorides. Studies of human populations have the advantage of evaluating pertinent concentrations in a physiologically relevant context, despite the limitations inherent in all epidemiologic observational studies of not controlling for all factors that might be pertinent. Relevant studies are summarized below according to route of exposure.

Ingestion Route

The most well-documented in vivo human study published was that of Y. Li et al. (1995), who assayed the fluoride concentrations in water, plasma, and urine in more than 700 individuals. Six groups of 120 subjects resided in different locales with average naturally occurring fluoride concentrations in drinking water varying between 0.2 and 5 mg/L. They observed that, although plasma and urine fluoride concentrations varied with water concentrations, the groups of subjects living in the regions with higher concentrations of fluoride had lower average SCEs per cell. The study controlled for the nutritional status of the subjects. Subsequently, Jackson et al. (1997) compared SCE occurrence in lymphocytes of residents of communities with water fluoride concentrations of 0.2, 1, and 4 mg/L. Residents of the 4-mg/L fluoride community had more average SCEs. In a follow-up study, there was no difference between the mean SCE level of a subsample of residents using the 4-mg/L community water and another sample of residents using 0.3-mg/L well water.

The following three less-well-documented studies reported associations between cytogenetic effects and residence in areas with high natural fluoride concentrations in drinking water. Sheth et al. (1994) published a preliminary investigation of SCEs in 100 residents of Gujarat, India, with fluorosis and 21 unaffected controls. They reported higher SCE rates among the fluorosis cases as well as higher fluoride concentrations in the cases' water. The design of this study was seriously deficient, particularly because of the possibility of selection bias; cases and controls were recruited from different areas (cases were from areas with higher naturally occurring fluoride in drinking water). Additionally, clinical criteria for case definition were not adequately documented. Wu and Wu (1995) examined peripheral blood lymphocytes in a small series (n = 53) of residents in a high-natural-fluoride area (4 to 15 mg/L) and 30 control residents from a low-fluoride area (<1 mg/L) of Inner Mongolia. SCEs and micronuclei were more frequent only among subjects with fluorosis and not among those with higher exposures who did not exhibit fluorosis. The report had a dearth of information on subject selection and on control of potential confounding factors. Joseph and Gadhia (2000) later compared residents of three villages that had drinking water concentrations of fluoride at 1.6 to 3.5 mg/L with residents of Gujarat, India, where there is fluoride in residential drinking water at 0.7 mg/L . Chromosome aberrations were strongly elevated in residents of all three of the villages. SCE rates were elevated only in residents of one of those, and the same village's residents also demonstrated higher chromosome aberrations in mitomycin-C-treated lymphocytes. Only 14 individuals were tested from each village, and the method of subject selection was not reported.

Van Asten et al. (1998) found no cytogenetic effects (aberrations, micronuclei, or cell cycle progression) on cultured lymphocytes in women who had been treated with fluoride (22.6 to 33.9 mg/day) for osteoporosis for 1 to 4 years.

Inhalation and/or Dermal Routes

Two articles published by Meng et al. (1995) and Meng and Zhang (1997) described cytogenetic assays in phosphate fertilizer workers. Inhalation of fluoride is the principal chemical exposure in these plants. The air concentrations of fluoride ranged from 0.5 to 0.8 mg/m^3 at the time of the study. Chromosomal aberrations, micronuclei, and SCEs were all elevated in exposed workers. The length of exposure did not show a dose-dependent relationship with these cytogenetic effects; those working at the plant for 5 to 10 years had the greatest effect compared with those working for more than 10 years or less than 5 years. It is not clear, however, whether length of employment is a pertinent exposure metric concerning the plausibility of cytogenetic risk of fluoride for this cohort.

Cell Transformation

Cell transformation is the conversion of normal cells to neoplastic cells in vitro. In the 1993 NRC report, the positive transformation results reported were largely in Syrian hamster embryo (SHE) cells for which results cannot be extrapolated to human systems or other cell types (NRC 1993). However, in the one study that included an additional system, BALB/3T3 mouse cells (Lasne et al. 1988), transformation was observed with NaF at 25 to 50 ppm primarily in a promotional model with a known carcinogen as an initiator, suggesting this mechanism for a potential carcinogenic effect of fluoride. Since that time, the only additional pertinent publication is by Matthews et al. (1993), who also used a BALB/3T3 system with assay modifications to increase sensitivity. They tested numerous chemicals including 1.2 to 4.6 mM NaF (19 to 193 ppm), which did not exhibit transformational activity according to their criteria.

DNA Synthesis and Repair

A report from India (Ramesh et al. 2001) described a case series of 20 osteosarcoma patients of which the two with the highest fluoride concentrations in their tumor tissue had mutations of the tumor-suppressor gene *p53* and the others did not. The normal *p53* allele appears to protect cells from some mutagenic exposures by enhancing DNA repair mechanisms, and the dominant, null mutation is often found in soft tissue and osteosarcomas (Wadayama et al. 1993; Hung and Anderson 1997; Semenza and Weasel 1997). However, it should be noted that the fluoride concentration reported in the tumors with *p53* mutations (i.e., 64,000 and 89,000 mg/kg versus 1,000-27,000 mg/kg in the remaining patients) exceed the theoretical maximum fluoride concentration of 37,700 mg/kg in bone (see Chapter 3). No data were presented regarding drinking water concentrations or other sources of fluoride exposures for those patients. The observations in this small case series are consistent with a role of fluoride in *p53* mutations that could influence the development of osteosarcoma.

No other studies on DNA synthesis or repair have been found since those reviewed in the 1993 NRC report. Previous results were inconsistent but suggested that a mechanism for genotoxicity might be secondary to inhibition of protein or DNA synthesis (NRC 1993).

Update on Genotoxicity Conclusions and Recommendations of NRC (1993)

Overall, the results in in vitro systems summarized above are inconsistent and do not strongly indicate the presence or absence of genotoxic potential for fluoride. In 1993, NRC concluded that the existing genotoxicity data probably were not of "genetic significance." There were no specific 1993 NRC recommendations regarding genotoxicity studies, although the report did mention the dearth of human in vivo assays. The more recent literature on in vitro assays does not resolve the overall inconsistencies in the earlier literature.

The human population in vivo studies published during the past 10 years comprise a new body of data that might be pertinent to evaluating the genotoxic potential of fluoride; those population studies by definition integrate the pharmacokinetic contexts and actual cell environment parameters resulting from external exposures, whether via water or other environmental media. However, the inconsistencies in the results of these in vivo studies do not enable a straightforward evaluation of fluoride's practical genotoxic potential in humans.

CARCINOGENICITY

Animal Cancer Studies

Two studies were judged in the 1993 NRC review as adequate for the consideration of carcinogenic evidence in animals: an NTP study in F344/N rats and $B_6C_3F_1$ mice (NTP 1990) and studies in Sprague-Dawley rats (Maurer et al. 1990) and in CD-1 mice (Maurer et al. 1993). The latter study in CD-1 mice was in press at the time of the NRC (1993) review. Two neoplasms were noted in the weight-of-evidence discussion:

1. Positive dose-related increase in the trend ($P = 0.027$) of osteosarcoma in male F344/N rats through drinking water route of exposure (NTP 1990)

2. Positive increase of osteoma in male and female CD-1 mice through dietary inclusion exposure (Maurer et al. 1993).

The review concluded that "the collective data from the rodent fluoride toxicological studies do not present convincing evidence of an association between fluoride and increased occurrence of bone cancer in animals" (NRC 1993).

Since the publication of the 1993 NRC review, the discussion on the uncertainties and overall weight of evidence in animals was further ex-

GENOTOXICITY AND CARCINOGENICITY 317

panded (WHO 2002; ATSDR 2003). Most of the uncertainties had already been highlighted in the NTP study. However, the nature of uncertainties in the existing data could also be viewed as supporting a greater precaution regarding the potential risk to humans. The key issues are presented in this section. In addition, the committee found another NTP study that adds to the database on fluoride.

NTP Studies

In the chronic bioassays by NTP (1990), F344/N rats and $B_6C_3F_1$ mice were administered NaF in drinking water at of 25, 100, and 175 mg/L, 7 days per week for 2 years. A summary of the neoplasms found is presented in Table 10-2. Osteosarcomas of the bone were found in male rats (1 of 50 and 3 of 80 in the mid- and high-dose groups, respectively) but not in female rats or in mice. An additional male rat in the 175-mg/L group had osteosarcoma of the subcutaneous tissue. Rats and mice exhibited tooth discoloration, and male rats had tooth deformities and attrition.

To adequately assess the oncogenicity of a chemical, it is important that the dose range used in the study is sufficiently high, attaining the maximum tolerated dose (MTD) or minimally toxic dose. There was a lack of significant toxicity of NaF in F344/N rats and $B_6C_3F_1$ mice, which suggested that higher doses could be tolerated (NTP 1990). Thus, it can be argued that the oncogenicity of fluoride in drinking water cannot be fully assessed on the basis of this study. Although this could be the case for the study in mice, given that rats at the high dose already showed various tooth abnormalities, higher-dose treatment might interfere with the rat's ability to eat (NTP 1990).

Increased incidence of osteosarcoma was reported in the high-dose male rats (Table 10-2). Opinion differs regarding the appropriateness of including the one case of extraskeletal osteosarcoma in the remaining incidence of osteosarcomas found in vertebrae and humerus (NTP 1990; PHS 1991; ATSDR 2003). The incidence from all sites gives stronger statistical significance than from the bone alone, lowering the P value from $P = 0.027$ to $P = 0.01$ for dose- related trend (logistic regression test) and from $P = 0.099$ to $P = 0.057$ for the pair-wise comparison with the controls (NTP 1990). A comparison with the historical control series was also presented, although its significance was compromised because of the higher fluoride in the standard diet used for the historical data, and because the radiograph used in the fluoride drinking water study was not routinely used in bone examinations (NTP 1990). Osteosarcoma is a rare tumor in rats. More recent historical data from Haseman et al. (1998) became available after the data from Haseman et al. (1985) that were used for the evaluation in the fluoride drinking water study. The data published in 1985 included studies

TABLE 10-2 Incidence of Neoplasms Highlighted in the NTP and Maurer et al. Studies

NaF in Drinking Water (NTP 1990)[a]

Site of Neoplasm	Control	25 mg/L	100 mg/L	175 mg/L
Male F344/N rats				
Osteosarcoma: bone	0/80 (0%)+	0/51 (0%)	1/50 (2%)	3/80 (4%)
Osteosarcoma: all sites	0/80 (0%)++	0/51 (0%)	1/50 (2%)	4/80 (5%)
Oral cavity[b]	0/80 (0%)	1/51 (2%)	2/50 (4%)	3/80 (4%)
Thyroid[c]	1/80 (1%)+	1/51 (2%)	1/50 (2%)	4/80 (5%)
Female F344/N rats				
Osteosarcoma: bone	0/80 (0%)	0/50 (0%)	0/50 (0%)	0/81 (0%)
Osteosarcoma: all sites	0/80 (0%)	0/50 (0%)	0/50 (0%)	0/81 (0%)
Oral cavity[b]	1/80 (1%)	1/50 (2%)	1/50 (2%)	3/81 (4%)
Thyroid[c]	2/80 (3%)	0/50 (0%)	2/50 (4%)	2/81 (2%)

NaF in Drinking Water (NTP 1992)

Site of Neoplasm	Control	250 mg/L		
Male F344 rats				
Osteosarcoma: bone	2/49 (4%)	1/49 (2%)		

NaF in Diet (Maurer et al. 1993)[d]

Site of neoplasm	Control	4 mg/kg/day	10 mg/kg/day	25 mg/kg/day
Male CD-1 mice				
Osteoma: bone	1/50 (2%)	0/42 (0%)	2/44 (5%)	13/50 (26%)***
Female CD-1 mice				
Osteoma: bone	2/50 (3%)	4/42 (10%)	2/44 (5%)	13/50 (26%)**

[a]Statistical significance: trend test at $P \leq 0.05$ (+); $P \leq 0.01$ (++). Fisher pair-wise comparison at $P \leq 0.01$ (**); $P \leq 0.001$ (***). The average daily dose for the male rat control, 25-, 100-, or 175-mg/L group was 0.2, 0.8, 2.5, or 4.1 mg of fluoride/kg/day.

[b]Included squamous papillomas and squamous cell carcinomas in oral mucosa, tongue, or pharynx.

[c]Follicular cell adenomas or carcinomas.

[d]The given dose is in NaF. Adjusted for the 45% weight difference between fluoride and NaF, the dose for the treatment group was 1.8, 4.5, or 11.3 mg of fluoride/kg/day. Fluoride intake for the control mice was 0.9 mg of NaF/kg/day (0.4 mg of fluoride/kg/day) for the males and 1.1 mg of NaF/kg/day (0.5 mg of fluoride/kg/day) for the females.

completed between 1979 and 1984, whereas the data published in 1998 were a 7-year collection up to January 1997. The 1990-1997 data showed a lower historical incidence of 0.1% (range 0% to 2%) each for bone and for all skin sites (Haseman et al. 1998). Ideally, historical data closer to the time frame of the bioassay of comparison would be more pertinent. On the basis of the 1990-1997 data, the incidence of osteosarcoma at the high dose appeared to exceed the historical range. Nevertheless, the same issues in making comparisons with historical data remain—historical control animals were not fed a low-fluoride diet and their bones were probably not examined with radiograph.

Additionally highlighted in the NTP report were the oral cavity squamous papillomas and squamous cell carcinomas (oral mucosa, tongue, pharynx) in male and female rats and thyroid follicular cell adenomas and carcinomas in high-dose male rats (Table 10-2). Both showed some increase with dose. The incidence at the high dose exceed the historical control but stayed within the high end of the historical range and was not statistically significant from the concurrent control. The marginal increase in these neoplasms might not provide additional weight to the overall evidence of oncogenicity, but their occurrence could serve as an additional guide for epidemiologic studies.

Among the other tumor sites and types highlighted in the NTP report as not statistically and biologically significant was the hepatocellular neoplasm (adenoma, carcinoma, hepatoblastoma, and hepatocholangiocarcinoma) in male and female mice (NTP 1990). Among these neoplasms, five in male and four in female treatment groups (unspecified) were reported by the contract laboratory as hepatocholangiocarcinoma (NTP 1990). All but one in the females were reclassified into hepatoblastoma by the NTP pathology working group (NTP 1990). The incidence of these rare neoplasms not seen in the concurrent controls (historical hepatoblastoma of 0/2,197 in male mice and 1/2,202 in female mice) was judged as not significant when grouped with the more common hepatocellular adenomas and carcinomas (NTP 1990).

Another study conducted by NTP (1992, released in 2005) that bears on the carcinogenicity evaluation of fluoride is one that investigated the interaction of fluoride on the development of osteosarcoma induced by ionizing radiation. Pertinent to the committee's evaluation was a group of nonirradiated male F344 rats that were administered NaF at 250 mg/L in drinking water for two years. Of the 49 rats per group that were examined, osteosarcoma of the bone occurred in one NaF treated rats and two nonirradiated controls. Thus, the results did not show an increase of osteosarcoma with NaF. However, this single data point does not have sufficient statistical power for detecting low level effects and rendered its observed results statistically compatible with those from the NTP (1990) bioassay. It is noteworthy that the study had the unexpected result that none of the irradiated animals developed osteosarcoma.

Maurer et al. Studies

Maurer et al. (1990, 1993) fed Sprague-Dawley rats and CD-1 mice diets containing NaF at doses of 4, 10, and 25 mg/kg/day for up to 99 weeks (rats) or 97 weeks (mice). Evidence of toxicity included decreased weight gain in the high-dose rats and non-neoplastic changes of the teeth (rats and mice), bones (rats and mice), joints (mice), and stomach (rats). In rats, no incidence of preneoplastic or neoplastic lesions was significantly different from that in controls. In mice, increased incidence of osteomas (noncancerous bone tumors) was reported (Table 10-2).

The many limitations of the studies in rats and mice were identified in the earlier NRC (1993) review. The histopathologic examination of bones was not performed for all test animals (PHS 1991; WHO 2002; ATSDR 2003). Data on neoplasm were reported only for the bone and stomach. Moreover, based on the joint review by the Carcinogenicity Assessment Committee, Center for Drug Evaluation and Research, and U.S. Food and Drug Administration, questions were raised about the adequacy of the histopathologic examinations (PHS 1991). In the original report, fibroblastic sarcoma with areas of osteoid formation, chordoma, and chondroma were found in the males and osteosarcoma and chondroma were found in the females. However, the joint review discovered additional osteosarcoma in males and females. Collectively, those discrepancies called into question the weight of this negative study in the overall weight-of-evidence consideration (PHS 1991).

In the study with CD-1 mice, increased osteoma was reported in males and females at the high dose (Maurer et al. 1993). The authors reported that retrovirus infection in mice from all test groups might have confounded the occurrence of osteoma. The earlier NRC (1993) review considered the impact of the infection and concluded that the fluoride exposure was the most obvious cause for the increase in osteoma. However, based on the view of the Armed Forces Institute of Pathology (AFIP) that the osteomas were more reminiscent of a hyperplastic lesion, NRC (1993) concluded that their relevance to humans was questionable.

Human Cancer Studies

General Issues

Inherent difficulties for conducting epidemiologic studies of the cancer potential of fluoride and drinking water are similar to those challenges of studying most environmental chemicals. The limitations severely affect the possibility of identifying relatively small effects on cancer incidence and, especially, cancer mortality. Chief among them are the latency of cancer diagnosis after exposure to causal factors, typically spanning more than 10 years and often reaching 30 years. Migrations into and out of fluoridated areas often lead to misclassification of exposures when individual residency histories are not known. The diversity of cancers, comprising many different diseases rather than a single entity, necessitates evaluating each type of cancer separately rather than all cancers combined. Even so, there are few cancers for which specific environmental chemicals impart high attributable risks for the overall population or even among exposed populations.

The basic criteria for evaluating studies are appropriate methodology, potential selection and information biases, statistical power to detect real associations, appropriate time windows for assessing exposures and potential effects, and control for potential confounding by sociodemographic and other factors. In addition, sufficiently specific end points (types of cancer) and adequate exposure estimation are necessary for any epidemiologic study of fluoride and cancer to be informative for the committee's task. A further issue is consideration of sensitive subpopulations based on a priori physiologic or previous epidemiologic data. Finally, it is necessary to apply biologic plausibility criteria and a weight-of-evidence approach to evaluate whether any observed associations should be interpreted as causal.

Many of the studies published before and since the 1993 NRC report are "ecologic studies." In these designs, populations rather than individuals are the units of observation. A typical ecologic study regresses disease rates in different areas against average exposures. Such studies are usually less expensive and less time-consuming to conduct because the component data are already available. Incidence data are often very reliable if they are derived from high-quality population-based registries and census data. However, ecologic studies are often insensitive to small effects because of their design. The Agency for Toxic Substances and Disease Registry (ATSDR 2003) estimated that the ecologic studies performed to date for fluoride and cancer did not have sensitivities to detect less than 10% to 20% increases in cancer risk. Ecologic studies can be subject to large amounts of bias. Confounding factors and limited ability to control for such factors can be particularly serious problems (see Appendix C for a more detailed discussion of ecologic bias).

In semi-individual (partially ecologic) designs, individual-level information is collected for outcome and important variables, but exposure is assigned at the group level (e.g., based on residence or job title). Although such studies can share some characteristics of fully ecologic studies, they have much better ability to control confounding (see Appendix C).

Individual-based studies are composed of (1) case-control studies in which a group of people with a disease are compared with a sample of the population giving rise to the cases (controls) with regard to exposures that occurred before diagnosis, (2) cohort studies in which exposed and nonexposed people are followed forward in time and the disease experience of the two groups are compared, and (3) hybrids of these case-control and cohort designs. In environmental epidemiology, generally hundreds of subjects are required to detect with statistical significance any less than a twofold increase in risk of disease associated with a particular exposure. If an environmental agent is a weak carcinogen, with risks as low as 1 per 100,000 or 1 per 1,000,000 of those affected, it is extremely difficult to detect such effects by standard epidemiologic methods. This is particularly true of cohort studies, which would need to enroll large numbers of subjects to detect differences between exposed and unexposed cohorts when the risks are low.

Epidemiology Data for Carcinogenicity of Fluoride

The weight of evidence for epidemiologic studies that NRC reviewed in 1993 did not indicate cancer risk to humans from fluoride exposure. However, the predominant methods used, particularly ecologic studies for which individual exposure histories could not be collected and confounding variables could not be controlled, were inadequate to rule out a weak effect. Some studies reported positive associations and some did not, but many of the studies were flawed in that adjustment for potential sociodemographic confounders was lacking or inadequate.

Epidemiologic studies published since the early 1990s and other pertinent studies not included in the 1993 NRC review are detailed in Table 10-3. The data are discussed below according to target sites for which associations with fluoride have been reported by at least one study.

Bone and Joint Cancers, Particularly Osteosarcoma

Osteosarcoma presents the greatest a priori plausibility as a potential cancer target site because of fluoride's deposition in bone, the NTP animal study findings of borderline increased osteosarcomas in male rats, and the known mitogenic effect of fluoride on bone cells in culture (see Chapter 5). Principles of cell biology indicate that stimuli for rapid cell division increase the risks for some of the dividing cells to become malignant, either by inducing random transforming events or by unmasking malignant cells that previously were in nondividing states. Osteosarcoma is a rare disease, with an overall annual incidence rate of approximately 0.3 per 100,000 in the United States (Schottenfeld and Fraumeni 1996). The age of diagnosis is bimodal with peaks before age 20 and after age 50.

The incidence and mortality studies of osteosarcoma reviewed by NRC 1993 were ecologic or semi-ecologic in design. Their results were contradictory and inconclusive. The incidence studies of Hoover et al. (1991) at the National Cancer Institute observed that osteosarcoma rates in young males increased in the fluoridated areas compared with the nonfluoridated areas of two SEER registries they analyzed (Iowa and Seattle). However, the authors concluded that an association of fluoridation and osteosarcoma was not supported by the data because there was no linear trend of increased rate of osteosarcoma with the duration of fluoridation of the pertinent water supplies. The Hrudey et al. (1990) osteosarcoma incidence study in Alberta, Canada, and the Freni and Gaylor (1992) mortality analysis of bone cancer

TABLE 10-3 Summary of Recent Studies of Fluoride and Cancer

Study Design & Location	Observations	Findings	Remarks	References
Individual-based studies (cohort or case control)				
Case-control study of pediatric osteosarcoma. NY State residents.	130 cases plus matched controls, patient and/or parent interviewed re residency history, fluoride ingestion sources, and other factors. 59 pairs of subjects were interviewed.	All data combined: odds ratios (ORs) for total fluoride ingestion decreased. ORs for water ingestion were elevated. Total fluoride protective for males. Reduced data for subjects (vs. parents) only: elevated ORs with dose response but wide confidence intervals.	Water ingestion alone not discussed by authors. No data or analysis of possible critical time window or latency. Paper contained some reversal of data columns in gender-specific tables and some misstatements regarding proportions of males and females with osteosarcoma. However, on the basis of information available to the committee, those specific errors do not appear to affect interpretation of this study.	Gelberg et al. 1995
Historical occupational cohort study of cryolite worker. SIRs. Denmark.	522 workers exposed.	Increased incidence or urinary bladder and respiratory cancers.	No smoking or drinking water data.	Grandjean et al. 1992; Grandjean and Olsen 2004
Case-control osteosarcoma analysis using public records only. Wisconsin.	167 cases and 989 cancer referents (brain, digestive system) from state cancer registry. Not interviewed.	No association with residential fluoridation, including ages 0 to 24. (Positive association with naturally occurring radiation in water.)	Lack of residential history via interview and use of cancer referents are limitations.	Moss et al. 1995

continued

TABLE 10-3 Continued

Study Design & Location	Observations	Findings	Remarks	References
Case control of osteosarcoma, age <20 years U.S. multi-site	91 cases, 188 controls, interviewed on residency history and other fluoride exposures. Hospital-based.	Associations of exposures to fluoride at approximately 1 mg/L in water with osteosarcoma during ages 6 to 8, particularly for males.	Exploratory dissertation, multiple limitations in design, analysis, and presentation of findings.	Bassin 2001
Ecologic studies				
Ecologic, correlations of cancer incidence (combined) for fluoride concentrations. Worldwide.	49 cities or countries on 5 continents, classified as high or low cancer incidence. Where fluoride concentrations were unavailable, used data from neighboring area.	Inverse relationship between cancer rates and fluoridation reported $R = -0.75$. Latitude and temperature also analyzed, but not together as covariates.	Averaged male and female rates, and combined all cancers.	Steiner 2002
Ecologic analysis using proportions of populations with estimated fluoride concentrations ≥ 0.7 mg/L. USA (6 cities, 3 states).	9 areas, 36 different sites of cancer, three 5-year periods. 15 years or 5 years (when different) coefficients and cancer incidence ratios. Used log-transformation of fluoride concentration and cancer rates.	Regression coefficient highest for females' 1990 oral/digestive and male bone cancers.	Large number of comparisons. Cancer rates not shown. Rate distribution stated to be Poisson. Results presented selectively. Statistical methods flawed.	Takahashi et al. 2001
Mortality trends or uterine cancer before and after fluoridation terminated. Multiple regression. Okinawa, Japan.	20 of the 53 towns included. Controlled for sociodemographics. All fluoride concentrations below 0.4 mg/L.	Positive correlation of fluoridation with mortality rates. Mortality rates among the towns converged after fluoridation terminated.	Up to 13 years of exposure data combined. Hypothesis generated and data analyzed further in same population.	Tohyama 1996

Comparison of cancer mortality rates for towns with high vs. low natural fluoride concentrations. Taiwan.	10 high and 10 low matched towns compared. Total populations exceeded 1 million. Rate ratios (SMRs) generated.	The only finding with 95% confidence interval excluding 1.0 = excess of bladder cancer in females. Also higher rate ratios in males for bone, females for uterus, colon, all sites; both genders.	Multiple comparisons. Hypothesis-generating. Controlled for urbanization and disinfection by-products. Osteosarcomas not distinguished from other bone cancers.	Yang et al. 2000
Incidence and mortality statistics; subtracted female from male osteosarcoma rates. Worldwide.	Used U.S. and NJ rates, among others.	Concluded that the difference between male and female rates between fluoridated and nonfluoridated areas indicates cancer associations.	Inappropriate to subtract one gender from other. Uses circular reasoning on causality.	Yiamouyiannis 1993

for 40 cancer registries worldwide found no evidence of association with fluoride.

Cohn (1992) in New Jersey had findings suggestive of an association of fluoride in public water with increased osteosarcoma in young males. The osteosarcoma rate ratio among males below age 20 in the Cohn analysis, based on 20 cases, was 3.4 (95% confidence interval [CI] 1.8 to 6). Mahoney et al. (1991) generated bone cancer and osteosarcoma incidence rate ratios for the years 1975-1987 for fluoridated and nonfluoridated counties of New York State (excluding New York City). The authors did not observe an association of fluoridation and osteosarcoma or other bone cancers for either gender, including for those younger than age 30.

As discussed above, strengths of all the ecologic studies included the largely complete ascertainment of cases through the population-based cancer registries; the chief limitation is the potential for large amounts of bias and poor ability to adjust for covariates.

Since the 1993 NRC report, Yang et al. (2000)[1] conducted an ecologic analysis of cancer mortality in 20 municipalities in Taiwan, half with measurable naturally occurring fluoride concentrations. They controlled for urbanization and sociodemographic variables. Bone cancers (not specifically osteosarcoma) were nonsignificantly elevated (rate ratio [RR] of 1.6, 95% CI 0.92 to 2.17) in males but decreased in females (RR of 0.87, 95% CI 0.52 to 1.44). The range of fluoride concentrations was not reported, but the median and mean were about 0.25 mg/L.

Also since 1993, four individual-based studies have been published. Gelberg (1994) and Gelberg et al. (1995) conducted a population-based case-control study of osteosarcoma before age 25 in New York State. It included 130 cases and one matched control for each case. Controls were drawn from birth certificates, with replacement for those that could not be located. Parents and/or patients were interviewed regarding residence history and exposure to fluoride through drinking water, consumer dental products, dental supplements, and fluoride treatments. Analyses were conducted according to estimated lifetime dose of fluoride in total milligrams from each source of potential exposure, both separately and combined. When data on all subjects were analyzed, total fluoride exposures showed an inverse relationship with osteosarcoma. Use of fluoride gels had strong negative associations with osteosarcoma. Based on the parents' interviews alone (97% of subjects), the authors found negative associations with total estimated fluoride intake from all sources, particularly due to a strong negative association of osteosarcoma with estimated quantities of fluoride ingested from toothpaste. Odds ratios (ORs) were above 1.0 for all catego-

---

[1]This study did not analyze age subgroups and, therefore, did not address particular risk for young males or females.

ries of lifetime fluoride intake from drinking water compared with those with zero estimated intake from that source, particularly among females. This distinction is particularly noteworthy because Gelberg et al. had higher estimates of the relative contributions of fluoride from toothpaste ingestion compared with drinking water than those reflected in Chapter 2 of this report (see Figure 2-1). The source of the study's estimates of toothpaste ingestion was not specified, but the relative proportions were most similar to those shown in Figure 2-1 for ages 2 to 6. If the relative contributions from toothpaste were exaggerated, then the findings regarding fluoride specifically from drinking water could arguably be given greater weight. Analyses of average annual fluoride exposure did not differ markedly from the observations on cumulative exposure estimates, thereby controlling to some degree for age of diagnoses.

A reduced set of 59 respondent pairs who were the actual patients or their controls (i.e, excluding proxies) showed positive associations, with very wide CIs, for both fluoridated water alone and for total fluoride exposure (only combined genders were analyzed in this smaller series). There were no analyses using lagged exposure estimates to consider hypothetical latencies between potential exposures and diagnosis of osteosarcoma, so it is possible that inclusion of nonpertinent exposures could lead to misclassification of relevant exposures.

Gelberg et al. concluded that their study showed no association of osteosarcoma and fluoride exposure. To date, this study is the closest to fulfilling the recommendation of the 1993 NRC report regarding conducting one or more analytic studies of osteosarcoma and fluoride exposure. However, no bone fluoride concentrations could be assessed through this design.

Moss et al. (1995) conducted a case-control analysis of osteosarcoma in Wisconsin by using only public records (without interviews). For the 167 cases, 989 cancer controls were selected from the state cancer registry among patients with other types of cancer (brain, digestive system). The study controlled for size of town, age at diagnosis, and radium levels in drinking water and did not observe an association of fluoridation at the time of case diagnosis with osteosarcoma. Because exposure classifications were assigned without interviews or other sources of residence history or water source data, this design is similar to that of a semiecologic study. The authors also examined young age groups specifically.

A pilot hospital-based case-control study of patients under age 40 was published by McGuire et al. (1991), indicating a nonsignificant negative association with a small series of osteosarcomas (34 cases and matched controls). A full-scale case-control study by this group (Douglass 2004) is now under way. Its design is described below because of its potential for future contribution to this issue.

Grandjean et al. (1992) and Grandjean and Olsen (2004) conducted a historical cohort study among cryolite production workers in Denmark who previously had been documented to suffer high rates of skeletal fluorosis. Cryolite is composed of about 50% fluoride, and the workers were not believed to be exposed to suspected carcinogens of any other type via their work. The authors did not control for smoking. There were no bone fluoride measurements. However, daily dose of fluoride to these workers during their time of employment could be estimated at about 30 mg/day. Over many years of employment, workers' exposure would tend to greatly exceed chronic exposures from ingestion of fluoride at the current MCL of 4 mg/L. No osteosarcoma incident or mortality cases were observed among their 522 subjects, and, given the rarity of osteosarcoma, the authors concluded an 18-fold upper bound on the relative risks of this disease from the exposures encountered by their cohort.

The central research chapter of an unpublished dissertation by Bassin (2001) on fluoride and osteosarcoma has recently become publicly available. The author described the work as exploratory. The report has important strengths and major deficits, some of which are described below.

The design is a case-control study of people under 20 years of age from 11 teaching hospitals in the United States. Cases (n = 91) were retrospectively ascertained and 188 controls were hospitalized patients in the same orthopedics departments. Controls were matched with cases according to distance of residence from the hospital. Hospital-based controls can introduce serious selection bias; osteosarcomas treated at the participating teaching hospitals are more likely to be representative of all osteosarcomas occurring in the surrounding populations, whereas patients treated for fractures or other common orthopedic ailments at these teaching hospitals may not be as representative of the overall population that gave rise to the cases. If fluoride exposure is either a risk factor or a protective factor for the group of hospitalized controls (e.g., fracture patients), the resulting relative risk estimates could be biased downward or upward, respectively. For example, the dissertation did not provide any data on what proportion of the controls comprised fracture patients.

All subjects or their surrogates were interviewed about lifetime residence history, a strength of the design. However, individual information on key socioeconomic factors such as education and income was not collected. Average income levels based on zip codes were used but might not reflect individual socioeconomic status. Lack of such information can be problematic if socioeconomic status, or factors for which it is a surrogate, introduce confounding.

The primary exposure metrics for fluoride in drinking water were based on a combination of data from the Centers for Disease Control and Prevention, states, locales, and purveyors on year-specific water system fluoride concentrations expressed as proportions of the recommended fluoride guidelines. Based on tertiles for the controls, three exposure categories were expressed as 100%, 30% to 99%, and <30% of the target concentrations for fluoridated water.

A unique feature of the analysis published in the literature so far was an exploratory analysis of ORs for each specific year of age. Bassin found elevated ORs for the highest tertile compared with the lowest centering on ages 6 to 8. At age 7, the respective ORs (and 95% confidence intervals) were 7.2 (1.7 to 30.0) for males and 2.0 (0.43 to 9.28) for females. For the highest tertile, graphed results for males indicated a gradual increase and then a decrease of estimated relative risk from exposure at ages 0 to 15 with peaks at age 7, with the middle tertile, compared with the lowest, showing stable ORs across all ages. For females, both the middle and highest tertiles of exposure showed relatively unchanging relative risk estimates across exposure ages.

There was no analysis of cumulative exposures to fluoride, and therefore it is difficult to compare the Gelberg study, which used only cumulative exposure indices, with the Bassin work. This dissertation had a paucity of data in the results section, hampering its interpretation; for example, the report did not provide numbers of subjects in the categories upon which the ultimate analyses were based. Also, there were no data on bias potential stemming from nonparticipation of subjects due to refusal to be included or inability to locate them.

Nevertheless, the higher ORs for males than for females, and the highest ORs at ages 6 to 8, during what the author describes as the "mid-childhood growth spurt for boys," are consistent with some previous ecologic or semiecologic studies (Hoover et al. 1991; Cohn 1992) and with a hypothesis of fluoride as an osteosarcoma risk factor operating during these ages. A publication based on the Bassin thesis is expected in the spring/summer of 2006 (E. Bassin, personal communication, Jan. 5, 2006). If this paper provides adequate documentation and analyses or the findings are confirmed by another study, more weight would be given to an assessment of fluoride as a human carcinogen.

A relatively large hospital-based case-control study of osteosarcoma and fluoride exposure is under way (Douglass 2004) and is expected to be reported in the summer of 2006 (C. Douglass, Harvard School of Dental Medicine, personal communication, January 3, 2006). Most of the incident cases are identified via eight participating medical centers in California, District of Columbia, Florida, Illinois, Massachusetts, Nebraska, and Ohio. The study has prospectively identified 189 incident cases of osteosarcoma and 289 hospital controls. Controls are orthopedic patients at the same hospitals as osteosarcoma patients and include patients diagnosed with malignancies other than osteosarcoma and other patients admitted for benign tumors, injuries, and inflammatory diseases. Matching criteria include gender, age, and geographic characteristics. The investigation includes residence histories and detailed interviews about water consumption as well as fluoride assays of bone specimens and toenails of all subjects. The ultimate analysis and validity of this study will depend partly on the degree to which control selection is not biased in such a way as to artificially increase or decrease the likelihood of fluoride exposure compared with the general population to which this study is intended to apply.

A preliminary retrospective recruitment phase of this investigation, including telephone interviews, residential history reconstruction, and an attempt to estimate dietary fluoride intakes, reported ORs of 1.2 to 1.4 that were not statistically significant (Douglass 2004). No confidence intervals were provided. The Douglass study may have limited statistical power to detect a small increase in osteosarcoma risk due to fluoride exposure, but the committee expects the forthcoming report is likely to be a useful addition to the weight of evidence regarding the presence or degree of carcinogenic hazard that fluoride ingestion might pose to osteosarcoma risk, particularly if it addresses some of the limitations of hospital-based studies that are mentioned above in the description and critique of the Bassin thesis.

Kidney and Bladder Cancers

The plausibility of the bladder as a target for fluoride is supported by the tendency of hydrogen fluoride to form under physiologically acid conditions, such as found in urine. Hydrogen fluoride is caustic and might increase the potential for cellular damage, including genotoxicity. The Hoover et al. (1991) analyses of the Iowa and Seattle cancer registries indicated a consistent, but not statistically significant, trend of kidney cancer incidence with duration of fluoridation. This trend has not been noted in other publications, although Yang et al. (2000) observed that the adjusted mortality rate ratios of kidney cancers among males in Taiwan was 1.55 (95% CI 0.84 to 2.84). The analogous rate for females was 1.37 (95% CI 0.51 to 3.70). Yang et al. noted statistically significant RRs in females for bladder cancer (RR = 2.79, 95% CI 1.41 to 5.55; for males RR = 1.27, 95% CI 0.75 to 2.15).

The Grandjean et al. (1992) and Grandjean and Olsen (2004) historical occupational cohort study of cryolite workers in Denmark (described earlier in the section on bone and joint cancers), who were followed from 1941 to 2002, observed an elevated standardized incidence ratio (SIR) for bladder cancers (SIR = 1.67, 95% CI 1.02 to 2.59). The SIR is the ratio of observed cases of cancer to the expected number of cases based on incidence rates of the general population. Higher SIRs were seen among males employed more than 10 years, males less than 35 years old when follow-up began, and among workers observed after a minimum latency of 30 years

(Grandjean and Olsen 2004). In the absence of data on smoking, the authors interpreted the higher SIRs for bladder cancer than for lung cancer to suggest that smoking was unlikely to be the major cause of the elevated bladder cancer incidence. The authors proposed (2002) that excretion of fluoride compounds entailed exposure of the pertinent target tissues. As noted above, the estimated exposures of the cryolite workers were about 4-fold greater than those estimated from ingestion of fluoridated water at the MCL of 4 mg/L. However, those workers were exposed for fewer years than those involved in lifetime residency.

Romundstad et al. (2000) reported on cancer among Norwegian aluminum workers exposed to polycyclic aromatic hydrocarbons and fluorides. SIRs for bladder and lung cancer were elevated among the exposed workers. However, separate effects from the two exposures could not be distinguished from this paper. Further, the authors review and compare earlier studies that used different aluminum plant processes, which support the role of polycyclic aromatic hydrocarbons in bladder cancer among the exposed cohort. It may be noteworthy that smoking did not appear to be a confounder for the risk of bladder or lung cancer among the exposed cohort. The authors state, but do not present data, that they found a "weak association" of bladder for fluoride exposures lagged less than 20 years.

Oral-Pharyngeal Cancer

The NCI analysis (Hoover et al. 1991) indicated an a priori interest in oral cancers. In Iowa, one of the two cancer registries they analyzed, the authors observed a trend among males in the incidence rates of oropharyngeal cancer with duration of fluoridation, but mortality analyses did not indicate an association with fluoridation. However, in an earlier study in England, oral-pharyngeal cancers among females constituted the only site-gender category for which standardized mortality ratios in England were found to be significantly elevated in areas with naturally occurring high fluoride concentrations, defined as more than 1.0 mg/L. Twenty-four site-gender combinations were examined for 67 small areas (Chilvers and Conway 1985).

Uterine Cancer

An association of uterine cancer (combination of cervical and corpus uteri) with fluoridation was reported by Tohyama (1996), who observed mortality rates in Okinawa before and after fluoridation was terminated, controlling for sociodemographics. This analysis is a follow-up of the positive results from a previous exploratory analysis that comprised a large number of comparisons conducted by this researcher with the same data set. The only other recent publication to report on uterine cancers is that of Yang et al. (2000), who observed a mortality rate ratio of 1.25 with 95% CI of 0.98 to 1.60.

Other Specific Cancers

Respiratory cancers were elevated among the cohort of Danish cryolite miners for whom exposure was by the inhalation route (Grandjean et al. 1992; Grandejan and Olsen 2004; see discussions above on this cohort study). SIRs of 1.51 (95% CI 1.11 to 2.01) were observed for the cohort as a whole, with higher SIRs among those after 30 years of exposure and among males younger than 35 when follow-up began. No smoking data on the cohort were collected. Also, except for mortality among females in Taiwan (Yang et al. 2000), there has not been corroborating data from other analyses for respiratory cancers.

No association between lung cancer and exposure to polycyclic aromatic hydrocarbons and fluorides was found in a study of the Norwegian aluminum industry (Romundstad et al. 2000).

The NCI incidence or mortality analyses conducted by Hoover et al. (1991) observed a few suggestive increases among some subgroups for soft tissue sarcoma, non-Hodgkin's lymphoma, colorectal cancer, and lip cancer, but those cancers were not a priori of concern as related to fluoride exposure based on biologic plausibility.

All Cancers Combined

A large number of mortality analyses for all cancers combined have been reported and reviewed previously (NRC 1993; McDonagh et al. 2000a), and most of those did not detect an association of combined cancer mortality with fluoridated water. Typically, studies that only report combined cancer rates are not informative for assessing possible associations between an environmental exposure and a specific cancer outcome, particularly an uncommon cancer. Thus, the committee did not use these types of studies as part of its evaluation.

Other Studies Evaluated

The following three studies were reviewed but were not included by the committee in the evaluation of weight of evidence of carcinogenicity of fluoride for the reasons summarized below.

Takahashi et al. (2001) conducted an ecologic analysis of data from nine U.S. cities for three 5-year intervals spanning 1978-1992 combined with fluoridation data. Their analysis involved regression of log-transformed cancer incidence rates on the log-transformed proportion of residents receiving fluoridated water. This paper is difficult to interpret and to compare with other studies in part because of its novel method of analysis. Unusual cancer subsites are included and major anatomical groupings typically appearing in cancer incidence reports (e.g., lymphocytic leukemia, breast, uterus) were omitted. Results were incompletely reported for subsets of data for particular cancer sites, creating issues of multiple comparisons and selective presentation. Another issue is that the ecologic exposure variable is the percentage of the population in each area with fluoridated water (or naturally occurring fluoride at 0.7 mg/L or higher). This is an aggregated form of a dichotomous variable on the individual level, which tends to bias results away from the null. There was inconsistent standardization of the outcome variable (which was age standardized) and the exposure variable (which was not), which can lead to bias. There was no adjustment for confounding by urbanization or other sociodemographic factors among the nine cities, which included widely different geographic, industrial, and demographic characteristics, and there was no population weighting by size. Finally, ecologic bias is best understood for linear or log-linear regression, making this study harder to interpret.

Steiner (2002) conducted an ecologic analysis of latitude, temperature, and fluoridated water in 49 cities worldwide. When fluoride concentrations were unavailable for these cities, he substituted data from neighboring areas. Average daily temperature and latitude were also included in his models, but not simultaneously. Steiner analyzed only all cancers combined. He found a negative association between cancer incidence and fluoridation.

Yiamouyiannis (1993) subtracted female from male cancer incidence rates for the United States and for New Jersey as an indication of fluoride's carcinogenic effect among males. This paper used circular reasoning to reach a conclusion of causality; that is, it concluded that higher cancer rates in males indicate an association with fluoride on the basis of a presumed causation by fluoride of cancers in males. Because most cancers do not occur at the same rates in each gender, the committee judges it is inappropriate to subtract rates of women from those of men as a means of evaluating factors that only affect bone cancer in males.

It has been suggested that differences in osteosarcoma rates found in provinces of Kenya could be related to fluoride exposure (C. Neurath, Fluoride Action Network, unpublished data, June 17, 2005). For eight provinces of Kenya, Neurath correlated enamel fluorosis prevalences reported by Chibole (1987) with osteosarcoma incidence rates reported by Bovill et al. (1985) and found a strong association. This type of fully ecologic analysis (see Appendix C) has its inherent advantages and limitations; in this instance, however, the underlying ratios of observed-to-expected osteosarcoma incidence are not reliable because Bovill et al. do not state that their incidence data were adjusted for differences in the age structure of various provincial populations. Bovill et al. state that Kenya is characterized by strong contrasts of ethnicity and other demographics among its geographic regions. The provincial summaries are weighted averages of the children examined, but it is not stated if they are also weighted averages of the underlying populations. Chibole does not state how the children examined in Kenyan schools and hospitals were selected (i.e., whether the fluorosis prevalence data collected were ascertained in a manner that would accurately reflect the populations of the component provinces). Chibole's detailed table indicates a wide range of prevalences of fluorosis within many of the provinces (e.g., from 3.7% to 69.5% in the Rift Valley province).

Summary of Cancer Epidemiology Findings

The combined literature described above does not clearly indicate that fluoride either is or is not carcinogenic in humans. The typical challenges of environmental epidemiology are magnified for the evaluation of whether fluoride is a risk factor for osteosarcoma. These challenges include: detection of relatively low risks, accurate exposure classification assessment of pertinent dose to target tissues, multiple causes for the effect of interest, and multiple effects of the exposure of interest. Assessing whether fluoride constitutes a risk factor for osteosarcoma is complicated by (1) how uncommon the disease is, so that cohort or semi-ecologic studies are not based on large numbers of outcomes, and (2) the difficulty of characterizing biologic dose of interest for fluoride because of the ubiquity of population exposure to fluoride and the difficulty of acquiring bone samples in nonaffected individuals.

In summary, there has been partial but incomplete fulfillment of NRC's recommendations on individual-based cancer studies in the intervening years since 1993; one analytic study of osteosarcoma has been published, but bone samples were not included. The alternative (hospital-based) design, including bone assays, from the Harvard group might be more useful in addressing this issue.

EPA GUIDELINES AND PRACTICE IN SETTING MCLGs REGARDING CARCINOGENICITY

The EPA Office of Drinking Water establishes MCLGs of zero for contaminants that are known or probable human carcinogens. Chemicals for which cancer hazard is judged to be absent are regulated via the reference dose (RfD) method (see Chapter 11). "Methodology for Deriving Ambient Water Quality Criteria for the Protection of Human Health (2000)" reviewed EPA's additional practice of applying an uncertainty factor between

"1 and 10" to an RfD derived from noncancer health effects (EPA 2000d). This procedure has been used for substances judged to be possibly carcinogenic in humans. That methodology document also stipulates that the water concentrations estimated to result in 10^{-6} to 10^{-5} excess cancer risks should also be assessed under the RfD scenario for comparison.

As of April 2005, EPA has adopted new "Guidelines for Carcinogen Risk Assessment," which has replaced the 1986 categories with weight-of-evidence descriptors, involving textual consideration and explanation of how each category was arrived at. In addition, the Guidelines provide for consideration of mode of action and sensitive subpopulations, especially children (EPA 2005a,b). In addition to mode of action, other factors for weighing human epidemiologic studies and lifetime whole animal bioassays include data on biomarkers (genotoxicity and other assays of exposure, susceptibility, and effect) and toxicokinetics. Thus, key decisions about cancer pertinent to a MCLG for drinking water include an assessment of whether an MCLG of zero is appropriate based on the current epidemiologic, animal bioassay, and additional contributing data. If not, EPA will need to decide whether an uncertainty (safety) factor greater than 1.0 and up to 10.0 should be applied to an RfD derived from a precursor response to tumors.

Some recent examples of the use by EPA of RfDs with additional safety factors imposed because of possible carcinogenic hazard, based on the July 1999 Cancer Guidelines, include the MCLG for disinfection by-products (EPA 2003c). For dibromochloromethane (DBCM), EPA imposed an additional uncertainty factor of 10 to account for possible carcinogenicity based on studies of DBCM by NTP in 1985 that showed an increase in liver tumors in both genders of mice but no increase in either gender of rats. Similarly for trichloroacetic acid (TCA), an additional uncertainty factor of 10 was added to the MCLG derived from the RfD; TCA induced liver tumors in mice but not in rats. The MCLGs for all regulated chemicals considered to be possible carcinogens has included the additional 10-fold risk management factor applied to the RfD (J. Donohue, EPA, personal commun., 2004).

FINDINGS

The 1993 NRC report recommended the following:

Conduct one or more highly focused, carefully designed analytical studies (case control or cohort) of the cancer sites that are most highly suspect, based on data from animal studies and the few suggestions of a carcinogenic effect reported in the epidemiological literature. Such studies should be designed to gather information on individual study subjects so that adjustments can be made for the potential confounding effects of other risk factors in analyses of individuals. Information on fluoride exposure from sources other than water must be obtained, and estimates of exposure from drinking water should be as accurate as possible. In addition, analysis of fluoride in bone samples from patients and controls would be valuable in inferring total lifetime exposures to fluoride. Among the disease outcomes that warrant separate study are osteosarcomas and cancers of the buccal cavity, kidney, and bones and joints.

As described above, some progress in those directions have been made, with the most comprehensive study still in progress (Douglass 2004).

Fluoride appears to have the potential to initiate or promote cancers, particularly of the bone, but the evidence to date is tentative and mixed (Tables 10-4 and 10-5). As noted above, osteosarcoma is of particular concern as a potential effect of fluoride because of (1) fluoride deposition in bone, (2) the mitogenic effect of fluoride on bone cells, (3) animal results described above, and (4) pre-1993 publication of some positive, as well as negative, epidemiologic reports on associations of fluoride exposure with osteosarcoma risk.

Several studies indicating at least some positive associations of fluoride with one or more types of cancer have been published since the 1993 NRC report. Several in vivo human studies of genotoxicity, although limited, suggest fluoride's potential to damage chromosomes. The human epidemiology study literature as a whole is still mixed and equivocal. As pointed out by Hrudey et al. (1990), rare diseases such as osteosarcoma are difficult to detect with good statistical power.

In animal studies, the overall incidence of osteosarcoma in male rats showed a positive trend. Based on the more recent historical control data (Haseman et al. 1998) that were closer to the time frame of the NTP study, the 4% to 5% incidence at the high dose might have exceeded the historical range. The relevance of rat osteosarcoma to humans was discussed based on the species differences in the development of long bone, the common site of human osteosarcoma (NTP 1990). Specifically, ossification of human long bones is completed by 18 years of age whereas it continues in rats throughout the first year of life (PHS 1991). Nevertheless, most of the osteosarcomas found in male rats were not in long bones.

In another study (NTP 1992), that used the same strain and sex of rats, no increase in osteosarcomas was reported, even though the animals were exposed to a higher concentration of fluoride than in the earlier study. However, the primary intent of the NTP (1992) study was to test the hypothesis that ionizing radiation is an initiator of osteosarcoma and that fluoride is a promoter, and the committee thought it was noteworthy that none of the irradiated animals developed osteosarcomas.

The 1993 NRC review concluded that the increase in osteoma in male and female mice (Maurer et al. 1993) was related to fluoride treatment.

GENOTOXICITY AND CARCINOGENICITY 337

TABLE 10-4 Evidence Summary for Carcinogenicity of Fluoride: Epidemiologic Studies and Rodent Lifetime Bioassays

Cancer Site/Type	Individual-Based Epidemiology Studies	Ecologic Epidemiology Studies	Animal Data
Osteosarcoma	Case-control studies ambiguous (additional comprehensive hospital-based case-control study including bone fluoride measurements is under way).	Mixed.	Male F344/N rats: Borderline positive. Male F344 rats: inconclusive
Oral cavity		NCI incidence elevated in males, but no mortality trends. Several other reports positive.	Nonstatistically significant increase in male rats.
Thyroid			Nonstatistically significant increase in male rats.
Kidney and/or bladder	Occupational cohort: positive finding, inhalation route, high exposures.	Some positive reports.	
Uterine		One positive report.	
Respiratory	Occupational cohort positive finding, inhalation route, high exposures.	One positive report.	

TABLE 10-5 Evidence Summary for Carcinogenicity of Fluoride: Genotoxicity and Mechanistic Assays

Type of Effect and Assay	Strength of Evidence
Mitogenesis	Well established.
Cytogenetic effects: human in vivo exposure, in vitro assay.	Inconsistent; and the positive findings were from weak papers.
Cytogenetic effects: human in vitro exposure, in vitro assay.	Inconsistent.
Cytogenetic effects: other mammalian systems.	Inconsistent.
Transformation.	Inconsistent; the positive results are consistent with a promotion mechanism.
DNA repair mechanism: human.	Suggestive positive finding regarding tumor suppressor gene, small case series.
Mutation: mammalian systems.	Inconsistent.
Mutation: microorganisms.	Negative.

Although the subsequent review by AFIP considered these mouse osteomas as more closely resembling hyperplasia than neoplasia, given that osteoma is widely recognized as neoplastic, the evidence of osteoma remains important in the overall weight-of-evidence consideration. The increased incidence and severity of osteosclerosis in high-dose female rats in the NTP study demonstrated the mitogenic effect of fluoride in stimulating osteoblasts and osteoid production (NTP 1990) (see also Chapter 5).

The genotoxicity data, particularly from in vivo human studies, are also conflicting; whereas three were positive on the basis of the ingestion route (Sheth et al. 1994; Wu and Wu 1995; Joseph and Gadhia 2000), all three of these reports had serious deficits in design and/or reporting, including the characterization of how the study populations were selected and whether the exposed and unexposed study subjects were comparable. Two studies (Meng et al. 1995; Meng and Zhang 1997) were positive for the inhalation route among workers in a phosphate fertilizer factory, although other contaminants cannot be ruled out as the causal factors. Contrasting negative observations by other investigators (Li et al. 1995; Jackson et al. 1997; Van Asten et al. 1998) must also be considered.

RECOMMENDATIONS

Carcinogenicity

- The results of the Douglass et al. multicenter osteosarcoma study (expected in the summer of 2006) could add important data to the current body of literature on fluoride risks for osteosarcoma because the study includes bone fluoride concentrations for cases and controls. When this study is published, it should be considered in context with the existing body of evidence to help determine what follow-up studies are needed.

- Further research on a possible effect of fluoride on bladder cancer risk should be conducted. Since bladder cancer is relatively common (compared with osteosarcoma), both cohort and case-control designs would be feasible to address this question. For example, valuable data might be yielded by analyses of cancer outcomes among the cohorts followed for other health outcomes, such as fractures (see Chapter 5).

Genotoxicity

- The positive in vivo genotoxicity studies described in the chapter were conducted in India and China, where fluoride concentrations in drinking water are often higher than those in the United States. Further, each had a dearth of information on the selection of subjects and was based on small numbers of participants. Therefore, in vivo human genotoxicity studies

GENOTOXICITY AND CARCINOGENICITY

in U.S. populations or other populations with nutritional and sociodemographic variables similar to those in the United States should be conducted. Documentation of subject enrollment with different fluoride concentrations would be useful for addressing the potential genotoxic hazards of fluoridated water in this country.

11

Drinking Water Standards for Fluoride

The U.S. Environmental Protection Agency (EPA) has three standards for fluoride in drinking water: a maximum-contaminant-level goal (MCLG), a maximum contaminant level (MCL), and a secondary maximum contaminant level (SMCL). In this chapter, the committee reviews the MCLG and SMCL for fluoride, the two nonenforceable standards, for their scientific basis and adequacy for protecting the public from adverse effects. First, an overview of current procedures for establishing exposure standards is provided, and risk assessment issues that have developed since the original MCLG and SMCL for fluoride were established are discussed.

CURRENT METHODS FOR SETTING STANDARDS FOR DRINKING WATER

To establish MCLGs for drinking water, EPA reviews studies of health effects of individual contaminants and uses the information to calculate an exposure level at which no known or anticipated adverse health effects would occur with an adequate margin of safety. MCLGs consider only public health and not the limits of detection or treatment technology, so they may be set at concentrations that water systems cannot achieve.

Noncarcinogenic Contaminants

For noncarcinogenic chemicals, the MCLG is based on the reference dose, which is defined as an estimate (with uncertainty spanning perhaps an order of magnitude or greater) of a daily dose to the human population (including susceptible subpopulations) that is likely to have no appreciable risk of deleterious health effects during a lifetime. The reference dose characterizes exposure conditions that are unlikely to cause noncancer health effects, which are typically assumed to have a threshold dose above which adverse health effects would be expected to occur.

Traditionally, reference doses are determined by identifying the most sensitive health effects that are relevant to the human, selecting a no-observed-adverse-effect level (NOAEL) or a lowest-observed-adverse-effect level (LOAEL), and dividing the NOAEL or LOAEL by one or more uncertainty factors to provide a margin of safety. Uncertainty factors are applied to address uncertainties with using experimental animal data for human effects (interspecies differences) to account for variable susceptibilities in the human population (intraspecies differences), to adjust for differences between the LOAEL and NOAEL when a LOAEL is used instead of a NOAEL (LOAEL-to-NOAEL extrapolation), to account for uncertainties with predicting chronic exposure effects on the basis of subchronic exposure studies (subchronic to chronic extrapolation), and to address uncertainties when the database on the chemical is inadequate. Sometimes a modifying factor is used to account for additional uncertainty not addressed by the standard uncertainty factors.

Typically, uncertainty factors are assigned values ranging from 1 to 10. If information about a factor is sparse and uncertainty is high, a default value of 10 is generally used. If information is available, the uncertainty factor might be reduced to 1. For an uncertainty factor that falls between 1 and 10, a factor of 3 is typically assigned, because 3 is the approximate logarithmic mean of 1 and 10, and it is assumed that the uncertainty factor is distributed lognormally (EPA 1994). To calculate a reference dose, the NOAEL or LOAEL is divided by the product of the uncertainty factors. EPA typically uses a maximum of 3,000 for the product of four uncertainty factors that individually are greater than 1 and a maximum of 10,000 with five uncertainty factors (Dourson 1994).

More recently, the benchmark dose is being used as the starting point for calculating reference doses. The benchmark dose is a dose with a specified low level of excess health risk, generally in the range of 1% to 10%, which can be estimated from data with little or no extrapolation outside the experimental dose range. Specifically, the benchmark dose is derived by modeling the data in the observed experimental range, selecting an incidence level within or near the observed range (e.g., the effective dose producing a 10% increased incidence of response), and determining the upper confidence limit on the model. To account for experimental variation, a lower confidence limit or uncertainty factors on the benchmark dose are used to ensure that the specified excess risk is not likely to be exceeded.

To derive an MCLG, the reference dose is multiplied by a typical adult body weight of 70 kg and divided by an assumed daily water consumption of 2 L to yield a drinking water equivalent level. That level is multiplied by a percentage of the total daily exposure contributed by drinking water (usually 20%) to calculate the MCLG. EPA then uses the MCLG to set an enforceable standard (the MCL). The MCL is set as close to the MCLG as feasible.

Carcinogenic Contaminants

EPA sets MCLGs of zero for contaminants that are known or probable human carcinogens. For chemicals judged to be possibly carcinogenic to humans, EPA has recently begun applying an uncertainty factor between 1 and 10 to the reference dose derived from noncancer health effects to determine some exposure standards, such as certain ambient water-quality criteria (EPA 2000d). EPA stipulates that the water concentrations estimated to result in 1×10^{-6} to 1×10^{-5} excess cancer risks should also be compared with the reference dose.

NEW RISK ASSESSMENT CONSIDERATIONS

Since the fluoride MCLG and SMCL were originally issued, there have been a number of developments in risk assessment. A few of those issues were described above in the discussion of current risk assessment practices (e.g., use of benchmark dose). Below, a few specific issues relevant to the committee's review of the drinking water standards for fluoride are discussed, including advances in carcinogenicity assessment, relative source contribution, special considerations for children, and explicit treatment of uncertainty and variability.

Carcinogenicity Assessment

In 2005, EPA issued its new *Guidelines for Carcinogen Risk Assessment* (EPA 2005a) as a replacement for its 1986 guidelines (EPA 1986). The revised guidelines were issued partly to address changes in the understanding of the variety of ways in which carcinogens can operate. For example, the guidelines provide a framework that allows all relevant biological information to be incorporated and the flexibility to consider future scientific advances.

The guidelines provide several options for constructing the dose-response relationship, in contrast to the single default dose-response relationship of the 1986 cancer guidelines. Biologically based extrapolation is the preferred approach for quantifying risk. It involves extrapolating from animals to humans based on a similar underlying mode of action. However, in the absence of data on the parameters used in such models, the guidelines allow for alternative quantitative methods. In the default approaches, response data are modeled in the range of observation and then the point of departure or the range of extrapolation below the range of observation is determined. In addition to modeling tumor data, other kinds of responses are modeled if they are considered measures of carcinogenic risk. Three default approaches—linear, nonlinear, and both—are provided. Curve fitting in the observed range provides the effective dose corresponding to the lower 95% limit on a dose associated with a low level of response (usually in the range of 1% to 10%). That dose is then used as a point of departure for extrapolating the origin as the linear default or for a margin of exposure as the nonlinear default.

Other modifications of interest in the new guidelines include the following:

- All biological information and not just tumor findings is considered in the hazard-assessment phase of risk assessment.
- Mode of action is emphasized to reduce the uncertainty in describing the likelihood of harm and in determining the dose-response approaches.
- A weight-of-evidence narrative replaces the 1986 alphanumeric classification categories. The narrative describes the key evidence, potential modes of action, conditions of hazard expression, and key default options used.
- Direction is provided on how the overall conclusion and the confidence about risk are presented and a call is made for assumptions and uncertainties to be clearly explained.

Relative Source Contribution

EPA has developed a relative source contribution policy for assessing total human exposure to a contaminant. Under this policy, nonwater sources of exposure are considered in development of the reference dose. The percentage of total exposure typically accounted for by drinking water is applied to the reference dose to determine the maximum amount of the reference dose "apportioned" to drinking water reflected by the MCLG value. In the drinking water program, the MCLG cannot account for more than 80% or for less than 20% of the reference dose (EPA 2000d). Typically, a conservative approach is used by applying a relative source contribution factor of 20% to the reference dose when exposure data are inadequate. It is assumed that the major portion (80%) of the total exposure comes from other sources, such as the diet. This policy contrasts with past "subtraction" methods of determining relative source contributions, in which sources of exposure other than drinking water were subtracted from the reference dose.

In EPA's *Methodology for Deriving Ambient Water Quality Criteria for the Protection of Human Health*, a process called the exposure decision tree (Figure 11-1) is proposed as another means for determining relative source

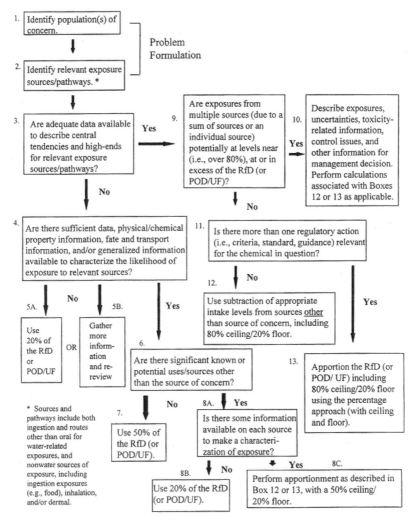

FIGURE 11-1 Exposure Decision Tree for Defining Proposed Reference Dose Apportionment. SOURCE: EPA 2000d. Abbreviations: POD, point of departure; UF, uncertainty factor

contributions (EPA 2000d). This method considers the adequacy of available exposure data, levels of exposure, relevant sources/media of exposure, and regulatory agendas. The exposure decision tree approach offers flexibility in the reference dose apportionment among sources of exposure and uses chemical information (e.g., chemical and physical properties, uses of the chemical, environmental fate and transformation, likelihood of occurrence in various media) when monitoring data are inadequate. The process also allows for use of either the subtraction or the percentage method to account for other exposures, depending on whether one or more health-based criterion is relevant for the chemical in question. The subtraction method can be used when only one criterion is relevant to a chemical. In those cases, other sources of exposure can be considered "background" and can be subtracted from the reference dose (EPA 2000d).

Risk to Children

In 1996, EPA's Office of the Administrator issued *Environmental Health Threats to Children* (EPA 1996b) and set an agenda that called for considering children's risks in all EPA actions. Children are considered a special subpopulation because their health risks can differ from those of adults as a result of their immature physiology, metabolism, and differing levels of exposure due to factors such as greater food consumption per unit of body weight and outdoor play activities. Different levels of exposure for children are typically considered in risk assessments, but the underlying toxicity database often does not specifically address effects on children. Such limitations in toxicity data are typically addressed by applying uncertainty factors to protect susceptible populations. In 2005, EPA issued special guidance for assessing susceptibility to carcinogens during early life stages (EPA 2005b).

FLUORIDE STANDARDS

Maximum-Contaminant-Level Goal

In 1986, EPA established an MCLG for fluoride of 4 mg/L to protect against "crippling" (clinical stage III) skeletal fluorosis. At that time, a reference dose for fluoride was not available, and the MCLG was calculated from a LOAEL of 20 mg/day estimated from case studies (Moller and Gudjonsson 1932), the assumption that adult water intake is 2 L per day, and the application of a safety factor of 2.5. EPA selected the safety factor to establish an MCLG that was in agreement with a recommendation from the U.S. Surgeon General (see Chapter 1).

The committee considered three toxicity end points for which there were sufficient relevant data for assessing the adequacy of the MCLG for fluoride to protect public health: severe enamel fluorosis, skeletal fluorosis, and bone fractures.

Severe Enamel Fluorosis

In the past, moderate to severe forms of enamel fluorosis were considered to be aesthetically displeasing but not adverse to health, largely because there was no direct evidence that moderate-to-severe enamel fluorosis, as observed in the United States, had resulted in tooth loss, loss of tooth function, or psychological problems. In reviewing the collective evidence, the committee considered moderate and severe forms of the condition separately. Severe enamel fluorosis is characterized by enamel loss and pitting. This damage compromises enamel's protective barrier and can make the teeth more susceptible to environmental stresses and to caries formation because it allows bacteria, plaque, and food particles to become entrapped in the enamel. Caries is dental decay caused by bacterial infection. When the infection goes unchecked, cavities may form that can cause toothache and tooth sensitivity to temperature and sweets. If cavities are untreated, the infection can lead to abscess, destruction of bone, and spread of the infection to other parts of the body (USDHHS 2000). While increased risk of caries has not been firmly established, the majority of the committee found that destruction of the enamel and the clinical practice of treating the condition even in the absence of caries provide additional lines of evidence for concluding that severe enamel fluorosis is an adverse health effect. Severe enamel fluorosis occurs at an appreciable frequency, approximately 10% on average, among children in U.S. communities with water fluoride concentrations at or near the current MCLG of 4 mg/L. Thus, the committee concludes that the MCLG of 4 mg/L is not protective against severe enamel fluorosis.

Two of the 12 members of the committee did not agree that severe enamel fluorosis should now be considered an adverse health effect. They agreed that it is an adverse dental effect but found that no new evidence has emerged to suggest a link between severe enamel fluorosis, as experienced in the United States, and a person's ability to function. They judged that demonstration of enamel defects alone from fluorosis is not sufficient to change the prevailing opinion that severe enamel fluorosis is an adverse cosmetic effect. Despite their disagreement on characterization of the condition, these two members concurred with the committee's conclusion that the MCLG should prevent the occurrence of this unwanted condition.

Strong evidence exits that the prevalence of severe enamel fluorosis is nearly zero at water fluoride concentrations to below 2 mg/L. For example, Horowitz et al. (1972) found that partial defluorination of drinking water from 6.7 mg/L to slightly below 2 mg/L prevented severe enamel fluorosis. Moderate forms of enamel fluorosis decreased from 42% to 3%.

Skeletal Fluorosis

Skeletal fluorosis is a bone and joint condition associated with prolonged exposure to high concentrations of fluoride. Fluoride increases bone density and appears to exacerbate the growth of osteophytes in the bone and joints, which leads to the radiological characteristics of the condition and associated pain. Crippling skeletal fluorosis (or clinical stage III) is the current basis of EPA's MCLG. The term crippling historically has been used to describe alterations in bone architecture and calcification of tissues that progress to the degree that they limit an individual's range of motion.

The committee judges that stage II skeletal fluorosis (the stage before mobility is significantly affected) should also be considered an adverse health effect. This stage is characterized by chronic joint pain, arthritic symptoms, slightly calcified ligaments, increased osteosclerosis/cancellous bones, and possibly osteoporosis of long bones (PHS 1991). No new studies and few clinical cases of skeletal fluorosis in healthy U.S. populations have been reported in recent decades. To determine whether EPA's MCLG protects the general public from stage II and stage III skeletal fluorosis, the committee compared pharmacokinetic predictions of bone-fluoride concentrations and historical data on iliac-crest bone-fluoride concentrations associated with the different stages of skeletal fluorosis. It found that bone-fluoride concentrations estimated to be achieved from lifetime exposure to fluoride at 4 mg/L (10,000 to 12,000 milligrams per kilogram [mg/kg] ash) fall within or exceed the ranges historically associated with stage II and stage III skeletal fluorosis (4,300 to 9,200 gm/kg ash and 4,200 to 12,700 mg/kg ash, respectively). This suggests that the MCLG might not protect all individuals from the adverse stages of the condition. However, stage III skeletal fluorosis appears to be a rare condition in the United States, and the existing epidemiologic evidence is insufficient for determining whether stage II skeletal fluorosis is occurring in U.S. residents. Thus, before any conclusions can be drawn, more research is needed to clarify the relationship between fluoride ingestion, fluoride concentrations in bone, and stage of skeletal fluorosis.

Bone Fractures

The database on fluoride's effects on bone fractures has expanded since the earlier National Research Council (NRC) review. A number of observational studies have compared bone fracture rates between populations exposed to different concentrations of fluoride in drinking water. The committee focused its review on studies involving exposure to fluoride near or within the range of 2 to 4 mg/L. Several strong studies (Sowers et al. 1991; Kurttio et al. 1999; Li et al. 2001) indicated an increased risk of bone fracture, and the results of other studies (Sowers et al. 1986; Alarcón-Herrera et al. 2001) were qualitatively consistent with that finding. The one study using serum fluoride concentrations found no appreciable relationship to fractures (Sowers et al. 2005). Because serum fluoride concentrations may not be a good measure of bone fluoride concentrations or long-term exposure, the ability to show an association might have been diminished.

A larger database on clinical trials of fluoride as an osteoporosis treatment was also reviewed. A meta-analysis of randomized clinical trials of fluoride reported an elevated risk of new nonvertebral fractures (1.85, 95% CI = 1.36, 2.50) and a slightly decreased risk of vertebral fractures (0.90, 95% CI = 0.71, 1.14) after 4 years (Haguenauer et al. 2000). An increased risk of bone fracture was found among those studies. Although the doses of fluoride were higher in the clinical trials than were experienced by people drinking water with fluoride at 4 mg/L, the length of exposure was shorter. Although comparison of these sets of data involves several assumptions, the ranges of estimated concentrations of bone fluoride were similar in the clinical trials (5,400 to 12,000 mg/kg ash) and observational studies (6,200 to >1,000 mg/kg ash). Pharmacokinetic modeling indicates that these concentrations of fluoride in bone could result from lifetime exposure to fluoride at 4 mg/L in drinking water.

Fracture risk and bone strength have been studied in animal models. The studies have shown that fluoride increases bone mass but results about its effect on the strength of bone are conflicting. Some investigators have reported a biphasic effect on bone strength (Beary 1969; Rich and Feist 1970; Turner et al. 1992), with lower concentrations of fluoride increasing strength and higher concentrations reducing it, but others have not found this effect (Turner et al. 1995). The weight of the evidence from laboratory studies indicates that, although fluoride might increase bone volume, strength per unit volume is lower. Studies of rats indicate that bone strength begins to decline when fluoride in bone ash reaches the range of 6,000 to 7,000 mg/kg (Turner et al. 1992). Studies in rabbits have shown that fluoride might decrease bone strength by altering the structural integrity of the bone microarchitecture (Turner et al. 1997; Chachra et al. 1999). However, more research is needed to address uncertainties associated with extrapolating animal data on bone strength and fractures to humans.

Overall, there was consensus among the committee that there is scientific evidence that under certain conditions fluoride can weaken bone and increase the risk of fractures. The majority of the committee concluded that lifetime exposure to fluoride at drinking water concentrations of 4 mg/L or higher is likely to increase fracture rates in the population, compared with exposure to 1 mg/L, particularly in some demographic subgroups that are prone to accumulate fluoride in their bones (e.g., people with renal disease). However, 3 of the 12 members judged that the evidence only supported a conclusion that the MCLG *might not* be protective against bone fracture.

These members judge that more evidence is needed that bone fractures occur at an appreciable frequency in human populations exposed to fluoride at 4 mg/L before drawing a conclusion that the MCLG is *likely* to be not protective.

Secondary Maximum Contaminant Level

EPA established an SMCL of 2 mg/L on the basis of cosmetically "objectionable" enamel fluorosis, defined as discoloration and/or pitting of teeth. The SMCL was selected to prevent objectionable enamel fluorosis in a significant portion of the population. EPA reviewed data on the prevalence of moderate and severe enamel fluorosis and found that, at a fluoride concentration of 2 mg/L in drinking water, the prevalence of moderate fluorosis ranged from 4% to 15% and that severe cases were observed at concentrations above 2.5 mg/L. Because of the anticaries properties of fluoride, EPA judged 2 mg/L to be an adequate upper-boundary guideline to limit the occurrence of objectionable enamel fluorosis and provide some anticaries benefit. The SMCL is not a recommendation to add fluoride to drinking water. The SMCL is a guideline for naturally occurring fluoride to be used by the states for reducing the occurrence and severity of enamel fluorosis, a condition considered by EPA to be a cosmetic condition. If fluoride in a community water system exceeds the SMCL but not the regulatory MCL, a notice about the potential risk of enamel fluorosis must be sent to all customers served by the system. The committee evaluated the SMCL only in terms of its protection against adverse cosmetic and health effects, including enamel fluorosis, skeletal fluorosis, and bone fracture. Prevention of caries was not evaluated.

Enamel Fluorosis

The committee considers moderate enamel fluorosis to be a cosmetic effect, because the available data are inadequate for categorizing the moderate form as adverse to health on the basis of structural or psychological effects. There are no studies since 1993 to assess the prevalence of enamel fluorosis at 2 mg/L, but previous reports have shown a distinct increase (approximately 15%) in moderate enamel fluorosis around 2 mg/L. Thus, the SMCL will not completely prevent the occurrence of moderate enamel fluorosis. As noted above, SMCL was intended to reduce the severity and occurrence of the condition to 15% or less of the exposed population. The available data indicates that less than 15% of children would experience moderate enamel fluorosis of aesthetic concern (discoloration of the front teeth). However, the degree to which moderate enamel fluorosis might go beyond a cosmetic effect to create an adverse psychological effect or an adverse effect on social functioning is not known.

While a few cases of severe enamel fluorosis occasionally have been reported in populations exposed at 2 mg/L, it appears that other sources of exposure to fluoride or other factors contributed to the condition. For example, similar rates of severe enamel fluorosis were reported in populations exposed to negligible amounts of fluoride in drinking water and in populations exposed at 2 mg/L (Selwitz et al. 1995; Kumar and Swango 1999; Nowjack-Raymer et al. 1995). Thus, the committee concludes that the SMCL of 2 mg/L adequately protects the public from the most severe stage of the condition (enamel pitting).

Skeletal Fluorosis

Few new data are available on skeletal fluorosis in populations exposed to fluoride in drinking water at 2 mg/L. Thus, the committee's evaluation was based on new estimates of the accumulation of fluoride into bone (iliac crest/pelvis) at that concentration (on average 4,000 to 5,000 mg/kg ash) and historical information on stage II skeletal fluorosis (4,300 to 9,200 mg/kg ash). A comparison of the bone concentrations indicates that lifetime exposure at the SMCL could lead to bone fluoride concentrations that historically have been associated with stage II skeletal fluorosis. However, as noted above, the existing epidemiologic evidence is insufficient for determining whether stage II skeletal fluorosis is occurring in U.S. residents, so no quantitative conclusions could be made about risks or safety at 2-mg/L exposures.

Bone Fracture

There were few studies to assess bone fracture risk in populations exposed to fluoride at 2 mg/L in drinking water. The best available study was from Finland, which provided data that suggested an increased rate of hip fracture in populations exposed to fluoride at >1.5 mg/L (Kurttio et al. 1999). However, this study alone is not sufficient to base judgment of fracture risk for people exposed to fluoride at 2 mg/L in drinking water. Thus, no quantitative conclusions could be drawn about fracture risk or safety at the SMCL.

Susceptible Subpopulations

Populations in need of special consideration when determining the MCLG and SMCL for fluoride include those at risk because their exposure to fluoride is greater than that of the average person or because they are particularly vulnerable to the effects of fluoride. The first category includes people who consume much larger volumes of water than assumed by EPA, such as athletes and outdoor workers, who consume large volumes of water to replace fluids lost because of strenuous activity, and people with medical conditions that cause them to consume excessive amounts of water (e.g., diabetes insipidus). Individuals who consume well over 2 L of water per day will accumulate more fluoride and reach critical bone concentrations before the average water drinker exposed to the same concentration of fluoride in drinking water. In Chapter 2, it was estimated that for high-water-intake individuals, drinking water would contribute 92% to 98% of the exposure to fluoride at 4 mg/L and 86% to 96% at 2 mg/L. Another consideration is individuals who are exposed to other significant sources of fluoride, such as occupational, industrial, and therapeutic sources.

There are also environmental, metabolic, and disease conditions that cause more fluoride to be retained in the body. For example, fluoride retention might be affected by environments or conditions that chronically affect urinary pH, including diet, drugs, altitude, and certain diseases (e.g., chronic obstructive pulmonary disease) (reviewed by Whitford 1996). It is also affected by renal function, because renal excretion is the primary route of fluoride elimination. Age and health status can affect renal excretion. Individuals with renal disease are of particular concern because their ability to excrete fluoride can be seriously inhibited, causing greater uptake of fluoride into their bones. However, the available data are insufficient to provide quantitative estimates of the differences between healthy individuals and people with renal disease.

Another category of individuals in need of special consideration includes those who are particularly susceptible or vulnerable to the effects of fluoride. For example, children are vulnerable for developing enamel fluorosis, because the condition occurs only when there is exposure while teeth are being formed (the pre-eruption stages). Thus, children up to the age of 8 are the susceptible subpopulation of concern for that end point. The elderly are another population of concern because of their long-term accumulation of fluoride into their bones. There are also medical conditions that can make people more susceptible to the effects of fluoride.

Relative Source Contribution

At the time the MCLG was established for fluoride, a reference dose was not available and the MCLG was calculated directly from available data rather than as an apportioned part of the reference dose. In Chapter 2, the committee shows that at 4 mg/L, drinking water is the primary contributor to total fluoride exposure, ranging from 72% to 94% for average-water-intake individuals and from 92% to 98% for high-water-intake individuals.

At 2 mg/L, drinking water contributes 57% to 90% for average-water-intake individuals and 86% to 96% for high-water-intake individuals. Thus, it is important that future revisions to the MCLG take into consideration that water is a significant, and sometimes the most significant, source of exposure to fluoride.

FINDINGS AND RECOMMENDATIONS

Maximum-Contaminant-Level Goal

In light of the collective evidence on various health end points and total exposure to fluoride, the committee concludes that EPA's MCLG of 4 mg/L should be lowered. Lowering the MCLG will prevent children from developing severe enamel fluorosis and will reduce the lifetime accumulation of fluoride into bone that the majority of the committee concluded is likely to put individuals at increased risk of bone fracture and possibly skeletal fluorosis, which are particular concerns for subpopulations that are prone to accumulating fluoride in their bone.

Recommendation: To develop an MCLG that is protective of severe enamel fluorosis, clinical stage II skeletal fluorosis, and bone fractures, EPA should update the risk assessment of fluoride to include new data on health risks and better estimates of total exposure (relative source contribution) in individuals and to use current approaches to quantifying risk, considering susceptible subpopulations, and characterizing uncertainties and variability.

Secondary Maximum Contaminant Level

The prevalence of severe enamel fluorosis is very low (near zero) at fluoride concentrations below 2 mg/L. However, from a cosmetic standpoint, the SMCL does not completely prevent the occurrence of moderate enamel fluorosis. EPA has indicated that the SMCL was intended to reduce the severity and occurrence of the condition to 15% or less of the exposed population. The available data indicates that fewer than 15% of children would experience moderate enamel fluorosis of aesthetic concern (discoloration of the front teeth). However, the degree to which moderate enamel fluorosis might go beyond a cosmetic effect to create an adverse psychological effect or an adverse effect on social functioning is not known.

Recommendations: Additional studies, including longitudinal studies, of the prevalence and severity of enamel fluorosis should be done in U.S. communities with fluoride concentrations greater than

1 mg/L. These studies should focus on moderate and severe enamel fluorosis in relation to caries and in relation to psychological, behavioral, and social effects among affected children, among their parents, and among affected children after they become adults.

To better define the aesthetics of enamel fluorosis, methods should be developed and validated to objectively assess enamel fluorosis. Staining and mottling of the anterior teeth should be distinguished from staining of the posterior teeth so that aesthetic consequences can be more easily assessed.

References

Aardema, M.J., and T. Tsutsui. 1995. Sodium fluoride-induced chromosome aberrations in different cell cycle stages. Mutat. Res. 331(1):171-172.

Abboud, T.K., L. D'Onofrio, A. Reyes, P. Mosaad, J. Zhu, M. Mantilla, H. Gangolly, D. Crowell, M. Cheung, A. Afrasiabi, N. Khoo, J. Davidson, Z. Steffens, and N. Zaki. 1989. Isoflurane or halothane for cesarean section: Comparative maternal and neonatal effects. Acta Anaesthesiol. Scand. 33(7):578-581.

ADA (American Dental Association). 2005. Fluoridation facts. Chicago, IL: American Dental Association.

Adachi, J.D., M.J. Bell, W.G. Bensen, F. Bianchi, A. Cividino, R.J. Sebaldt, M. Gordon, G. Ioannidis, and C. Goldsmith. 1997. Fluoride therapy in prevention of rheumatoid arthritis induced bone loss. J. Rheumatol. 24(12):2308-2313.

Adair, S.M. 1999. Overview of the history and current status of fluoride supplementation schedules. J. Public Health Dent. 59(4):252-258.

Adair, S.M., W.P. Piscitelli, and C. McKnight-Hanes. 1997. Comparison of the use of a child and an adult dentifrice by a sample of preschool children. Pediatr. Dent. 19(2):99-103.

Adams, G.R. 1977. Physical attractiveness, personality, and social reactions to peer pressure. J. Psychol. 96(Part 2):287-296.

Adams, G.R., and T. Huston. 1975. Social perceptions of middle-aged persons varying in physical attractiveness. Dev. Psychol. 11:657-658.

Ahmad, R., and J.M. Hammond. 2004. Primary, secondary, and tertiary hyperparathyroidism. Otolaryngol. Clin. N. Am. 37(4):701-713.

Ahn, H.W., B. Fulton, D. Moxon, and E.H. Jeffery. 1995. Interactive effects of fluoride and aluminum uptake and accumulation in bones of rabbits administered both agents in their drinking water. J. Toxicol. Environ. Health 44(3):337-350.

Akano, A., and S.W. Bickler. 2003. Pineal gland calcification in sub-Saharan Africa. Clin. Radiol. 58(4):336-337.

Akpata, E.S., Z. Fakiha, and N. Khan. 1997. Dental fluorosis in 12-15-year-old rural children exposed to fluorides from well drinking water in the Hail region of Saudi Arabia. Community Dent. Oral Epidemiol. 25(4):324-327.

REFERENCES

Al-Alousi, W., D. Jackson, G. Crompton, and O.C. Jenkins. 1975. Enamel mottling in a fluoride and in a non-fluoride community. Br. Dent. J. 138(1):9-15.
Alarcón-Herrera, M.T., I.R. Martín-Domínguez, R. Trejo-Vázquez, and S. Rodriguez-Dozal. 2001. Well water fluoride, dental fluorosis, and bone fractures in the Guadiana Valley of Mexico. Fluoride 34(2):139-149.
Albino, J.E., J.J. Cunat, R.N. Fox, E.A. Lewis, M.J. Slakter, and L.A. Tedesco. 1981. Variables discriminating individuals who seek orthodontic treatment. J. Dent. Res. 60(9):1661-1667.
Aleandri, V., V. Spina, and A. Ciardo. 1997. The role of the pineal body in the endocrine control of puberty [in Italian]. Minerva Ginecol. 49(1-2):43-48.
Alhava, E.M., H. Olkkonen, P. Kauranen, and T. Kari. 1980. The effect of drinking water fluoridation on the fluoride content, strength and mineral density of human bone. Acta Orthop. Scand. 51(3):413-420.
Al-Hiyasat, A.S., A.M. Elbetieha, and H. Darmani. 2000. Reproductive toxic effects of ingestion of sodium fluoride in female rats. Fluoride 33(2):79-84.
Allegood, J. 2005. Water treatment process called potential risk. Chemicals' mix with plumbing could put lead in tap water. The News & Observer. May 18, 2005 [online]. Available: http://www.newsobserver.com/news/health_science/story/2417101p-8794959c.html [accessed Sept. 20, 2005]
Allolio, B., and R. Lehmann. 1999. Drinking water fluoridation and bone. Exp. Clin. Endocrinol. Diabetes 107(1):12-20.
Almond, F.W. 1923. Letter from F.W. Almond, Director, Public Health Service, Boise, ID, to the Surgeon General, U.S. Public Health Service. November 5, 1923 (From the H. Trendley Dean Papers, MS C 468, The History of Medicine Division, National Library of Medicine).
Alonge, O.K., D.D. Williamson, and S. Narendran. 2000. Dental fluorosis among third graders in Harris County, Texas--1998 study findings. Tex. Dent. J. 117(9):22-29.
al-Wakeel, J.S., A.H. Mitwalli, S. Huraib, S. al-Mohaya, H. Abu-Aisha, A.R. Chaudhary, S.A. al-Majed, and N. Memon. 1997. Serum ionic fluoride levels in haemodialysis and continuous ambulatory peritoneal dialysis patients. Nephrol. Dial. Transplant. 12(7):1420-1424.
American Diabetes Association. 2004. Basic Diabetes Information [online]. Available: http://www.diabetes.org. [accessed Sept. 10, 2004].
Anbar, M., S. Guttmann, and Z. Lewitus. 1959. Effect of monofluorosulphonate, difluorophosphate and fluoroborate ions on the iodine uptake of the thyroid gland. Nature 183(4674):1517-1518.
Andersen, L., A. Richards, A.D. Care, H.M. Andersen, J. Kragstrup, and O. Fejerskov. 1986. Parathyroid glands, calcium, and vitamin D in experimental fluorosis in pigs. Calcif. Tissue Int. 38(4):222-226.
Anderson, R.E., D.M. Woodbury, and W.S. Jee. 1986. Humoral and ionic regulation of osteoclast activity. Cacif. Tissue Int. 39(4):252-258.
Anderson, S.E., G.E. Dallal, and A. Must. 2003. Relative weight and race influence average age at menarche: Results from two nationally representative surveys of U.S. girls studied 25 years apart. Pediatrics 111(4 Pt 1):844-850.
Ando, M., M. Tadano, S. Yamamoto, K. Tamura, S. Asanuma, T. Watanabe, T. Kondo, S. Sakurai, R. Ji, C. Liang, X. Chen, Z. Hong, and S. Cao. 2001. Health effects of fluoride pollution caused by coal burning. Sci. Total Environ. 271(1-3):107-116.
Angelillo, I.F., I. Torre, C.G. Nobile, and P. Villari. 1999. Caries and fluorosis prevalence in communities with different concentrations of fluoride in water. Caries Res. 33(2):114-122.
Angmar-Mansson, B., and G.M. Whitford. 1990. Environmental and physiological factors affecting dental fluorosis. J. Dent. Res. 69(Spec.):706-713.

Anisimov, V.N. 2003. The role of pineal gland in breast cancer development. Crit. Rev. Oncol. Hematol. 46(3):221-234.
Antich, P.P., C.Y. Pak, J. Gonzales, J. Anderson, K. Sakhaee, and C. Rubin. 1993. Measurement of intrinsic bone quality in vivo by reflection ultrasound: Correction of impaired quality with slow-release sodium fluoride and calcium citrate. J. Bone Miner. Res. 8(3):301-311.
Antonny, B., and M. Chabre. 1992. Characterization of the aluminum and beryllium fluoride species which activate transducin. Analysis of the binding and dissociation kinetics. J. Biol. Chem. 267(10):6710-6718.
Aoba, T., and O. Fejerskov. 2002. Dental fluorosis: Chemistry and biology. Crit. Rev. Oral. Biol. Med. 13(2):155-170.
Arendt, J. 2003. Importance and relevance of melatonin to human biological rhythms. J. Neuroendocrinol. 15(4):427-431.
Arnala, I., E.M. Alhava, and P. Kauranen. 1985. Effects of fluoride on bone in Finland. Histomorphometry of cadaver bone from low and high fluoride areas. Acta Orthop. Scand. 56(2):161-166.
Arnala, I., E.M. Alhava, R. Kivivuori, and P. Kauranen. 1986. Hip fracture incidence not affected by fluoridation. Osteofluorosis studied in Finland. Acta Orthop. Scand. 57(4):344-348.
Arnow, P.M., L.A. Bland, S. Garcia-Houchins, S. Fridkin, and S.K. Fellner. 1994. An outbreak of fatal fluoride intoxication in a long-term hemodialysis unit. Ann. Intern. Med. 121(5):339-344.
Arola, D., and R.K. Reprogel. 2005. Effects of aging on the mechanical behavior of human dentin. Biomaterials 26(18):4051-4061.
Assem, E.S., and B.Y. Wan. 1982. Stimulation of H$^+$ ion secretion from the isolated mouse stomach by sodium fluoride. Experientia 38(3):369-370.
ATSDR (Agency for Toxic Substances and Disease Registry). 1999. Toxicological Profile for Aluminum. U.S. Department of Health and Human Services, Public Health Service, Atlanta, GA. July 1999.
ATSDR (Agency for Toxic Substances and Disease Registry). 2002. Toxicological Profile for Beryllium. U.S. Department of Health and Human Services, Public Health Service, Atlanta, GA. September 2002.
ATSDR (Agency for Toxic Substances and Disease Registry). 2003. Toxicological Profile for Fluorides, Hydrogen Fluoride, and Fluorine. U.S. Department of Health and Human Services, Public Health Service, Atlanta, GA. September 2003.
Austen, K.F., M. Dworetzky, R.S. Farr, G.B. Logan, S. Malkiel, E. Middleton Jr., M.M. Miller, R. Patterson, C.E. Reed, S.C. Siegel, and P.P. Van Arsdel Jr. 1971. A statement on the question of allergy to fluoride as used in the fluoridation of community water supplies. J. Allergy 47(6):347-348.
Avorn, J., and L.C. Niessen. 1986. Relationship between long bone fractures and water fluoridation. Gerodontics 2(5):175-179.
Awadia, A.K., J.M. Birkeland, O. Haugejorden, and K. Bjorvatn. 2000. An attempt to explain why Tanzanian children drinking water containing 0.2 or 3.6 mg fluoride per liter exhibit a similar level of dental fluorosis. Clin. Oral Investig. 4(4):238-244.
Awadia, A.K., J.M. Birkeland, O. Haugejorden, and K. Bjorvatn. 2002. Caries experience and caries predictors—a study of Tanzanian children consuming drinking water with different fluoride concentrations. Clin. Oral Invest. 6(2):98-103.
Bachinskii, P.P., O.A. Gutsalenko, N.D. Naryzhniuk, V.D. Sidora, and A.I. Shliakhta. 1985. Action of the body fluorine of healthy persons and thyroidopathy patients on the function of hypophyseal-thyroid the system [translated from Russian by Ralph McElroy. Translation Company, Austin, TX]. Probl. Endokrinol. 31(6):25-29.

REFERENCES

Baelum, V., O. Fejerskov, F. Manji, and M.J. Larsen. 1987. Daily dose of fluoride and dental fluorosis. Tandlaegebladet 91(10):452-456.

Baker, M.T., and W.C. Ronnenberg Jr. 1992. Acute stimulation of trifluoroethene defluorination and cytochrome P450 inactivation in the rat by exposure to isoflurane. Toxicol. Appl. Pharmacol. 114(1):25-30.

Balabolkin, M.I., N.D. Mikhailets, R.N. Lobovskaia, and N.V. Chernousova. 1995. The interrelationship of the thyroid and immune statuses of workers with long-term fluorine exposure [in Russian]. Ter. Arkh. 67(1):41-42.

Bang, S., G. Boivin, J.C. Gerster, and C.A. Baud. 1985. Distribution of fluoride in calcified cartilage of a fluoride-treated osteoporosis patient. Bone 6(4):207-210.

Barnhart, W.E., L K. Hiller, G.J. Leonard, and S.E. Michaels. 1974. Dentifrice usage and ingestion among four age groups. J. Dent. Res. 53(4):1317-1322.

Bartels, D., K. Haney, and S.S. Khajotia. 2000. Fluoride concentrations in bottled water. J. Okla. Dent. Assoc. 91(1):18-22.

Bassin, E.B. 2001. Pp. 68-83 and 92-100 in Association Between Fluoride in Drinking Water During Growth and Development and the Incidence of Osteosarcoma for Children and Adolescents. D.M.S. Thesis, Harvard School of Dental Medicine, Boston, Massachusetts [online]. Available: http://www.fluoridealert.org/health/cancer/bassin-2001.pdf [accessed Oct. 10, 2005].

Baud, C.A., R. Lagier, G. Boivin, and M.A. Boillat. 1978. Value of bone biopsy in the diagnosis of industrial fluorosis. Virchows. Arch. A Pathol. Anat. Histol. 380(4):283-297.

Baum, K., W. Börner, C. Reiners, and E. Moll. 1981. Bone density and thyroid gland function in adolescents in relation to fluoride content of drinking water [in German]. Fortschr. Med. 99(36):1470-1472.

Bayley, T.A., J.E. Harrison, T.M. Murray, R.G. Josse, W. Sturtridge, K.P. Pritzker, A. Strauss, R. Vieth, and S. Goodwin. 1990. Fluoride-induced fractures: Relation to osteogenic effect. J. Bone Miner. Res. 5(Suppl. 1):S217-S222.

Baylis, P.H., and T. Cheetham. 1998. Diabetes insipidus. Arch. Dis. Child. 79(1):84-89.

Beary, D.F. 1969. The effects of fluoride and low calcium on the physical properties of the rat femur. Anat. Rec. 164(3):305-316.

Becaria, A., S.C. Bondy, and A. Campbell. 2003. Aluminum and copper interact in the promotion of oxidative but not inflammatory events: Implications for Alzheimer's disease. J. Alzheimers Dis. 5(1):31-38.

Beers, M.H., and R. Berkow, eds. 1999. The Merck Manual of Diagnosis and Therapy, 17th Ed. Whitehouse Station, NJ: Merck Research Laboratories.

Behrendt, A., V. Oberste, and W.E. Wetzel. 2002. Fluoride concentration and pH of iced tea products. Caries Res. 36(6):405-410.

Belchetz, P.E., and P.J. Hammond. 2003. Mosby's Color Atlas and Text of Diabetes and Endocrinology. Edinburgh: Mosby.

Bello, V.A., and H.J. Gitelman. 1990. High fluoride exposure in hemodialysis patients. Am. J. Kidney Dis. 15(4):320-324.

Bellows, C.G., J.N. Heersche, and J.E. Aubin. 1990. The effects of fluoride on osteoblast progenitors in vitro. J. Bone Miner. Res. 5(Suppl. 1.):S101-S105.

Beltran-Aguilar, E.D., S.O. Griffin, and S.A. Lockwood. 2002. Prevalence and trends in enamel fluorosis in the United States from the 1930s to the 1980s. J. Am. Dent. Assoc. 133(2):157-165.

Bentley, E.M., R.P. Ellwood, and R.M. Davies. 1999. Fluoride ingestion from toothpaste by young children. Br. Dent. J. 186(9):460-462.

Bergan, T. 1989. Pharmacokinetics of ciprofloxacin with reference to other fluorinated quinolones. J. Chemother. 1(1):10-17.

Berry, W.T. 1958. A study of the incidence of mongolism in relation to the fluoride content of water. Am. J. Ment. Defic. 62(4):634-636.
Berscheid, E., and E. Walster. 1974. Physical attractiveness. Adv. Exp. Soc. Psychol. 7: 157-215.
Bhat, M., and K.B. Nelson. 1989. Developmental enamel defects in primary teeth in children with cerebral palsy, mental retardation, or hearing defects: A review. Adv. Dent. Res. 3(2):132-142.
Bhatnagar, M., P. Rao, J. Sushma, and R. Bhatnagar. 2002. Neurotoxicity of fluoride: Neurodegeneration in hippocampus of female mice. Indian J. Exp. Biol. 40(5):546-554.
Biggerstaff, R.H., and J.C. Rose. 1979. The effects of induced prenatal hypothyroidism on lamb mandibular third primary molars. Am. J. Phys. Anthropol. 50(3):357-362.
Bigsby, R., R.E. Chapin, G.P. Daston, B.J. Davis, J. Gorski, L.E. Gray, K.L. Howdeshell, R.T. Zoeller, and F.S. vom Saal. 1999. Evaluating the effects of endocrine disruptors on endocrine function during development. Environ. Health Perspect. 107(Suppl. 4):613-618.
Binder, K. 1973. Comparison of the effects of fluoride drinking water on caries frequency and mottled enamel in three similar regions of Austria over a 10-year period. Caries Res. 7(2):179-183.
Biondi, B., E.A. Palmieri, G. Lombardi, and S. Fazio. 2002. Effects of subclinical thyroid dysfunction on the heart. Ann. Intern. Med. 137(11):904-914.
Björk, J., and U. Strömberg. 2002. Effects of systematic exposure assessment errors in partially ecologic case-control studies. Int. J. Epidemiol. 31(1):154-160.
Blanco, E., M.I. Vidal, J. Blanco, S. Fagundo, O. Campana, and J. Alvarez. 1995. Comparison of maintenance and recovery characteristics of sevoflurane-nitrous oxide and enflurane-nitrous oxide anaesthesia. Eur. J. Anaesthesiol. 12(5):517-523.
Bobek, S., S. Kahl, and Z. Ewy. 1976. Effect of long-term fluoride administration on thyroid hormones level blood in rats. Endocrinol. Exp. 10(4):289-295.
Boillat, M.A., J. Garcia, and L. Velebit. 1980. Radiological criteria of industrial fluorosis. Skeletal. Radiol. 5(3):161-165.
Boivin, G., P. Chavassieux, M.C. Chapuy, C.A. Baud, and P.J. Meunier. 1986. Histomorphometric profile of bone fluorosis induced by prolonged ingestion of Vichy Saint-Yorre water. Comparison with bone fluorine levels [in French]. Pathol. Biol. 34(1):33-39.
Boivin, G., M.C. Chapuy, C.A. Baud, and P.J. Meunier. 1988. Fluoride content in human iliac bone: Results in controls, patients with fluorosis, and osteoporotic patients treated with fluoride. J. Bone Miner. Res. 3(5):497-502.
Boivin, G., P. Chavassieux, M.C. Chapuy, C.A. Baud, and P.J. Meunier. 1989. Skeletal fluorisis: Histomorphometric analysis of bone changes and bone fluoride content in 29 patients. Bone 10(2):89-99.
Boivin, G., P. Chavassieux, M.C. Chapuy, C.A. Baud, and P.J. Meunier. 1990. Skeletal fluorosis: Histomorphometric findings. J. Bone Miner. Res. 5(Suppl. 1):S185-S189.
Boivin, G., J. Duriez, M.C. Chapuy, B. Flautre, P. Hardouin, and P.J. Meunier. 1993. Relationship between bone fluoride content and histological evidence of calcification defects in osteoporotic women treated long term with sodium fluoride. Osteoporos. Int. 3(4):204-208.
Borke, J.L., and G.M. Whitford. 1999. Chronic fluoride ingestion decreases 45Ca uptake by rat kidney membranes. J. Nutr. 129(6):1209-1213.
Boros, I., P. Keszler, G. Csikós, and H. Kalász. 1998. Fluoride intake, distribution, and bone content in diabetic rats consuming fluoridated drinking water. Fluoride 31(1):33-42.
Bottled Water Web. 2004. The Definitive Bottled Water Site [online]. Available: http://www.bottledwaterweb.com. [accessed Feb. 20, 2004].
Bovill, E.G., Jr., A. Kung'u, A. Bencivenga, M.K. Jeshrani, B.S. Mbindyo, and P.M. Heda. 1985.

An epidemiological study of osteogenic sarcoma in Kenya: The variations in incidence between ethnic groups and geographic regions, 1968-1978. Int. Orthop. 9(1):59-63.

Brambilla, E., G. Belluomo, A. Malerba, M. Buscaglia, and L. Strohmenger. 1994. Oral administration of fluoride in pregnant women, and the relation between concentration in maternal plasma and in amniotic fluid. Arch. Oral Biol. 39(11):991-994.

Brenner, H., D.A. Savitz, K.H. Jöckel, and S. Greenland. 1992. Effects of nondifferential exposure misclassification in ecologic studies. Am. J. Epidemiol. 135(1):85-95.

Briancon, D., and P.J. Meunier. 1981. Treatment of osteoporosis with fluoride, calcium and vitamin D. Orthop. Clin. North Am. 12(3):629-648.

Bringhurst, F.R., M.B. Demay, and H.M. Kronenberg. 2002. Hormones and disorders of mineral metabolism. Pp. 1303-1371 in Williams Textbook of Endocrinology, 10th Ed., P.R. Larsen, H.M. Kronenberg, S. Melmed, and K.S. Polonsky, eds. Philadelphia, PA: Saunders.

Bronckers, A.L., D.M. Lyaruu, T.J. Bervoets, and J.H. Woltgens. 2002. Fluoride enhances intracellular degradation of amelogenins during secretory phase of amelogenesis of hamster teeth in organ culture. Connect. Tissue Res. 43(2-3):456-465.

Brown, A.S., E. Feingold, K.W. Broman, and S.L. Sherman. 2000. Genome-wide variation in recombination in female meiosis: A risk factor for non-disjunction of chromosome 21. Hum. Mol. Genet. 9(4):515-523.

Brownlee, M., L.P. Aiello, E. Friedman, A.I. Vinik, R.W. Nesto, and A.J.M. Boulton. 2002. Complications of diabetes mellitus. Pp. 1509-1583 in Williams Textbook of Endocrinology, 10th Ed., P.R. Larsen, H.M. Kronenberg, S. Melmed, and K.S. Polonsky, eds. Philadelphia, PA: Saunders.

Brucker-Davis, F., K. Thayer, and T. Colborn. 2001. Significant effects of mild endogenous hormonal changes in humans: Considerations for low-dose testing. Environ. Health Perspect. 109(Suppl. 1):21-26.

Bucher, J.R., M.R. Hejtmancik, J.D. Toft, II, R.L. Persing, S.L. Eustis, and J.K. Haseman. 1991. Results and conclusions of the National Toxicology Program's rodent carcinogenicity studies with sodium fluoride. Int. J. Cancer. 48(5):733-737.

Bürgi, H., L. Siebenhüner, and E. Miloni. 1984. Fluorine and thyroid gland function: A review of the literature. Klin. Wochenschr. 62(12):564-569.

Burgstahler, A.W., and M.A. Robinson. 1997. Fluoride in California wines and raisins. Fluoride 30(3):142-146.

Burt, B.A. 1992. The changing patterns of systemic fluoride intake. J. Dent. Res. 71(5):1228-1237.

Burt, B.A., and S.A. Eklund. 1999. Dentistry, Dental Practice, and the Community, 5th Ed. Phildelphia, PA: WB Saunders Co.

Burt, B.A., M.A. Keels, and K.E. Heller. 2003. Fluorosis development in seven age cohorts after an 11-month break in water fluoridation. J. Dent. Res. 82(1):64-68.

Buse, J.B., K.S. Polonsky, and C.F. Burant. 2002. Type 2 diabetes mellitus. Pp. 1427-1483 in Williams Textbook of Endocrinology, 10th Ed., P.R. Larsen, H.M. Kronenberg, S. Melmed, and K.S. Polonsky, eds. Philadelphia, PA: Saunders.

Butler, J.E., M. Satam, and J. Ekstrand. 1990. Fluoride: An adjuvant for mucosal and systemic immunity. Immunol. Lett. 26(3):217-220.

Cajochen, C., K. Kräuchi, and A. Wirz-Justice. 2003. Role of melatonin in the regulation of human circadian rhythms and sleep. J. Neuroendocrinol. 15(4):432-437.

Caldera, R., J. Chavinie, J. Fermanian, D. Tortrat, and A.M. Laurent. 1988. Maternal-fetal transfer of fluoride in pregnant women. Biol. Neonate 54(5):263-269.

Calderon, J., B. Machado, M. Navarro, L. Carrizales, M.D. Ortiz, and F. Diaz-Barriga. 2000. Influence of fluoride exposure on reaction time and visuospatial organization in children. Epidemiology 11(4):S153.

Call, R.A., D.A. Greenwood, W.H. Lecheminant, J.L. Shupe, H.M. Nielsen, L.E. Olson, R.E. Lamborn, F.L. Mangelson, and R.V. Davis. 1965. Histological and chemical studies in man on effects of fluoride. Public Health Rep. 80:529-538.

Calvo, M.S., R. Eastell, K.P. Offord, E.J. Bergstralh, and M.F. Burritt. 1991. Circadian variation in ionized calcium and intact parathyroid hormone: Evidence for sex differences in calcium homeostasis. J. Clin. Endocrinol. Metab. 72(1):69-76.

Cardinali, D.P., M.G. Ladizesky, V. Boggio, R.A. Cutrera, and C. Mautalen. 2003. Melatonin effects on bone: Experimental facts and clinical perspectives. J. Pineal. Res. 34(2):81-87.

Carlson, C.H., L. Singer, and W.D. Armstrong. 1960. Radiofluoride distribution in tissues of normal and nephrectomized rats. Proc. Soc. Exp. Biol. Med. 103:418-420.

Carmona, R.H. 2004. Surgeon General's Statement on Community Water Fluoridation. U.S. Department of Health and Human Services, Public Health Service, Office of the Surgeon General, Rockville, MD. July 28, 2004 [online]. Available: http://www.cdc.gov/oralhealth/waterfluoridation/fact_sheets/sg04.htm [accessed Sept. 15, 2005].

Cauley, J.A., P.A. Murphy, T.J. Riley, and A.M. Buhari. 1995. Effects of fluoridated drinking water on bone mass and fractures: The study of osteoporotic fractures. J. Bone Miner. Res. 10(7):1076-1086.

Caverzasio, J., T. Imai, P. Ammann, D. Burgener, and J.P. Bonjour. 1996. Aluminum potentiates the effect of fluoride on tyrosine phosphorylation and osteoblast replication in vitro and bone mass in vivo. J. Bone Miner. Res. 11(1):46-55.

Caverzasio, J., G. Palmer, A. Suzuki, and J.P. Bonjour. 1997. Mechanism of the mitogenic effect of fluoride on osteoblast-like cells: Evidence of a G protein-dependent tyrosine phosphorylation process. J. Bone Miner. Res. 12(12):1975-1983.

Caverzasio, J., G. Palmer, and J.P. Bonjour. 1998. Fluoride: Mode of action. Bone 22(6):585-589.

Cawson, R.A., A.W. McCracken, and P.B. Marcus. 1982. Pathologic Mechanisms and Human Disease, 2nd Ed. St. Louis, MO: The C.V. Mosby Company.

CDC (Centers for Disease Control and Prevention). 1993. Fluoridation Census 1992. Atlanta, GA: U.S. Department of Health and Human Services, Public Health Service, Centers for Disease Control and Prevention, National Center for Prevention Services, Division of Oral Health.

CDC (Centers for Disease Control and Prevention). 1995. Engineering and Administrative Recommendations for Water Fluoridation, 1995. Morbidity and Mortality Weekly Report, Recommendations and Reports 44(RR-13). Atlanta, GA: U.S. Department of Health and Human Services, Public Health Service, Centers for Disease Control and Prevention.

CDC (Centers for Disease Control and Prevention). 1999. Ten great public health achievements—United States, 1900-1999. MMWR 48(12):241-243.

CDC (Centers for Disease Control and Prevention). 2001. Recommendations for Using Fluoride to Prevent and Control Dental Caries in the United States. Morbidity and Mortality Weekly Report 50(RR-14). Atlanta, GA: U.S. Department of Health and Human Services, Centers for Disease Control and Prevention [online]. Available: http://www.cdc.gov/mmwr/preview/mmwrhtml/rr5014a1.htm [accessed Sept. 13, 2004].

CDC (Centers for Disease Control and Prevention). 2002a. Populations receiving optimally fluoridated public drinking water-United States, 2000. MMWR 51(7):144-147 [online]. Available: http://www.cdc.gov/mmwr/preview/mmwrhtml/mm5107a2.htm [accessed July 7, 2004].

CDC (Centers for Disease Control and Prevention). 2002b. Fluoridation Statistics 2000: Status of Water Fluoridation in the United States. Fact Sheet. Oral Health Resources. U.S. Department of Health and Human Services, Centers for Disease Control and Prevention, Na-

tional Center for Chronic Disease Prevention and Health Promotion [online]. Available: http://www.cdc.gov/OralHealth/factsheets/fl-stats-us2000.htm [accessed Sept. 9, 2004].
CDC (Centers for Disease Control and Prevention). 2002c. Fluoridation Statistics 1992: Status of Water Fluoridation in the United States. Fact Sheet. Oral Health Resources. U.S. Department of Health and Human Services, Centers for Disease Control and Prevention, National Center for Chronic Disease Prevention and Health Promotion [online]. Available: http://www.cdc.gov/OralHealth/factsheets/fl-stats1992.htm [accessed Sept. 9, 2004].
CDC (Centers for Disease Control). 2002d. Iodine Level, United States, 2000. U.S. Department of Health and Human Services, Centers for Disease Control and Prevention, National Center for Health Statistics, Hyattsville, MD [online]. Available: http://www.cdc.gov/nchs/products/pubs/pubd/hestats/iodine.htm [accessed Nov. 9, 2004].
CDC (Centers for Disease Control and Prevention). 2003. Second National Report on Human Exposure to Environmental Chemicals. U.S. Department of Health and Human Services, Centers for Disease Control and Prevention, Atlanta, GA [online]. Available: http://www.cdc.gov/exposurereport/2nd/ [accessed Sept. 9, 2004].
CDC (Centers for Disease Control and Prevention). 2005. Third National Report on Human Exposure to Environmental Chemicals. U.S. Department of Health and Human Services, Centers for Disease Control and Prevention, Atlanta, GA [online]. Available: http://www.cdc.gov/exposurereport/ [accessed July 26, 2005].
CEPA (Canadian Environmental Protection Act). 1996. National Ambient Air Quality Objectives for Hydrogen Fluoride (HF): Science Assessment Document. Ontario: CEPA/FPAC Working Group on Air Quality Objectives and Guidelines, Canadian Environmental Protection Act (as cited in ATSDR 2003).
Cettour-Rose, P., C. Theander-Carrillo, C. Asensio, M. Klein, T.J. Visser, A.G. Burger, C.A. Meier, and F. Rohner-Jeanrenaud. 2005. Hypothyroidism in rats decreases peripheral glucose utilisation, a defect partially corrected by central leptin infusion. Diabetologia 48(4):624-633.
Chabre, M. 1990. Aluminofluoride and berylofluoride complexes: A new phosphate analogs in enzymology. Trends Biochem. Sci. 15(1):6-10.
Chachra, D., C.H. Turner, A.J. Dunipace, and M.D. Grynpas. 1999. The effect of fluoride treatment on bone mineral in rabbits. Calcif. Tissue Int. 64(4):345-351.
Chadha, M., and S. Kumar. 2004. Fluorosis-induced hyperparathyroidism mimicking a giant-cell tumour of the femur. J. Bone Joint Surg. Br. 86(4):594-496.
Challacombe, S.J. 1996. Does fluoridation harm immune function? Community Dent. Health 13(Suppl. 2):69-71.
Chan, J.T., and S.H. Koh. 1996. Fluoride content in caffeinated, decaffeinated and herbal teas. Caries Res. 30(1):88-92.
Chan, J.T., C. Stark, and A.H. Jeske. 1990. Fluoride content of bottled waters: Implications for dietary fluoride supplementation. Tex. Dent. J. 107(4):17-21.
Chang, C.Y., P.H. Phillips, E.B. Hart, and G. Bonstedt. 1934. The effect of feeding raw rock phosphate on the fluorine content of the organs and tissues of dairy cows. J. Dairy Sci. 17:695-700 (as cited in Galletti and Joyet 1958).
Chang, W.K., S.A. McClave, M.S. Lee, and Y.C. Chao. 2004. Monitoring bolus nasogastric tube feeding by the Brix value determination and residual volume measurement of gastric contents. JPEN J. Parenter. Enteral. Nutr. 28(2):105-112.
Charen, J., D.R. Taves, J.W. Stamm, and F.M. Parkins. 1979. Bone fluoride concentrations associated with fluoridated drinking water. Calcif. Tissue Int. 27(2):95-99.
Chavassieux, P., P. Pastoureau, G. Boivin, M.C. Chapuy, P.D. Delmas, and P.J. Meunier. 1991. Dose effects on ewe bone remodeling of short-term sodium fluoride administration--a histomorphometric and biochemical study. Bone 12(6):421-427.

Cheetham, T., and P.H. Baylis. 2002. Diabetes insipidus in children: Pathophysiology, diagnosis and management. Pediatr. Drugs 4(12):785-796.
Chen, B.C. 1989. Epidemiological study on dental fluorosis and dental caries prevalence in communities with negligible, optimal, and above-optimal fluoride concentrations in drinking water supplies. Zhonghua Ya Yi Xue Hui Za Zhi. 8(3):117-127.
Chen, C.J., C.S. Anast, and E.M. Brown. 1988. Effects of fluoride on parathyroid hormone secretion and intracellular second messengers in bovine parathyroid cells. J. Bone Miner. Res. 3(3):279-288.
Chen, H., S. Shoumura, S. Emura, M. Utsumi, T. Yamahira, and H. Isono. 1990. Effects of pinealectomy on the ultrastructure of the golden hamster parathyroid gland. Histol. Histopathol. 5(4):477-484.
Chen, H., S. Shoumura, S. Emura, M. Utsumi, T. Yamahira, and H. Isono. 1991. Effects of melatonin on the ultrastructure of the golden hamster parathyroid gland. Histol. Histopathol. 6(1):1-7.
Chibole, O. 1987. Epidemiology of dental fluorosis in Kenya. J. R. Soc. Health 107(6): 242-243.
Chilvers, C., and D. Conway. 1985. Cancer mortality in England in relation to levels of naturally occurring fluoride in water supplies. J. Epidemiol. Community Health 39(1):44-47.
Chinoy, N.J., and M.V. Narayana. 1994. In vitro fluoride toxicity in human spermatozoa. Reprod. Toxicol. 8(2):155-159.
Chinoy, N.J., and D. Patel. 1998. Influences of fluoride on biological free radicals in ovary of mice and its reversal. Environ. Sci. 6(3):171-184.
Chinoy, N.J., and T.N. Patel. 2001. Effects of sodium fluoride and aluminum chloride on ovary and uterus of mice and their reversal by some antidotes. Fluoride 34(1):9-20.
Chinoy, N.J., and E. Sequeira. 1989. Effects of fluoride on the histoarchitecture of reproductive organs of the male mouse. Reprod. Toxicol. 3(4):261-267.
Chinoy, N.J., and E. Sequeira. 1992. Reversible fluoride induced fertility impairment in male mice. Fluoride 25(2):71-76.
Chinoy, N.J., and A. Sharma. 1998. Amelioration of fluoride toxicity by vitamins E and D in reproductive functions of male mice. Fluoride 31(4):203-216.
Chinoy, N.J., and A. Sharma. 2000. Reversal of fluoride-induced alteration in cauda epididymal spermatozoa and fertility impairment in male mice. Environ. Sci. 7(1):29-38.
Chinoy, N.J., M.V. Rao, M.V. Narayana, and E. Neelakanta. 1991a. Microdose vasal injection of sodium fluoride in the rat. Reprod. Toxicol. 5(6):505-512.
Chinoy, N.J., E. Sequeira, and M.V. Narayana. 1991b. Effects of vitamin C and calcium on the reversibility of fluoride-induced alterations in spermatozoa of rabbits. Fluoride 24(1):29-39.
Chinoy, N.J., P.K. Pradeep, and E. Sequeira. 1992. Effect of fluoride ingestion on the physiology of reproductive organs of male rats. J. Environ. Biol. 13(1):55-61.
Chinoy, N.J., A.S. Walimbe, H.A. Vyas, and P. Mangla. 1994. Transient and reversible fluoride toxicity in some soft tissues of female mice. Fluoride 27(4):205-214.
Chinoy, N.J., M.V. Narayana, V. Dalal, M. Rawat, and D. Patel. 1995. Amelioration of fluoride toxicity in some accessory reproductive glands and spermatozoa of rat. Fluoride 28(2):75-86.
Chinoy, N.J., B.C. Patel, D.K. Patel, and A.K. Sharma. 1997. Fluoride toxicity in the testis and cauda epididymis of guinea pig and reversal by ascorbate. Med. Sci. Res. 25(2):97-100.
Choubisa, S.L. 1999. Some observations on endemic fluorosis in domestic animals in Southern Rajasthan (India). Vet. Res. Commun. 23(7):457-465.
Christensen, G.J. 2005. The advantages of minimally invasive dentistry. J. Am. Dent. Assoc. 136(11):1563-1565.

Chumlea, W.C., C.M. Schubert, A.F. Roche, H.E. Kulin, P.A. Lee, J.H. Himes, and S.S. Sun. 2003. Age at menarche and racial comparisons in U.S. girls. Pediatrics 111(1):110-113.
Cinar, A., and M. Selcuk. 2005. Effects of chronic fluorosis on thyroxine, triiodothyronine, and protein-bound iodine in cows. Fluoride 38(1):65-68.
Cittanova, M.L., B. Lelongt, M.C. Verpont, M. Geniteau-Legendre, F. Wahbe, D. Prie, P. Coriat, and P.M. Ronco. 1996. Fluoride ion toxicity in human kidney collecting duct cells. Anesthesiology 84(2):428-435.
Cittanova, M.L., L. Estepa, R. Bourbouze, O. Blanc, M.C. Verpont, E. Wahbe, P. Coriat, M. Daudon, and P.M. Ronco. 2002. Fluoride ion toxicity in rabbit kidney thick ascending limb cells. Eur. J. Anaesthesiol. 19(5):341-349.
Clabby, C. 2005. Water tests to plumb extent of lead problem. Durham among cities where chemical could cause toxic taint in tap water. The News and Observer, Raleigh, NC, September 3, 2005 [online]. Available: http://www.newsobserver.com [accessed Nov. 7, 2005].
Clark, D.C. 1995. Evaluation of aesthetics for the different classifications of the Tooth Surface Index of Fluorosis. Community Dent. Oral. Epidemiol. 23(2):80-83.
Clark, D.C., H.J. Hann, M.F. Williamson, and J. Berkowitz. 1993. Aesthetic concerns of children and parents in relation to different classifications of the Tooth Surface Index of Fluorosis. Community Dent. Oral. Epidemiol. 21(6):360-364.
Clarkson, J. 1989. Review of terminology, classifications, and indices of developmental defects of enamel. Adv. Dent. Res. 3(2):104-109.
Clarkson, J., and J. McLoughlin. 2000. Role of fluoride in oral health promotion. Int. Dent. J. 50(3):119-128.
Clarkson, J., and D. O'Mullane. 1989. A modified DDE index for use in epidemiological studies of enamel defects. J. Dent. Res. 68(3):445-450.
Clarkson, J., K. Hardwick, D. Barmes, and L.M. Richardson. 2000. International collaborative research on fluoride. J. Dent. Res. 79(4):893-904.
Clay, A.B., and J.W. Suttie. 1987. Effect of dietary fluoride on dairy cattle: Growth of young heifers. J. Dairy Sci. 70(6):1241-1251.
Clovis, J., and J.A. Hargreaves. 1988. Fluoride intake from beverage consumption. Community Dent. Oral. Epidemiol. 16(1):11-15.
Cohen, M., M. Lippman, and B. Chabner. 1978. Role of pineal gland in aetiology and treatment of breast cancer. Lancet 2(8094):814-816.
Cohen-Solal, M.E., F. Augry, Y. Mauras, C. Morieux, P. Allain, and M.C. de Vernejoul. 2002. Fluoride and strontium accumulation in bone does not correlate with osteoid tissue in dialysis patients. Nephrol. Dial. Transplant. 17(3):449-454.
Cohn, P.D. 1992. A Brief Report on the Association of Drinking Water Fluoridation and the Incidence of Osteosarcoma among Young Males. New Jersey Department of Health, November 8, 1992. 17pp.
Collins, T.F., R.L. Sprando, M.E. Shackelford, T.N. Black, M.J. Ames, J.J. Welsh, M.F. Balmer, N. Olejnik, and D.I. Ruggles. 1995. Developmental toxicity of sodium fluoride in rats. Food Chem. Toxicol. 33(11):951-960.
Collins, T.F., R.L. Sprando, T.N. Black, M.E. Shakelford, M.A. Bryant, N. Olejnik, M.J. Ames, J.I Rorie, and D.I. Ruggles. 2001a. Multigenerational evaluation of sodium fluoride in rats. Food Chem. Toxicol. 39(6):601-613.
Collins, T.F., R.L. Sprando, T.N. Black, M.E. Shackelford, N. Olejnik, M.J. Ames, J.I Rorie, and D.I. Ruggles. 2001b. Developmental toxicity of sodium fluoride measured during multiple generations. Food Chem. Toxicol. 39(8):867-876.
Colquhoun, J. 1997. Why I changed my mind about water fluoridation. Perspect. Biol. Med. 41(1):29-44.
Cone, R.D., M.J. Low, J.K. Elmquist, and J.L. Cameron. 2002. Neuroendocrinology. Pp. 81-

176 in Williams Textbook of Endocrinology, 10th Ed., P.R. Larsen, H.M. Kronenberg, S. Melmed, and K.S. Polonsky, eds. Philadelphia, PA: Saunders.

Convertino, V.A., L.E. Armstrong, E.F. Coyle, G.W. Mack, M.N. Sawka, L.C. Senay, Jr., and W.M. Sherman. 1996. American College of Sports Medicine position stand: Exercise and fluid replacement. Med. Sci. Sports Exerc. 28(1):1-7.

Conzen, P.F., M. Nuscheler, A. Melotte, M. Verhaegen, T. Leupolt, H. Van Aken, and K. Peter. 1995. Renal function and serum fluoride concentrations in patients with stable renal insufficiency after anesthesia with sevoflurane or enflurane. Anesth. Analg. 81(3):569-575.

Cooper, C., C. Wickham, R.F. Lacey, and D.J. Barker. 1990. Water fluoride concentration and fracture of the proximal femur. J. Epidemiol. Community Health 44(1):17-19.

Cooper, C., C.A. Wickham, D.J. Barker, and S.J. Jacobsen. 1991. Water fluoridation and hip fracture. JAMA 266(4):513-514.

Coplan, M.J., and R.D. Masters. 2001. Silicofluorides and fluoridation. Fluoride 34(3): 161-164.

Coris, E.E., A.M. Ramirez, and D.J. Van Durme. 2004. Heat illness in athletes: The dangerous combination of heat, humidity and exercise. Sports Med. 34(1):9-16.

Corrêa Rodrigues, M.A., J.R. de Magalhães Bastos, and M.A.R. Buzalaf. 2004. Fingernails and toenails as biomarkers of subchronic exposure to fluoride from dentifrice in 2- to 3-year-old children. Caries Res. 38(2):109-114.

Cortes, D.F., R.P. Ellwood, D.M. O'Mullane, and J.R. Bastos. 1996. Drinking water fluoride levels, dental fluorosis, and caries experience in Brazil. J. Public Health Dent. 56(4):226-228.

Cousins, M.J., and R.I. Mazze. 1973. Methoxyflurane nephrotoxicity: A study of the dose response in man. JAMA 225(13):1611-1616.

Cousins, M.J., L.R. Greenstein, B.A. Hitt, and R.I. Mazze. 1976. Metabolism and renal effects of enflurane in man. Anesthesiology 44(1):44-53.

Cousins, M.J., G.K. Gourlay, K.M. Knights, P.D. Hall, C.A. Lunam, and P. O'Brien. 1987. A randomized prospective controlled study of the metabolism and hepatotoxicity of halothane in humans. Anesth. Analg. 66(4):299-308.

Cox, G.R., E.M. Broad, M.D. Riley, and L.M. Burke. 2002. Body mass changes and voluntary fluid intakes of elite level water polo players and swimmers. J. Sci. Med. Sport. 5(3):183-193.

Coyle, E.F. 2004. Fluid and fuel intake during exercise. J. Sports Sci. 22(1):39-55.

Crapper, D.R., and A.J. Dalton. 1973. Aluminum induced neurofibrillary degeneration, brain electrical activity and alterations in acquisition and retention. Physiol. Behav. 10(5):935-945.

Cross, D.W., and R.J. Carton. 2003. Fluoridation: A violation of medical ethics and human rights. Int. J. Occup. Environ. Health 9(1):24-29.

Crossman, M. 2003. Inside a diet plan. Sporting News (Dec. 29, 2003-Jan. 5, 2004):26-27. [online]. Available: http://www.findarticles.com/p/articles/mi_m1208/is_52_227/ai_112168690 [accessed Sept. 10, 2004].

Curzon, M.E., and P.C. Spector. 1977. Enamel mottling in a high strontium area of the U.S.A. Community Dent. Oral. Epidemiol. 5(5):243-247.

Cutress, T.W., and G.W. Suckling. 1990. Differential diagnosis of dental fluorosis. J. Dent. Res. 69(Spec.):714-720.

Cutress, T.W., G.W. Suckling, G.E. Coote, and J. Gao. 1996. Fluoride uptake into the developing enamel and dentine in sheep incisors following daily ingestion of fluoridated milk or water. N. Z. Dent. J. 92(409):68-72.

Czarnowski, W., and J. Krechniak. 1990. Fluoride in the urine, hair, and nails of phosphate fertiliser workers. Br. J. Ind. Med. 47(5):349-351.

Czarnowski, W., J. Krechniak, B. Urbanska, K. Stolarska, M. Taraszewska-Czarnowska, and

A. Muraszko-Klaudel. 1999. The impact of water-borne fluoride on bone density. Fluoride 32(2):91-95.
Dabeka, R.W., and A.D. McKenzie. 1987. Lead, cadmium, and fluoride levels in market milk and infant formulas in Canada. J. Assoc. Off. Anal. Chem. 70(4):754-757.
Dabeka, R.W., and A.D. McKenzie. 1995. Survey of lead, cadmium, fluoride, nickel, and cobalt in food composites and estimation on dietary intakes of these elements by Canadians in 1986-1988. J. AOAC Int. 78(4):897-909.
Dabeka, R.W., K.F. Karpinski, A.D. McKenzie, and C.D. Bajdik. 1986. Survey of lead, cadmium and fluoride in human milk and correlation of levels with environmental and food factors. Food Chem. Toxicol. 24(9):913-921.
Dandona, P., A. Coumar, D.S. Gill, J. Bell, and M. Thomas. 1988. Sodium fluoride stimulates osteocalcin in normal subjects. Clin. Endocrinol. 29(4):437-441.
Daniels, T.C., and E.C. Jorgensen. 1977. Physicochemical properties in relation to biological action. Pp. 5-62 in Textbook of Organic Medicinal and Pharmaceutical Chemistry, 7th Ed., C.O. Wilson, O. Gisvold, and R.F. Doerge, eds. Philadelphia: Lippincott.
Danielson, C., J.L. Lyon, M. Egger, and G.K. Goodenough. 1992. Hip fractures and fluoridation in Utah's elderly population. JAMA 268(6):746-748.
Das, T.K., A.K. Susheela, I.P. Gupta, S. Dasarathy, and R.K. Tandon. 1994. Toxic effects of chronic fluoride ingestion on the upper gastrointestinal tract. J. Clin. Gastroenterol. 18(3):194-199.
Dasarathy, S., T.K. Das, I.P. Gupta, A.K. Susheela, and R.K. Tandon. 1996. Gastroduodenal manifestations in patients with skeletal fluorosis. J. Gastroenterol. 31(3):333-337.
Daston, G.P., B.F. Rehnberg, B. Carver, and R.J. Kavlock. 1985. Toxicity of sodium fluoride to the postnatally developing rat kidney. Environ. Res. 37(2):461-474.
Davies, T.F., and P.R. Larsen. 2002. Thyrotoxicosis. Pp. 374-421 in Williams Textbook of Endocrinology, 10th Ed., P.R. Larsen, H.M. Kronenberg, S. Melmed, and K.S. Polonsky, eds. Philadelphia, PA: Saunders.
Day, T.K., and P.R. Powell-Jackson. 1972. Fluoride, water hardness, and endemic goitre. Lancet 1(7761):1135-1138.
Dean, H.T. 1934. Classification of mottled enamel diagnosis. J. Am. Dent. Assoc. 21: 1421-1426.
Dean, H.T. 1942. The investigation of physiological effects by the epidemiological method. Pp. 23-31 in Fluorine and Dental Health, F.R. Mouton, ed. AAAS No. 19. Washington, DC: American Association for the Advancement of Science.
de Camargo, A.M., and J. Merzel. 1980. Histological and histochemical appearance of livers and kidneys of rats after long-term treatment with different concentrations of sodium fluoride in drinking water. Acta Anat. 108(3):288-294.
Delagrange, P., J. Atkinson, J.A. Boutin, L. Casteilla, D. Lesieur, R. Misslin, S. Pellissier, L. Penicaud, and P. Renard. 2003. Therapeutic perspectives for melatonin agonists and antagonists. J. Neuroendocrinol. 15(4):442-448.
de la Sota, M., R. Puche, A. Rigalli, L.M. Fernandez, S. Benassati, and R. Boland. 1997. Changes in bone mass and in glucose homeostasis in subjects with high spontaneous fluoride intake. Medicina (B Aires) 57(4):417-420.
DeLucia, M.C., M.E. Mitnick, and T.O. Carpenter. 2003. Nutritional rickets with normal circulating 25-hydroxyvitamin D: A call for reexamining the role of dietary calcium intake in North American infants. J. Clin. Endocrinol. Metab. 88(8):3539-3545.
Demole, V. 1970. Toxic effects on the thyroid. Pp. 255-262 in Fluorides and Human Health, Monograph Series No. 59. Geneva: World Health Organization.
DenBesten, P.K. 1994. Dental fluorosis: Its use as a biomarker. Adv. Dent. Res. 8(1): 105-110.

DenBesten, P.K. 1999. Biological mechanisms of dental fluorosis relevant to the use of fluoride supplements. Community Dent. Oral Epidemiol. 27(1):41-47.

DenBesten, P.K., and H. Thariani. 1992. Biological mechanisms of fluorosis and level and timing of systemic exposure to fluoride with respect to fluorosis. J. Dent. Res. 71(5): 1238-1243.

DenBesten, P.K., Y. Yan, J.D. Featherstone, J.F. Hilton, C.E. Smith, and W. Li. 2002. Effects of fluoride on rat dental enamel matrix proteinases. Arch. Oral Biol. 47(11):763-770.

Desai, V.K., D.M. Solanki, and R.K. Bansal. 1993. Epidemiological study on goitre in endemic fluorosis district of Gujarat. Fluoride 26(3):187-190.

Dick, A.E., R.P. Ford, P.J. Schluter, E.A. Mitchell, B.J. Taylor, S.M. Williams, A.W. Stewart, D.M. Becroft, J.M. Thompson, R. Scragg, I.B. Hassall, D.M. Barry, and E.M. Allen. 1999. Water fluoridation and the sudden infant death syndrome. N.Z. Med. J. 112(1093):286-289.

DIF (The Diabetes Insipidus Foundation, Inc.). 2004. Question # 0007 FAQ Keywords: prevalence, nocturnal enuresis, incidence. General Question, Frequently Asked Questions. The Diabetes Insipidus Foundation, Inc. [online]. Available: http://www.diabetesinsipidus.org/faqs5.htm#GENERAL%20QUESTIONS [accessed Sept. 10, 2004].

Dillenberg, J.S., S.M. Levy, D.C. Schroeder, E.N. Gerston, and C.J. Andersen. 1992. Arizona providers' use and knowledge of fluoride supplements. Clin. Prev. Dent. 14(5):15-26.

Dominok, G., K. Siefert, J. Frege, and B. Dominok. 1984. Fluoride content of bones of retired fluoride workers. Fluoride 17(1):23-26.

Dost, F.N., R.M. Knaus, D.E. Johnson, and C.H. Wang. 1977. Fluoride impairment of glucose utilization: Nature of effect in rats during and after continuous NaF infusion. Toxicol. Appl. Pharmacol. 41(3):451-458.

Douglass, C. 2004. Fluoride Exposure and Osteosarcoma. Grant No. 5 ROI ES06000. National Institute of Environmental Health Sciences.

Dourson, M.L. 1994. Methods for establishing oral reference doses (RfDs). Pp. 51-61 in Risk Assessment of Essential Elements, W. Mertz, C.O. Abernathy, and S.S. Olin, eds. Washington, DC: ILSI Press.

Driessen, B., L. Zarucco, E.P. Steffey, C. McCullough, F. Del Piero, L. Melton, B. Puschner, and S.M. Stover. 2002. Serum fluoride concentrations, biochemical and histopathological changes associated with prolonged sevoflurane anaesthesia in horses. J. Vet. Med. A Physiol. Pathol. Clin. Med. 49(7):337-347.

Driscoll, W.S., H.S. Horowitz, R.J. Meyers, S.B. Heifetz, A. Kingman, and E.R. Zimmerman. 1983. Prevalence of dental caries and dental fluorosis in areas with optimal and above-optimal water fluoride concentrations. J. Am. Dent. Assoc. 107(1):42-47.

Driscoll, W.S., H.S. Horowitz, R.J. Meyers, S.B. Heifetz, A. Kingman, and E.R. Zimmerman. 1986. Prevalence of dental caries and dental fluorosis in areas with negligible, optimal, and above-optimal fluoride concentrations in drinking water. J. Am. Dent. Assoc. 113(1):29-33.

Duchassaing, D., B. Rigat, J.P. Barberousse, and M.J. Laisne. 1982. The elimination of inorganic fluoride after enflurane anesthesia--transitory action on parathyroid tissue. Int. J. Clin. Pharmacol. Ther. Toxicol. 20(8):366-372.

Duell, P.B., and C.H. Chesnut III. 1991. Exacerbation of rheumatoid arthritis by sodium fluoride treatment of osteoporosis. Arch. Int. Med. 151(4):783-784.

Dunipace, A.J., E.J. Brizendine, W. Zhang, M.E. Wilson, L.L. Miller, B.P. Katz, J.M. Warrick, and G.K. Stookey. 1995. Effect of aging on animal response to chronic fluoride exposure. J. Dent. Res. 74(1):358-368.

Dunipace, A.J., C.A. Wilson, M.E. Wilson, W. Zhang, A.H. Kafrawy, E.J. Brizendine, L.L. Miller, B.P. Katz, J.M. Warrick, and G.K. Stookey. 1996. Absence of detrimental effects of fluoride exposure in diabetic rats. Arch. Oral Biol. 41(2):191-203.

Dunipace, A.J., E.J. Brizendine, M.E. Wilson, W. Zhang, B.P. Katz, and G.K. Stookey. 1998. Chronic fluoride exposure does not cause detrimental, extraskeletal effects in nutritionally deficient rats. J. Nutr. 128(8):1392-1400.
Duperon, D.F., J.R. Jedrychowski, and J. Kong. 1995. Fluoride content of Los Angeles County water. J. Calif. Dent. Assoc. 23(2):45-46, 48.
Durand, M., and J. Grattan. 2001. Effects of volcanic air pollution on health. Lancet 357(9251):164.
Dure-Smith, B.A., S.M. Farley, S.G. Linkhart, J.R. Farley, and D.J. Baylink. 1996. Calcium deficiency in fluoride-treated osteoporotic patients despite calcium supplementation. J. Clin. Endocrinol. Metab. 81(1):269-275.
Duursma, S.A., J.H. Glerum, A. van Dijk, R. Bosch, H. Kerkhoff, J. van Putten, and J.A. Raymakers. 1987. Responders and non-responders after fluoride therapy in osteoporosis. Bone 8(3):131-136.
Easmann, R.P., D.E. Steflik, D.H. Pashley, R.V. McKinney, and G.M. Whitford. 1984. Surface changes in rat gastric mucosa induced by sodium fluoride: A scanning electron microscopic study. J. Oral Pathol. 13(3):255-264.
Eastell, R., M.S. Calvo, M.F. Burritt, K.P. Offord, R.G. Russell, and B.L. Riggs. 1992. Abnormalities in circadian patterns of bone resorption and renal calcium conservation in type I osteoporosis. J. Clin. Endocrinol. Metab. 74(3):487-494.
Eble, D.M., T.G. Deaton, F.C. Wilson Jr., and J.W. Bawden. 1992. Fluoride concentrations in human and rat bone. J. Public Health Dent. 52(5):288-291.
EC (European Commission). 2003. Directive 2003/40/EC of 16 May 2003 establishing the list, concentration limits and labelling requirements for the constituents of natural mineral waters and the conditions for using ozone-enriched air for the treatment of natural mineral waters and spring waters. Official Journal of the European Union (22.5.2003)L 126/34-39 [online]. Available: http://europa.eu.int/comm/food/food/labellingnutrition/water/index_en.htm [accessed Nov. 1, 2005].
Eckerlin, R.H., L. Krook, G.A. Maylin, and D. Carmichael. 1986a. Toxic effects of food-borne fluoride in silver foxes. Cornell Vet. 76(4):395-402.
Eckerlin, R.H., G.A. Maylin, and L. Krook. 1986b. Milk production of cows fed fluoride contaminated commercial feed. Cornell Vet. 76(4):403-414.
Edwards, S.L., T.L. Poulos, and J. Kraut. 1984. The crystal structure of fluoride-inhibited cytochrome c peroxidase. J. Biol. Chem. 259(21):12984-12988.
Eichner, R., W. Börner, D. Henschler, W. Köhler, and E. Moll. 1981. Osteoporosis therapy and thyroid function. Influence of 6 months of sodium fluoride treatment on thyroid function and bone density [in German]. Fortschr. Med. 99(10):342-348.
Eisenbarth, G.S., K.S. Polonsky, and J.B. Buse. 2002. Type 1 diabetes mellitus. Pp. 1485-1508 in Williams Textbook of Endocrinology, 10th Ed., P.R. Larsen, H.M. Kronenberg, S. Melmed, and K.S. Polonsky, eds. Philadelphia, PA: Saunders.
Ekambaram, P., and V. Paul. 2001. Calcium preventing locomotor behavioral and dental toxicities of fluoride by decreasing serum fluoride level in rats. Environ. Toxicol. Pharmacol. 9(4):141-146.
Ekambaram, P., and V. Paul. 2002. Modulation of fluoride toxicity in rats by calcium carbonate and by withdrawal of fluoride exposure. Pharmacol. Toxicol. 90(2):53-58.
Eklund, S.A., B.A. Burt, A.I. Ismail, and J.J. Calderone. 1987. High-fluoride drinking water, fluorosis, and dental caries in adults. J. Am. Dent. Assoc. 114(3):324-328.
Eklund, S.A., J.L. Pittman, and K.E. Heller. 2000. Professionally applied topical fluoride and restorative care in insured children. J. Public Health Dent. 60(1):33-38.
Ekstrand, J. 1978. Relationship between fluoride in the drinking water and the plasma fluoride concentration in man. Caries Res. 12(3):123-127.

Ekstrand, J., and C.J. Spak. 1990. Fluoride pharmacokinetics: Its implications in the fluoride treatment of osteoporosis. J. Bone Miner. Res. 5(Suppl.1):S53-S61.
Ekstrand, J., M. Ehrnebo, and L.O. Boreus. 1978. Fluoride bioavailability after intravenous and oral administration: Importance of renal clearance and urine flow. Clin. Pharmacol. Ther. 23(3):329-337.
Ekstrand, J., S.J. Fomon, E.E. Ziegler, and S.E. Nelson. 1994. Fluoride pharmacokinetics in infancy. Pediatr. Res. 35(2):157-163.
Elbetieha, A., H. Darmani, and A.S. Al-Hiyasat. 2000. Fertility effects of sodium fluoride in male mice. Fluoride 33(3):128-134.
el-Hajj Fuleihan, G., E.B. Klerman, E.N. Brown, Y. Choe, E.M. Brown, and C.A. Czeisler. 1997. The parathyroid hormone circadian rhythm is truly endogenous—a general clinical research center study. J. Clin. Endocrinol. Metab. 82(1):281-286.
Ellwood, R., D. O'Mullane, J. Clarkson, and W. Driscoll. 1994. A comparison of information recorded using the Thylstrup Fejerskov index, Tooth Surface index of Fluorosis and Developmental Defects of Enamel index. Int. Dent. J. 44(6):628-636.
Emsley, J., D.J. Jones, J.M. Miller, R.E. Overill, and R.A. Waddilove. 1981. An unexpectedly strong hydrogen bond: Ab initio calculations and spectroscopic studies of amide-fluoride systems. J. Am. Chem. Soc. 103:24-28.
Englander, H.R., and P.F. DePaola. 1979. Enhanced anticaries action from drinking water containing 5 ppm fluoride. J. Am. Dent. Assoc. 98(1):35-39.
EPA (U.S. Environmental Protection Agency). 1986. Guidelines for Carcinogen Risk Assessment. EPA/630/R-00/004. Risk Assessment Forum, U.S. Environmental Protection Agency, Washington, DC. September 1986 [online]. Available: http://www.epa.gov/ncea/raf/car2sab/guidelines_1986.pdf [accessed Jan. 25, 2005].
EPA (U.S. Environmental Protection Agency). 1988. Summary Review of Health Effects Associated with Hydrogen Fluoride and Related Compounds. Health Issue Assessment. EPA/600/8-89/002F. Environmental Criteria and Assessment Office, Office of Health and Environmental Assessment, Office of Research and Development, U.S. Environmental Protection Agency, Research Triangle Park, NC. December 1988.
EPA (U.S. Environmental Protection Agency). 1989. Fluorine (Soluble Fluoride) (CASRN 7782-41-4). Integrated Risk Information System, U.S. Environmental Protection Agency [online]. Available: http://www.epa.gov/iris/subst/0053.htm [accessed Sept. 10, 2004].
EPA (U.S. Environmental Protection Agency). 1992. Sulfuryl Fluoride. R.E.D. (Registration Eligibility Decision) Facts. EPA-738-R-93-012. Office of Prevention, Pesticide and Toxic Substances, U.S. Environmental Protection Agency [online]. Available: http://www.fluorideaction.org/pesticides/ Sulfuryl_fluoride_RED.1992.pdf [accessed Sept. 10, 2004].
EPA (U.S. Environmental Protection Agency). 1994. Methods for Derivation of Inhalation Reference Concentrations and Application of Inhalation Dosimetry. EPA/600/8-90/066F. Environmental Criteria and Assessment Office, Office of Health and Environmental Assessment, Office of Research and Development, U.S. Environmental Protection Agency, Research Triangle Park, NC. October 1994.
EPA (U.S. Environmental Protection Agency). 1996a. Cryolite. R.E.D. (Registration Eligibility Decision) Facts. EPA-738-R-96-016. Office of Prevention, Pesticide and Toxic Substances, U.S. Environmental Protection Agency [online]. Available: www.epa.gov/oppsrrd1/REDs/factsheets/0087fact.pdf [accessed Sept. 10, 2004].
EPA (U.S. Environmental Protection Agency). 1996b. Environmental Health Threats to Children. EPA175-F-96-001. Office of the Administrator, U.S. Environmental Protection Agency. September 1996.
EPA (U.S. Environmental Protection Agency). 1997. Exposure Factors Handbook,Vol. I, II, III. EPA/600/P-95/002Fa-c. National Center for Environmental Assessment, Office of

REFERENCES

Research and Development, U.S. Environmental Protection Agency [online]. Available: http://www.epa.gov/ncea/exposfac.htm [accessed Oct. 13, 2004].

EPA (U.S. Environmental Protection Agency). 2000a. Estimated Per Capita Water Ingestion in the United States: Based on Data Collected by the United States Department of Agriculture's 1994-96 Continuing Survey of Food Intakes by Individuals. EPA-822-R-00-008. Office of Water, U.S. Environmental Protection Agency. April 2000.

EPA (U.S. Environmental Protection Agency). 2000b. Report to Congress. EPA Studies on Sensitive Subpopulations and Drinking Water Contaminants. EPA 815-R-00-015. Office of Water, U.S. Environmental Protection Agency, Washington, DC. December 2000.

EPA (U.S. Environmental Protection Agency). 2000c. Food Commodity Intake Database (FCID). Computer data file. NTIS PB2000-500101. Office of Pesticide Programs, U.S. Environmental Protection Agency, Washington, DC.

EPA (U.S. Environmental Protection Agency). 2000d. Methodology for Deriving Ambient Water Quality Criteria for the Protection of Human Health (2000), Final Report. EPA-822-B-00-004. Office of Science and Technology, Office of Water, U.S. Environmental Protection Agency, Washington, DC [online]. Available: http://www.epa.gov/waterscience/humanhealth/method/complete.pdf [accessed Oct. 13, 2004].

EPA (U.S. Environmental Protection Agency). 2003a. Occurrence Estimation Methodology and Occurrence Findings Report for the Six-Year Review of Existing National Primary Drinking Water Regulations. EPA-815-R-03-006. Office of Water, U.S. Environmental Protection Agency [online]. Available: http://www.epa.gov/safewater/standard/review/pdfs/support_6yr_occurancemethods_final.pdf [accessed Sept. 10, 2004].

EPA (U.S. Environmental Protection Agency). 2003b. Preliminary Risk Assessment of the Developmental Toxicity Associated with Exposure to Perfluorooctanoic Acid and Its Salts. Risk Assessment Division, Office of Pollution Prevention and Toxics, U.S. Environmental Protection Agency. April 10, 2003 [online]. Available: http://www.nicnas.gov.au/publications/pdf/pfospreliminaryriskassessment.pdf [accessed Sept. 10, 2004].

EPA (U.S. Environmental Protection Agency). 2003c. Health Risks to Fetuses, Infants and Children (Proposed Stage 2 Disinfectant/Disinfection Byproducts): A Review. EPA-822-R-03-010. Office of Water, Office of Science and Technology, U.S. Environmental Protection Agency. March 2003.

EPA (U.S. Environmental Protection Agency). 2004. Human Health Risk Assessment for Sulfuryl Fluoride and Fluoride Anion Addressing the Section 3 Registration of Sulfuryl Fluoride Post-Harvest Fumigation of Stored Cereal Grains, Dried Fruits and Tree Nuts and Pest Control in Grain Processing Facilities. PP# 1F6312. Memorandum to Dennis McNeilly/Richard Keigwin, Fungicide Branch, Registration Division, from Michael Doherty, Edwin Budd, Registration Action Branch 2, and Becky Daiss, Registration Action Branch 4, Health Effects Division, Office of Prevention, Pesticides and Toxic Substances, U.S. Environmental Protection Agency, Washington, DC. January 20, 2004 [online]. Available: http://www.fluorideaction.org/pesticides/fr.sulfuryl.fluoride.htm [accessed Sept. 10, 2004].

EPA (U.S. Environmental Protection Agency). 2005a. Guidelines for Carcinogen Risk Assessment. EPA/630/P-03/001F. Risk Assessment Forum, U.S. Environmental Protection Agency, Washington, DC [online]. Available: http://cfpub.epa.gov/ncea/cfm/recordisplay.cfm?deid=116283 [accessed Oct. 10, 2005].

EPA (U.S. Environmental Protection Agency). 2005b. Supplemental Guidance for Assessing Susceptibility from Early-Life Exposure to Carcinogens. EPA/630/R-03/003F. Risk Assessment Forum, U.S. Environmental Protection Agency, Washington, DC [online]. Available: http://cfpub.epa.gov/ncea/cfm/recordisplay.cfm?deid=116283 [accessed Oct. 10, 2004].

Erben, J., B. Hajakova, M. Pantucek, and L. Kubes. 1984. Fluoride metabolism and renal

osteodystrophy in regular dialysis treatment. Proc. Eur. Dial. Transplant Assoc. Eur. Ren. Assoc. 21:421-425.
Erdal, S., and S.N. Buchanan. 2005. A quantitative look at fluorosis, fluoride exposure, and intake in children using a health risk assessment approach. Environ. Health Perspect. 113(1):111-117.
Erickson, J.D. 1980. Down syndrome, water fluoridation, and maternal age. Teratology 21(2):177-180.
Erickson, J.D., G.P. Oakley, J.W. Flynt, and S. Hay. 1976. Water fluoridation and congenital malformation: No association. J. Am. Dent. Assoc. 93(5):981-984.
Ericsson, Y., K. Gydell, and T. Hammarskioeld. 1973. Blood plasma fluoride: An indicator of skeletal fluoride content. J. Int. Res. Commun. Syst. 1:33-35.
Erlacher, L., H. Templ, and D. Magometschnigg. 1995. A comparative bioavailability study on two new sustained-release formulations of disodiummonoflu9orophosphate versus a non-sustained-release formulation in healthy volunteers. Calcif. Tissue. Int. 56(3):196-200.
Ermiş, R.B., F. Koray, and B.G. Akdeniz. 2003. Dental caries and fluorosis in low- and high-fluoride areas in Turkey. Quintessence Int. 34(5):354-360.
Ernst, P., D. Thomas, and M.R. Becklake. 1986. Respiratory survey of North American Indian children living in proximity to an aluminum smelter. Am. Rev. Respir. Dis. 133(2):307-312.
Eto, B., M. Boisset, and J.F. Desjeux. 1996. Sodium fluoride inhibits the antisecretory effect of peptide YY and its analog in rabbit jejunum. Arch. Physiol. Biochem. 104(2):180-184.
Evans, R.W., and J.W. Stamm. 1991. An epidemiologic estimate of the critical period during which human maxillary central incisors are most susceptible to fluorosis. J. Public Health Dent. 51(4):251-259.
Everett, E.T., M.A. McHenry, N. Reynolds, H. Eggertsson, J. Sullivan, C. Kantmann, E.A. Martinez-Mier, J.M. Warrick, and G.K. Stookey. 2002. Dental fluorosis: Variability among different inbred mouse strains. J. Dent. Res. 81(11):794-798.
Fabiani, L., V. Leoni, and M. Vitali. 1999. Bone-fracture incidence rate in two Italian regions with different fluoride concentration levels in drinking water. J. Trace Elem. Med. Biol. 13(4):232-237.
Façanha, A.R., and A.L. Okorokova-Façanha. 2002. Inhibition of phosphate uptake in corn roots by aluminum-fluoride complexes. Plant Physiol. 129(4):1763-1772.
Faccini, J.M. 1967. Inhibition of bone resorption in the rabbit by fluoride. Nature 214(94): 1269-1271.
Faccini, J.M. 1969. Fluoride-induced hyperplasia of the parathyroid glands. Proc. R Soc. Med. 62(3):241.
Faccini, J.M., and A.D. Care. 1965. Effect of sodium fluoride on the ultrastructure of the parathyroid glands of the sheep. Nature 207(4):1399-1401.
Farkas, G., A. Fazekas, and E. Szekeres. 1983. The fluoride content of drinking water and menarcheal age. Acta Univ. Szeged. Acta Biol. 29(1-4):159-168.
Farley, J.R., J.R. Wergedal, and D.J. Baylink. 1983. Fluoride directly stimulates proliferation and alkaline phosphatase activity of bone forming cells. Science 222(4621):330-332.
Farley, J.R., N. Tarbaux, S. Hall, and D.J. Baylink. 1988. Evidence that fluoride-stimulated 3[H]- thymidine incorporation in embryonic chick calvarial cell cultures is dependent on the presence of a bone cell mitogen, sensitive to changes in the phosphate concentration, and modulated by systemic skeletal effectors. Metabolism 37(10):988-995.
Farley, J.R., S.L. Hall, S. Herring, and M.A. Tanner. 1993. Fluoride increase net calcium uptake by SaOS-2 cells: The effect is phosphate dependent. Calcif. Tissue Int. 53(3):187-192.
FDI (Fédération Dentaire Internationale). 1982. An epidemiological index of developmental defects of dental enamel (DDE index). Int. Dent. J. 32(2):159-167.

REFERENCES

Featherstone, J.D. 1999. Prevention and reversal of dental caries: Role of low level fluoride. Community Dent. Oral Epidemiol. 27(1):31-40.
Fein, N.J., and F.L. Cerklewski. 2001. Fluoride content of foods made with mechanically separated chicken. J. Agric. Food Chem. 49(9):4284-4286.
Fejerskov, O. 2004. Changing paradigms in concepts on dental caries: Consequences for oral health care. Caries Res. 38(3):182-191.
Fejerskov, O., K.W. Stephen, A. Richards, and R. Speirs. 1987. Combined effect of systemic and topical fluoride treatments on human deciduous teeth—case studies. Caries Res. 21(5):452-459.
Fejerskov, O., F. Manji, and V. Baelum. 1990. The nature and mechanisms of dental fluorosis in man. J. Dent. Res. 69(Spec. Iss.):692-700.
Fejerskov, O., M.J. Larsen, A. Richards, and V. Baelum. 1994. Dental tissue effects of fluoride. Adv. Dent. Res. 8(1):15-31.
Felsenfeld, A.J., and M.A. Roberts. 1991. A report of fluorosis in the United States secondary to drinking well water. JAMA 265(4):486-488.
Feltman, R., and G. Kosel. 1961. Prenatal and postnatal ingestion of fluorides—Fourteen years of investigation. Final report. J. Dent. Med. 16(Oct.):190-198.
Feskanich, D., W. Owusu, D.J. Hunter, W. Willett, A. Ascherio, D. Spiegelman, S. Morris, V.L. Spate, and G. Colditz. 1998. Use of toenail fluoride levels as an indicator for the risk of hip and forearm fractures in women. Epidemiology 9(4):412-416.
Fleischer, M. 1962. Fluoride Content of Ground Water in the Conterminous United States. U.S. Geological Survey Miscellaneous Geological Investigation I-387. Washington, DC: U.S. Geological Survey (as cited in ATSDR 2003).
Fleischer, M., R.M. Forbes, R.C. Harriss, L. Krook, and J. Kubots. 1974. Fluorine. Pp. 22-25 in Geochemistry and the Environment, Vol. I: The Relation of Selected Trace Elements to Health and Disease. Washington, DC: National Academy of Sciences (as cited in ATSDR 2003).
Fomon, S.J., and J. Ekstrand. 1999. Fluoride intake by infants. J. Public Health Dent. 59(4):229-234.
Fomon, S.J., J. Ekstrand, and E.E. Ziegler. 2000. Fluoride intake and prevalence of dental fluorosis: Trends in fluoride intake with special attention to infants. J. Public Health Dent. 60(3):131-139.
Forsman, B. 1974. Dental fluorosis and caries in high-fluoride districts in Sweden. Community Dent. Oral Epidemiol. 2(3):132-148.
Foster, N.L., T.N. Chase, P. Fedio, N.J. Patronas, R.A. Brooks, and G. Di Chiro. 1983. Alzheimer's disease: Focal cortical changes shown by positron emission tomography. Neurology 33(8):961-965.
Franke, J., and E. Auermann. 1972. Significance of iliac crest puncture with histological and microanalytical examination of the obtained bone material in the diagnosis of fluorosis [in German]. Int. Arch. Arbeitsmed. 29(2):85-94.
Franke, J., F. Rath, H. Runge, F. Fengler, E. Auermann, and G.L. Lenart. 1975. Industrial fluorosis. Fluoride 8(2):61-85.
Fraser, W.D., F.C. Logue, J.P. Christie, S.J. Gallacher, D. Cameron, D.S. O'Reilly, G.H. Beastall, and I.T. Boyle. 1998. Alteration of the circadian rhythm of intact parathyroid hormone and serum phosphate in women with established postmenopausal osteoporosis. Osteoporos Int. 8(2):121-126.
Freni, S.C. 1994. Exposure to high fluoride concentrations in drinking water is associated with decreased birth rates. J. Toxicol. Environ. Health 42(1):109-121.
Freni, S.C., and D.W. Gaylor. 1992. International trends in the incidence of bone cancer are not related to drinking water fluoridation. Cancer 70(3):611-618.
Frink, E.J. Jr., H. Ghantous, T.P. Malan, S. Morgan, J. Fernando, A.J. Gandolfi, and B.R. Brown

Jr. 1992. Plasma inorganic fluoride with sevoflurane anesthesia: Correlation with indices of hepatic and renal function. Anesth. Analg. 74(2):231-235.

Fujii, A., and T. Tamura. 1989. Deleterious effect of sodium fluoride on gastrointestinal tract. Gen. Pharmacol. 20(5):705-710.

Fujita, T., and G.M. Palmieri. 2000. Calcium paradox disease: Calcium deficiency prompting secondary hyperparathyroidism and cellular calcium overload. J. Bone Miner. Metab. 18(3):109-125.

Gabler, W.L., and P.A. Leong. 1979. Fluoride inhibition of polymorphonuclear leukocytes. J. Dent. Res. 58(9):1933-1939.

Gadhia, P.K., and S. Joseph. 1997. Sodium fluoride induced chromosome aberrations and sister chromatid exchange in cultured human lymphocytes. Fluoride 30(3):153-156.

Galletti, P.M., and G. Joyet. 1958. Effect of fluorine on thyroidal iodine metabolism in hyperthyroidism. J. Clin. Endocrinol. Metab. 18(10):1102-1110.

García-Patterson, A., M. Puig-Domingo, and S.M. Webb. 1996. Thirty years of human pineal research: Do we know its clinical relevance? J. Pineal. Res. 20(1):1-6.

Gedalia, I., and N. Brand. 1963. The relationship of fluoride and iodine in drinking water in the occurrence of goiter. Arch. Int. Pharmacodyn. Ther. 142(April 1):312-315.

Gedalia, I., J. Gross, S. Guttmann, J.E. Steiner, F.G. Sulman, and M.M. Weinreb. 1960. The effects of water fluorination on thyroid function, bones and teeth of rats on a low iodine diet. Arch. Int. Pharmacodyn. Ther. 129(Dec.1):116-124.

Gelberg, K.H. 1994. Case-control Study of Childhood Osteosarcoma. Ph.D. Dissertation, Yale University.

Gelberg, K.H., E.F. Fitzgerald, S.A. Hwang, and R. Dubrow. 1995. Fluoride exposure and childhood osteosarcoma: A case-control study. Am. J. Public Health 85(12):1678-1683.

Gerster, J.C., S.A. Charhon, P. Jaeger, G. Boivin, D. Briancon, A. Rostan, C.A. Baud, P.J. Meunier. 1983. Bilateral fractures of femoral neck in patients with moderate renal failure receiving fluoride for spinal osteoporosis. Br. Med. J. 287(6394):723-725.

Gessner, B.D., M. Beller, J.P. Middaugh, and G.M. Whitford. 1994. Acute fluoride poisoning from a public water system. N. Engl. J. Med. 330(2):95-99.

Gharzouli, K., S. Amira, S. Khennouf, and A. Gharzouli. 2000. Effects of sodium fluoride on water and acid secretion, soluble mucus and adherent mucus of the rat stomach. Can. J. Gastroenterol. 14(6):493-498.

Ghosh, D., S. Das Sarkar, R. Maiti, D. Jana, and U.B. Das. 2002. Testicular toxicity in sodium fluoride treated rats: Association with oxidative stress. Reprod. Toxicol. 16(4):385-390.

Gibson, S.L. 1992. Effects of fluoride on immune system function. Complement. Med. Res. 6:111-113.

Gill, D.S., A. Coumar, and P. Dandona. 1989. Effect of fluoride on parathyroid hormone. Clin. Sci. 76(6):677-678.

Gispen, W.H., and R.L. Isaacson. 1986. Excessive grooming in response to ACTH. Pp. 273-312 in Neuropeptides and Behavior, Vol. 1. CNS Effects of ACTH, MSH, and Opioid Peptides, D. de Weid, W.H. Gispen, and T.B. van Wimersma Greidanus, eds. New York: Pergamon Press.

Goh, E.H., and A.W. Neff. 2003. Effects of fluoride on Xenopus embryo development. Food Chem. Toxicol. 41(11):1501-1508.

Gold, M.S., A.L. Pottash, and I. Extein. 1981. Hypothyroidism and depression. Evidence from complete thyroid function evaluation. JAMA 245(19):1919-1922.

Goldberg, M.E., J. Cantillo, G.E. Larijani, M. Torjman, D. Vekeman, and H. Schieren. 1996. Sevoflurane versus isoflurane for maintenance of anesthesia: Are serum inorganic fluoride ion concentrations of concern? Anesth. Analg. 82(6):1268-1272.

Goldman, R., Y. Granot, and U. Zor. 1995. A pleiotropic effect of fluoride on signal trans-

duction in macrophages: Is it mediated by GPT-binding proteins? J. Basic Clin. Physiol. Pharmacol. 6(1):79-94.
Gomez-Ubric, J.L., J. Liebana, J. Gutierrez, and A. Castillo A. 1992. In vitro immune modulation of polymorphonuclear leukocyte adhesiveness by sodium fluoride. Eur. J. Clin. Invest. 22(10):659-661.
Goodman, H.M. 2003. Basic Medical Endocrinology, 3rd Ed. San Diego, CA: Academic Press.
Gopalakrishnan, P., R.S. Vasan, P.S. Sarma, K.S. Nair, and K.R. Thankappan. 1999. Prevalence of dental fluorosis and associated risk factors in Alappuzha district, Kerala. Natl. Med. J. India 12(3):99-103.
Goward, P.E. 1982. Mottling on deciduous incisor teeth. A study of 5-year-old Yorkshire children from districts with and without fluoridation. Brit. Dent. J. 153(10):367-369.
Goyer, R.A. 1995. Nutrition and metal toxicity. Am. J. Clin. Nutr. 61(3 Suppl.):646S-650S.
Grandjean, P., and J.H. Olsen. 2004. Extended follow-up of cancer incidence in fluoride-exposed workers. J. Natl. Cancer Inst. 96(10):802-803.
Grandjean, P., J.H. Olsen, O.M. Jensen, and K. Juel. 1992. Cancer incidence and mortality in workers exposed to fluoride. J. Natl. Cancer Inst. 84(24):1903-1909.
Grattan, J., M. Durand, and S. Taylor. 2003. Illness and elevated human mortality in Europe coincident with the Laki Fissure eruption. Pp. 401-414 in Volcanic Degassing, C. Oppenheimer, D.M. Pyle, and J. Barclay, eds. Geological Society Special Publication No. 213. London: Geological Society.
Greenberg, L.W., C.E. Nelsen, and N. Kramer. 1974. Nephrogenic diabetes insipidus with fluorosis. Pediatrics 54(3):320-322.
Greenland, S. 1992. Divergent biases in ecologic and individual-level studies. Stat. Med. 11(9):1209-1223.
Greenland, S. 1998. Meta-analysis. Pp. 643-674 in Modern Epidemiology, 2nd Ed, K.J. Rothman, and S. Greenland, eds. Philadelphia, PA: Lippincott-Raven.
Greenland, S., and J. Robins. 1994. Invited commentary: Ecologic studies—biases, misconceptions and counterexamples. Am. J. Epidemiol. 139(8):747-760.
Griffin, S.O., E.D. Beltran, S.A. Lockwood, and L.K. Barker. 2002. Esthetically objectionable fluorosis attributable to water fluoridation. Community Dent. Oral. Epidemiol. 30(3):199-209.
Griffith-Jones, W. 1977. Fluorosis in dairy cattle. Vet. Rec. 100(5):84-89.
Grimbergen, G.W. 1974. A double blind test for determination of intolerance to fluoridated water. Preliminary report. Fluoride 7(3):146-152.
Grobler, S.R., A.J. Louw, and T.J. van Kotze. 2001. Dental fluorosis and caries experience in relation to three different drinking water fluoride levels in South Africa. Int. J. Paediatr. Dent. 11(5):372-379.
Grossman, J. 2002. Bottled Water not Affecting Tooth Decay. UPI Science News, May 30, 2002 [online]. Available: http://www.nrwa.org/2001/frontpage/front%20page%20cells/toothdecay.htm [accessed Sept. 13, 2004].
Groudine, S.B., R.J. Fragen, E.D. Kharasch, T.S. Eisenman, E J. Frink, and S. McConnell. 1999. Comparison of renal function following anesthesia with low-flow sevoflurane and isoflurane. J. Clin. Anesth. 11(3):201-207.
Grucka-Mamczar, E., M. Machoy, R. Tarnawski, E. Birkner, and A. Mamczar. 1997. Influence of long-term sodium fluoride administration on selected parameters of rat blood serum and liver function. Fluoride 30(3):157-164.
Grucka-Mamczar, E., E. Birkner, J. Zalejska-Fiolka, and Z. Machoy. 2005. Disturbances of kidney function in rats with fluoride-induced hyperglycemia after acute poisoning by sodium fluoride. Fluoride 38(1):48-51.
Gruebbel, A.O. 1952. Summarization of the subject. J. Am. Dent. Assoc. 44(2):151-155.

Grumbach, M.M., and D.M. Styne. 2002. Puberty: Ontogeny, neuroendocrinology, physiology, and disorders. Pp. 1115-1286 in Williams Textbook of Endocrinology, 10th Ed., P.R. Larsen, H.M. Kronenberg, S. Melmed, and K.S. Polonsky, eds. Philadelphia, PA: Saunders.

Guan, Z.Z., Z.J. Zhuang, P.S. Yang, and S. Pan. 1988. Synergistic action of iodine-deficiency and fluorine-intoxication on rat thyroid. Chin. Med. J. 101(9):679-684.

Guan, Z.Z., P.S. Yang, N.D. Yu, and Z.J. Zhuang. 1989. An experimental study of blood biochemical diagnostic indices for chronic fluorosis. Fluoride 22(3):112-118.

Guan, Z.Z., Y.N. Wang, K.Q. Xiao, D.Y. Dai, Y.H, Chen, J.L. Liu, P. Sindelar, and G. Dallner. 1998. Influence of chronic fluorosis on membrane lipids in rat brain. Neurotoxicol. Teratol. 20(5):537-542.

Guan, Z.Z., X. Zhang, K. Blennow, and A. Nordberg. 1999. Decreased protein level of nicotinic receptor alpha7 subunit in the frontal cortex from schizophrenic brain. NeuroReport 10(8):1779-1782.

Guan, Z.Z., K.Q. Xiao, X.Y. Zeng, Y.G. Long, Y.H. Cheng, S.F. Jiang, and Y.N. Wang. 2000. Changed cellular membrane lipid composition and lipid peroxidation of kidney in rats with chronic fluorosis. Arch. Toxicol. 74(10):602-608.

Guna Sherlin, D.M., and R.J. Verma. 2001. Vitamin D ameliorates fluoride-induced embryotoxicity in pregnant rats. Neurotoxicol. Teratol. 23(2):197-201.

Gupta, I.P., T.K. Das, A.K. Susheela, S. Dasarathy, and R.K. Tandon. 1992. Fluoride as a possible aetiological factor in non-ulcer dyspepsia. J. Gastroenterol. Hepatol. 7(4):355-359.

Gupta, S., A.K. Seth, A. Gupta, and A.G. Gavane. 1993. Transplacental passage of fluorides. J. Pediatr. 123(1):139-141.

Gupta, S.K., R.C. Gupta, and A.K. Seth. 1994. Increased incidence of spina bifida occulta in fluorosis prone areas. Indian Pediatr. 31(11):1431-1432.

Gupta, S.K., R.C. Gupta, A.K. Seth, and C.S. Chaturvedi. 1995. Increased incidence of spina bifida occulta in fluorosis prone areas. Acta Paediatr. Jpn. 37(4):503-506.

Gupta, S.K., T.I. Khan, R.C. Gupta, A.B. Gupta, K.C. Gupta, P. Jain, and A. Gupta. 2001. Compensatory hyperparathyroidism following high fluoride ingestion—a clinicobiochemical correlation. Indian Pediatr. 38(2):139-146.

Gutteridge, D.H., R.I. Price, G.N. Kent, R.L. Prince, and P.A. Michell. 1990. Spontaneous hip fractures in fluoride-treated patients: Potential causative factors. J. Bone Miner. Res. 5 (Suppl. 1):S205-S215.

Hac, E., W. Czarnowski, T. Gos, and J. Krechniak. 1997. Lead and fluoride content in human bone and hair in the Gdansk region. Sci. Total Environ. 206(2-3):249-254.

Haftenberger, M., G. Viergutz, V. Neumeister, and G. Hetzer. 2001. Total fluoride intake and urinary excretion in German children aged 3-6 years. Caries Res. 35(6):451-457.

Haguenauer, D., V. Welch, B. Shea, P. Tugwell, J.D. Adachi, and G. Wells. 2000. Fluoride for the treatment of postmenopausal osteoporotic fractures: A meta-analysis. Osteoporosis Int. 11(9):727-738.

Haimanot, R.T., A. Fekadu, and B. Bushra. 1987. Endemic fluorosis in the Ethiopian Rift Valley. Trop. Geogr. Med. 39(3):209-217.

Hanhijärvi, H. 1974. The effect of renal diseases on the free ionized plasma fluoride concentrations in patients from anb artificially fluoridated and non-fluoridated drinking water community. Proc. Finn. Dent. Soc. 70(Suppl. 1-3):35-43.

Hanhijärvi, H. 1982. The effect of renal impairment of fluoride retention of patients hospitalized in a low-fluoride community. Proc. Finn. Dent. Soc. 78(1):13-19.

Hanhijärvi, H., and I. Penttilä. 1981. The relationship between human ionic plasma fluoride and serum creatinine concentrations in cases of renal and cardiac insufficiency in a fluoridated community. Proc. Finn. Dent. Soc. 77(6):330-335.

Hanhijärvi, H., V.M. Anttonen, A. Pekkarinen, and A. Penttila. 1972. The effects of artificially

fluoridated drinking water on the plasma of ionized fluoride content in certain clinical disease and in normal individuals. Acta Pharmacol. Toxicol. 31(1):104-110.
Hanhijärvi, H., I. Penttilä, and A. Pekkarinen. 1981. Human ionic plasma fluoride concentrations and age in a fluoridated community. Proc. Finn. Dent. Soc. 77(4):211-221.
Hansson, T., and B. Roos. 1987. The effect of fluoride and calcium on spinal bone mineral content: A controlled prospective (3year) study. Calcif. Tissue Int. 40(6):315-317.
Hara, K. 1980. Studies on fluorosis, especially effects of fluoride on thyroid metabolism [in Japanese]. Koku Eisei Gakkai Zasshi 30(1):42-57.
Hara, T., M. Fukusaki, T. Nakamura, and K. Sumikawa. 1998. Renal function in patients during and after hypotensive anesthesia with sevoflurane. J. Clin. Anesth. 10(7):539-545.
Harbrow, D.J., M.G. Robinson, and P.A. Monsour. 1992. The effect of chronic fluoride administration on rat condylar cartilage. Aust. Dent. J. 37(1):55-62.
Hardin, J.A., M.H. Kimm, M. Wirasinghe, and D.G. Gall. 1999. Macromolecular transport across the rabbit proximal and distal colon. Gut 44(2):218-225.
Harris, N.O., and R.L. Hayes. 1955. A tracer study of the effect of acute and chronic exposure to sodium fluoride on the thyroid iodine metabolism of rats. J. Dent. Res. 34(4):470-477.
Harrison, J.E., K.G. McNeill, W.C. Sturtridge, T.A. Bayley, T.M. Murray, C. Williams, C. Tam, and V. Fornasier. 1981. Three year changes in bone mineral mass of osteoporotic patients based on neutron activation analysis of the central third of the skeleton. J. Clin. Endocrinol. Metab. 52(4):751-758.
Hartfield, P.J., and J.M. Robinson. 1990. Fluoride-mediated activation of the respiratory burst in electropermeabilized neutrophils. Biochim. Biophys. Acta. 1054(2):176-180.
Hase, K., K. Meguro, and T. Nakamura. 2000. Effects of sevoflurane anesthesia combined with epidural block on renal function in the elderly: Comparison with isoflurane. J. Anesth. 14(2):53-60.
Haseman, J.K., J.E. Huff, G.N. Rao, J.E. Arnold, G.A. Boorman, and E.E. McConnell. 1985. Neoplasms observed in untreated and corn oil gavage control groups of F344/N rats and (C57BL/6N X C_3H/HeN)F_1 (B6C3F_1) mice. J. Natl. Cancer Inst. 75(5):975-984.
Haseman, J.K., J.R. Hailey, and R.W. Morris. 1998. Spontaneous neoplasm incidences in Fischer 344 rats and $B_6C_3F_1$ mice in two-year carcinogenicity studies: A National Toxicology Program update. Toxicol. Pathol. 26(3):428-441.
Hasling, C., H.E. Nielsen, F. Melsen, and L. Mosekilde. 1987. Safety of osteoporosis treatment with sodium fluoride, calcium phosphate and vitamin D. Miner. Electrolyte Metab. 13(2):96-103.
Hassold, T., and S. Sherman. 2000. Down syndrome: Genetic recombination and the origin of the extra chromosome 21. Clin. Genet. 57(2):95-100.
Hassold, T., S. Sherman, and P. Hunt. 2000. Counting cross-overs: Characterizing meiotic recombination in mammals. Hum. Mol. Genet. 9(16):2409-2419.
Hattner, R., B.N. Epker, and H.M. Frost. 1965. Suggested sequential mode of control of changes in cell behaviour in adult bone remodeling. Nature 206(983):489-490.
Hawley, G.M., R.P. Ellwood, and R.M. Davies. 1996. Dental caries, fluorosis and the cosmetic implications of different TF scores in 14-year-old adolescents. Community Dent. Health 13(4):189-192.
He, H., V. Ganapathy, C.M. Isales, and G.M. Whitford. 1998. pH-dependent fluoride transport in intestinal brush border membrane vesicles. Biochim. Biophys. Acta. 1372(2):244-254.
Heath, K., V. Singh, R. Logan, and J. McIntyre. 2001. Analysis of fluoride levels retained intraorally or ingested following routine clinical applications of topical fluoride products. Aust. Dent. J. 46(1):24-31.

Hefti, A., and T.M. Marthaler. 1981. Bone fluoride concentrations after 16 years of drinking water fluoridation. Caries Res. 15(1):85-89.

Heifetz, S.B., W.S. Driscoll, H.S. Horowitz, and A. Kingman. 1988. Prevalence of dental caries and dental fluorosis in areas with optimal and above-optimal water-fluoride concentrations: A 5-year follow-up survey. J. Am. Dent. Assoc. 116(4):490-495.

Heilman, J.R., M.C. Kiritsy, S.M. Levy, and J.S. Wefel. 1997. Fluoride concentrations of infant foods. J. Am. Dent. Assoc. 128(7):857-863.

Heilman, J.R., M.C. Kiritsy, S.M. Levy, and J.S. Wefel. 1999. Assessing fluoride levels of carbonated soft drinks. J. Am. Dent. Assoc. 130(11):1593-1599.

Hein, J.W., F.A. Smith, and F. Brudevold. 1954. Distribution of 1 ppm fluoride as radioactively tagged NaF in soft tissues of adult female albino rats. J. Dent. Res. 33(Oct.):709-710.

Hein, J.W., Bonner, J.F., F. Brudevold, F.A. Smith, and H.C. Hodge. 1956. Distribution in the soft tissue of the rat of radioactive fluoride administered as sodium fluoride. Nature 178(4545):1295-1296.

Heindel, J.J., H.K. Bates, C.J. Price, M.C. Marr, C.B. Myers, and B.A. Schwetz. 1996. Developmental toxicity evaluation of sodium fluoride administered to rats and rabbits in drinking water. Fundam. Appl. Toxicol. 30(2):162-177.

Helfand, M. 2004. Screening for subclinical thyroid dysfunction in nonpregnant adults: A summary of the evidence for the U.S. Preventive Services Task Force. Ann. Intern. Med. 140(2):128-141.

Heller, K.E., S.A. Eklund, and B.A. Burt. 1997. Dental caries and dental fluorosis at varying water fluoride concentrations. J. Public Health Dent. 57(3):136-143.

Heller, K.E., W. Sohn, B.A. Burt, and S.A. Eklund. 1999. Water consumption in the United States in 1994-96 and implications for water fluoridation policy. J. Public Health Dent. 59(1):3-11.

Heller, K.E., W. Sohn, B.A. Burt, and R.J. Feigal. 2000. Water consumption and nursing characteristics of infants by race and ethnicity. J. Public Health Dent. 60(3):140-146.

Higuchi, H., S. Arimura, H. Sumikura, T. Satoh, and M. Kanno. 1994. Urine concentrating ability after prolonged sevoflurane anaesthesia. Br. J. Anaesth. 73(2):239-240.

Higuchi, H., H. Sumikura, S. Sumita, S. Arimura, F. Takamatsu, M. Kanno, and T. Satoh. 1995. Renal function in patients with high serum fluoride concentrations after prolonged sevoflurane anesthesia. Anesthesiology 83(3):449-458.

Hileman, B. 1988. Fluoridation of water: Questions about health risks and benefits remain after more than 40 years. Chem. Eng. News (August 1):26-42 [online]. Available: http://www.fluoridealert.org/hileman.htm [accessed Sept. 9, 2004].

Hill, A.B. 1965. The environment and disease: Association or causation? Proc. R. Soc. Med. 58(May):295-300.

Hillier, S., C. Cooper, S. Kellingray, G. Russell, H. Hughes, and D. Coggon. 2000. Fluoride in drinking water and risk of hip fracture in the UK: A case-control study. Lancet 355(9200):265-269.

Hillman, D., D.L. Bolenbaugh, and E.M. Convey. 1979. Hypothyroidism and anemia related to fluoride in dairy cattle. J. Dairy Sci. 62(3):416-423.

Hinkle, A.J. 1989. Serum inorganic fluoride levels after enflurane in children. Anesth. Analg. 68(3):396-399.

Hinrichs, E.H., Jr. 1966. Dental changes in juvenile hypothyroidism. J. Dent. Child. 33(3):167-173.

Hirano, S., M. Ando, and S. Kanno. 1999. Inflammatory responses of rat alveolar macrophages following exposure to fluoride. Arch. Toxicol. 73(6):310-315.

Hirano, T., D.B. Burr, C.H. Turner, M. Sato, R.L. Cain, and J.M. Hock. 1999. Anabolic effects of human biosynthetic parathyroid hormone fragment (1-34), LY333334, on remodeling and mechanical properties of cortical bone in rabbits. J. Bone Miner. Res. 14(4):536-545.

Hirayama, T., K. Niho, O. Fujino, and M. Murakami. 2003. The longitudinal course of two cases with cretinism diagnosed after adolescence. J. Nippon Med. Sch. 70(2):175-178.

Hirschauer, S.C. 2004. Too much fluoride; Parts of state don't meet drinking water standards. The Daily Press. October 10, 2004 [online]. Available: http://www.fluoridealert.org/news/2066.html [accessed Oct. 12, 2005].

Hodge, H.C., and F.A. Smith. 1965. Fluorine Chemistry, Vol. 4, J.H. Simons, ed. New York: Academic Press.

Hodsman, A.B., and D.J. Drost. 1989. The response of vertebral bone mineral density during the treatment of osteoporosis with sodium fluoride. J. Clin. Endocrinol. Metab. 69(5):932-938.

Hoffman, R., J. Mann, J. Calderone, J. Trumbull, and M. Burkhart. 1980. Acute fluoride poisoning in a New Mexico elementary school. Pediatrics 65(5):897-900.

Hong, L., S.M. Levy, J.J. Warren, G.R. Bergus, D.V. Dawson, J.S. Wefel, and B. Broffitt. 2004. Primary tooth fluorosis and amoxicillin use during infancy. J. Public Health Dent. 64(1):38-44.

Honkanen, K., R. Honkanen, L. Heikkinen, H. Kröger, and D. Saarikoski. 1999. Validity of self-reports of fractures in perimenopausal women. Am. J. Epidemiol. 150(5):511-516.

Hoover, R.N., S.S. Devesa, K.P. Cantor, J.H. Lubin, and J.F. Fraumeni. 1991. Fluoridation of Drinking Water and Subsequent Cancer Incidence and Mortality. Appendix E in Review of Fluoride Benefits and Risks: Report of the Ad Hoc Subcommittee on Fluoride Committee of the Committee to Coordinate Environmental Health and Related Programs. Public Health Service, U.S. Department of Health and Human Services, Washington, DC.

Horowitz, H.D., S.B. Heifetz, and W.S. Driscoll. 1972. Partial defluoridation of a community water supply and dental fluorosis. Health Serv. Rep. 87(5):451-455.

Horowitz, H.S. 1996. The effectiveness of community water fluoridation in the United States. J. Public Health Dent. 56(5 Spec. No.):253-258.

Horowitz, H.S., W.S. Driscoll, R.J. Meyers, S.B. Heifetz, and A. Kingman. 1984. A new method for assessing the prevalence of dental fluorosis: The Tooth Surface index of Fluorosis. J. Am. Dent. Assoc. 109(1):37-41.

Horswill, C.A. 1998. Effective fluid replacement. Int. J. Sport Nutr. 8(2):175-195.

Hossny, E., S. Reda, S. Marzouk, D. Diab, and H. Fahmy. 2003. Serum fluoride levels in a group of Egyptian infants and children from Cairo city. Arch. Environ. Health 58(5):306-315.

Hotchkiss, C.E., R. Brommage, M. Du, and C.P. Jerome. 1998. The anesthetic isoflurane decreases ionized calcium and increases parathyroid hormone and osteocalcin in cynomolgus monkeys. Bone 23(5):479-484.

Hrudey, S.E., C.L. Soskolne, J. Berkel, and S. Fincham. 1990. Drinking water fluoridation and osteosarcoma. Can. J. Public Health 81(6):415-416.

Huang, C.C. 1987. Bone resorption in experimental otosclerosis in rats. Am. J. Otolaryngol. 8(5):332-341.

Huang, Z., K. Li, G. Hou, Z. Shen, C. Wang, K. Jiang, and X. Luo. 2002. Study on the correlation of the biochemical indexes in fluoride workers [in Chinese]. Zhonghua Lao Dong Wei Sheng Zhi Ye Bing Za Zhi 20(3):192-194.

Hudak, P.F. 1999. Fluoride levels in Texas groundwater. J. Environ. Sci. Health Part A 34(8):1659-1676.

Hull, W.E., R.E. Port, R. Herrmann, B. Britsch, and W. Kunz. 1988. Metabolites of 5-fluorouracil in plasma and urine, as monitored by 19F nuclear magnetic resonance spectroscopy, for patients receiving chemotherapy with or without methotrexate pretreatment. Cancer Res. 48(6):1680-1688.

Hung, J., and R. Anderson. 1997. p53: Functions, mutations and sarcomas. Acta Orthop. Scand. 273(Suppl.):68-73.
Huraib, S., M.Z. Souqqiyeh, S. Aswad, and A.R. al-Swailem. 1993. Pattern of renal osteodystrophy in haemodialysis patients in Saudi Arabia. Nephrol. Dial Transplant. 8(7):603-608.
Husdan, H., R. Vogl, D. Oreopoulos, C. Gryfe, and A. Rapoport. 1976. Serum ionic fluoride: Normal range and relationship to age and sex. Clin. Chem. 22(11):1884-1888.
Imbeni, V., J.J. Kruzic, G.W. Marshall, S.J. Marshall, and R.O. Ritchie. 2005. The dentin-enamel junction and the fracture of human teeth. Nat. Mater. 4(3):229-232.
Inkielewicz, I., and J. Krechniak. 2003. Fluoride content in soft tissues and urine of rats exposed to sodium fluoride in drinking water. Fluoride 36(4):263-266.
IOM (Institute of Medicine). 1997. Dietary Reference Intakes: For Calcium, Phosphorus, Magnesium, Vitamin D, and Fluoride. Washington, DC: National Academy Press.
IOM (Institute of Medicine). 2004. Dietary Reference Intakes for Water, Potassium, Sodium, Chloride, and Sulfate. Washington, DC: The National Academies Press.
Ionescu, O., E. Sonnet, N. Roudaut, F. Prédine-Hug, and V. Kerlan. 2004. Oral manifestations of endocrine dysfunction [in French]. Ann. Endocrinol. (Paris) 65(5):459-465.
Irie, T., T. Aizawa, and S. Kokubun. 2005. The role of sex hormones in the kinetics of chondrocytes in the growth plate. A study in the rabbit. J. Bone Joint Surg. Br. 87(9): 1278-1284.
Isaacson, R.L., J.M. Fahey, and F.A. Mughairbi. 2003. Environmental conditions unexpectedly affect the long-term extent of cell death following a hypoxic episode. Ann. NY Acad. Sci. 993(May):179-194.
Ishii, T., and G. Suckling. 1991. The severity of dental fluorosis in children exposed to water with a high fluoride content for various periods of time. J. Dent. Res. 70(6):952-956.
Ismail, A.I., J. Shoveller, D. Langille, W.A. MacInnis, and M. McNally. 1993. Should the drinking water of Truro, Nova Scotia, be fluoridated? Water fluoridation in the 1990s. Community Dent. Oral. Epidemiol. 21(3):118-125.
Jackson, D., and S.M. Weidmann. 1958. Fluorine in human bone related to age and the water supply of different regions. J. Pathol. Bacteriol. 76(2):451-459.
Jackson, P.J., P.W. Harvey, and W.F. Young. 2002. Chemistry and Bioavailability Aspects of Fluoride in Drinking Water. Report No. CO 5037. WRc-NSF Ltd., Marlow, Bucks.
Jackson, R., S. Kelly, T. Noblitt, W. Zhang, A. Dunipace, Y. Li, G. Stookey, B. Katz, E. Brizendine, S. Farley, and D. Baylink. 1994. The effect of fluoride therapy on blood chemistry parameters in osteoporotic females. Bone Miner. 27(1):13-23.
Jackson, R.D., S.A. Kelly, B.P. Katz, J.R. Hull, and G.K. Stookey. 1995. Dental fluorosis and caries prevalence in children residing in communities with different levels of fluoride in the water. J. Public Health Dent. 55(2):79-84.
Jackson, R.D., S.A. Kelly, T.W. Noblitt, W. Zhang, M.E. Wilson, A.J. Dunipace, Y. Li, B.P. Katz, E.J. Brizendine, and G.K. Stookey. 1997. Lack of effect of long-term fluoride ingestion on blood chemistry and frequency of sister chromatid exchange in human lymphocytes. Environ. Mol. Mutagen. 29(3):265-271.
Jackson, R.D., S.A. Kelly, B. Katz, E. Brizendine, and G.K. Stookey. 1999. Dental fluorosis in children residing in communities with different water fluoride levels: 33-month follow-up. Pediatr. Dent. 21(4):248-254.
Jackson, R.D., E.J. Brizendine, S.A. Kelly, R. Hinesley, G.K. Stookey, and A.J. Dunipace. 2002. The fluoride content of foods and beverages from negligibly and optimally fluoridated communities. Community Dent. Oral Epidemiol. 30(5):382-391.
Jacobsen, S.J., J. Goldberg, T.P. Miles, J.A. Brody, W. Stiers, and A.A. Rimm. 1990. Regional variation in the incidence of hip fracture. U.S. white women aged 65 years and older. JAMA 264(4):500-502.

Jacobsen, S.J., J. Goldberg, C. Cooper, and S.A. Lockwood. 1992. The association between water fluoridation and hip fracture among white women and men aged 65 years and older: A national ecologic study. Ann. Epidemiol. 2(5):617-626.

Jacobsen, S.J., W.M. O'Fallon, and L.J. Melton, III. 1993. Hip fracture incidence before and after the fluoridation of public water supply, Rochester, Minnesota. Am. J. Public Health 83(5):743-745.

Jacqmin, H., D. Commenges, L. Letenneur, P. Barberger-Gateau, and J.F. Dartigues. 1994. Components of drinking water and risk of cognitive impairment in the elderly. Am. J. Epidcmiol. 139(1).48-57.

Jacqmin-Gadda, H., D. Commenges, and J.F. Dartigues. 1995. Fluorine concentrations in drinking water and fractures in the elderly [letter]. JAMA 273(10):775-776.

Jacqmin-Gadda, H., A. Fourrier, D. Commenges, and J.F. Dartigues. 1998. Risk factors for fractures in the elderly. Epidemiology 9(4):417-423.

Jagiello, G., and J.S. Lin. 1974. Sodium fluoride as potential mutagen in mammalian eggs. Arch. Environ. Health 29(4):230-235.

Jain, S.K., and A.K. Susheela. 1987. Effect of sodium fluoride on antibody formation in rabbits. Environ. Res. 44(1):117-125.

Jenkins, G.N. 1991. Fluoride intake and its safety among heavy tea drinkers in a British fluoridated city. Proc. Finn. Dent. Soc. 87(4):571-579.

Jenny, J., and J.M. Proshek. 1986. Visibility and prestige of occupations and the importance of dental appearance. J. Can. Dent. Assoc. 52(12):987-989.

Jeschke, M., G.J. Standke, and M. Scaronuscarona. 1998. Fluoroaluminate induces activation and association of Src and Pyk2 tyrosine kinases in osteoblastic $MC_3T_3-E_1$ cells. J. Biol. Chem. 273(18):11354-11361.

Jobson, M.D., S.E. Grimm, III, K. Banks, and G. Henley. 2000. The effects of water filtration systems on fluoride: Washington, D.C. metropolitan area. J. Dent. Child. 67(5):350-354.

Johnson, K.A., B.L. Holman, S.P. Mueller, T.J. Rosen, R. English, J.S. Nagel, and J.H. Growdon. 1988. Single photon emission computed tomography in Alzheimer's disease. Abnormal iofetamine I 123 uptake reflects dementia severity. Arch. Neurol. 45(4):392-396.

Johnson, S.A., and C. DeBiase. 2003. Concentration levels of fluoride in bottled drinking water. J. Dent. Hyg. 77(3):161-167.

Johnson, W.J., D.R. Taves, and J. Jowsey. 1979. Fluoridation and bone disease. Pp. 275-293 in Continuing Evaluation of the Use of Fluorides. E. Johansen, D.R. Taves, and T.O. Olsen, eds. AAAS Selected Symposium. Boulder, CO: Westview Press.

Jonderko, G., K. Kita, J. Pietrzak, B. Primus-Slowinska, B. Ruranska, M. Zylka-Wloszczyk, and J. Straszecka. 1983. Effect of subchronic poisoning with sodium fluoride on the thyroid gland of rabbits with normal and increased supply of iodine [in Polish]. Endokrynol. Pol. 34(3):195-203.

Jones, K.F., and J.H. Berg. 1992. Fluoride supplementation. A survey of pediatricians and pediatric dentists. Am. J. Dis. Child. 146(12):1488-1491.

Jooste, P.L., M.J. Weight, J.A. Kriek, and A.J. Louw. 1999. Endemic goitre in the absence of iodine deficiency in schoolchildren of the Northern Cape Province of South Africa. Eur. J. Clin. Nutr. 53(1):8-12.

Jope, R.S. 1988. Modulation of phosphoinositide hydrolysis by NaF and aluminum in rat cortical slices. J. Neurochem. 51(6):1731-1736.

Joseph, S., and P.K. Gadhia. 2000. Sister chromatid exchange frequency and chromosome aberrations in residents of fluoride endemic regions of south Gujarat. Fluoride 33(4): 154-158.

Jubb, T.F., T.E. Annand, D.C. Main, and G.M. Murphy. 1993. Phosphorus supplements and fluorosis in cattle—a northern Australian experience. Aust. Vet. J. 70(10):379-383.

Juncos, L.I., and J.V. Donadio Jr. 1972. Renal failure and fluorosis. JAMA 222(7):783-785.

Juuti, M., and O.P. Heinonen. 1980. Incidence of urolithiasis and composition of household water in southern Finland. Scand. J. Urol. Nephrol. 14(2):181-187.

Kameyama, Y., S. Nakane, H. Maeda, T. Saito, S. Konishi, and N. Ito. 1994. Effect of fluoride on root resorption caused by mechanical injuries of the periodontal soft tissues in rats. Endod. Dent. Traumatol. 10(5):210-214.

Kapoor, V., T. Prasad, and K.C. Bhatia. 1993. Effect of dietary fluorine on histopathological changes in calves. Fluoride 26(2):105-110.

Karagas, M.R., J.A. Baron, J.A. Barrett, and S.J. Jacobsen. 1996. Patterns of fracture among the United States elderly: Geographic and fluoride effects. Ann. Epidemiol. 6(3):209-216.

Kassem, M., L. Mosekilde, and E.F. Eriksen. 1994. Effects of fluoride on human bone cells in vitro: Differences in responsiveness between stromal osteoblast precursors and mature osteoblasts. Eur. J. Endocrinol. 130(4):381-386.

Kato, S., H. Nakagaki, Y. Toyama, T. Kanayama, M. Arai, A. Togari, S. Matsumoto, M. Strong, and C. Robinson. 1997. Fluoride profiles in the cementum and root dentine of human permanent anterior teeth extracted from adult residents in a naturally fluoridated and a non-fluoridated area. Gerodontology 14(1):1-8.

Kaur, P., A. Tewari, and H.S. Chawla. 1987. Changing trends of dental cries and enamel mottling after change of fluoride content in drinking water in endemic fluoride belt. J. Indian Soc. Pedod. Prev. Dent. 5(1):37-44.

Kawase, T., and A. Suzuki. 1989. Studies on the transmembrane migration of fluoride and its effects on proliferation of L-929 fibroblasts (L cells) in vitro. Arch. Oral Biol. 34(2):103-107.

Kawase, T., A. Oguro, M. Orikasa, and D.M. Burns. 1996. Characteristics of NaF-induced differentiation of HL-60 cells. J. Bone Miner. Res. 11(11):1676-1687.

Kedryna, T., M.B. Stachurska, J. Ignacak, and M. Guminska. 1993. Effect of environmental fluorides on key biochemical processes in humans. Folia Med. Cracov 34(1-4):49-57.

Kekki, M., E. Lampainen, P. Kauranen, F. Hoikka, M. Alhava, and A. Pasternack. 1982. The nonlinear tissue binding characteristics of fluoride kinetics in normal and anephric subjects. Nephron 31(2):129-134.

Kernan, W.J., and P.J. Mullenix. 1991. Stability and reproducibility of time structure in spontaneous behavior of male rats. Pharmacol. Biochem. Behav. 39(3):747-754.

Kernan, W.J., P.J. Mullenix, and D.L. Hopper. 1987. Pattern recognition of rat behavior. Pharmacol. Biochem. Behav. 27(3):559-564.

Kernan, W.J., P.J. Mullenix, R. Kent, D.L. Hopper, and N.A. Cressie. 1988. Analysis of the time distribution and time sequence of behavioral acts. Int. J. Neurosci. 43(1-2):35-51.

Kertesz, P., T. Kerenyi, J. Kulka, and J. Banoczy. 1989. Comparison of the effects of NaF and CaF_2 on rat gastric mucosa. A light-, scanning- and transmission electron microscopic study. Acta Morphol. Hung. 37(1-2):21-28.

Khalil, A.M. 1995. Chromosome aberrations in cultured rat bone marrow cells treated with inorganic fluorides. Mutat. Res. 343(1):67-74.

Khalil, A.M., and A.A. Da'dara. 1994. The genotoxic and cytotoxic activities of inorganic fluoride in cultured rat bone marrow cells. Arch. Environ. Contam. Toxicol. 26(1):60-63.

Kharasch, E.D., and D.C. Hankins. 1996. P450-dependent and nonenzymatic human liver microsomal defluorination of fluoromethyl-2,2-difluoro-1-(trifluoromethyl)vinyl ether (compound A), a sevoflurane degradation product. Drug Metab. Dispos. 24(6):649-654.

Kingman, A. 1994. Current techniques for measuring dental fluorosis: Issues in data analysis. Adv. Dent. Res. 8(1):56-65.

Kinney, J.H., R.K. Nalla, J.A. Pople, T.M. Breunig, and R.O. Ritchie. 2005. Age-related transparent root dentin: Mineral concentration, crystallite size, and mechanical properties. Biomaterials 26(16):3363-3376.

Kiritsy, M.C., S.M. Levy, J.J. Warren, N. Guha-Chowdhury, J.R. Heilman, and T. Marshall.

1996. Assessing fluoride concentrations of juices and juice-flavored drinks. J. Am. Dent. Assoc. 127(7):895-902.
Kishi, K., and T. Ishida. 1993. Clastogenic activity of sodium fluoride in great ape cells. Mutat. Res. 301(3):183-188.
Kleerekoper, M., and R. Balena. 1991. Fluoride and osteoporosis. Ann. Rev. Nut. 11: 309-324.
Kleerekoper, M., and D.B. Mendlovic. 1993. Sodium fluoride therapy of postmenopausal osteoporosis. Endocr. Rev. 14(3):312-323.
Kleerekoper, M., E.L. Peterson, D.A. Nelson, E. Phillips, M.A. Schork, B.C. Tilley, and A.M. Parfitt. 1991. A randomized trial of sodium fluoride as a treatment for postmenopausal osteoporosis. Osteoporosis Int. 1(3):155-161.
Klein, H. 1975. Dental fluorosis associated with hereditary diabetes insipidus. Oral Surg. Oral Med. Oral Pathol. 40(6):736-741.
Klein, R.Z., J.D. Sargent, P.R. Larsen, S.E. Waisbren, J.E. Haddow, and M.L. Mitchell. 2001. Relation of severity of maternal hypothyroidism to cognitive development of offspring. J. Med. Screen. 8(1):18-20.
Knappwost, A., and J. Westendorf. 1974. Inhibition of cholinesterases caused by fluorine complex of silicon and of iron [in German]. Naturwissenschaften 61(6):275.
Kolka, M.A., W.A. Latzka, S.J. Montain, W.P. Corr, K.K. O'Brien, and M.N. Sawka. 2003. Effectiveness of revised fluid replacement guidelines for military training in hot weather. Aviat. Space Environ. Med. 74(3):242-246.
Kragstrup, J., A. Richards, and O. Fejerskov. 1984. Experimental osteo-fluorosis in the domestic pig: A histomorphometric study of vertebral trabecular bone. J. Dent. Res. 63(6):885-889.
Krasowska, A., and T. Włostowski. 1992. The effect of high fluoride intake on tissue trace elements and histology of testicular tubules in the rat. Comp. Biochem. Physiol. C 103(1):31-34.
Krishnamachari, K.A. 1982. Trace elements in serum and bone in endemic Genu valgum: Manifestation of fluorosis. Fluoride 15(1):25-31.
Krishnamachari, K.A. 1986. Skeletal fluorosis in humans: A review of recent progress in the understanding of the disease. Prog. Food Nutr. Sci. 10(3-4):279-314.
Kroger, H., E. Alhava, R. Honkanen, M. Tuppurainen, and S. Saarkoski. 1994. The effect of fluoridated drinking water on axial bone mineral density—a population based study. Bone Miner. 27(1):33-41.
Kudo, N., and Y. Kawashima. 2003. Toxicity and toxicokinetics of perfluorooctanoic acid in humans and animals. J. Toxicol. Sci. 28(2):49-57.
Kumar, A., and A.K. Susheela. 1994. Ultrastructural studies of spermiogenesis in rabbit exposed to chronic fluoride toxicity. Int. J. Fertil. Menopausal. Stud. 39(3):164-171.
Kumar, A., and A.K. Susheela. 1995. Effects of chronic fluoride toxicity on the morphology of ductus epididymis and the maturation of spermatozoa of rabbit. Int. J. Exp. Pathol. 76(1):1-11.
Kumar, J.V., and P.A. Swango. 1999. Fluoride exposure and dental fluorosis in Newburgh and Kingston, New York: Policy implications. Community Dent. Oral. Epidemiol. 27(3):171-180.
Kunz, D., S. Schmitz, R. Mahlberg, A. Mohr, C. Stöter, K.J. Wolf, and W.M. Herrmann. 1999. A new concept for melatonin deficit: On pineal calcification and melatonin excretion. Neuropsychopharmacol. 21(6):765-772.
Kunzel, V.W. 1976. Cross sectional comparison of the median eruption time for permanent teeth in children from fluoride poor and optimally fluoridated areas [in German]. Stomatol. DDR 5:310-321.

Kunzel, W. 1980. Caries and dental fluorosis in high-fluoride districts under sub-tropical conditions. J. Int. Assoc. Dent. Child. 11(1):1-6.

Kuo, H.C., and J.W. Stamm. 1974. Fluoride levels in human rib bone: A preliminary study. Can. J. Public Health. 65(5):359-361.

Kuo, H.C., and J.W. Stamm. 1975. The relationship of creatinine clearance to serum fluoride concentration and urinary fluoride excretion in man. Arch. Oral Biol. 20(4):235-238.

Kurttio, P., N. Gustavsson, T. Vartianinen, and J. Pekkanen. 1999. Exposure to natural fluoride in well water and hip fracture: A cohort analysis in Finland. Am. J. Epidemiol. 150(9):817-824.

Lafage, M.H., R. Balena, M.A. Battle, M. Shea, J.G. Seedor, H. Klein, W.C. Hayes, and G.A. Rodan. 1995. Comparison on alendronate and sodium fluoride effects on cancellous and cortical bone in minipigs. A one-year study. J. Clin. Invest. 95(5):2127-2133.

Laisalmi, M., A. Soikkeli, H. Kokki, H. Markkanen, A. Yli-Hankala, P. Rosenberg, and L. Lindgren. 2003. Fluoride metabolism in smokers and non-smokers following enflurane anaesthesia. Br. J. Anaesth. 91(6):800-804.

Lalumandier, J.A., and L.W. Ayers. 2000. Fluoride and bacterial content of bottled water vs. tap water. Arch. Fam. Med. 9(3):246-250.

Lalumandier, J.A., and R.G. Rozier. 1998. Parents' satisfaction with children's tooth color: Fluorosis as a contributing factor. J. Am. Dent. Assoc. 129(7):1000-1006.

Lamb, N.E., S.B. Freeman, A. Savage-Austin, D. Pettay, L. Taft, J. Hersey, Y. Gu, J. Shen, D. Saker, K.M. May, D. Avramopoulos, M.B., Petersen, A. Hallberg, M. Mikkelsen, T.J. Hassold, and S.L. Sherman. 1996. Susceptible chiasmate configurations of chromosome 21 predispose to non-disjunction in both maternal meiosis I and meiosis II. Nat. Genet. 14(4):400-405.

Lamb, N.E., E. Feingold, A. Savage, D. Avramopoulos, S. Freeman, Y. Gu, A. Hallberg, J. Hersey, G. Karadima, D. Pettay, D. Saker, J. Shen, L. Taft, M. Mikkelsen, M.P. Petersen, T. Hassold, and S.L. Sherman. 1997. Characterization of susceptible chiasma configurations that increase the risk for maternal nondisjunction of chromosome 21. Hum. Mol. Genet. 6(9):1391-1399.

Lantz, O., M.H. Jouvin, M.C. De Vernejoul, and P. Druet. 1987. Fluoride induced chronic renal failure. Am. J. Kidney Dis. 10(2):136-137.

Larsen, M.J., F. Melsen, L. Mosekilde, and M.S. Christensen. 1978. Effects of a single dose of fluoride on calcium metabolism. Calcif. Tissue Res. 26(3):199-202.

Larsen, P.R., and T.F. Davies. 2002. Hypothyroidism and thyroiditis. Pp. 423-455 in Williams Textbook of Endocrinology, 10th Ed., P.R. Larsen, H.M. Kronenberg, S. Melmed, and K.S. Polonsky, eds. Philadelphia, PA: Saunders.

Larsen, P.R., T.F. Davies, M.J. Schlumberger, and I.D. Hay. 2002. Thyroid physiology and diagnostic evaluation of patients with thyroid disorders. Pp. 331-373 in Williams Textbook of Endocrinology, 10th Ed., P.R. Larsen, H.M. Kronenberg, S. Melmed, and K.S. Polonsky, eds. Philadelphia, PA: Saunders.

Lasne, C., Y.P. Lu, and I. Chouroulinkov. 1988. Transforming activities of sodium fluoride in cultured Syrian hamster embryo and BALB/3T3 cells. Cell Biol. Toxicol. 4(3):311-324.

Lau, K.H., and D.J. Baylink. 1998. Molecular mechanism of action of fluoride on bone cells. J. Bone Miner. Res. 13(11):1660-1667.

Lau, K.H., J.R. Farley, and D.J. Baylink. 1985. Phosphotyrosyl-specific protein phosphatase activity of a bovine skeletal acid phosphatase isoenzyme. Comparison with the phosphotyrosyl protein phosphatase activity of skeletal alkaline phosphatase. J. Biol. Chem. 260(8):4653-4660.

Lau, K.H., T.K. Freeman, and D.J. Baylink. 1987. Purification and characterization of an acid phosphatase that displays phosphotyrosyl-protein phosphatase activity from bovine cortical bone matrix. J. Biol. Chem. 262(3):1389-1397.

Lau, K.H., J.R. Farley, T.K. Freeman, and D.J. Baylink. 1989. A proposed mechanism of the mitogenic action of fluoride on bone cells: Inhibition of the activity of an osteoblastic acid phosphatase. Metabolism 38(9):858-868.

Lee, Z.H., and H.H. Kim. 2003. Signal transduction by receptor activator of nuclear factor kappa B in osteoclasts. Biochem. Biophys. Res. Commun. 305(2):211-214.

Lees, S., and D.B. Hanson. 1992. Effect of fluoride dosage on bone density, sonic velocity, and longitudinal modulus of rabbit femurs. Calcif. Tissue Int. 50(1):88-92.

Lehmann, R., M. Wapniarz, B. Hofmann, B. Pieper, I. Haubitz, and B. Allolio. 1998. Drinking water fluoridation: Bone mineral density and hip fracture incidence. Bone 22(3) 273-278.

Lejus, C., O. Delaroche, C. Le Roux, E. Legendre, O. Rivault, H. Floch, M. Renaudin, and M. Pinaud. 2002. Does sevoflurane inhibit serum cholinesterase in children? Anaesthesia 57(1):44-48.

Leone, N.C., M.B. Shimkin, F.A. Arnold, Jr., C.A. Stevenson, E.R. Zimmermann, P.B. Geiser, and J.E. Lieberman. 1954a. Medical aspects of excessive fluoride in a water supply. A ten-year study. Pp. 110-130 in Fluoridation as a Public Health Measure, J.H. Shaw, ed. Washington, DC: American Association for the Advancement of Science.

Leone, N.C., M.B. Shimkin, F.A. Arnold, Jr., C.A. Stevenson, E.R. Zimmermann, P.B. Geiser, and J.E. Lieberman. 1954b. Medical aspects of excessive fluoride in a water supply. Public Health Rep. 69(10):925-936.

Leone, N.C., C.A. Stevenson, T.F. Hilbish, and M.C. Sosman. 1955a. A roentgenologic study of a human population exposed to high fluoride domestic water: A 10 year study. Am. J. Roentgenol. Radium Ther. Nucl. Med. 74(5):874-875.

Leone, N.C., F.A. Arnold, Jr., E.R. Zimmermann, P.B. Geiser, and J.E. Lieberman. 1955b. Review of the Bartlett-Cameron survey: A ten year fluoride study. J. Am. Dent. Assoc. 50(3):277-281.

Leone, N.C., E.C. Leatherwood, I.M. Petrie, and L. Lieberman. 1964. Effect of fluoride on thyroid gland: Clinical study. J. Am. Dent. Assoc. 69(Aug.):179-180.

Leroy, R., K. Bogaerts, E. Lesaffre, and D. Declerck. 2003. The effect of fluorides and caries in primary teeth on permanent tooth emergence. Community Dent. Oral Epidemiol. 31(6):463-470.

Levy, B.M., S. Dreizen, S. Bernick, and J.K. Hampton Jr. 1970. Studies on the biology of the periodontium of marmosets. IX. Effect of parathyroid hormone on the alveolar bone of marmosets pretreated with fluoridated and nonfluoridated drinking water. J. Dent. Res. 49(4):816-821.

Levy, F.M., J.R. de Magalhães Bastos, and M.A.R. Buzalaf. 2004. Nails as biomarkers of fluoride in children of fluoridated communities. J. Dent. Child. 71(2):121-125.

Levi, J.E., and H.E. Silberstein. 1955. Lack of effect of fluorine ingestion on uptake of iodine 131 by the thyroid gland. J. Lab. Clin. Med. 45(3):348-351.

Levy, S.M. 1993. A review of fluoride intake from fluoride dentifrice. J. Dent. Child. 60(2): 115-124.

Levy, S.M. 1994. Review of fluoride exposures and ingestion. Community Dent. Oral Epidemiol. 22(3):173-180.

Levy, S.M. 2003. An update on fluorides and fluorosis. J. Can. Dent. Assoc. 69(5):286-291.

Levy, S.M., and N. Guha-Chowdhury. 1999. Total fluoride intake and implications for dietary fluoride supplementation. J. Public Health Dent. 59(4):211-223.

Levy, S.M., and G. Muchow. 1992. Provider compliance with recommended dietary fluoride supplement protocol. Am. J. Public Health 82(2):281-283.

Levy, S.M., and D.A. Shavlik. 1991. The status of water fluoride assay programs and implications for prescribing of dietary fluoride supplements. J. Dent. Child. 58(1):23-26.

Levy, S.M., and Z. Zarei-M. 1991. Evaluation of fluoride exposures in children. J. Dent. Child. 58(6):467-473.

Levy, S.M., T.J. Maurice, and J.R. Jakobsen. 1992. A pilot study of preschoolers' use of regular-flavored dentifrices and those flavored for children. Pediatr. Dent. 14(6):388-391.

Levy, S.M., T.J. Maurice, and J.R. Jakobsen. 1993. Dentifrice use among preschool children. J. Am. Dent. Assoc. 124(9):57-60.

Levy, S.M., M.C. Kiritsy, and J.J. Warren. 1995a. Sources of fluoride intake in children. J. Public Health Dent. 55(1):39-52.

Levy, S.M., F.J. Kohout, N. Guha-Chowdhury, M.C. Kiritsy, J.R. Heilman, and J.S. Wefel. 1995b. Infants' fluoride intake from drinking water alone and from water added to formula, beverages, and food. J. Dent. Res. 74(7):1399-1407.

Levy, S.M., F.J. Kohout, M.C. Kiritsy, J.R. Heilman, and J.S. Wefel. 1995c. Infants' fluoride ingestion from water, supplements and dentifrice. J. Am. Dent. Assoc. 126(12):1625-1632.

Levy, S.M., M.C. Kiritsy, S.L. Slager, J.J. Warren, and F.J. Kohout. 1997. Patterns of fluoride dentifrice use among infants. Pediatr. Dent. 19(1):50-55.

Levy, S.M., J.A. McGrady, P. Bhuridej, J.J. Warren, J.R. Heilman, and J.S. Wefel. 2000. Factors affecting dentifrice use and ingestion among a sample of U.S. preschoolers. Pediatr. Dent. 22(5):389-394.

Levy, S., K. Furst, and W. Chern. 2001a. A pharmacokinetic evaluation of 0.5% and 5% fluorouracil topical cream in patients with actinic keratosis. Clin. Ther. 23(6):908-920.

Levy, S.M., J.J. Warren, C.S. Davis, H.L. Kirchner, M.J. Kanellis, and J.S. Wefel. 2001b. Patterns of fluoride intake from birth to 36 months. J. Public Health Dent. 61(2):70-77.

Levy, S.M., S.L. Hillis, J.J. Warren, B.A. Broffitt, A.K. Mahbubul Islam, J.S. Wefel, and M.J. Kanellis. 2002a. Primary tooth fluorosis and fluoride intake during the first year of life. Community Dent. Oral Epidemiol. 30(4):286-295.

Levy, S.M., J.J. Warren, and J.R. Jakobsen. 2002b. Follow-up study of dental students' esthetic perceptions of mild dental fluorosis. Community Dent. Oral. Epidemiol. 30(1):24-28.

Levy, S.M., B. Broffitt, R. Slayton, J.J. Warren, and M.J. Kanellis. 2003a. Dental visits and professional fluoride applications for children ages 3 to 6 in Iowa. Pediatr. Dent. 25(6):565-571.

Levy, S.M., J.J. Warren, and B. Broffitt. 2003b. Patterns of fluoride intake from 36 to 72 months of age. J. Public Health Dent. 63(4):211-220.

Lewis, D.W., and H. Limeback. 1996. Comparison of recommended and actual mean intakes of fluoride by Canadians. J. Can. Dent. Assoc. 62(9):708-715.

Lewis, H.A., U. M. Chikte, and A. Butchart. 1992. Fluorosis and dental caries in school children from rural areas with about 9 and 1 ppm F in the water supplies. Community Dent. Oral Epidemiol. 20(1):53-54.

Li, G., and L. Ren. 1997. Effects of excess fluoride on bone turnover under conditions of diet with different calcium contents [in Chinese]. Zhonghua Bing Li Xue Za Zhi 26(5):277-280.

Li, L. 2003. The biochemistry and physiology of metallic fluoride: Action, mechanism, and implications. Crit. Rev. Oral Biol. 14(2):100-114.

Li, L.C., Y.S. Zhang, R.Z. Hu, and X.C. Zhou. 1992. Inhibitory effect of fluoride on renal stone formation in rats. Urol. Int. 48(3):336-341.

Li, X.S., J.L. Zhi, and R.O. Gao. 1995. Effect of fluoride exposure on intelligence in children. Fluoride 28(4):189-192.

Li, Y., C.K. Liang, B.P. Katz, E.J. Brizendine, and G.K. Stookey. 1995. Long-term exposure to fluoride in drinking water and sister chromatid exchange frequency in human blood lymphocytes. J. Dent. Res. 74(8):1468-1474.

Li, Y., C. Liang, C.W. Slemenda, R. Ji, S. Sun, J. Cao, C.L. Emsley, F. Ma, Y. Wu, P. Ying, Y. Zhang, S. Gao, W. Zhang, B.P. Katz, S. Niu, S. Cao, and C.C. Johnston, Jr. 2001. Effects

of long-term exposure to fluoride in drinking water on risks of bone fractures. J. Bone Miner. Res. 16(5):932-939.
Likins, R.C., F.J. McClure, and A.C. Steere. 1956. Urinary excretion of fluoride following defluoridation of a water supply. Public Health Rep. 71(3):217-220.
Limeback, H. 1999a. A re-examination of the pre-eruptive and post-eruptive mechanism of the anti-caries effects of fluoride: Is there any anti-caries benefit from swallowing fluoride? Community Dent. Oral Epidemiol. 27(1):62-71.
Limeback, H. 1999b. Appropriate use of fluoride supplements for the prevention of dental caries. Consensus Conference of the Canadian Dental Association, Toronto, Canada, 28-29 November 1997. Introduction. Community Dent. Oral Epidemiol. 27(1):27-30.
Limeback, H., A. Ismail, D. Banting, P. DenBesten, J. Featherstone, and P.J. Riordan. 1998. Canadian Consensus Conference on the appropriate use of fluoride supplements for the prevention of dental caries in children. J. Can. Dent. Assoc. 64(9):636-639.
Lin, F.F., Aihaiti, H.X. Zhao, J. Lin, J.Y. Jiang, Maimaiti, and Aiken. 1991. The relationship of a low-iodine and high-fluoride environment to subclinical cretinism in Xinjiang. IDD Newsletter 7(3):24-25 [online]. Available: http://www.people.virginia.edu/~jtd/iccidd/newsletter/idd891.htm#Relationship [accessed Oct. 5, 2004].
Lindskog, S., M.E. Flores, E. Lilja, and L. Hammarstrom. 1989. Effect of a high dose of fluoride on resorbing osteoclasts in vivo. Scand. J. Dent. Res. 97(6):483-487.
Lindstrom, J. 1997. Nicotinic acetylcholine receptors in health and disease. Mol. Neurobiol. 15(2):193-222.
Liptrot, G.F. 1974. Modern Inorganic Chemistry. London: Mills and Boon, Ltd.
Liu, C., L.E. Wyborny, and J.T. Chan. 1995. Fluoride content of dairy milk from supermarket: A possible contributing factor to dental fluorosis. Fluoride 28(1):10-16.
Liu, C.C., and D.J. Baylink. 1977. Stimulation of bone formation and bone resorption by fluoride in thyroparathyroidectomized rats. J. Dent. Res. 56(3):304-311.
Liu, J.L., T. Xia, Y.Y. Yu, X.Z. Sun, Q. Zhu, W. He, M. Zhang, and A. Wang. 2005. The dose-effect relationship of water fluoride levels and renal damage in children [in Chinese]. Wei Sheng Yan Jiu 34(3):287-288.
Locker, D. 1999. Benefits and Risks of Water Fluoridation. An Update of the 1996 Federal-Provincial Sub-committee Report. Prepared under contract for Public Health Branch, Ontario Ministry of Health, First Nations and Inuit Health Branch, Health, Canada, by David Locker, Community Dental Health Services Research Unit, Faculty of Dentistry, University of Toronto. November 15, 1999 [online]. Available: http://www.health.gov.on.ca/english/public/pub/ministry_reports/fluoridation/fluor.pdf [accessed Oct. 8, 2004].
Loevy, H.T., H. Aduss, and I.M. Rosenthal. 1987. Tooth eruption and craniofacial development in congenital hypothyroidism: Report of case. J. Am. Dent. Assoc. 115(3):429-431.
Loftenius, A., B. Andersson, J. Butler, and J. Ekstrand. 1999. Fluoride augments the mitogenic and antigenic response of human blood lymphocytes in vitro. Caries Res. 33(2):148-155.
Long, Y.G., Y.N. Wang, J. Chen, S.F. Jiang, A. Nordberg, and Z.Z. Guan. 2002. Chronic fluoride toxicity decreases the number of nicotinic acetylcholine receptors in rat brain. Neurotoxicol. Teratol. 24(6):751-757.
Lu, Y., Z.R. Sun, L.N. Wu, X. Wang, W. Lu, and S.S. Liu. 2000. Effect of high-fluoride water on intelligence in children. Fluoride 33(2):74-78.
Luke, J. 2001. Fluoride deposition in the aged human pineal gland. Caries Res. 35(2): 125-128.
Luke, J.A. 1997. The Effect of Fluoride on the Physiology of the Pineal Gland. Ph.D. Thesis, University of Surrey, Guildford. 278 pp.
Lundy, M.W., M. Stauffer, J.E. Wergedal, D.J. Baylink, J.D. Featherstone, S.F. Hodgson, and

B.L. Riggs. 1995. Histomorphometric analysis of iliac crest bone biopsies in placebo-treated versus fluoride-treated subjects. Osteoporos. Int. 5(2):115-129.
Maas, R.P., S.C. Patch, and A.M. Smith. 2005. Effects of Fluorides and Chloramines on Lead Leaching from Leaded-Brass Surfaces. Technical Report 05-142. Environmental Quality Institute, University of North Carolina, Asheville, NC. June 2005.
Macek, M.D., T.D. Matte, T. Sinks, and D.M. Malvitz. 2006. Blood lead concentrations in children and method of water fluoridation in the United States, 1988-1994. Environ. Health Perspect. 114(1):130-134.
Machaliński, B., M. Zejmo, I. Stecewicz, A. Machalinska, Z. Machoy, and M.Z. Ratajczak. 2000. The influence of sodium fluoride on the clonogenecity of human hematopoietic progenitor cells: Preliminary report. Fluoride 33(4):168-173.
Machaliński, B., M. Baskiewicz-Masiuk, B. Sadowska, A. Machalinska, M. Marchlewicz, B. Wiszniewska, and I. Stecewicz. 2003. The influence of sodium fluoride and sodium hexafluorosilicate on human leukemic cell lines: Preliminary report. Fluoride 36(4):231-240.
Madans, J., J.C. Kleinman, and J. Cornoni-Huntley. 1983. The relationship between hip fracture and water fluoridation: An analysis of national data. Am. J. Public Health 73(3):296-298.
Maguire, A., P.J. Moynihan, and V. Zohouri. 2004. Bioavailability of Fluoride in Drinking Water—A Human Experimental Study, June 2004. Prepared for the UK Department of health, by School of Dental Sciences, University of Newcastle [online]. Available: http://www.ncl.ac.uk/dental/assets/docs/fluoride_bioavailability [accessed Sept. 16, 2004].
Mahoney, M.C., P.C. Nasca, W.S. Burnett, and J.M. Melius. 1991. Bone cancer incidence rates in New York State: Time trends and fluoridated drinking water. Am. J. Public Health 81(4):475-479.
Maier, F.J. 1953. Defluoridation of municipal water supplies. J. Am. Water Works Assoc. 45:879-888.
Maier, N.R.F. 1929. Reasoning in White Rats. Comparative Psychology Monographs 6(29). Baltimore: John Hopkins Press.
Maier, N.R.F. 1932. Cortical destruction of the posterior part of the brain and its effect on reasoning in rats. J. Comp. Neurol. 56(1):179-214.
Malhotra, A., A. Tewari, H.S. Chawla, K. Gauba, and K. Dhall. 1993. Placental transfer of fluoride in pregnant women consuming optimum fluoride in drinking water. J. Indian Soc. Pedod. Prev. Dent. 11(1):1-3.
Mamelle, N., P.J. Meunier, R. Dusan, M. Guillaume, J.L Martin, A. Gaucher, A. Prost, G. Zeigler, and P. Netter. 1988. Risk-benefit ratio of sodium fluoride treatment in primary vertebral osteoporosis. Lancet 2(8607):361-365.
Mann, J., M. Tibi, and H.D. Sgan-Cohen. 1987. Fluorosis and caries prevalence in a community drinking above-optimal fluoridated water. Community Dent. Oral Epidemiol. 15(5):293-295.
Mann, J., W. Mahmoud, M. Ernest, H. Sgan-Cohen, N. Shoshan, and I. Gedalia. 1990. Fluorosis and dental caries in 6-8-year-old children in a 5 ppm fluoride area. Community Dent. Oral Epidemiol. 18(2):77-79.
Marie, P.J., and M. Hott. 1986. Short term effects of fluoride and strontium on bone formation and resorption in the mouse. Metabolism 35(6):547-551.
Marier, J.R. 1977. Some current aspects of environmental fluoride. Sci. Total Environ. 8(3):253-265.
Martin, R.B. 1988. Ternary hydroxide complexes in neutral solutions of Al^{3+} and F^-. Biochem. Biophys. Res. Commun. 155(3):1194-1200.
Martínez, O.B., C. Díaz, T.M. Borges, E. Díaz, and J.P. Pérez. 1998. Concentrations of fluoride in wines from the Canary Islands. Food Addit. Contam. 15(8):893-897.

Masters, R.D., and M. Coplan. 1999. Water treatment with silicofluorides and lead toxicity. Int. J. Environ. Sci. 56:435-449.
Masters, R.D., M.J. Coplan, B.T. Hone, and J.E. Dykes. 2000. Association of silicofluoride treated water with elevated blood lead. Neurotoxicology 21(6):1091-1100.
Mathias, R.S., U. Amin, C.H. Mathews, and P. DenBesten. 2000. Increased fluoride content in the femur growth plate and cortical bone of uremic rats. Pediatr. Nephrol. 14(10-11):935-939.
Matsumura, C., O. Kemmotsu, Y. Kawano, K. Takita, H. Sugimoto, and T. Mayumi. 1994. Serum and urine inorganic fluoride levels following prolonged low-dose sevoflurane anesthesia combined with epidural block. J. Clin. Anesth. 6(5):419-424.
Matsuo, S., K. Kiyomiya, and M. Kurebe. 1998. Mechanism of toxic action of fluoride in dental fluorosis: Whether trimeric G proteins participate in the disturbance of intracellular transport of secretory ameloblast exposed to fluoride. Arch. Toxicol. 72(12):798-806.
Matthews, E.J., J.W. Spalding, and R.W. Tennant. 1993. Transformation of BALB/c-3T3 cells: V. Transformation responses of 168 chemicals compared with mutagenicity in Salmonella and carcinogenicity in rodent bioassays. Environ Health Perspect. 101(Suppl. 2):347-482.
Maumené, E. 1854. Experience pour determiner l'action des fluores sur l'economie animale. Compt. Rend. Acad. Sci. Paris 39:538 (as cited in Gedalia and Brand 1963).
Maumené, E. 1866. Recherches expérimentales sur les causes du goitre. Compt. Rend. Acad. Sci. Paris 62:381 (as cited in Murray et al. 1948).
Maurer, J.K., M.C. Cheng, B.G. Boysen, and R.L. Anderson. 1990. Two-year carcinogenicity study of sodium fluoride in rats. J. Natl. Cancer Inst. 82(13):1118-1126.
Maurer, J.K., M.C. Cheng, B.G. Boysen, R.A. Squire, J.D. Strandberg, S.E. Weisbrode, J.L. Seymour, and R.L. Anderson. 1993. Confounded carcinogenisity study of sodium fluoride in CD-1 mice. Regul. Toxicol. Pharmacol. 18(2):154-168.
Mazze, R.I., R.K. Calverley, and N.T. Smith. 1977. Inorganic fluoride nephrotoxicity: Prolonged enflurane and halothane anesthesia in volunteers. Anesthesiology 46(4):265-271.
McCarty, M.F., and C.A. Thomas. 2003. PTH excess may promote weight gain by impeding catecholamine-induced lipolysis: Implications for the impact of calcium, vitamin D, and alcohol on body weight. Med. Hypotheses 61(5-6):535-542.
McClure, F.J. 1943. Ingestion of fluoride and dental caries. Quantitative relations based on food and water requirements of children one to twelve years old. Am. J. Dis. Child. 66:362-369 (as cited in IOM 1997).
McClure, F.J. 1970. Water Fluoridation, the Search and the Victory. Bethesda, MD: U.S. Department of Health, Education, and Welfare, National Institutes of Health, National Institute of Dental Research.
McClure, F.J., H.H. Mitchell, T.S. Hamilton, and C.A. Kinser. 1945. Balances of fluorine ingested from various sources in food and water by five young men. Excretion of fluorine through the skin. J. Ind. Hyg. Toxicol. 27:159-170 (as cited in Singer et al. 1985).
McDonagh, M., P. Whiting, M. Bradley, J. Cooper, A. Sutton, I. Chestnutt, K. Misso, P. Wilson, E. Treasure, and J. Kleijnen. 2000a. A Systematic Review of Public Water Fluoridation. NHS Centre for Reviews and Dissemination, University of York, York, UK [online]. Available: http://www.york.ac.uk/inst/crd/fluorid.pdf [accessed Sept. 28, 2004].
McDonagh, M.S., P.F. Whiting, P.M. Wilson, A.J. Sutton, I. Chestnutt, J. Cooper, K. Misso, M. Bradley, E. Treasure, and J. Kleijnen. 2000b. Systematic review of water fluoridation. Br. Med. J. 321(7265):855-859.
McDonnell, S.T., D. O'Mullane, M. Cronin, C. MacCormac, and J. Kirk. 2004. Relevant factors when considering fingernail clippings as a fluoride biomarker. Community Dent. Health 21(1):19-24.
McGuire, S.M., E.D. Vanable, M.H. McGuire, J.A. Buckwalter, and C.W. Douglass. 1991.

Is there a link between fluoridated water and osteosarcoma? J. Am. Dent. Assoc. 122(4):38-45.
McInnes, P.M., B.D. Richardson, and P.E. Cleaton-Jones. 1982. Comparison of dental fluorosis and caries in primary teeth of preschool-children living in arid high and low fluoride villages. Community Dent. Oral Epidemiol. 10(4):182-186.
McKay, F.S. 1933. Mottled enamel: The prevention of its further production through a change of the water supply at Oakley, Ida. J. Am. Dental Assoc. 20(7):1137-1149.
McLaren, J.R. 1976. Possible effects of fluorides on the thyroid. Fluoride 9(2):105-116.
McNall, P.E., and J.C. Schlegel. 1968. Practical thermal environmental limits for young adult males working in hot, humid environments. ASHRAE Trans. 74:225-235 (as cited in EPA 1997).
Mehta, M.N., K. Raghavan, V.P. Gharpure, and R. Shenoy. 1998. Fluorosis: A rare complication of diabetes insipidus. Indian Pediatr. 35(5):463-467.
Mella, S.O., X.M. Molina, and E.S. Atalah. 1994. Prevalence of dental fluorosis and its relation with fluoride content of communal drinking water [in Spanish]. Rev. Med. Chile 122(11):1263-1270.
Meng, Z., and B. Zhang. 1997. Chromosomal aberrations and micronuclei in lymphocytes of workers at a phosphate fertilizer factory. Mutat. Res. 393(3):283-288.
Meng, Z., H. Meng, and X. Cao. 1995. Sister-chromatid exchanges in lymphocytes of workers at a phosphate fertilizer factory. Mutat. Res. 334(2):243-246.
Menon, A., and K.R. Indushekar. 1999. Prevalence of dental caries and co-relation with fluorosis in low and high fluoride areas. J. Indian Soc. Pedod. Prev. Dent. 17(1):15-20.
Mernagh, J.R., J.E. Harrison, R. Hancock, and K.G. McNeill. 1977. Measurement of fluoride in bone. Int. J. Appl. Radiat. Isot. 28(6):581-583.
Messer, H.H., W.D. Armstrong, and L. Singer. 1973a. Fluoride, parathyroid hormone and calcitonin: Inter-relationships in bone calcium metabolism. Calcif. Tissue Res. 13(3):217-224.
Messer, H.H., W.D. Armstrong, and L. Singer. 1973b. Fluoride, parathyroid hormone and calcitonin: Effects on metabolic processes involved in bone resorption. Calcif. Tissue Res. 13(3):227-233.
Mg'ang'a, P.M., and M.L. Chindia. 1990. Dental and skeletal changes in juvenile hypothyroidism following treatment: Case report. Odontostomatol. Trop. 13(1):25-27.
Michael, M., V.V. Barot, and N.J. Chinoy. 1996. Investigations of soft tissue functions in fluorotic individuals of north Gujarat. Fluoride 29(2):63-71.
Mihashi, M., and T. Tsutsui. 1996. Clastogenic activity of sodium fluoride to rat vertebral body-derived cells in culture. Mutat. Res. 368(1):7-13.
Mikhailets, N.D., M.I. Balabolkin, V.A. Rakitin, and I.P. Danilov. 1996. Thyroid function during prolonged exposure to fluorides [in Russian]. Probl. Endocrinol. 42(1):6-9.
Millan-Plano, S., J.J. Garcia, E. Martinez-Ballarin, R.J. Reiter, S. Ortega-Gutierrez, R.M. Lazaro, and J.F. Escanero. 2003. Melatonin and pinoline prevent aluminium-induced lipid peroxidation in rat synaptosomes. J. Trace Elem. Med. Biol. 17(1):39-44.
Miller-Ihli, N.J., P.R. Pehrsson, R.L. Cutrifelli, and J.M. Holden. 2003. Fluoride content of municipal water in the United States: What percentage is fluoridated? J. Food Compos. Anal. 16(5):621-628.
Mitchell, E.A., J.M. Thompson, and B. Borman. 1991. No association between fluoridation of water supplies and sudden infant death syndrome. N.Z. Med. J. 104(924):500-501.
Miu, A.C., C.E. Andreescu, R. Vasiu, and A.L. Olteanu. 2003. A behavioral and histological study of the effects of long-term exposure of adult rats to aluminum. Int. J. Neurosci. 113(9):1197-1211.
Mohr, H. 1990. Fluoride effect on bone formation—an overview [in Danish]. Tandlaegebladet 94(18):761-763.

Mokrzynski, S., and Z. Machoy. 1994. Fluoride incorporation into fetal bone. Fluoride 27(3):151-154.
Moller, P.F., and S.V. Gudjonsson. 1932. Massive fluorosis of bones and ligaments. Acta Radiol. 13:269-294.
Moonga, B.S., M. Pazianas, A.S. Alam, V.S. Shankar, C.L. Huang, and M. Zaidi. 1993. Stimulation of a Gs-like G protein in the osteoclast inhibits bone resorption by enhances tartrate-resistant acid phosphatase secretion. Biochem. Biophys. Res. Commun. 190(2):496-501.
Morgan, L., E. Allred, M. Tavares, D. Bellinger, and H. Needleman. 1998. Investigation of the possible associations between fluorosis, fluoride exposure, and childhood behavior problems. Pediatr. Dent. 20(4):244-252.
Morgenstern, H. 1998. Ecologic studies. Pp. 459-480 in Modern Epidemiology, 2nd Ed., K.J. Rothman, and S. Greenland, eds. Philadelphia, PA: Lippincott-Raven.
Morris, M.D. 2004. The Chemistry of Fluorosilicate Hydrolysis in Municipal Water Supplies. A Review of the Literature and a Summary of University of Michigan Studies. Report to the National Academy of Science, by M.D. Morris, University of Michigan, Ann Arbor, MI. January 23, 2004.
Moss, M.E., M.S. Kanarek, H.A. Anderson, L.P. Hanrahan, and P.L. Remington. 1995. Osteosarcoma, seasonality, and environmental factors in Wisconsin, 1979-1989. Arch. Environ. Health 50(3):235-241.
Moss, S.J. 1999. The case for retaining the current supplementation schedule. J. Public Health Dent. 59(4):259-262.
Mukai, M., M. Ikeda, T. Yanagihara, G. Hara, K. Kato, K. Ishiguro, H. Nakagaki, and C. Robinson. 1994. Fluoride distribution in dentine and cementum in human permanent teeth with vital and non-vital pulps. Arch. Oral Biol. 39(3):191-196.
Mukhopadhyay, D., L. Gokulkrishnan, and K. Mohanaruban. 2001. Lithium-induced nephrogenic diabetes insipidus in older people. Age Ageing 30(4):347-350.
Mullenix, P.J., P.K. DenBesten, A. Schunior, and W.J. Kernan. 1995. Neurotoxicity of sodium fluoride in rats. Neurotoxicol. Teratol. 17(2):169-177.
Munday, I.T., P.A. Stoddart, R.M. Jones, J. Lytle, and M.R. Cross. 1995. Serum fluoride concentration and urine osmolality after enflurane and sevoflurane anesthesia in male volunteers. Anesth. Analg. 81(2):353-359.
Murao, H., N. Sakagami, T. Iguchi, T. Murakami, and Y. Suketa. 2000. Sodium fluoride increases intracellular calcium in rat renal epithelial cell line NRK-52E. Biol. Pharm. Bull. 23(5):581-584.
Murcia García, J., A. Muñoz Hoyos, A. Molina Carballo, J.M. Fernández García, E. Narbona López, and J. Uberos Fernández. 2002. Puberty and melatonin [in Spanish]. An. Esp. Pediatr. 57(2):121-126.
Murray, J.M., M.B. Frarcsi, and T.R. Trinick. 1992. Plasma fluoride concentrations during and after prolonged anesthesia: A comparison of halothane and isoflurane. Anesth. Analg. 74(2):236-240.
Murray, M.M., J.A. Ryle, B.W. Simpson, and D.C. Wilson. 1948. Thyroid Enlargement and Other Changes Related to the Mineral Content of Drinking Water, with a Note on Goitre Prophylaxis. Medical Research Council Memorandum 18. London: His Majesty's Stationary Office.
Myers, B.S., J.S. Martin, D.T. Dempsey, H.P. Parkman, R.M. Thomas, and J.P. Ryan. 1997. Acute experimental colitis decreases colonic circular smooth muscle contractility in rats. Am. J. Physiol. 273(4 Pt 1):G928-G936.
Naccache, H., P.L. Simard, L. Trahan, J.M. Brodeur, M. Demers, D. Lachapelle, and P.M. Bernard. 1992. Factors affecting the ingestion of fluoride dentifrice by children. J. Public Health Dent. 52(4):222-226.

Naguib, M., A.H. Samarkandimb, Y. Al-Hattab, A. Turkistani, M.B. Delvi, W. Riad, and M. Attia. 2001. Metabolic, hormonal and gastric fluid and pH changes after different preoperative feeding regimens. Can. J. Anaesth. 48(4):344-350.

Nakano, O., C. Sakamoto, H. Nishisaki, Y. Konda, K. Matsuda, K. Wada, M. Nagao, and T. Matozaki. 1990. Difference in effects of sodium fluoride and cholecystokinin on diacylglycerol accumulation and calcium increase in guinea pig gastric chief cells. Life Sci. 47(7):647-654.

Narayana, M.V., and N.J. Chinoy. 1994a. Reversible effects of sodium fluoride on spermatozoa of the rat. Int. J. Fertil. Menopausal. Stud. 39(6):337-346.

Narayana, M.V., and N.J. Chinoy. 1994b. Effect of fluoride on rat testicular steroidogenesis. Fluoride 27(1):7-12.

Narchi, P., D. Edouard, P. Bourget, J. Otz, and I. Cattaneo. 1993. Gastric fluid pH and volume in gynaecologic out-patients. Influences of cimetidine and cimetidine-sodium citrate combination. Eur. J. Anaesthesiol. 10(5):357-361.

NCDC (National Climatic Data Center). 2002a. State, Regional, and National Monthly Temperature Weighted by Area 1971-2000 (and Previous Normals Periods). Historical Climatography Series No. 4-1. Asheville, NC: National Oceanic and Atmospheric Administration, National Climatic Data Center [online]. Available: http://lwf.ncdc.noaa.gov/oa/climate/normals/usnormals.html [accessed Sept. 15, 2004].

NCDC (National Climatic Data Center). 2002b. Divisional Normals and Standard Deviations of Temperature, Precipitation, and Heating and Cooling Degree Days 1971-2000 (and Previous Normals Periods), Climatography of the United States No. 85. Asheville, NC: National Oceanic and Atmospheric Administration, National Climatic Data Center [online]. Available: http://lwf.ncdc.noaa.gov/oa/climate/normals/usnormals.html [accessed Sept. 15, 2004].

Needleman, H.L., S.M. Pueschel, and K.J. Rothman. 1974. Fluoridation and the occurrence of Down's syndrome. N. Engl. J. Med. 291(Oct. 17):821-823.

Nesin, B.C. 1956. A water supply perspective of the fluoridation discussion. J. Maine Water Util. Assoc. 32:33-47.

Nevitt, M.C., S.R. Cummings, W.S. Browner, D.G. Seeley, J.A. Cauley, T.M. Vogt, and D.M. Black. 1992. The accuracy of self-report of fractures in elderly women: Evidence from a prospective study. Am. J. Epidemiol. 135(5):490-499.

Newbrun, E. 1986. Fluorides and Dental Caries, 3rd Ed. Springfield, IL: Charles C. Thomas.

Newbrun, E. 1989. Effectiveness of water fluoridation. J. Public Health Dent. 49(5):279-289.

Newbrun, E. 1992. Current regulations and recommendations concerning water fluoridation, fluoride supplements, and topical fluoride agents. J. Dent. Res. 71(5):1255-1265.

Newbrun, E. 1999. The case for reducing the current Council on Dental Therapeutics fluoride supplementation schedule. J. Public Health Dent. 59(4):263-268.

Newhouse, P.A., A. Potter, and E.D. Levin. 1997. Nicotinic system involvement in Alzheimer's and Parkinson's diseases. Implications for therapeutics. Drugs Aging. 11(3):206-228.

Newman, P.J., A.C. Quinn, G.M. Hall, and R.M. Grounds. 1994. Circulating fluoride changes and hepatorenal function following sevoflurane anaesthesia. Anaesthesia 49(11):936-939.

Newton, J.T., N. Prabhu, and P.G. Robinson. 2003. The impact of dental appearance on the appraisal of personal characteristics. Int. J. Prosthodont. 16(4):429-434.

Ng, A.H., G. Hercz, R. Kandel, and M.D. Grynpas. 2004. Association between fluoride, magnesium, aluminum and bone quality in renal osteodystrophy. Bone 34(1):216-224.

Nicolay, A., P. Bertocchio, E. Bargas, and J.P. Reynier. 1997. Long-term follow up of ionic plasma fluoride level in patients receiving hemodialysis treatment. Clin. Chim. Acta 263(1):97-104.

Nicolay, A., P. Bertocchio, E. Bargas, F. Coudore, G. Al Chahin, and J.P. Reynier. 1999. Hyperkalemia risks in hemodialysed patients consuming fluoride-rich water. Clin. Chim. Acta 281(1-2):29-36.
Noren, J.G., and J. Alm. 1983. Congenital hypothyroidism and changes in the enamel of deciduous teeth. Acta Paediatr. Scand. 72(4):485-489.
Nourjah, P., A.M. Horowitz, and D.K. Wagener. 1994. Factors associated with the use of fluoride supplements and fluoride dentifrice by infants and toddlers. J. Public Health Dent. 54(1):47-54.
Nowak, A., and M.V. Nowak. 1989. Fluoride concentration of bottled and processed waters. Iowa Dent. J. 7(4):28.
Nowjack-Raymer, R.E., R.H. Selwitz, A. Kingman, and W.S. Driscoll. 1995. The prevalence of dental fluorosis in a school-based program of fluoride mouthrinsing, fluoride tablets, and both procedures combined. J. Public Health Dent. 55(3):165-170.
NRC (National Research Council). 1977. Drinking Water and Health, Vol. 1. Washington, DC: National Academy Press.
NRC (National Research Council). 1989a. Biologic Markers in Reproductive Toxicology. Washington, DC: National Academy Press.
NRC (National Research Council). 1989b. Recommended Dietary Allowances, 10th Ed. Washington, DC: National Academy Press.
NRC (National Research Council). 1993. Health Effects of Ingested Fluoride. Washington, DC: National Academy Press.
NTP (National Toxicology Program). 1990. NTP Technical Report on the Toxicology and Carcinogenesis Studies of Sodium Fluoride (CAS no. 7682-49-4) in F344/N Rats and B6C3F$_1$ (Drinking Water Studies) /. Technical Report 393. NIH Publ. No. 91-2848. National Institutes of Health, Public Health Service, U.S. Department of Health and Human Services, Research Triangle Park, NC.
NTP (National Toxicology Program). 1992. NTP Supplemental 2-Year Study of Sodium Fluoride in Male F344 Rats (CAS No. 7681-49-4). Study No. C55221D. National Toxicology Program, National Institute of Environmental Health Sciences, Research Triangle Park, NC [online]. Available: http://ntp.niehs.nih.gov/index.cfm?objectid=16577B88-F1F6-975E-750961B2062D514E [accessed August 23, 2005].
NTP (National Toxicology Program). 2002. The National Toxicology Program. Annual Plan, Fiscal Year 2002. NIH Publication No. 03-5309. National Toxicology Program, Public Health Service, U.S. Department of Health and Human Services, Research Triangle Park, NC [online]. Available: http://ntp-server.niehs.nih.gov/ntp/htdocs/2002AP/AP2002.pdf [accessed Oct. 6, 2005].
Nunn, J.H., J.J. Murray, P. Reynolds, D. Tabari, and J. Breckon. 1992. The prevalence of developmental defects of enamel in 15-16-year-old children residing in three districts (natural fluoride, adjusted fluoride, low fluoride) in the north east of England. Community Dent. Health 9(3):235-247.
Obata, R., H. Bito, M. Ohmura, G. Moriwaki, Y. Ikeuchi, T. Katoh, and S. Sato. 2000. The effects of prolonged low flow sevoflurane anesthesia on renal and hepatic function. Anesth. Analg. 91(5):1262-1268.
Obel, A.O. 1982. Goitre and fluorosis in Kenya. East Afr. Med. J. 59(6):363-365.
Ogilvie, A.L. 1953. Histologic findings in the kidney, liver, pancreas, adrenal, and thyroid glands of the rat following sodium fluoride administration. J. Dent. Res. 32(3):386-397.
Oguro, A., J. Cervenka, and K. Horii. 1995. Effect of sodium fluoride on chromosomal ploidy and breakage in cultured human diploid cells (IMR-90): An evaluation of continuous and short-time treatment. Pharmacol. Toxicol. 76(4):292-296.
Oguro, A., T. Kawase, and M. Orikasa. 2003. NaF induces early differentiation of murine bone

marrow cells along the granulocytic pathway but not the monocytic or preosteoclastic pathway in vitro. In Vitro Cell Dev. Biol. Anim. 39(5-6):243-248.

Oikkonen, M., and O. Meretoja. 1989. Serum fluoride in children anaesthetized with enflurane. Eur. J. Anaesthesiol. 6(6):401-407.

OIV (Office International de la Vigne et du Vin). 1990. Recueil des Méthodes Internationales de' Analyse des Vins et des Mouts. Paris: OIV Édition officelle (as cited in Martínez et al. 1998).

Okuda, A., J. Kanehisa, and J.N. Heersche. 1990. The effects of sodium fluoride on the resorptive activity of osteoclasts. J. Bone Miner. Res. 5(Suppl. 1):S115-S120.

Oliveby, A., S. Twetman, and J. Ekstrand. 1990. Diurnal fluoride concentration in whole saliva in children living in a high- and a low-fluoride area. Caries Res. 24(1):44-47.

Olsen, G.W., T.R. Church, E.B. Larson, G. van Belle, J.K. Lundberg, K.J. Hansen, J.M. Burris, J.H. Mandel, and L.R. Zobel. 2004. Serum concentrations of perfluorooctanesulfonate and other fluorochemicals in an elderly population from Seattle, Washington. Chemosphere 54(11):1599-1611.

Olsen, G.W., T.R. Church, J.P. Miller, J.M. Burris, K.J. Hansen, J.K. Lundberg, J.B. Armitage, R.M. Herron, Z. Medhdizadehkashi, J.B. Nobiletti, E.M. O'Neill, J.H. Mandel, and L.R. Zobel. 2003. Perfluorooctanesulfonate and other fluorochemicals in the serum of American Red Cross adult blood donors. Environ. Health Perspect. 111(16):1900.

Olson, S.C., S.R. Tyagi, and J.D. Lambeth. 1990. Fluoride activates diradylglycerol and superoxide generation in human neutrophils via PLD/PA phosphohydrolase-dependent and -independent pathways. FEBS Lett. 272(1-2):19-24.

Olsson, B. 1979. Dental findings in high-fluoride areas in Ethiopia. Community Dent. Oral Epidemiol. 7(1):51-56.

Ophaug, R.H., L. Singer, and B.F. Harland. 1980. Estimated fluoride intake of average two-year-old children in four dietary regions of the United States. J. Dent. Res. 59(5):777-781.

Ophaug, R.H., L. Singer, and B.F. Harland. 1985. Dietary fluoride intake of 6-months and 2-year-old children in four dietary regions of the United States. J. Clin. Nutr. 42(4):701-707.

Opinya, G.N., N. Bwibo, J. Valderhaug, J.M. Birkeland, and P. Lökken. 1991. Intake of fluoride and excretion in mothers' milk in a high fluoride (9 ppm) area in Kenya. Eur. J. Clin. Nutr. 45(1):37-41.

Orcel, P., M.C. de Vernejoul, A. Prier, L. Miravet, D. Kuntz, and G. Kaplan. 1990. Stress fractures of the lower limbs in osteoporotic patients treated with fluoride. J. Bone Miner. Res. 5(Suppl. 1):S191-S194.

Ortiz-Perez, D., M. Rodriguez-Martinez, F. Martinez, V.H. Borja-Aburto, J. Castelo, J.I. Grimaldo, E. de la Cruz, L. Carrizales, and F. Diaz-Barriga. 2003. Fluoride-induced disruption of reproductive hormones in men. Environ. Res. 93(1):20-30.

Osborne, M.A., J.M. Eddleston, and W. McNicoll. 1996. Inorganic fluoride concentration after long-term sedation with isoflurane. Intensive Care Med. 22(7):677-682.

O'Shea, J.J., K.B. Urdahl, H.T. Luong, T.M. Chused, L.E. Samelson, and R.D. Klausner. 1987. Aluminum fluoride induces phosphatidylinositol turnover, elevation of cytoplasmic free calcium, and phosphorylation of the T cell antigen receptor in murine T cells. J. Immunol. 139(10):3463-3469.

Padez, C., and M.A. Rocha. 2003. Age at menarche in Coimbra (Portugal) school girls: A note on the secular changes. Ann. Hum. Biol. 30(5):622-632.

Pak, C.Y., K. Sakhaee, B. Adams-Huet, V. Piziak, R.D. Peterson, and J.R. Poindexter. 1995. Treatment of postmenopausal osteoporosis with slow-release sodium fluoride. Ann. Intern. Med. 123(6):401-408.

Paloyan Walker, R., E. Kazuko, C. Gopalsami, J. Bassali, A.M. Lawrence, and E. Paloyan.

REFERENCES

1997. Hyperparathyroidism associated with a chronic hypothyroid state. Laryngoscope 107(7):903-909.

Pang, D.T., C.L. Phillips, and J.W. Bawden. 1992. Fluoride intake from beverage consumption in a sample of North Carolina children. J. Dent. Res. 71(7):1382-1388.

Panzer, A. 1997. Melatonin in osteosarcoma: An effective drug? Med. Hypotheses 48(6):523-525.

Parkins, F.M., N. Tinanoff, M. Moutinho, M.B. Anstey, and M.H. Waziri. 1974. Relationships of human plasma fluoride and bone fluoride to age. Calcif. Tissue Res. 16(4):335-338.

Parodi, S., D. Malacarne, and M. Taningher. 1991. Examples of uses of databases for quantitative and qualitative correlation studies between genotoxicity and carcinogenicity. Environ. Health Perspect. 96:61-66.

Parsons, V., A.A. Choudhury, J.A. Wass, and A. Vernon. 1975. Renal excretion of fluoride in renal failure and after renal transplantation. Br. Med. J. 1(5950):128-130.

Partanen, S. 2002. Inhibition of human renal acid phosphatases by nephrotoxic micromolar concentrations of fluoride. Exp. Toxicol. Pathol. 54(3):231-237.

Pashley, D.H., N.B. Allison, R.P. Easmann, R.V. McKinney, J.A. Horner, and G.M. Whitford. 1984. The effects of fluoride on the gastric mucosa of the rat. J. Oral Pathol. 13(5):535-545.

Paul, V., P. Ekambaram, and A.R. Jayakumar. 1998. Effects of sodium fluoride on locomotor behavior and a few biochemical parameters in rats. Environ. Toxicol. Pharmacol. 6(3):187-191.

Pellestor, F., B. Andréo, F. Arnal, C. Humeau, and J. Demaille. 2003. Maternal aging and chromosomal abnormalities: New data drawn from in vitro unfertilized human oocytes. Hum. Genet. 112(2):195-203.

Pendrys, D.G. 1990. The fluorosis risk index: A method for investigating risk factors. J. Public Health Dent. 50(5):291-298.

Pendrys, D.G., and D.E. Morse. 1995. Fluoride supplement use by children in fluoridated communities. J. Public Health Dent. 55(3):160-164.

Penman, A.D., B.T. Brackin, and R. Embrey. 1997. Outbreak of acute fluoride poisoning caused by a fluoride overfeed, Mississippi, 1993. Public Health Rep. 112(5):403-409.

Peres, K.G., R. Latorre Mdo, M.A. Peres, J. Traebert, and M. Panizzi. 2003. Impact of dental caries and dental fluorosis on 12-year-old schoolchildren's self-perception of appearance and chewing [in Portuguese]. Cad. Saude Publica 19(1):323-330.

Pessan, J.P., M.L. Pin, C.C. Martinhon, S.M. de Silva, J.M. Granjeiro, and M.A. Buzalaf. 2005. Analysis of fingernails and urine as biomarkers of fluoride exposure from dentifrice and varnish in 4- to 7-year-old children. Caries Res. 39(5):363-370.

Petersen, L.R., D. Denis, D. Brown, J.L. Hadler, and S.D. Helgerson. 1988. Community health effects of a municipal water supply hyperfluoridation accident. Am. J. Public Health. 78(6):711-713.

Petersen, M.B., and M. Mikkelsen. 2000. Nondisjunction in trisomy 21: Origin and mechanisms. Cytogenet. Cell. Genet. 91(1-4):199-203.

Petraborg, H.T. 1977. Hydrofluorosis in the fluoridated Milwaukee area. Fluoride 10(4):165-168.

Pettifor, J.M., C.M. Schnitzler, F.P. Ross, and G.P. Moodley. 1989. Endemic skeletal fluorosis in children: Hypocalcemia and the presence of renal resistance to parathyroid hormone. Bone Miner. 7(3):275-288.

Phipps, K.R., E.S. Orwoll, J.D. Mason, and J.A. Cauley. 2000. Community water fluoridation, bone mineral density, and fractures: Prospective study of effects in older women. BMJ 321(7265):860-864.

PHS (Public Health Service). 1991. Review of Fluoride Benefits and Risks: Report of the Ad Hoc Subcommittee on Fluoride Committee of the Committee to Coordinate Environmental Health and Related Programs. Public Health Service, U.S. Department of Health and Human Services, Washington, DC.

Pirinen, S. 1995. Endocrine regulation of craniofacial growth. Acta Odontol. Scand. 53(3): 179-185.

Pivonello, R., A. Colao, C. Di Somma, G. Facciolli, M. Klain, A. Faggiano, M. Salvatore, and G. Lombardi. 1998. Impairment of bone status in patients with central diabetes insipidus. J. Clin. Endocrinol. Metab. 83(7):2275-2280.

Poole, C. 1985. Exception to the rule about nondifferential misclassification [abstract]. Am. J. Epidemiol. 122(3):508.

Powell, J.J., and R.P. Thompson. 1993. The chemistry of aluminum in the gastrointestinal lumen and its uptake and absorption. Proc. Nutr. Soc. 52(1):241-253.

Pradhan, K.M., N.K. Arora, A. Jena, A.K. Susheela, and M.K. Bhan. 1995. Safety of ciprofloxacin therapy in children: Magnetic resonance images, body fluid levels of fluoride and linear growth. Acta Paediatr. 84(5):555-560.

Procopio, M., and G. Borretta. 2003. Derangement of glucose metabolism in hyperparathyroidism. J. Endocrionol. Invest. 26(11):1136-1142.

Qin, L.S., and S.Y. Cui. 1990. The Influence of drinking water fluoride on pupils IQ, as measured by Rui Wen's standards [in Chinese]. Chinese J. Control Endemic Dis. 5:203-204.

Raisz, L.G., B.E. Kream, and J.A. Lorenzo. 2002. Metabolic bone disease. Pp. 1373-1410 in Williams Textbook of Endocrinology, 10th Ed., P.R. Larsen, H.M. Kronenberg, S. Melmed, and K.S. Polonsky, eds. Philadelphia, PA: Saunders.

Ramesh, N., A.S. Vuayaraghavan, B.S. Desai, M. Natarajan, P.B. Murthy, and K.S. Pillai. 2001. Low levels of p53 mutations in Indian patients with osteosarcoma and the correlation with fluoride levels in bone. J. Environ. Pathol. Toxicol. Oncol. 20(3):237-243.

Rantanen, N.W., J.E. Alexander, and G.R. Spencer. 1972. Interaction of fluoride, calcium, phosphorus, and thyroidectomy on porcine bone. Am. J. Vet. Res. 33(7):1347-1358.

Rao, H.V., R.P. Beliles, G.M. Whitford, and C.H. Turner. 1995. A physiologically based pharmacokinetic model for fluoride uptake by bone. Regul. Toxicol. Pharmacol. 22(1):30-42.

Rapaport, I. 1956. Contribution to the study of mongolism: Pathogenicity of fluorine [in French]. Bull. Acad. Nat. Med. Paris 140(28-29):529-531.

Rapaport, I. 1963. Oligophrenie mongolienne et caries dentaires. Rev. Stomatol. 64: 207-218.

Ray, S.K., S. Ghosh, I.C. Tiwari, J. Nagchaudhuri, P. Kaur, and D.C. Reddy. 1982. Prevalence of dental fluorosis in relation to fluoride in drinking water in two villages of Varanasi (U.P.). Indian J. Public Health 26(3):173-178.

Read, S.G. 1982. The distribution of Down's syndrome. J. Ment. Defic. Res. 26(Pt 4): 215-227.

Ream, L.J., and R. Principato. 1981a. Ultrastructural observations on the mechanism of secretion in the rat parathyroid after fluoride ingestion. Cell Tissue Res. 214(3):569-573.

Ream, L.J., and R. Principato. 1981b. Glycogen accumulation in the parathyroid gland of the rat after fluoride ingestion. Cell Tissue Res. 220(1):125-130.

Ream, L.J., and R. Principato. 1981c. Fluoride stimulation of the rat parathyroid gland: An ultrastructural study. Am. J. Anat. 162(3):233-241.

Record, S., D.F. Montgomery, and M. Milano. 2000. Fluoride supplementation and caries prevention. J. Pediatr. Health Care 14(5):247-249.

Reed, B.Y., J.E. Zerwekh, P.P. Antich, and C.Y. Pak. 1993. Fluoride-stimulated [3H]thymidine uptake in a human osteoblastic osteosarcoma cell line is dependent of transforming growth factor beta. J. Bone Miner. Res. 8(1):19-25.

Reginster, J.Y., L. Meurmans, B. Zegels, L.C. Rovati, H.W. Minne, G. Giacovelli, A.N. Taquet, I. Setnikar, J. Collett, and C. Gosset. 1998. The effect of sodium monofluorophosphate plus calcium on vertebral fracture rate in postmenopausal women with moderate osteoporosis. A randomized, controlled trial. Ann. Intern. Med. 129(1):1-8.

Reiter, R.J. 1998. Melatonin and human reproduction. Ann. Med. 30(1):103-108.

Reiter, R.J., D.X. Tan, J.C. Mayo, R.M. Sainz, J. Leon, and D. Bandyopadhyay. 2003a. Neurally-mediated and neurally-independent beneficial actions of melatonin in the gastrointestinal tract. J. Physiol. Pharmacol. 54(Suppl. 4):113-125.

Reiter, R.J., D.X. Tan, J.C. Mayo, R.M. Sainz, J. Leon, and Z. Czarnocki. 2003b. Melatonin as an antioxidant: Biochemical mechanisms and pathophysiological implications in humans. Acta Biochim. Pol. 50(4):1129-1146.

Retief, D.H., E.L. Bradley, F.H. Barbakow, M. Friedman, E.H. van der Merwe, and J.I. Bischoff. 1979. Relationships among fluoride concentration in enamel, degree of fluorosis and caries incidence in a community residing in a high fluoride area. J. Oral Pathol. 8(4):224-236.

Ribeiro, D.A., M.E. Marques, G.F. de Assis, A. Anzai, M.L. Poleti, and D.M. Salvadori. 2004a. No relationship between subchronic fluoride intake and DNA damage in Wistar rats. Caries Res. 38(6):576-579.

Ribeiro, D.A., C. Scolastici, M.E. Marques, and D.M. Salvadoir. 2004b. Fluoride does not induce DNA breakage in Chinese hamster ovary cells in vitro. Pesqui Odontol. Bras. 18(3):192-196.

Rice-Evans, C., and P. Hoschstein. 1981. Alterations in erythrocyte membrane fluidity by phenylhydrazine-induced peroxidation of lipids. Biochem. Biophys. Res. Commun. 100(4):1537-1541.

Rich, C., and J. Ensinck. 1961. Effect of sodium fluoride on calcium metabolism of human beings. Nature 191(July 8):184-185.

Rich, C., and E. Feist. 1970. The action of fluoride on bone. Pp. 70-87 in Fluoride in Medicine, T.L. Vischer, ed. Bern: Hans Huber.

Richards, A., L. Mosekilde, and C.H. Sogaard. 1994. Normal age-related changes in fluoride content of vertebral trabecular bone—relation to bone quality. Bone 15(1):21-26.

Ridefelt, P., P. Hellman, J. Rastad, R. Larsson, G. Akerstrom, and E. Gylfe. 1992. Fluoride interactions with stimulus-secretion coupling of normal and pathological parathyroid cells. Acta Physiol. Scand. 145(3):275-285.

Rigalli, A., J.C. Ballina, E. Roveri, and R.C. Puche. 1990. Inhibitory effect of fluoride on the secretion of insulin. Calcif. Tissue Int. 46(5):333-338.

Rigalli, A., J.C. Ballina, and R.C. Puche. 1992. Bone mass increase and glucose tolerance in rats chronically treated with sodium fluoride. Bone Miner. 16(2):101-108.

Rigalli, A., R. Alloatti, I. Menoyo, and R.C. Puche. 1995. Comparative study of the effect of sodium fluoride and sodium monofluorophosphate on glucose homeostasis in the rat. Arzneimittel-Forsch. 45(3):289-292.

Riggs, B.L., S.F. Hodgson, W.M. O'Fallon, E.Y. Chao, H.W. Wahner, J.M. Muhs, S.L. Cedel, and L.J. Melton III. 1990. Effect of fluoride treatment on the fracture rate in post-menopausal women with osteoporosis. N. Engl. J. Med. 322(12):802-809.

Rimoli, C., C.N. Carducci, C. Dabas, C. Vescina, M. E. Quindimil, and A. Mascaro. 1991. Relationship between serum concentrations of flecainide and fluoride in humans. Boll. Chim. Farm. 130(7):279-282.

Riordan, P.J. 1993. Perceptions of dental fluorosis. J. Dent. Res. 72(9):1268-1274.

Ripa, L.W. 1993. A half-century of community water fluoridation in the United States: Review and commentary. J. Public Health Dent. 53(1):17-44.

Robinson, A.G., and J.G. Verbalis. 2002. Posterior pituitary gland. Pp. 281-329 in Williams Textbook of Endocrinology, 10th Ed., P.R. Larsen, H.M. Kronenberg, S. Melmed, and K.S. Polonsky, eds. Philadelphia, PA: Saunders.

Robinson, C., J. Kirkham, and J.A. Weatherell. 1996. Fluoride in teeth and bone. Pp. 69-87 in Fluoride in Dentistry, 2nd Ed., O. Fejerskov, J. Ekstrand, and B.A. Burt, eds. Copenhagen: Munksgaard.

Roholm, K. 1937. Fluorine Intoxication: A Clinical-Hygienic Study, with a Review of the Literature and Some Experimental Investigations. London: H.K. Lewis & Co.

Rojas-Sanchez, F., S.A. Kelly, K.M. Drake, G.J. Eckert, G.K. Stookey, and A.J. Dunipace. 1999. Fluoride intake from foods, beverages and dentifrice by young children in communities with negligibly and optimally fluoridated water: A pilot study. Community Dent. Oral Epidemiol. 27(4):288-297.

Rölla, G., and J. Ekstrand. 1996. Fluoride in oral fluids and dental plaque. Pp. 215-229 in Fluoride in Dentistry, 2nd Ed, O. Fejerskov, J. Ekstrand, and B.A. Burt, eds. Copenhagen: Munksgaard.

Romundstad, P., A. Andersen, and T. Haldorsen. 2000. Cancer incidence among workers in six Norwegian aluminum plants. Scand. J. Work Environ. Health 26(6):461-469.

Rosenquist, J., and L. Boquist. 1973. Effects of supply and withdrawal of fluoride. Experimental studies on growing and adult rabbits. 2. Parathyroid morphology and function. Acta Pathol. Microbiol. Scand. A 81(5):637-644.

Rosenquist, J.B., P.R. Lorentzon, and L.L. Boquist. 1983. Effect of fluoride on parathyroid activity of normal and calcium-deficient rats. Calcif. Tissue Int. 35(4-5):533-537.

Ross, J.F., and G.P. Daston. 1995. Neurotoxicity of sodium fluoride in rats [letter]. Neurotoxicol. Teratol. 17(6):685-688.

Ross, P.D. 1996. Osteoporosis: Frequency, consequences and risk factors. Ann. Intern. Med. 156(13):1399-1411.

Roth, J.A., B.G. Kim, W.L. Lin, and M.I. Cho. 1999. Melatonin promotes osteoblast differentiation and bone formation. J. Biol. Chem. 274(31):22041-22047.

Rothman, K.J., and S. Greenland, eds. 1998. Modern Epidemiology, 2nd Ed. Philadelphia, PA: Lippincott-Raven.

Rozier, R.G. 1994. Epidemiologic indices for measuring the clinical manifestations of dental fluorosis: Overview and critique. Adv. Dent. Res. 8(1):39-55.

Rozier, R.G., and G.G. Dudney. 1981. Dental fluorosis in children exposed to multiple sources of fluoride: Implications for school fluoridation programs. Public Health Rep. 96(6):542-546.

Rozman, K.K., and J. Doull. 2000. Dose and time as variables of toxicity. Toxicology 144(1-3):169-178.

Russell, A.L., and E. Elvove. 1951. Domestic water and dental caries. VII. A study of the fluoride-dental caries relationship in an adult population. Public Health Rep. 66(43): 1389-1401.

Rwenyonyi, C.M., K. Bjorvatn, J. Birkeland, and O. Haugejorden. 1999. Altitude as a risk indicator of dental fluorosis in children residing in areas with 0.5 and 2.5 mg fluoride per liter in drinking water. Caries Res. 33(4):267-274.

Rwenyonyi, C.M., J.M. Birkeland, O. Haugejorden, and K. Bjorvatn. 2001. Dental caries among 10- to 14-year-old children in Ugandan rural areas with 0.5 and 2.5 mg fluoride per liter in drinking water. Clin. Oral Investig. 5(1):45-50.

Saka, O., P. Hallac, and I. Urgancioglu. 1965. The effect of fluoride on the thyroid of the rat. New Istanbul Contrib. Clin. Sci. 8(2):87-90.

Sakai, T., and M. Takaori. 1978. Biodegradation of halothane, enflurane, and methoxyflurane. Br. J. Anaesth. 50(8):785-791.

Salti, R., F. Galluzzi, G. Bindi, F. Perfetto, R. Tarquini, F. Halberg, and G. Cornelissen. 2000. Nocturnal melatonin patterns in children. J. Clin. Endocrinol. Metab. 85(6):2137-2144.

Saltzman, W.M. 2004. Pp. 91-93 in Tissue Engineering: Engineering Principles for the Design of Replacement Organs and Tissues. New York, NY: Oxford University Press.

Sampaio, F.C., and P. Arneberg. 1999. Dental plaque fluoride and pH in children exposed to different water fluoride levels. Acta Odontol. Scand. 57(2):65-71.
Sandyk, R., P.G. Anastasiadis, P.A. Anninos, and N. Tsagas. 1992. Is postmenopausal osteoporosis related to pineal gland functions? Int. J. Neurosci. 62(3-4):215-225.
Sankaran, B., and N.G. Gadekar. 1964. Skeletal fluorosis. Pp. 357-362 in Bone and Tooth: Proceedings of the First European Symposium held at Somerville College, Oxford, April 1963, H.J.J. Blackwood, ed. New York: Pergamon Press.
Sapov, K., I. Gedalia, S. Grobler, I. Lewinstein, I. Roman, L. Shapira, Z. Hirschfeld, and S. Teotia. 1999. A laboratory assessment of enamel hypoplasia of teeth with varying severities of dental fluorosis. J. Oral Rehabil. 26(8):672-677.
Sarner, J.B., M. Levine, P.J. Davis, J. Lerman, D.R. Cook, and E.K. Motoyama. 1995. Clinical characteristics of sevoflurane in children. Anesthesiology 82(1):38-46.
Sauer, G.R., L.N. Wu, M. Iijima, and R.E. Wuther. 1997. The influence of trace elements on calcium phosphate formation by matrix vesicles. J. Inorg. Biochem. 65(1):57-65.
Sauerbrunn, B.J., D.M. Ryan, and J.F. Shaw. 1965. Chronic fluoride intoxication with fluorotic radiculomyelopathy. Ann. Intern. Med. 63(6):1074-1078.
Savage, L.M. 2001. In search of the neurobiological underpinnings of the differential outcomes effect. Integr. Physiol. Behav. Sci. 36(3):182-195.
Savas, S., M. Cetin, M. Akdogan, and N. Heybeli. 2001. Endemic fluorosis in Turkish patients: Relationships with knee osteoarthritis. Rheumatol. Int. 21(1):30-35.
SCDHEC (South Carolina Department of Health and Environmental Control). 2004. South Carolina Public Water System Annual Compliance Report for Calendar Year 2003. [online]. Available: http://www.scdhec.gov/eqc/water/pubs/dwreport.doc.
Schamschula, R.G., E. Sugár, P.S. Un, K. Tóth, D.E. Barmes, and B.L. Adkins. 1985. Physiological indicators of fluoride exposure and utilization: An epidemiological study. Community Dent. Oral Epidemiol. 13(2):104-107.
Scheffrahn, R.H., R.C. Hsu, and N.Y. Su. 1989. Fluoride residues in frozen foods fumigated with sulfuryl fluoride. Bull. Environ. Contam. Toxicol. 43(6):899-903.
Scheffrahn, R.H., L. Bodalbhai, and N.Y. Su. 1992. Residues of methyl bromide and sulfuryl fluoride in manufacturer-packaged household foods following fumigation. Bull. Environ. Contam. Toxicol. 48(6):821-827.
Schellenberg, D., T.A. Marks, C.M. Metzler, J.A. Oostveen, and M.J. Morey. 1990. Lack of effect of fluoride on reproductive performance and development in Shetland sheepdogs. Vet. Hum. Toxicol. 32(4):309-314.
Schiffl, H.H., and U. Binswanger. 1980. Human urinary fluoride excretion as influenced by renal functional impairment. Nephron 26(2):69-72.
Schlegel, H.H. 1974. Industrial skeletal fluorosis: Preliminary report on 61 cases from aluminum smelter [in German]. Soz. Praventiv. Med. 19:269-274.
Schlesinger, E.R., D.E. Overton, H.C. Chase, and K.T. Cantwell. 1956. Newburgh-Kingston caries-fluorine study. XIII. Pediatric findings after ten years. J. Am. Dent. Assoc. 52(3):296-306.
Schlumberger, M.-J., S. Filetti, and I.D. Hay. 2002. Nontoxic goiter and thyroid neoplasia. Pp. 457-490 in Williams Textbook of Endocrinology, 10th Ed., P.R. Larsen, H.M. Kronenberg, S. Melmed, and K.S. Polonsky, eds. Philadelphia, PA: Saunders.
Schneider, M.J., S.N. Fiering, S.E. Pallud, A.F. Parlow, D.L. St Germain, and V.A. Galton. 2001. Targeted disruption of the type 2 selenodeiodinase gene (DIO_2) results in a phenotype of pituitary resistance to T4. Mol. Endocrinol. 15(12):2137-2148.
Schottenfeld, D., and J.F. Fraumeni. 1996. Cancer Epidemiology and Prevention, 2nd Ed. New York: Oxford University Press.
Schulze-Specking, A., J. Duyster, P.J. Gebicke-Haerter, S. Wurster, and P. Dieter. 1991. Effect of fluoride, pertussis and cholera toxin on the release of arachidonic acid and the formation

of prostaglandin E2, D2, superoxide and inositol phosphates in rat liver macrophages. Cell Signal 3(6):599-606.
Selwitz, R.H. 1994. Strategies for improving methods of assessing fluoride accumulation in body fluids and tissues. Adv. Dent. Res. 8(1):111-112.
Selwitz, R.H., R.E. Nowjack-Raymer, A. Kingman, and W.S. Driscoll. 1995. Prevalence of dental caries and dental fluorosis in areas with optimal and above-optimal water fluoride concentrations: A 10-year follow-up survey. J. Public Health Dent. 55(2):85-93.
Selwitz, R.H., R.E. Nowjack-Raymer, A. Kingman, and W.S. Driscoll. 1998. Dental caries and dental fluorosis among schoolchildren who were lifelong residents of communities having either low or optimal levels of fluoride in drinking water. J. Public Health Dent. 58(1):28-35.
Semenza, J.C., and L.H. Weasel. 1997. Molecular epidemiology in environmental health: The potential of tumor suppressor gene p53 as a biomarker. Environ. Health Perspect. 105(Suppl. 1):155-163.
Seow, W.K. 1993. Clinical diagnosis and management strategies of amelogenesis imperfecta-variants. Pediatr. Dent. 15(6):384-393.
Seow, W.K., and M.J. Thomsett. 1994. Dental fluorosis as a complication of hereditary diabetes insipidus: Studies of six affected patients. Pediatr. Dent. 16(2):128-132.
Shashi, A. 1992a. Biochemical effects of fluoride on lipid metabolism in the reproductive organs of male rabbits. Fluoride 25(3):149-154.
Shashi, A. 1992b. Studies on alterations in brain lipid metabolism following experimental fluorosis. Fluoride 25(2):77-84.
Shashi, A. 2002. Histopathological effects of sodium fluoride on the duodenum of rabbits. Fluoride 35(1):28-37.
Shashi, A., and S.P. Thapar. 2001. Histopathology of fluoride-induced hepatotoxicity in rabbits. Fluoride 34(1):34-42.
Shashi, A., J.P. Singh, and S.P. Thapar. 1994. Effect of long-term administration of fluoride on levels of protein, free amino acids and RNA in rabbit brain. Fluoride 27(3):155-159.
Shashi, A., J.P. Singh, and S.P. Thapar. 2002. Toxic effects of fluoride on rabbit kidney. Fluoride 35(1):38-50.
Shayiq, R.M., H. Raza, and A.M. Kidwai. 1984. Alteration in gastric secretion of rats administered NaF. Fluoride 17(3):178-182.
Sheth, F.J., A.S. Multani, and N.J. Chinoy. 1994. Sister chromatid exchanges: A study in fluorotic individuals of North Gujarat. Fluoride 27(4):215-219.
Shimonovitz, S., D. Patz, P. Ever-Hadani, L. Singer, D. Zacut, G. Kidroni, and M. Ron. 1995. Umbilical cord fluoride serum levels may not reflect fetal fluoride status. J. Perinat. Med. 23(4):279-282.
Shivarajashankara, Y.M., A.R. Shivashankara, P.G. Bhat, S.M. Rao, and S.H. Rao. 2002. Histological changes in the brain of young fluoride-intoxicated rats. Fluoride 35(1):12-21.
Shoback, D.M., and J.M. McGhee. 1988. Fluoride stimulates the accumulation of inositol phosphates, increases intracellular free calcium, and inhibits parathyroid hormone release in dispersed bovine parathyroid cells. Endocrinology 122(6):2833-2839.
Short, E.M. 1944. Domestic water and dental caries: VI. The relation of fluoride domestic waters to permanent tooth eruption. J. Dent. Res. 23:247-255. [Reprinted as pp. 137-141 in McClure, F.J. (Ed.) 1962. Fluoride Drinking Waters. A selection of Public Health Service papers on dental fluorosis and dental caries; physiological effects, analysis and chemistry of fluoride. Bethesda, MD: U.S. Department of Health, Education, and Welfare, Public Health Service.]
Shortt, H.M., G.R. McRobert, T.W. Bernard, and A.S.M. Nayar. 1937. Endemic fluorosis in Madras Presidency. Indian Med. Gazette 72:396-398.
Shoumura, S., H. Chen, S. Emura, M. Utsumi, D. Hayakawa, T. Yamahira, K. Terasawa,

A. Tamada, M. Arakawa, and H. Isono. 1992. An in vitro study on the effects of melatonin on the ultrastructure of the hamster parathyroid gland. Histol. Histopathol. 7(4):715-718.

Shu, W.S., Z.Q. Zhang, C.Y. Lan, and M.H. Wong. 2003. Fluoride and aluminum concentrations of tea plants and tea products from Sichuan Province, PR China. Chemosphere 52(9):1475-1482.

Shulman, J.D., and L.M. Wells. 1997. Acute fluoride toxicity from ingesting home-use dental products in children, birth to 6 years of age. J. Public Health Dent. 57(3):150-158.

Shulman, J.D., J.A. Lalumandier, and J.D. Grabenstein. 1995. The average daily dose of fluoride: A model based on fluid consumption. Pediatr. Dent. 17(1):13-18.

Siddiqui, A.H. 1955. Fluorosis in Nalgonda district, Hyderabad-Deccan. Br. Med. J. 2(4953):1408-1413.

Siddiqui, A.H. 1960. Incidence of simple goitre in areas of endemic fluorosis. J. Endocrinol. 20(Apr.):101-105.

Sidhu, K.S., and R.O. Kimmer. 2002. Fluoride overfeed at a well site near an elementary school in Michigan. J. Environ. Health 65(3):16-21, 38.

Sidora, V.D., A.I. Shliakhta, V.K. Iugov, A.S. Kas'ianenko, and V.G. Piatenko. 1983. Indices of the pituitary-thyroid system in residents of cities with various fluorine concentrations in drinking water [in Russian]. Probl. Endokrinol. 29(4):32-35.

Siebenhüner, L., E. Miloni, and H. Bürgi. 1984. Effects of fluoride on thyroid hormone biosynthesis. Studies in a highly sensitive test system. Klin. Wochenschr. 62(18):859-861.

Silverman, D.H., and G.W. Small. 2002. Prompt identification of Alzheimer's disease with brain PET imaging of a woman with multiple previous diagnoses of other neuropsychiatric conditions. Am. J. Psychiatry 159(9):1482-1488.

Silverman, S., Jr. 1971. Oral changes in metabolic diseases. Postgrad. Med. 49(1):106-110.

Simard, P.L., H. Naccache, D. Lachapelle, and J.M. Brodeur. 1991. Ingestion of fluoride from dentifrices by children aged 12 to 24 months. Clin. Pediar. 30(11):614-617.

Simonen, O., and O. Laitinen. 1985. Does fluoridation of drinking water prevent bone fragility and osteoporosis? Lancet 2(8452):432-433.

Singer, L., R.H. Ophaug, and B.F. Harland. 1985. Dietary fluoride intake of 15-19-year-old male adults residing in the United States. J. Dent. Res. 64(11):1302-1305.

Singh, A., and S.S. Jolly. 1961. Endemic fluorosis. Q. J. Med. 30(Oct.):357-372.

Singh, A., S.S. Jolly, and B.C. Bansal. 1961. Skeletal fluorosis and its neurological complications. Lancet 1:197-200.

Singh, A., B.M. Singh, I.D. Singh, S.S. Jolly, and K.C. Malhotra. 1966. Parathyroid function in endemic fluorosis. Indian J. Med. Res. 54(6):591-597.

Singh, P.P., M.K. Barjatiya, S. Dhing, R. Bhatnagar, S. Kothari, and V. Dhar. 2001. Evidence suggesting that high intake of fluoride provokes nephrolithiasis in tribal populations. Urol. Res. 29(4):238-244.

Sjögren, K., and N.H. Melin. 2001. The influence of rinsing routines on fluoride retention after toothbrushing. Gerodontology 18(1):15-20.

Smith, M.C., and H.V. Smith. 1940. Observations on the durability of mottled teeth. Am. J. Public Health 30:1050-1052.

Snow, G.R., and C. Anderson. 1986. Short term fluoride administration and trabecular bone remodeling in beagles: A pilot study. Calcif. Tissue Int. 38(4):217-221.

Søgaard, C.H., L. Mosekilde, A. Richards, and L. Mosekilde. 1994. Marked decrease in trabecular bone quality after five years of sodium fluoride therapy—assessed by biomechanical testing of iliac crest bone biopsies in osteoporotic patients. Bone 15(4): 393-399.

Søgaard, C.H., L. Mosekilde, W. Schwartz, G. Leidig, H.W. Minne, and R. Ziegler. 1995. Effects of fluoride on rat vertebral body biomechanical competence and bone mass. Bone 16(1):163-169.

Søgaard, C.H.. L. Mosekilde, J.S. Thomsen, A. Richards, and J.E. McOsker. 1997. A comparison of the effects of two anabolic agents (fluoride and PTH) on ash density and bone strength assessed in an osteopenic rat model. Bone 20(5):439-449.

Sohn, W., K.H. Heller, and B.A. Burt. 2001. Fluid consumption related to climate among children in the United States. J. Public Health Dent. 61(2):99-106.

Sokoloff, L. 1966. Cerebral circulatory and metabolic changes associated with aging. Res. Publ. Assoc. Res. Nerv. Ment. Dis. 41:237-254.

Sondhi, H., M.L. Gupta, and G.L. Gupta. 1995. Intestinal effects of sodium fluoride in Swiss albino mice. Fluoride 28(1):21-24.

Sowers, M.F., G.M. Whitford, M.K. Clark, and M.L. Jannausch. 2005. Elevated serum fluoride concentrations in women are not related to fractures and bone mineral density. J. Nutr. 135(9):2247-2252.

Sowers, M.F.R., R.B. Wallace, and J.H. Lemke. 1986. The relationship of bone mass and fracture history to fluoride and calcium intake: A study of three communities. Am. J. Clin. Nutr. 44(6):889-898.

Sowers, M.F.R., M.K. Clark, M.L. Jannausch, and R.B. Wallace. 1991. A prospective study of bone mineral content and fracture in communities with differential fluoride exposure. Am. J. Epidemiol. 133(7):649-660.

Spak, C.J., L.I. Hardell, and P. De Chateau. 1983. Fluoride in human milk. Acta Paediatr. Scand. 72(5):699-701.

Spak, C.J., S. Sjostedt, L. Eleborg, B. Veress, L. Perbeck, and J. Ekstrand. 1990. Studies of human gastric mucosa after application of 0.42% fluoride gel. J. Dent. Res. 69(2):426-429.

Spencer, H., L. Kramer, C. Norris, and E. Wiatrowski. 1980. Effect of aluminum hydroxide on fluoride metabolism. Clin. Pharmacol. Ther. 28(4):529-535.

Spira, L. 1962. Fluorine-induced endocrine disturbances in mental illness. Folia Psychiatr. Neurol. Jpn. 16(Apr.):4-14.

Spittle, B. 1994. Psychopharmacology of fluoride: A review. Int. Clin. Psychopharmacol. 9(2):79-82.

Sprando, R.L., T.N. Black, M.J. Ames, J.I. Rorie, and T.F. Collins. 1996. Effect of intratesticular injection of sodium fluoride on spermatogenesis. Food Chem. Toxicol. 34(4):377-384.

Sprando, R.L., T.F. Collins, T.N. Black, J. Rorie, M.J. Ames, and M. O'Donnell. 1997. Testing the potential of sodium fluoride to affect spermatogenesis in the rat. Food Chem. Toxicol. 35(9):881-890.

Sprando, R.L., T.F. Collins, T. Black, N. Olejnik, and J. Rorie. 1998. Testing the potential of sodium fluoride to affect spermatogenesis: A morphometric study. Food Chem. Toxicol. 36(12):1117-1124.

Srinivasan, V. 2002. Melatonin oxidative stress and neurodegenerative diseases. Indian J. Exp. Biol. 40(6):668-679.

Srivastava, R.N., D.S. Gill, A. Moudgil, R.K. Menon, M. Thomas, and P. Dandona. 1989. Normal ionized calcium, parathyroid hypersecretion, and elevated osteocalcin in a family with fluorosis. Metabolism 38(2):120-124.

Stamp, T.C., M.V. Jenkins, N. Loveridge, P.W. Saphier, M. Katakity, and S.E. MacArthur. 1988. Fluoride therapy in osteoporosis: Acute effects on parathyroid and mineral homoeostasis. Clin. Sci. 75(2):143-146.

Stamp, T.C., P.W. Saphier, N. Loveridge, C.R. Kelsey, A.J. Goldstein, M. Katakity, M.V. Jenkins, and G.A. Rose. 1990. Fluoride therapy and parathyroid hormone activity in osteoporosis. Clin. Sci. 79(3):233-238.

Stannard, J., J. Rovero, A. Tsamtsouris, and V. Gavris. 1990. Fluoride content of some bottled waters and recommendations for fluoride supplementation. J. Pedod. 14(2):103-107.

Stannard, J.G., Y.S. Shim, M. Kritsineli, P. Labropoulou, and A. Tsamtsouris. 1991. Fluoride levels and fluoride contamination of fruit juices. J. Clin. Pediatr. Dent. 16(1):38-40.
Starkstein, S.E., S. Vazquez, R. Migliorelli, A. Teson, L. Sabe, and R. Leiguarda. 1995. A single-photon emission computed tomographic study of anosognosia in Alzheimer's disease. Arch. Neurol. 52(4):415-420.
Stehle, J.H., C. von Gall, and H.W. Korf. 2003. Melatonin: A clock-output, a clock-input. J. Neuroendocrinol. 15(4):383-389.
Steiner, G.G. 2002. Cancer incidence rates and environmental factors: An ecological study. J. Environ. Pathol. Toxicol. Oncol. 21(3):205-212.
Sternweis, P.C., and A.G. Gilman. 1982. Aluminum: A requirement for activation of the regulatory component of adenylate cyclase by fluoride. Proc. Natl. Acad. Sci. USA 79(16):4888-4891.
Stevenson, C.A., and A.R. Watson. 1957. Fluoride osteosclerosis. Am. J. Roentgenol. Radium Ther. Nucl. Med. 78(1):13-18.
Steyn, D.G. 1948. Fluorine and endemic goitre. S.A. Med. J. 22(16):525-526.
Steyn, D.G., J. Kieser, W.A. Odendaal, H. Malherbe, H.W. Snyman, W. Sunkel, C.P. Naude, H. Klintworth, and E. Fisher. 1955. Endemic goitre in the Union of South Africa and some neighbouring territories [excerpts]. Union of South Africa, Department of Nutrition. March 1955 [online]. Available: http://www.slweb.org/south-africa.goitre.html [accessed Oct. 7, 2004].
Stoffer, S.S., W.E. Szpunar, and M. Block. 1982. Hyperparathyroidism and thyroid disease: A study of their association. Postgrad. Med. 71(6):91-94.
Stokinger, H.E., N.J. Ashenburg, J. DeVoldre, J.K. Scott, and F.A. Smith. 1949. The Enhancing Effect of the Inhalation of Hydrogen Fluoride Vapor on Beryllium Sulfate Poisoning in Animals. Atomic Energy Project, UR-68, University of Rochester, NY.
Stolc, V., and J. Podoba. 1960. Effect of fluoride on the biogenesis of thyroid hormones. Nature 188(4753):855-856.
Stone, R. 2004. Iceland's doomsday scenario? Science 306(5700):1278-1281.
Stormont, J., F.L. Kozelka, and M.H. Seevers. 1931. The iodine content of the thyroid following chronic fluoride administration. J. Pharmacol. Exp. Ther. 57:143-144.
Strunecka, A., and J. Patocka. 2002. Aluminofluoride complexes: A useful tool in laboratory investigations, but a hidden danger for living organisms? Pp. 271-282 in Group 13 Chemistry: From Fundamentals to Applications, P.J. Shapiro, and D.A. Atwood, eds. ACS Symposium Series 822. Washington, DC: American Chemical Society.
Strunecka, A., O. Strunecky, and J. Patocka. 2002. Fluoride plus aluminum: Useful tools in laboratory investigations, but messengers of false information. Physiol. Res. 51(6):557-564.
Suarez-Almazor, M.E., G. Flowerdew, S.D. Saunders, C.L. Soskolne, and AS. Russell. 1993. The fluoridation of drinking water and hip fracture hospitalization rates in two Canadian communities. Am. J. Public Health 83(5):689-693.
Suay, L.L., and D.F. Ballester. 2002. Review of studies on exposure to aluminum and Alzheimer's disease [in Spanish]. Rev. Esp. Salud. Publica. 76(6):645-658.
Subbareddy, V.V., and A. Tewari. 1985. Enamel mottling at different levels of fluoride in drinking water: In an endemic area. J. Indian Dent. Assoc. 57(6):205-212.
Sugimoto, T., C. Ritter, E. Slatopolsky, and J. Morrissey. 1990. Role of guanine nucleotide binding protein, cytosolic calcium and cAMP in fluoride-induced suppression of PTH secretion. Miner. Electrolyte Metab. 16(4):224-231.
Suketa, Y., and Y. Kanamoto. 1983. A role of thyroid-parathyroid function in elevation of calcium content in kidney of rats after a single large dose of fluoride. Toxicology 26(3-4):335-345.
Suketa, Y., Y. Asao, Y. Kanamoto, T. Sakashita, and S. Okada. 1985. Changes in adrenal

function as a possible mechanism for elevation of serum glucose by a single large dose of fluoride. Toxicol. Appl. Pharmacol. 80(2):199-205.
Sundström B. 1971. Thyroidal C-cells and short term, experimental fluorosis in the rat. Acta Pathol. Microbiol. Scand. A 79(4):407-409.
Susa, M., G.J. Standke, M. Jeschke, and D. Rohner. 1997. Fluoroaluminate induces pertussis toxin-sensitive protein phosphorylation: Differences in $MC_3T_3-E_1$ osteoblastic and NIH3T3 fibroblastic cells. Biochem. Biophys. Res. Commun. 235(3):680-684.
Susheela, A.K., and T.K. Das. 1988. Chronic fluoride toxicity: A scanning electron microscopic study of duodenal mucosa. J. Toxicol. Clin. Toxicol. 26(7):467-476.
Susheela, A.K., and P. Jethanandani. 1996. Circulating testosterone levels in skeletal fluorosis patients. J. Toxicol. Clin. Toxicol. 34(2):183-189.
Susheela, A.K., and A. Kumar. 1991. A study of the effect of high concentrations of fluoride on the reproductive organs of male rabbits, using light and scanning electron microscopy. J. Reprod. Fertil. 92(2):353-360.
Susheela, A.K., and A. Kumar. 1997. Ultrastructural studies on the Leydig cells of rabbits exposed to chronic fluoride toxicity. Environ. Sci. 5(2):79-94.
Susheela, A.K., A. Kumar, M. Bhatnagar, and R. Bahadur. 1993. Prevalence of endemic fluorosis with gastrointestinal manifestations in people living in some North-Indian villages. Fluoride 26(2):97-104.
Susheela, A.K., M. Bhatnagar, K. Vig, and N.K. Mondal. 2005. Excess fluoride ingestion and thyroid hormone derangements in children living in Delhi, India. Fluoride 38(2):98-108.
Swarup, D., S.K. Dwivedi, S. Dey, and S.K. Ray. 1998. Fluoride intoxication in bovines due to industrial pollution. Indian J. Anim. Sci. 68(7):605-608.
Szpunar, S.M., and B.A. Burt. 1988. Dental caries, fluorosis, and fluoride exposure in Michigan schoolchildren. J. Dent. Res. 67(5):802-806.
Szpunar, S.M., and B.A. Burt. 1990. Fluoride exposure in Michigan schoolchildren. J. Public Health Dent. 50(1):18-23.
Taivainen, T., P. Tiainen, O.A. Meretoja, L. Raiha, and P.H. Rosenberg. 1994. Comparison of the effects of sevoflurane and halothane on the quality of anaesthesia and serum glutathione transferase alpha and fluoride in paediatric patients. Br. J. Anaesth. 73(5):590-595.
Takahashi, K. 1998. Fluoride-linked Down syndrome births and their estimated occurrence due to water fluoridation. Fluoride 31(2):61-73.
Takahashi, K., K. Akiniwa, and K. Narita. 2001. Regression analysis of cancer incidence rates and water fluoride in the U.S.A. based on IACR/IARC (WHO) data (1978-1992). International Agency for Research on Cancer. J. Epidemiol. 11(4):170-179.
Talbot, J.R., M.M. Fischer, and S.M. Farley, C. Libanati, J. Farley, A. Tabuenca, and D.J. Baylink. 1996. The increase in spinal bone density that occurs in response to fluoride therapy for osteoporosis is not maintained after therapy is discontinued. Osteoporosis Int. 6(6):442-447.
Taves, D.R., and W.S. Guy. 1979. Distribution of fluoride among body compartments. Pp. 159-186 in Continuing Evaluation of the Use of Fluorides, E. Johansen, D.R. Taves, and T.O. Olsen, eds. AAAS Selected Symposium 11. Boulder, CO: Westview Press.
Taylor, M.L., A. Boyde, and S.J. Jones. 1989. The effect of fluoride on the patterns of adherence of osteoclasts cultured on and resorbing dentine: A 3-D assessment of vinculin-labelled cells using confocal microscopy. Anat. Embryol. 180(5):427-435.
Taylor, M.L., E. Maconnachie, K. Frank, A. Boyde, and S.J. Jones. 1990. The effect of fluoride on the resorption of dentine by osteoclasts in vitro. J. Bone Miner. Res. 5(Suppl. 1):S121-S130.
Tennant, R.W. 1987. Issues in biochemical applications to risk assessment: Are short-term tests predictive of in vivo tumorigenicity? Environ. Health Perspect. 76:163-167.

Tennant, R.W., B.H. Margolin, M.D. Shelby, E. Zeiger, J.K. Haseman, J. Spalding, W. Caspary, M. Resnick, S. Stasiewicz, and B. Anderson. 1987. Prediction of chemical carcinogenicity in rodents from in vitro genetic toxicity assays. Science 236(4804):933-941.

Teotia, M., A. Rodgers, S.P. Teotia, A.E. Wandt, and M. Nath. 1991. Fluoride metabolism and fluoride content of stones from children with endemic vesical stones. Br. J. Urol. 68(4):425-429.

Teotia, M., S.P. Teotia, and K.P. Singh. 1998. Endemic chronic fluoride toxicity and dietary calcium deficiency interaction syndromes of metabolic bone disease and deformities in India: Year 2000. Indian J. Pediatr. 65(3):371-381.

Teotia, S.P., and M. Teotia. 1973. Secondary hyperparathyroidism in patients with endemic skeletal fluorosis. Br. Med. J. 1(5854):637-640.

Teotia, S.P., M. Teotia, R.K. Singh, D.R. Taves, and S. Heels. 1978. Endocrine aspects of endemic skeletal fluorosis. J. Assoc. Physicians India 26(11):995-1000.

Thaper, R., A. Tewari, H.S. Chawla, and V. Sachdev. 1989. Prevalence and severity of dental fluorosis in primary and permanent teeth at varying fluoride levels. J. Indian Soc. Pedod. Prev. Dent. 7(1):38-45.

Thylstrup, A., and O. Fejerskov. 1978. Clinical appearance of dental fluorosis in permanent teeth in relation to histologic changes. Community Dent. Oral. Epidemiol. 6(6):315-328.

Tice, R.R., H.F. Stack, and M.D. Waters. 1996. Human exposures to mutagens—an analysis using the genetic activity profile database. Environ. Health Perspect. 104(Suppl 3):585-589.

Tiwari, S., S.K. Gupta, K. Kumar, R. Trivedi, and M.M. Godbole. 2004. Simultaneous exposure of excess fluoride and calcium deficiency alters VDR, CaR, and Calbindin D 9 k mRNA levels in rat duodenal mucosa. Calcif. Tissue Int. 75(4):313-320.

Tohyama, E. 1996. Relationship between fluoride concentration in drinking water and mortality rate from uterine cancer in Okinawa prefecture, Japan. J. Epidemiol. 6(4):184-191.

Tokar', V.I., and O.N. Savchenko. 1977. Effect of inorganic fluorine compounds on the functional state of the pituitary-testis system [in Russian]. Probl. Endokrinol. 23(4):104-107.

Tokar', V.I., V.V. Voroshnin, and S.V. Sherbakov. 1989. Chronic effects of fluorides on the pituitary-thyroid system in industrial workers [in Russian]. Gig. Tr. Prof. Zabol. 1989(9):19-22.

Tormanen, C.D. 2003. Substrate inhibition of rat liver and kidney arginase with fluoride. J. Inorg. Biochem. 93(3-4):243-246.

Torra, M., M. Rodamilans, and J. Corbella. 1998. Serum and urine fluoride concentration: Relationships to age, sex, and renal function in a non-fluoridated population. Sci. Total Environ. 220(1):81-85.

Toumba, K.J., S. Levy, and M.E. Curzon. 1994. The fluoride content of bottled drinking waters. Br. Dent. J. 176(7):266-268.

Trabelsi, M., F. Guermazi, and N. Zeghal. 2001. Effect of fluoride on thyroid function and cerebellar development in mice. Fluoride 34(3):165-173.

Train, T.E., A.G. McWhorter, N.S. Seale, C.F. Wilson, and I.Y. Guo. 1996. Examination of esthetic improvement and surface alteration following microabrasion in fluorotic human incisors in vivo. Pediatr. Dent. 18(5):353-362.

TRI (Toxic Release Inventory). 2003. TRI Explorer. Release Reports: Chemical Report for Fluorine and Hydrogen Fluoride, 2001 Data Update as of July 25, 2003. Toxic Release Inventory, U.S. Environmental Protection Agency, Washington, DC [online]. Available: http://www.epa.gov/triexplorer/trends.htm [accessed Sept. 14, 2004].

Trivedi, N., A. Mithal, S.K. Gupta, and M.M. Godbole. 1993. Reversible impairment of glucose tolerance in patients with endemic fluorosis. Fluoride Collaborative Study Group. Diabetologia 36(9):826-828.

Truman, B.I., B.F. Gooch, I. Sulemana, H.C. Gift, A.M. Horowitz, C.A. Evans, S.O. Griffin, and V.G. Carande-Kulis. 2002. Reviews of evidence on interventions to prevent dental caries, oral and pharyngeal cancers, and sports-related craniofacial injuries. Am. J. Prev. Med. 23(1 Suppl.):21-54.

TSO (The Stationery Office Limited). 2004. The Natural Mineral Water, Spring Water and Bottled Drinking Water (Amendment) England Regulations 2004. Statutory Instrument 2004 No. 656 [online]. Available: http://www.opsi.gov.uk/si/si2004/20040656.htm [accessed Nov. 1, 2005].

Tsutsui, T., Y. Tanaka, Y. Matsudo, A. Uehama, T. Someya, F. Hamaguchi, H. Yamamoto, and M. Takahashi. 1995. No increases in chromosome aberrations in human diploid fibroblasts following exposure to low concentrations of sodium fluoride for long times. Mutat. Res. 335(1):15-20.

Turner, C.H., M.P. Akhter, and R.P. Heaney. 1992. The effects of fluoridated water on bone strength. J. Orthop. Res. 10(4):581-587.

Turner, C.H., G. Boivin, and P.J. Meunier. 1993. A mathematical model for fluoride uptake by the skeleton. Calcif. Tissue Int. 52(2):130-138.

Turner, C.H., K. Hasegawa, W. Zhang, M. Wilson, Y. Li, and A.J. Dunipace. 1995. Fluoride reduces bone strength in older rats. J. Dent. Res. 74(8):1475-1481.

Turner, C.H., I. Owan, E.J. Brizendine, W. Zhang, M.E. Wilson, and A.J. Dunipace. 1996. High fluoride intakes cause osteomalacia and diminished bone strength in rats with renal deficiency. Bone 19(6):595-601.

Turner, C.H., L.P. Garetto, A.J. Dunipace, W. Zhang, M.E. Wilson, M.D. Grynpas, D. Chachra, R. McClintock, M. Peacock, and G.K. Stookey. 1997. Fluoride treatment increased serum IGF-1, bone turnover, and bone mass, but not bone strength, in rabbits. Calcif. Tissue Int. 61(1):77-83.

Urbansky, E.T. 2002. Fate of fluorosilicate drinking water additives. Chem. Rev. 102(8): 2837-2854.

Urbansky, E.T., and M.R. Schock. 2000. Can fluoride affect lead (II) in potable water? Hexafluorosilicate and fluoride equilibria in aqueous solution. Int. J. Environ. Studies 57:597-637.

Urbansky, E.T., and M.R. Schock. Undated. Can Fluoride Affect Water Lead (II) Levels and Lead (II) Neurotoxicity? National Risk Management Research Laboratory, Office of Research and Development, U.S. Environmental Protection Agency, Cincinnati, OH. [online]. Available: http:/fluoride.oralhealth.org/papers/pdf/urbansky.pdf [accessed Oct. 1, 2004].

U.S. Army. 1983. Water Consumption Planning Factors Study. Directorate of Combat Developments, U. S. Army Quartermaster School, Fort Lee, VA (as cited in EPA 1997).

USASMA (U.S. Army Sergeants Major Academy). 2003. Revised Fluid Replacement Policy for Warm Weather Training. Health Promotion Office, U.S. Army Sergeants Major Academy [online]. Available: http://usasma.bliss.army.mil/HPO/Fluid.htm [accessed Sept. 20, 2004].

U.S. Census Bureau. 2000. Census 2000. Population Division, U.S. Census Bureau [online]. Available: http://factfinder.census.gov [accessed August 17, 2005].

USDA (U.S. Department of Agriculture). 2004. USDA National Fluoride Database of Selected Beverages and Foods. Nutrient Data Laboratory, Beltsville Human Nutrition Research Center, Agricultural Research Service, U.S. Department of Agriculture [online]. Available: http://www.nal.usda.gov/fnic/foodcomp/Data/Fluoride/Fluoride.html [accessed Sept. 21, 2005].

USDHHS (U.S. Department of Health and Human Services). 2000. Oral Health in America: A Report of the Surgeon General. U.S. Department of Health and Human Services, National Institute of Dental and Craniofacial Research, National Institutes of Health, Rockville, MD.

Usuda, K., K. Kono, and Y. Yoshida. 1997. The effect of hemodialysis upon serum levels of fluoride. Nephron 75(2):175-178.
Uz, T., N. Dimitrijevic, M. Akhisaroglu, M. Imbesi, M. Kurtuncu, and H. Manev. 2004. The pineal gland and anxiogenic-like action of fluoxetine in mice. Neuroreport 15(4): 691-694.
van Asten, P., F. Darroudi, A.T. Natarajan, I.J. Terpstra, and S.A. Duursma. 1998. Cytogenetic effects on lymphocytes in osteoporotic patients on long-term fluoride therapy. Pharm. World Sci. 20(5):214-218.
Vanden Heuvel, J.P., B.I. Kuslikis, M.J. Van Rafelghem, and R.E. Peterson. 1991. Tissue distribution, metabolism, and elimination of perfluorooctanoic acid in male and female rats. J. Biochem. Toxicol. 6(2):83-92.
van Kesteren, R.G., S.A. Duursma, W.J. Visser, J. van der Sluys Veer, and O. Backer Dirks. 1982. Fluoride in serum and bone during treatment of osteoporosis with sodium fluoride, calcium and vitamin D. Metab. Bone Dis. Relat. Res. 4(1):31-37.
Van Winkle, S., S.M. Levy, M.C. Kiritsy, J.R. Heilman, J.S. Wefel, and T. Marshall. 1995. Water and formula fluoride concentrations: Significance for infant fed formula. Pediatr. Dent. 17(4):305-310.
Varner, J.A., C.W. Huie, W.J. Horvath, K.F. Jensen, and R.L. Isaacson. 1993. Chronic AIF_3 administration: II. Selected histological observations. Neurosci. Res. Commun. 13(2):99-104.
Varner, J.A., K.F. Jensen, W. Horvath, and R.L. Isaacson. 1998. Chronic administration of aluminum-fluoride or sodium fluoride to rats in drinking water: Alterations in neuronal and cerebrovascular integrity. Brain Res. 784(1-2):284-298.
Varner, J.A., W.J. Horvath, C.W. Huie, H.R. Naslund, and R.L. Isaacson. 1994. Chronic aluminum fluoride administration. Behav. Neural. Biol. 61(3):233-241.
Venkateswarlu, P., D.N. Rao, and K.R. Rao. 1952. Studies in endemic fluorosis: Visakhapatnam and suburban areas. Fluorine mottled enamel and dental caries. Ind. J. Med. Res. 40(4):535-548.
Verma, R.J., and D.M. Guna Sherlin. 2001. Vitamin C ameliorates fluoride-induced embryotoxicity in pregnant rats. Hum. Exp. Toxicol. 20(12):619-623.
Verma, R.J., and D.M. Guna Sherlin. 2002a. Sodium fluoride-induced hypoproteinemia and hypoglycemia in parental and F_1 generation rats and amelioration by vitamins. Food Chem. Toxicol. 40(12):1781-1788.
Verma, R.J., and D.M. Guna Sherlin. 2002b. Hypocalcaemia in parental and F_1 generation rats treated with sodium fluoride. Food Chem. Toxicol. 40(4):551-554.
Vieira, A.P., R. Hancock, H. Limeback, R. Maia, and M.D. Grynpas. 2004. Is fluoride concentration in dentin and enamel a good indicator of dental fluorosis? J. Dent. Res. 83(1):76-80.
Vieira, A.P.G.F., M. Mousny, R. Maia, R. Hancock, E.T. Everett, and M.D. Grynpas. 2005. Assessment of teeth as biomarkers for skeletal fluoride exposure. Osteoporos. Int. 16(12):1576-1582.
Vígh, B., A. Szél, K. Debreceni, Z. Fejér, M.J. Manzano e Silva, and I. Vígh-Teichmann. 1998. Comparative histology of pineal calcification. Histol. Histopathol. 13(3):851-870.
Vine, M.F. 1994. Biological markers: Their use in quantitative assessments. Adv. Dent. Res. 8(1):92-99.
Virtanen, J.I., R.S. Bloigu, and M.A. Larmas. 1994. Timing of eruption of permanent teeth: Standard Finnish patient documents. Community Dent. Oral Epidemiol. 22(5 Part 1):286-288.
Vogt, R.L., L. Witherell, D. LaRue, and D.N. Klaucke. 1982. Acute fluoride poisoning associated with an on-site fluoridator in a Vermont elementary school. Am. J. Public Health 72(10):1168-1169.

von Tirpitz, C., J. Klaus, M. Steinkamp, L.C. Hofbauer, W. Kratzer, R. Mason, B.O. Boehm, G. Adler, and M. Reinshagen. 2003. Therapy of osteoporosis in patients with Crohn's disease: A randomized study comparing sodium fluoride and ibandronate. Aliment. Pharmacol. Ther. 17(6):807-816.
Wadayama, B., J. Toguchida, T. Yamaguchi, M.S. Sasaki, and T. Yamamuro. 1993. p53 expression and its relationship to DNA alterations in bone and soft tissue sarcomas. Br. J. Cancer 68(6):1134-1139.
Wagener, D.K., P. Nourjah, and A.M. Horowitz. 1992. Trends in Childhood Use of Dental Care Products Containing Fluoride: United States, 1983-1989. Hyattsville, MD: U.S. Department of Health and Human Services, Public Health Service, Centers for Disease Control.
Waldbott, G.L. 1956. Incipient chronic fluoride intoxication from drinking water. II. Distinction between allergic reactions and drug intolerance. Int. Arch. Allergy Appl. Immunol. 9(5):241-249.
Waldbott, G.L., A.W. Burgstahler, and H.L. McKinney. 1978. Fluoridation: The Great Dilemma. Lawrence, KS: Coronado Press.
Wang, A.G., T. Xia, Q.L. Chu, M. Zhang, F. Liu, X.M. Chen, and K.D. Yang. 2004. Effects of fluoride on lipid peroxidation, DNA damage and apoptosis in human embryo hepatocytes. Biomed. Environ. Sci. 17(2):217-222.
Wang, D., and M. Qian. 1989. Report on the village use of the recension Chinese combined Raven's Test. Inform. Phychol. Sci. 5:23-27.
Wang, R. 2005. Anisotropic fracture in bovine root and coronal dentin. Dent. Mater. 21(5):429-436.
Wang, Y.N., K.Q. Xiao, J.L. Liu, G. Dallner, and Z.Z. Guan. 2000. Effect of long term fluoride exposure on lipid composition in rat liver. Toxicology 146(2-3):161-169.
Warnakulasuriya, K.A., S. Balasuriya, P.A. Perera, and L.C. Peiris. 1992. Determining optimal levels of fluoride in drinking water for hot, dry climates—a case study in Sri Lanka. Community Dent. Oral Epidemiol. 20(6):364-367.
Warnakulasuriya, S., C. Harris, S. Gelbier, J. Keating, and T. Peters. 2002. Fluoride content of alcoholic beverages. Clin. Chim. Acta 320(1-2):1-4.
Warren, D.P., H.A. Henson, and J.T. Chan. 1996. Comparison of fluoride content in caffeinated, decaffeinated and instant coffee. Fluoride 29(3):147-150.
Warren, J.J., and S.M. Levy. 1999. Systemic fluoride. Sources, amounts, and effects of ingestion. Dent. Clin. North Am. 43(4):695-711.
Waterhouse, C., D. Taves, and A. Munzer. 1980. Serum inorganic fluoride: Changes related to previous fluoride intake, renal function and bone resorption. Clin. Sci. 58(2):145-152.
Webster, T. 2000. Bias in Ecologic and Semi-Individual Studies. D.Sc. dissertation, Boston University School of Public Health.
Webster, T. 2002. Commentary: Does the spectre of ecologic bias haunt epidemiology? Int. J. Epidemiol. 31(1):161-162.
Weetman, A.P. 1997. Hypothyroidism: Screening and subclinical disease. Br. Med. J. 314(7088): 1175-1178.
Weinberger, S.J. 1991. Bottled drinking waters: Are the fluoride concentrations shown on the labels accurate? Int. J. Pediatr. Dent. 1(3):143-146.
Weiner, M.F., W.H. Wighton-Benn, R. Risser, D. Svetlik, R. Tintner, J. Hom, R.N. Rosenberg, and F.J. Bonte. 1993. Xenon-133 SPECT-determined regional cerebral blood flow in Alzheimer's disease: What is typical? J. Neuropsychiatry Clin. Neurosci. 5(4):415-418.
Weiss, B. 1969a. Similarities and differences in the norepinephrine-and sodium fluoride-sensitive adenyl cyclase system. J. Pharmacol. Exp. Ther. 166(2):330-338.
Weiss, B. 1969b. Effects of environmental lighting and chronic denervation on the activation of

adenyl cyclase of rat pineal gland by norepinephrine and sodium fluoride. J. Pharmacol. Exp. Ther. 168(1):146-152.

Wergedal, J.E., K.H. Lau, and D.J. Baylink. 1988. Fluoride and bovine bone extract influence cell proliferation and phosphatase activities in human bone cell cultures. Clin. Orthop. Relat. Res. 233:274-282.

Wergedal, J.W., and K.H. Lau. 1992. Human bone cells contain a fluoride sensitive acid phosphatase: Evidence that this enzyme functions at neutral pH as a phosphotyrosyl protein phosphatase. Clin. Biochem. 25(1):47-53.

Westendorf, J. 1975. The Kinetics of Actylcholinesterase Inhibition and the Influence of Fluoride and Fluoride Complexes on the Permeability of Erythrocyte Membranes [in German]. Ph.D. Thesis, University of Hamburg, Hamburg, Germany (as cited in Masters et al. 2000).

Whitford, G.M. 1994. Intake and metabolism of fluoride. Adv. Dent. Res. 8(1):5-14.

Whitford, G.M. 1996. The Metabolism and Toxicity of Fluoride, 2nd Rev. Ed. Monographs in Oral Science Vol. 16. New York: Karger.

Whitford, G.M. 1999. Fluoride metabolism and excretion in children. J. Pub. Health Dent. 59(4):224-228.

Whitford, G.M., and N.L. Birdsong-Whitford. 2000. Plasma as a biomarker for bone fluoride concentrations in rats. Abstract No. 2283. J. Dent Res. 79(IADR Abstracts):429.

Whitford, G.M., and E.R. Taves. 1973. Fluoride-induced diuresis. Anesthesiology 39(4): 416-427.

Whitford, G.M., D.H. Pashley, and K.E. Reynolds. 1979. Fluoride tissue distribution: short-term kinetics. Am. J. Physiol. 236(2):F141-F148.

Whitford, G.M., E.D. Biles, and N.L. Birdsong-Whitford. 1991. A comparative study of fluoride pharmacokinetics in five species. J. Dent. Res. 70(6):948-951.

Whitford, G.M., J.W. Bawden, W.H. Bowen, L.J. Brown, J.E. Ciardi, T.W. Clarkson, P.B. Imrey, M. Kleerekoper, T.M. Marthaler, S. McGuire, R.H. Ophaug, C. Robinson, J.S. Schultz, G.K. Stookey, M.S. Tochman, P. Venkateswarlu, and D.T. Zero. 1994. Report for Working Group I: Strategies for improving the assessment of fluoride accumulation in body fluids and tissues. Adv. Dent. Res. 8(1):113-115.

Whitford, G.M., D.H. Pashley, and R.H. Garman. 1997. Effects of fluoride on structure and function of canine gastric mucosa. Dig. Dis. Sci. 42(10):2146-2155.

Whitford, G.M., F.C. Sampaio, P. Arneberg, and F.R. von der Fehr. 1999a. Fingernail fluoride: A method for monitoring fluoride exposure. Caries Res. 33(6):462-467.

Whitford, G.M., J.E. Thomas, and S.M Adair. 1999b. Fluoride in whole saliva, parotid ductal saliva and plasma in children. Arch. Oral Biol. 44(10):785-788.

Whitford, G.M., M.A. Buzalaf, M.F. Bijella, and J.L. Waller. 2005. Plaque fluoride concentrations in a community without water fluoridation: Effects of calcium and use of a fluoride or placebo dentifrice. Caries Res. 39(2):100-107.

Whiting, P., M. MacDonagh, and J. Kleijnen. 2001. Association of Down's syndrome and water fluoride level: A systematic review of the evidence. BMC Public Health 1(1):6.

WHO (World Health Organization). 2002. Fluorides. Environmental Health Criteria 227. Geneva, Switzerland: International Programme on Chemical Safety, World Health Organization [online]. Available: http://www.inchem.org/documents/ehc/ehc/ehc227.htm [accessed Oct. 8, 2004].

Whyte, M.P., K. Essmyer, F.H. Gannon, and W.R. Reinus. 2005. Skeletal fluorosis and instant tea. Am. J. Medicine 118(1):78-82.

Wilkinson, P.C. 1983. Effects of fluoride on locomotion of human blood leucocytes in vitro. Arch. Oral Biol. 28(5):415-418.

Wilson, D.C. 1941. Fluorine in the aetiology of endemic goitre. Lancet 1(Feb. 15):211-212.

Winterer, G., and D. Goldman. 2003. Genetics of human prefrontal function. Brain Res. Rev. 43(1):134-163.
Wolff, W.A., and E.G. Kerr. 1938. The composition of human bone in chronic fluoride poisoning. Am. J. Med. Sci. 195:493-497.
Wollan, M. 1968. Controlling the potential hazards of government-sponsored technology. George Wash. Law Rev. 36(5):1105-1137.
Wolstenholme, J., and R.R. Angell. 2000. Maternal age and trisomy—A unifying mechanism of formation. Chromosoma 109(7):435-438.
Wondwossen , F., A.N. Astrom, K. Bjorvatn, and A. Bardsen. 2004. The relationship between dental caries and dental fluorosis in areas with moderate- and high-fluoride drinking water in Ethiopia. Community Dent. Oral Epidemiol. 32(5):337-344.
Wong, F.S., and G.B. Winter. 2002. Effectiveness of microabrasion technique for improvement of dental aesthetics. Br. Dent. J. 193(3):155-158.
Wong, M.H., K.F. Fung, and H.P. Carr. 2003. Aluminum and fluoride content in tea, with emphasis on brick tea, and their health implications. Toxicol. Lett. 137(1-2):111-120.
Woolf, A.D., and K. Åkesson. 2003. Preventing fractures in elderly people. Br. Med. J. 327(7406):89-95.
Wu, D.Q., and Y. Wu. 1995. Micronucleus and sister chromatid exchange frequency in endemic fluorosis. Fluoride 28(3):125-127.
Wu, T., P. Mendola, and G.M. Buck. 2002. Ethnic differences in the presence of secondary sex characteristics and menarche among US girls: The Third National Health and Nutrition Examination Survey, 1988-1994. Pediatrics 110(4):752-757.
Xiang, Q., Y. Liang, L. Chen, C. Wang, B. Chen, X. Chen, and M. Zhou. 2003a. Effect of fluoride in drinking water on children's intelligence. Fluoride 36(2):84-94.
Xiang, Q., Y. Liang, M. Zhou, and H. Zang. 2003b. Blood lead of children in Wamiao-Xinhuai intelligence study [letter]. Fluoride 36(3):198-199.
Yamamoto, S., K. Katagiri, M. Ando, and X.Q. Chen. 2001. Suppression of pulmonary antibacterial defenses mechanisms and lung damage in mice exposed to fluoride aerosol. J. Toxicol. Environ. Health A 62(6):485-494.
Yang, C.Y., M.F. Cheng, S.S. Tsai, and C.F. Hung. 2000. Fluoride in drinking water and cancer mortality in Taiwan. Environ. Res. 82(3):189-193.
Yang, Q., S.L. Sherman, T.J. Hassold, K. Allran, L. Taft, D. Pettay, M.J. Khoury, J.D. Erickson, and S.B. Freeman. 1999. Risk factors for trisomy 21: Maternal cigarette smoking and oral contraceptive use in a population-based case-control study. Genet. Med. 1(3):80-88.
Yang, Y., X. Wang, and X. Guo. 1994. Effects of high iodine and high fluorine on children's intelligence and the metabolism of iodine and fluorine [in Chinese]. Zhonghua Liu Xing Bing Xue Za Zhi 15(5):296-298.
Yatani, A., and A.M. Brown. 1991. Mechanism of fluoride activation of G protein-gated muscarinic atrial K+ channels. J. Biol. Chem. 266(34):22872-22877.
Yates, C., S. Doty, and R.V. Talmage. 1964. Effects of sodium fluoride on calcium homeostasis. Proc. Soc. Exp. Biol. Med. 115(Apr.):1103-1108.
Yiamouyiannis, J.A. 1993. Fluoridation and cancer: The biology and epidemiology of bone and oral cancer related to fluoridation. Fluoride 26(2):83-96.
Yoder, K.M., L. Mabelya, V.A. Robison, A.J. Dunipace, E.J. Brizendine, and G.K. Stookey. 1998. Severe dental fluorosis in a Tanzanian population consuming water with negligible fluoride concentration. Community Dent. Oral Epidemiol. 26(6):382-393.
Yokel, R.A., S.S. Rhineheimer, R.D. Brauer, P. Sharma, D. Elmore, and P.J. McNamara. 2001. Aluminum bioavailability from drinking water is very low and not appreciably influenced by stomach contents or water hardness. Toxicology 161(1-2):93-101.
Yuan, S., K. Song, Q. Xie, and F. Lu. 1994. Experimental study of inhibition of lactation due to fluorosis in rat. Environ. Sci. 2(4):179-187.

Zaleske, D.J., M.G. Ehrlich, C. Piliero, J.W. May, Jr., and H.J. Mankin. 1982. Growth-plate behavior in whole joint replantation in the rabbit. J. Bone Joint Surg. Am. 64)2):249-258

Zatz, M. 1977. Effects of cholera toxin on supersensitive and subsensitive rat pineal glands: Regulation of sensitivity at multiple sites. Life Sci. 21(9):1267-1276.

Zatz, M. 1979. Low concentrations of lithium inhibit the synthesis of cyclic AMP and cyclic GMP in the rat pineal gland. J. Neurochem. 32(4):1315-1321.

Zeiger, E., D.K. Gulati, P. Kaur, A.H. Mohamed, J. Revazova, and T.G. Deaton. 1994. Cytogenetic studies of sodium fluoride in mice. Mutagenesis 9(5):467-471.

Zero, D.T., R.F. Raubertas, J. Fu, A.M. Pedersen, A.L. Hayes, and J.D. Featherstone. 1992. Fluoride concentrations in plaque, whole saliva, and ductal saliva after application of home-use topical fluorides. J. Dent. Res. 71(11):1768-1775.

Zerwekh, J., A. Morris, P. Padalino, F. Gottschalk, and C.Y. Pak. 1990. Fluoride rapidly and transiently raises intracellular Ca in human osteoblasts. J. Bone Miner. Res. 5(Suppl. 1): S131-S136.

Zerwekh, J.E., P.P. Antich, S. Mehta, K. Sakhaee, F. Gottschalk, and C.Y. Pak. 1997a. Reflection ultrasound velocities and histomorphometric and connectivity analyses: Correlations and effect of slow-release sodium fluoride. J. Bone Miner. Res. 12(12):2068-2075.

Zerwekh, J.E., P. Padalino, and C.Y. Pak. 1997b. The effect of intermittent slow-release sodium fluoride and continuous calcium citrate therapy on calcitropic hormones, biochemical markers of bone metabolism, and blood chemistry in postmenopausal osteoporosis. Calcif. Tissue Int. 61(4):272-278.

Zhao, L.B., G.H. Liang, D.N. Zhang, and X.R. Wu. 1996. Effect of a high fluoride water supply on children's intelligence. Fluoride 29(4):190-192.

Zhao, W., H. Zhu, Z. Yu, K. Aoki, J. Misumi, and X. Zhang. 1998. Long-term effects of various iodine and fluorine doses on the thyroid and fluorosis in mice. Endocr. Regul. 32(2):63-70.

Zhao, Z.L., N.P. Wu, and W.H. Gao. 1995. The influence of fluoride on the content of testosterone and cholesterol in rat. Fluoride 28(3):128-130.

Zipkin, I., F.J. McClure, N.C. Leone, and W.A. Lee. 1958. Fluoride deposition in human bones after prolonged ingestion of fluoride in drinking water. Public Health Rep. 73(8):732-740.

Zohouri, F.V., A. Maguire, and P.J. Moynihan. 2003. Fluoride content of still bottled waters available in the North-East of England, UK. Br. Dent. J. 195(9):515-518.

APPENDIX
A

Biographical Information on the Committee on Fluoride in Drinking Water

JOHN DOULL (*Chair*) is professor emeritus of pharmacology and toxicology at the University of Kansas Medical School. His distinguished career in toxicology includes service in a variety of leadership positions and on numerous scientific advisory committees. Most notably, he is past president of the Society of Toxicology and the American Board of Toxicology. Dr. Doull is the recipient of many awards, including the International Achievement Award from the International Society for Regulatory Toxicology and Pharmacology, the Commanders Award for Public Service from the Department of the Army, and the Stockinger Award from the American Conference of Governmental Industrial Hygienists. He was the first recipient of the John Doull Award, which was established by the Central States Chapter of the Society of Toxicology to recognize his contributions to the discipline of toxicology. He is former chair of the NRC Committee on Toxicology and former vice chair of the Board on Environmental Studies and Toxicology. He is a national associate of the National Academies. Dr. Doull received his M.D. and Ph.D. in pharmacology from the University of Chicago.

KIM BOEKELHEIDE is professor and acting chair of the Department of Pathology and Laboratory Medicine at Brown University. His research interests are in male reproductive biology and toxicology, particularly the potential roles of germ-cell proliferation and apoptosis and local paracrine growth factors in regulating spermatogenesis after toxicant-induced injury. Dr. Boekelheide serves on the NRC Committee on Toxicity Testing and Assessment of Environmental Agents and has served on the Committee on Gender Differences in Susceptibility to Environmental Factors: A Priority

Assessment. He is a past member of the Board of Scientific Counselors of the National Toxicology Program (NTP), currently serves on the NTP Center for the Evaluation of Risks to Human Reproduction expert panel that is evaluating di-(2-ethlyhexyl)phthalate, and was chair of the National Institutes of Health Center for Scientific Review Special Emphasis Panel, Fetal Basis of Adult Disease: Role of the Environment. Dr. Boekelheide received his M.D. and Ph.D. in pathology from Duke University and is board certified in anatomic and clinical pathology.

BARBARA FARISHIAN is a practicing dentist in Washington, DC, and is on the faculty of the University of Maryland Dental School. She is a fellow of the Academy of General Dentistry, past president of the Capitol Academy of Dentistry, and a member of the Board of Directors of the District of Columbia Dental Society, an affiliate of the American Dental Association. Before attending dental school, Dr. Farishian was a toxicologist at the U.S. Environmental Protection Agency and was on the biomedical research staff of the Wistar Institute of the University of Pennsylvania. She received her D.D.S. from the Georgetown University Dental School.

ROBERT L. ISAACSON is a distinguished professor of psychology at Binghamton University. His research interests are in behavioral neuroscience, particularly the study of recovery from brain damage, functions of the limbic system, mechanisms responsible for neuronal cell death, and the neurotoxic effects of certain fluoride complexes. He is a past president of the International Behavioral Neuroscience Society and is a recipient of the Society's Lifetime Achievement Award. He serves on a number of editorial boards, including that of *Brain Research*. He has received fellow status in several scientific societies. He has served as chairperson and member of several committees of the Society for Neuroscience. In the past he has served as a member of grant review panels for the National Institutes of Health, the National Institute of Mental Health, and the National Science Foundation. He received his Ph.D. from the University of Michigan.

JUDITH B. KLOTZ is an adjunct associate professor at the University of Medicine and Dentistry of New Jersey School of Public Health. Previously, she was program manager of the cancer surveillance and environmental epidemiology programs at the New Jersey Department of Health and Senior Services. Her research interests are in epidemiological studies of cancer incidence and reproductive outcomes, gene-environment interactions, evaluation of biological exposures to environmental contaminants, and the application of health risk assessment and epidemiology to public policy. She received her M.S. in genetics from the University of Michigan and her

Dr.P.H. in environmental health sciences from Columbia University School of Public Health.

JAYANTH V. KUMAR is director of the Oral Health Surveillance & Research Unit, Bureau of Dental Health, at the New York State Department of Health. He also holds an appointment as an associate professor in the Department of Health Policy, Management, and Behavior at the School of Public Health of the University at Albany, State University of New York. He is a diplomate and former president of the American Board of Dental Public Health. His research interests are in exposure to fluoride, its effects on oral health, and health promotion and disease prevention strategies. Dr. Kumar received his dental degree from Bangalore University, M.P.H. from Johns Hopkins University, and postdoctoral certificate in dental public health from the New York State Department of Health.

HARDY LIMEBACK is an associate professor and head of preventive dentistry at the University of Toronto; he is also a part-time practicing dentist. His research interests are in tooth development, enamel proteins, caries, and prevention of dental fluorosis. Dr. Limeback is a former president of the Canadian Association of Dental Research. He has been involved for many years in reviewing the scientific literature related to fluoridation of drinking water. He received his Ph.D. in collagen biochemistry and his D.D.S. from the University of Toronto.

CHARLES POOLE is an associate professor in the Department of Epidemiology at the University of North Carolina School of Public Health. Previously, he was with the Boston University School of Public Health. Dr. Poole's work currently focuses on the development and utilization of epidemiologic methods and principles, including problem definition, study design, data collection, statistical analysis, and interpretation and application of research results, including systematic review and meta-analysis. His research experience includes studies in environmental and occupational epidemiology and other substantive areas. Dr. Poole was an epidemiologist in the Office of Pesticides and Toxic Substances of the U.S. Environmental Protection Agency for 5 years and worked for a decade as an epidemiologic consultant, both with a firm and independently. He received his M.P.H in health administration from the University of North Carolina School of Public Health and his Sc.D. in epidemiology from the Harvard School of Public Health. Dr. Poole was a member of the Institute of Medicine Committee on Gulf War and Health: Review of the Literature on Pesticides and Solvents and the National Research Council Committee on Estimating the Health-Risk-Reduction Benefits of Proposed Air Pollution Regulations.

J. EDWARD PUZAS is the Donald and Mary Clark Professor of Orthopaedics at the University of Rochester School of Medicine and Dentistry. He also holds faculty appointments in biochemistry, biomedical engineering, oncology, and pathology and laboratory medicine. He is director of the university's Osteoporosis Center and Center for Musculoskeletal Research. His research interests are in all aspects of bone, cartilage, orthopaedic, and dental biology, with a particular interest in diseases of the skeleton, such as osteoporosis and some skeletal cancers. He also directs the osteotoxicology research core at the university's National Institutes of Environmental Health Sciences center program at the University of Rochester Medical Center, where he conducts research on adverse impacts of environmental agents on skeletal tissue. He has won several awards for his research, including the Kappa Delta Prize for Outstanding Orthopaedic Research and the Kroc Foundation Award for Excellence in Cartilage and Bone Research. Dr. Puzas is president of the Orthopaedic Research Society. He received his M.S. and Ph.D. in radiation biology and biophysics from the University of Rochester.

NU-MAY RUBY REED is a staff toxicologist with the California Environmental Protection Agency's (Cal/EPA) Department of Pesticide Regulation, where she is the lead person on risk assessment issues in the health assessment section. Her research interests are in evaluating health risks and developing dietary assessment guidelines for pesticides. She has been on several Cal/EPA working groups that initiate, research, and revise risk assessment guidelines and policies, and she represented her department in task forces on community concerns and emergency response, risk management guidance, and public education. Dr. Reed is also a lecturer on health risk assessment at the University of California at Davis. She received her Ph.D. from the University of California at Davis and is a diplomate of the American Board of Toxicology.

KATHLEEN M. THIESSEN is a senior scientist at SENES Oak Ridge, Inc., Center for Risk Analysis. She has extensive experience in evaluating exposures, doses, and risks to human health from environmental contaminants and in using uncertainty analysis for environmental and health risk assessment. More recently, Dr. Thiessen has led a working group on dose reconstruction for the International Atomic Energy Agency's Biosphere Modeling and Assessment Methods program. She received her Ph.D. in genetics from the University of Tennessee-Oak Ridge Graduate School of Biomedical Sciences.

THOMAS WEBSTER is assistant professor in the Department of Environmental Health at the Boston University School of Public Health. His research interests include methods in environmental epidemiology (particularly spatial epidemiology and ecologic bias), applications of mathematical modeling to toxicology and epidemiology, and persistent organic pollutants, particularly brominated fire retardants. He received his D.Sc. in environmental health from the Boston University School of Public Health.

APPENDIX
B

Measures of Exposure to Fluoride in the United States: Supplementary Information

U.S. DATA ON ARTIFICIAL AND NATURAL FLUORIDE IN DRINKING WATER

The recommended "optimal" fluoride concentrations for community public water supply systems and school public water supply systems are shown in Table B-1. Both sets of recommendations are based on the "annual average of maximum daily air temperatures" (CDC 1995, based on two studies in the 1950s). Table B-2 provides the approximate number of persons receiving artificially fluoridated public water in 1992, by fluoride concentration. In practice, most states seem to use a single fluoride concentration for the whole state. Figure B-1 shows the fluoride concentration by state with respect to annual average temperature for that state over the period 1971-2000. Table B-3 presents the approximate number of persons receiving naturally fluoridated public water in 1992, by fluoride concentration.

The number of persons served with public water supplies exceeding 4 milligrams (mg) of fluoride per liter (L) is expected to be substantially lower now than in 1992. For example, South Carolina, which had more than half of the persons in that category in 1992 (Table B-3), now has only occasional violations of the maximum contaminant level (MCL) (e.g., two water systems with 10 violations in calendar year 2003; SCDHEC 2004[1]). On the other hand, a recent news article indicates that some areas in Virginia

---

[1]See also local drinking water information by state at http://www.epa.gov/safewater/dwinfo.htm.

TABLE B-1 Recommended Optimal Fluoride Concentrations for Public Water Supply Systems

Annual Average of Maximum Daily Air Temperatures[a]		Recommended Fluoride Concentrations, mg/L	
°F	°C	Community Water Systems	School Water Systems[b]
50.0-53.7	10.0-12.0	1.2	5.4
53.8-58.3	12.1-14.6	1.1	5.0
58.4-63.8	14.7-17.7	1.0	4.5
63.9-70.6	17.8-21.4	0.9	4.1
70.7-79.2	21.5-26.2	0.8	3.6
79.3-90.5	26.3-32.5	0.7	3.2

[a]Based on temperature data obtained for a minimum of 5 years.
[b]Based on 4.5 times the optimal fluoride level for communities. School water fluoridation is recommended only when the school has its own source of water and is not connected to a community water system. Several other criteria are also considered; for example, if >25% of the children attending the school already receive optimally fluoridated water at home, the school's water should not be fluoridated.

SOURCE: CDC 1995.

are still served by water systems with fluoride exceeding 4 mg/L (Hirschauer 2004).

Miller-Ihli et al. (2003) reported on fluoride concentrations in water samples collected in 1999 from 24 locations nationwide; these locations were expected to provide nationally representative samples for the National Food and Nutrient Analysis Program.[2] Not unexpectedly, their findings indicate a bimodal distribution of fluoride concentrations in public drinking water: either water was fluoridated at approximately 1 mg/L or it was not fluoridated, with concentrations bordering on undetectable.

WATER INGESTION AND FLUORIDE INTAKES

Tables B-4 to B-7 summarize recent estimates by the U.S. Environmental Protection Agency (EPA) of the mean and selected percentiles of water ingestion by source (community supplies, bottled water, "other" sources, and all sources combined) and subpopulation (EPA 2000a); Tables B-8 and B-9

---

[2]Miller-Ihli et al. (2003) reported that 40% of the samples were fluoridated and suggested that, rather than using an average fluoride concentration for the country, an individual should be assumed to have a 40% probability of ingesting fluoridated water and a 60% probability of ingesting nonfluoridated water. However, CDC (2002a) estimates that about two-thirds of the U.S. population served by public water supplies receives fluoridated water. Thus, the sampling reported by Miller-Ihli et al. was probably not sufficiently representative on a population-weighted basis.

TABLE B-2 Population Sizes by Level of Artificial Fluoridation in 1992

Fluoride, mg/L	Number of States[a]	Population	Percentage	States
0.7	1	149,290	0.11	Hawaii
0.7-0.9	1	8,014,583	5.88	Texas
0.7-1.0	1	1,282,425	0.94	Arizona
0.8	4	12,886,396	9.46	Florida, Louisiana, Oklahoma, South Carolina
0.8-1.0	1	432,700	0.32	Delaware
0.9	2	7,177,525	5.27	Kentucky,[b] Virginia[c]
0.9-1.2	1	1,921,525	1.41	Colorado
1.0	29	93,060,026	68.30	Alabama, California, Connecticut, District of Columbia, Georgia, Idaho, Illinois, Indiana,[c] Kansas, Maryland, Massachusetts, Michigan, Mississippi, Missouri, Nebraska, Nevada, New Jersey, New Mexico, New York, North Carolina,[c] Ohio, Oregon, Pennsylvania, Rhode Island, Tennessee, Utah, Washington, West Virginia,[c] Wisconsin
1.0-1.1	2	1,931,337	1.42	Iowa, Wyoming
1.0-1.2	2	214,865	0.16	Montana, New Hampshire
1.1	1	233,447	0.17	Vermont[d]
1.2	5	5,026,243	3.69	Alaska, Maine, Minnesota,[e] North Dakota, South Dakota
No data[f]	2	3,911,884	2.87	Arkansas, Puerto Rico
Total	52	136,242,246	100	

[a]Includes the 50 states, the District of Columbia, and Puerto Rico.
[b]A few small water supplies have artificial fluoride concentrations of 4.0 mg/L.
[c]A few small water supplies have artificial fluoride concentrations of 4.5 mg/L.
[d]A few small water supplies have artificial fluoride concentrations of 4.9 mg/L.
[e]A few small water supplies have artificial fluoride concentrations of 5.4 mg/L.
[f]Data for Arkansas were not provided (the table for Arkansas contained a duplication of the Alaska data). The water fluoridation data were not provided for Puerto Rico.
SOURCE: CDC 1993.

give the corresponding estimates for consumption of community water or all water as a function of body weight. The data in Tables B-4 through B-9 are for those persons who actually consume water from the indicated source, rather than per capita estimates for the entire population. Estimates include plain (noncarbonated) drinking water and indirect water (water added to foods and beverages during preparation at home or by local food service establishments). Water in processed foods (commercial water) or naturally contained in foods (biological water) was not included.

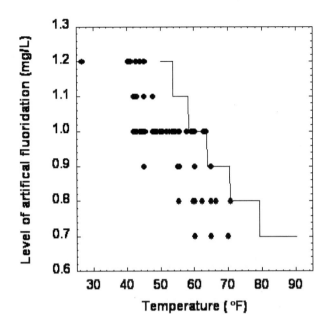

FIGURE B-1 Level of artificial fluoridation in 1992 by state (Table B-2; CDC 1993) versus area-weighted annual average temperature (°F) for that state over the period 1971-2000 (NCDC 2002a). Temperature for the District of Columbia is for Climate District 4 of the state of Maryland (NCDC 2002b). States with a range of artificial fluoride levels (Arizona, Colorado, Delaware, Iowa, Montana, New Hampshire, Texas, and Wyoming) are included at each relevant fluoride level. Arkansas and Puerto Rico are not included because of the lack of information on fluoride levels. Thin line indicates the "recommended optimal fluoride levels" for the given range of "annual average of *maximum* daily air temperatures" (emphasis added; Table B-1; CDC 1995).

EPA's estimates are based on U.S. Department of Agriculture surveys taken in 1994, 1995, and 1996 of food ingestion data for two nonconsecutive days for a sample of more than 15,000 individuals in the 50 states and the District of Columbia selected to represent the entire U.S. population based on 1990 census data (EPA 2000a). (An additional survey of children in 1998 was included in the estimates used in Chapter 2.) Because these estimates were developed for the purpose of estimating people's exposures to substances in drinking water and also are based on relatively recent data,

TABLE B-3 Population Sizes by Level of Natural Fluoridation in 1992

State[a]	Reported Range, mg/L	Reported Level of Natural Fluoride, mg/L				Not given[b]	Reported Total[c]
		≤1.2	1.3-1.9	2.0-3.9	≥4.0		
Alabama	0.7-3.6	27,368	25,195	6,827	0	—	54,283
Arizona	0.7-7.4	242,309	63,132	39,259	516	—	345,266
Arkansas	NA[d]	—	—	—	—	—	17,239
California	0.7-3.5	389,715	24,583	500	0	—	414,798
Colorado	0.1-11.2	363,905	75,755	361,969	1,926	—	801,224
Connecticut	0.7-1.9	870	160	0	0	—	1,030
Delaware	0.6-0.9	7,171	0	0	0	—	7,171
Florida	0.5-3.6	890,443	37,435	1,227	0	—	929,105
Georgia	0.7-2.0	16,039	878	1,200	0	7,475	25,592
Hawaii	0.7	354	0	0	0	—	354
Idaho	0.6-15.9	293,127	8,275	2,650	500	—	304,552
Illinois	0.7-4.0	291,600	91,237	56,481	500	6,658	446,050
Indiana	0.7-4.4	177,890	36,254	5,541	5,790	31,928	264,233
Iowa	0.7-7.0	186,936	90,182	28,484	1,445	—	302,652
Kansas	0.5-2.6	81,884	14,958	22,846	0	41,558	161,515
Kentucky	NA[e]	0	0	0	0	1,899	1,899
Louisiana	0.7-3.8	302,520	44,787	12,599	0	—	357,210
Maryland	0.3-5.1	36,583	11,705	100	225	—	48,613
Massachusetts	1.0-1.1	122	0	0	0	—	122
Michigan	0.7-1.9	114,605	9,968	0	0	—	124,623
Minnesota	0.7-3.2	2,386	908	367	0	—	4,000
Mississippi	0.8-3.5	93,120	9,965	1,560	0	—	104,645
Missouri	0.7-5.0	74,412	58,168	16,906	180	—	143,603
Montana	0.1-7.3	85,452	3,923	7,171	1,814	492	82,985
Nebraska	0.3-1.4	31,246	4,352	0	0	—	35,598
Nevada	0.5-2.6	16,440	3,628	5,187	0	—	25,255
New Hampshire	1.0-3.9	12,612	3,749	11,190	0	—	27,551

New Jersey	0.7–2.5	32,344	56,450	24,651	0	—	113,445
New Mexico	0.7–13	178,754	45,619	58,556	4,295	261	287,485
New York	NA[e]	0	0	0	0	1,536	1,216
North Carolina	0.0–2.7	0	7,200	325	0	183,076	190,601
North Dakota	0.5–7.0	5,205	6,002	6,024	3,793	—	20,421
Ohio	0.8–2.8	131,963	104,558	13,450	0	1,010	249,755
Oklahoma	0.7–12.0	62,353	20,803	8,966	18,895	—	111,017
Oregon	0.7–2.4	39,865	2,320	680	0	—	42,865
South Carolina	0.1–5.9	62,924	27,968	190,430	105,618	—	378,995
South Dakota	0.7–6.0	10,097	14,053	41,038	692	—	37,758
Texas	0.7–8.8	2,234,504	426,341	233,326	36,863	25,200	2,955,395
Utah	0.7–2.0	8,240	2,560	0	0	—	10,800
Virginia	0.7–6.3	8,418	11,423	207,924	18,726	408	246,694
Washington	0.7–2.7	54,460	3,117	4,916	0	—	62,493
West Virginia	1.2	659	0	0	0	—	659
Wisconsin	0.7–2.7	90,713	36,570	50,140	0	—	174,850
Wyoming	0.7–4.5	14,694	21,984	2,144	120	—	38,942
Totals		6,674,302	1,406,165	1,424,634	201,898	301,501	9,954,559

[a]Alaska, the District of Columbia, Maine, Pennsylvania, Rhode Island, Tennessee, and Vermont reported no water systems with natural fluoridation.

[b]Reported as 0.0 or some other number suspected to be a misprint.

[c]Total given in the summary table for each state. Because of apparent internal inconsistencies, the numbers in the preceding columns do not necessarily give the same total.

[d]Data for Arkansas were not provided (the table for Arkansas contained a duplication of the Alaska data).

[e]Reported as 0.0 for all systems with natural fluoride.

SOURCE: CDC 1993.

TABLE B-4 Estimated Average Daily Water Ingestion (mL/day) from Community Sources During 1994-1996, by People Who Consume Water from Community Sources

Population	Mean	50th Percentile	90th Percentile	95th Percentile	99th Percentile	Sample Size	Population
All consumers	1,000	785	2,069	2,600	4,273	14,012	242,641,675
<0.5 year	529	543	943	1,064	1,366	111	1,062,136
0.5-0.9 year	502	465	950	1,122	1,529	135	1,449,698
1-3 years	351	267	719	952	1,387	1,625	10,934,001
4-6 years	454	363	940	1,213	1,985	1,110	11,586,632
7-10 years	485	377	995	1,241	1,999	884	14,347,058
11-14 years	641	473	1,415	1,742	2,564	759	14,437,898
15-19 years	817	603	1,669	2,159	3,863	777	16,735,467
20-24 years	1,033	711	2,175	3,082	5,356	644	17,658,027
25-54 years	1,171	965	2,326	2,926	4,735	4,599	106,779,569
55-64 years	1,242	1,111	2,297	2,721	4,222	1,410	19,484,112
≥ 65 years	1,242	1,149	2,190	2,604	3,668	1,958	28,167,077
Males (all)	1,052	814	2,164	2,733	4,616	7,082	118,665,763
<1 year	462	441	881	1,121	1,281	118	1,191,526
1-10 years	444	355	934	1,155	1,731	1,812	18,847,070
11-19 years	828	595	1,673	2,058	3,984	768	15,923,625
≥ 20 years	1,242	1,038	2,387	3,016	4,939	4,384	82,703,542
Females (all)	951	747	2,005	2,482	3,863	6,930	123,975,912
<1 year	560	542	967	1,122	1,584	128	1,320,308
1-10 years	426	329	940	1,109	2,014	1,807	18,020,621
11-19 years	638	457	1,382	1,774	2,598	768	15,249,740
≥ 20 years	1,116	943	2,165	2,711	4,268	4,227	89,385,243
Lactating women	1,665	1,646	2,959	3,588	4,098	34	971,057
Pregnant women	872	553	1,844	2,588	3,448	65	1,645,565
Women aged 15-44 years	984	756	2,044	2,722	4,397	2,176	55,251,477

SOURCE: EPA 2000a.

TABLE B-5 Estimated Average Daily Water Ingestion (mL/day) from Bottled Water During 1994-1996, by People Who Consume Bottled Water

Population	Mean	50th Percentile	90th Percentile	95th Percentile	99th Percentile	Sample Size	Population
All consumers	737	532	1,568	1,967	3,316	3,078	57,316,806
<0.5 year	411	349	896	951	1,193	51	538,267
0.5-0.9 year	437	361	802	808	1,578	37	456,103
1-3 years	302	232	649	819	1,175	368	2,532,201
4-6 years	390	315	794	922	1,319	213	2,336,873
7-10 years	416	323	828	985	1,767	164	2,808,756
11-14 years	538	361	1,099	1,420	2,192	148	2,896,893
15-19 years	665	468	1,503	1,777	3,149	163	3,528,434
20-24 years	786	532	1,640	2,343	3,126	179	5,089,216
25-54 years	822	621	1,773	1,981	3,786	1,174	28,487,354
55-64 years	860	685	1,833	2,306	2,839	279	3,987,578
≥ 65 years	910	785	1,766	2,074	2,548	302	4,655,131
Males (all)	749	523	1,626	2,097	3,781	1,505	26,298,392
<1 year	414	317	805	1,012	1,397	48	575,019
1-10 years	365	266	767	847	1,685	376	3,755,220
11-19 years	682	464	1,423	1,822	2,802	144	2,969,950
≥ 20 years	845	592	1,774	2,303	3,855	937	18,998,203
Females (all)	727	532	1,542	1,893	3,031	1,573	31,018,414
<1 year	436	428	895	896	1,301	40	419,351
1-10 years	375	289	765	993	1,347	369	3,922,610
11-19 years	544	357	1,116	1,537	3,143	167	3,455,377
≥ 20 years	819	690	1,747	1,975	3,060	997	23,221,076
Lactating women	749	608	1,144	1,223	1,286	7	278,308
Pregnant women	891	683	1,910	1,957	2,198	27	698,645
Women aged 15-44 years	766	592	1,598	1,922	3,093	611	16,279,438

SOURCE: EPA 2000a.

TABLE B-6 Estimated Average Daily Water Ingestion (mL/day) from Other Sources (e.g., Wells and Cisterns) During 1994-1996, by People Who Consume Water from Those Sources

Population	Mean	50th Percentile	90th Percentile	95th Percentile	99th Percentile	Sample Size	Population
All consumers	965	739	1,971	2,475	3,820	2,129	34,693,744
<0.5 year	306	188	637	754	878	15	117,444
0.5-0.9 year	265	172	552	560	567	14	198,639
1-3 years	347	291	710	761	1,190	206	1,243,498
4-6 years	390	285	778	1,057	1,332	137	1,382,002
7-10 years	485	399	992	1,093	1,623	134	2,121,832
11-14 years	733	553	1,561	1,884	3,086	121	2,243,452
15-19 years	587	395	1,221	1,721	2,409	109	2,372,842
20-24 years	640	472	1,305	1,648	1,937	67	1,809,825
25-54 years	1,124	917	2,175	2,834	4,728	731	15,480,754
55-64 years	1,276	1,110	2,365	2,916	5,152	272	3,504,576
≥65 years	1,259	1,188	2,136	2,470	3,707	323	4,218,880
Males (all)	1,031	785	2,107	2,821	4,734	1,155	17,880,530
<1 year	243	148	554	567	773	16	198,829
1-10 years	426	320	884	1,077	1,630	259	2,566,652
11-19 years	702	564	1,366	1,753	2,787	103	2,011,715
≥20 years	1,212	1,001	2,286	3,017	4,883	777	13,103,334
Females (all)	894	710	1,826	2,225	3,035	974	16,813,214
<1 year	344	256	537	579	759	13	117,254
1-10 years	416	352	865	1,039	1,165	218	2,180,680
11-19 years	624	406	1,394	1,873	2,489	127	2,604,579
≥ 20 years	1,046	941	1,925	2,371	3,123	616	11,910,701
Lactating women	1,248	915	2,148	2,410	2,620	7	182,414
Pregnant women	1,066	660	1,676	1,807	3,374	7	168,433
Women aged 15-44 years	904	666	1,863	2,319	3,056	283	6,759,992

SOURCE: EPA 2000a.

TABLE B-7 Estimated Average Daily Water Ingestion (mL/day) from All Sources During 1994-1996 by Consumers of Water

Population	Mean	50th Percentile	90th Percentile	95th Percentile	99th Percentile	Sample Size	Population
All consumers	1,241	1,045	2,345	2,922	4,808	15,172	259,972,235
<0.5 year	544	545	947	1,078	1,365	156	1,507,727
0.5-0.9 year	580	563	1,130	1,273	1,672	154	1,732,993
1-3 years	422	351	807	993	1,393	1,814	12,143,483
4-6 years	548	468	1,019	1,268	2,031	1,193	12,438,322
7-10 years	608	514	1,131	1,425	2,172	937	15,248,676
11-14 years	815	651	1,625	1,962	3,033	812	15,504,627
15-19 years	1,006	776	1,897	2,414	4,027	814	17,697,092
20-24 years	1,283	1,013	2,508	3,632	5,801	678	18,544,787
25-54 years	1,486	1,273	2,638	3,337	5,259	4,906	113,011,204
55-64 years	1,532	1,378	2,557	2,999	4,395	1,541	21,145,387
≥65 years	1,453	1,345	2,324	2,708	3,750	2,167	30,997,937
Males (all)	1,300	1,070	2,483	3,149	5,212	7,689	126,998,276
<1 year	549	538	1,121	1,278	1,567	151	1,560,310
1-10 years	536	451	1,024	1,254	1,817	1,993	20,495,833
11-19 years	1,001	761	1,898	2,434	4,011	809	16,887,932
≥20 years	1,549	1,331	2,740	3,524	5,526	4,736	88,054,201
Females (all)	1,185	1,021	2,221	2,703	4,252	7,483	132,973,959
<1 year	577	559	950	1,131	1,654	159	1,680,410
1-10 years	528	445	993	1,226	2,035	1,951	19,334,648
11-19 years	830	664	1,652	1,955	3,083	817	16,313,787
≥20 years	1,389	1,221	2,416	2,928	4,512	4,556	95,645,114
Lactating women	1,806	1,498	3,021	3,767	4,024	41	1,171,868
Pregnant women	1,318	1,228	2,339	2,674	3,557	70	1,751,888
Women aged 15-44 years	1,265	1,065	2,366	2,952	4,821	2,314	58,549,659

SOURCE: EPA 2000a.

TABLE B-8 Estimated Average Daily Water Ingestion (mL/kg of Body Weight per Day) from Community Sources during 1994-1996, by People Who Consume Water from Community Sources

Population	Mean	50th Percentile	90th Percentile	95th Percentile	99th Percentile	Sample Size	Population
All consumers	17	13	33	44	79	13,593	236,742,834
<0.5 year	88	85	169	204	240	106	1,034,566
0.5-0.9 year	56	52	116	127	170	128	1,405,128
1-3 years	26	20	53	68	112	1,548	10,417,368
4-6 years	23	18	45	65	95	1,025	10,751,616
7-10 years	16	12	33	39	60	820	13,427,986
11-14 years	13	10	27	36	54	736	14,102,256
15-19 years	12	9	26	32	62	771	16,646,551
20-24 years	15	11	31	39	80	637	17,426,127
25-54 years	16	13	32	40	65	4,512	104,816,948
55-64 years	17	14	32	38	58	1,383	19,011,778
≥65 years	18	16	32	37	53	1,927	27,702,510
Males (all)	16	13	32	43	81	6,935	117,076,195
<1 year	66	60	139	175	235	115	1,180,289
1-10 years	21	16	43	55	87	1,705	17,865,064
11-19 years	14	10	27	38	67	755	15,717,364
≥20 years	15	13	30	38	62	4,360	82,313,478
Females (all)	17	14	35	45	77	6,658	119,666,639
<1 year	72	69	139	169	203	119	1,259,405
1-10 years	21	17	45	61	98	1,688	16,731,906
11-19 years	12	9	26	32	48	752	15,031,443
≥20 years	17	14	33	41	63	4,099	86,643,885
Lactating women	26	20	54	55	57	33	940,375
Pregnant women	14	9	33	43	47	65	1,645,565
Women aged 15-44 years	15	12	32	39	66	2,126	54,000,618

SOURCE: EPA 2000a.

TABLE B-9 Estimated Average Daily Water Ingestion (mL/kg of Body Weight per Day) from All Sources During 1994-1996 by Consumers of Water

Population	Mean	50th Percentile	90th Percentile	95th Percentile	99th Percentile	Sample Size	Population
All consumers	21	17	38	50	87	14,726	253,667,688
<0.5 year	92	87	169	196	239	149	1,465,837
0.5-0.9 year	65	58	120	164	185	147	1,688,423
1-3 years	31	26	60	74	118	1,732	11,603,245
4-6 years	27	23	51	68	97	1,103	11,556,872
7-10 years	20	17	36	44	70	873	14,329,604
11-14 years	16	14	33	40	60	786	15,116,291
15-19 years	15	12	29	38	66	806	17,564,502
20-24 years	18	14	34	44	86	658	18,224,524
25-54 years	20	17	37	46	69	4,813	110,938,819
55-64 years	20	18	35	42	59	1,513	20,646,201
≥65 years	21	19	34	39	54	2,136	30,533,370
Males (all)	20	16	38	49	86	7,532	125,266,552
<1 year	77	66	164	173	233	147	1,538,210
1-10 years	25	20	48	62	91	1,832	19,480,513
11-19 years	16	13	32	42	69	794	16,642,651
≥20 years	19	16	34	43	67	4,709	87,605,178
Females (all)	22	18	39	50	88	7,194	128,401,136
<1 year	79	72	158	170	200	149	1,616,050
1-10 years	26	21	50	66	104	1,826	18,009,208
11-19 years	15	13	29	36	56	798	16,038,142
≥20 years	21	18	37	45	69	4,421	92,737,736
Lactating women	28	25	53	57	70	40	1,141,186
Pregnant women	21	19	39	44	61	69	1,729,947
Women aged 15-44 years	20	16	36	46	77	2,258	57,164,907

SOURCE: EPA 2000a.

they are appropriate for the present purpose of estimating the range of current exposures to fluoride. These estimates are based on a 2-day average, whereas for fluoride exposure, long-term averages of intake are usually more important. However, given the size of the population sampled, the likelihood that the entire sample represents days of unusually high or unusually low water intake is small. Thus, these values are considered reasonable indicators both of typical water consumption and of the likely range of water consumption from various sources on a long-term basis. However, they should not be used by themselves to estimate the number of individuals or percentage of the population that consumes a given amount of water on a long-term basis, especially not at the extremes of the range. Water intakes at the low end are not of major importance for the present report, and water intakes at the high end are considered separately (Chapter 2), with additional information beyond what is provided by EPA.

It may be helpful to compare the water intakes (all sources, Table B-7) with values for adequate intake[3] (AI) of water recently published by the Institute of Medicine (IOM 2004; Table B-10). The AI for total water (drinking water, other beverages, and moisture contained in food) is set "to prevent deleterious, primarily acute, effects of dehydration, which include metabolic and functional abnormalities" (IOM 2004). "Given the extreme variability in water needs which are not solely based on differences in metabolism, but also in environmental conditions and activity, there is not a single level of water intake that would ensure adequate hydration and optimal health for half[4] of all apparently healthy persons in all environmental conditions" (IOM 2004). The AI for total water is based on the median total water intake from U.S. survey data (NHANES III, 1988-1994; described by IOM 2004). Daily consumption below the AI is not necessarily a concern "because a wide range of intakes is compatible with normal hydration. Higher intakes of *total* water will be required for those who are physically active or who are exposed to [a] hot environment" (IOM 2004). For the intake values shown in Table B-10, approximately 80% of the intake comes from drinking water and other beverages (including caffeinated and alcoholic beverages).

Use of bottled water in the United States has at least doubled since 1990 (Grossman 2002), suggesting that more people use bottled water now than in 1994-1996 and/or that individuals use more bottled water per person.

---

[3]"Adequate intake" is defined as "the recommended average daily intake level based on observed or experimentally determined approximations or estimates of nutrient intake by a group (or groups) of apparently healthy people that are assumed to be adequate—used when an RDA [recommended dietary allowance] cannot be determined" (IOM 2004).

[4]The estimated average requirement (EAR) on which a recommended dietary allowance is based is defined as "the average daily nutrient intake level estimated to meet the requirement of half the healthy individuals in a particular life stage and gender group" (IOM 2004).

TABLE B-10 Adequate Intake Values (L/day) for Total Water

	Males			Females		
Group	From Foods	From Beverages	Total Water	From Foods	From Beverages	Total Water
0-6 months	0	0.7	0.7	0	0.7	0.7
7-12 months	0.2	0.6	0.8	0.2	0.6	0.8
1-3 years	0.4	0.9	1.3	0.4	0.9	1.3
4-8 years	0.5	1.2	1.7	0.5	1.2	1.7
9-13 years	0.6	1.8	2.4	0.5	1.6	2.1
14-18 years	0.7	2.6	3.3	0.5	1.8	2.3
>19 years	0.7	3.0	3.7	0.5	2.2	2.7
Pregnancy[a]	—	—	—	0.7	2.3	3.0
Lactation[a]	—	—	—	0.7	3.1	3.8

[a]Women aged 14-50 years.
SOURCE: IOM 2004.

However, total water consumption per person from all sources combined probably has not changed substantially. Information for a few groups in the tables (children < 1 year of age, pregnant and lactating women) is based on relatively small sample sizes, and the confidence to be placed in specific percentile values is therefore lower. Sample sizes for some other population subgroups of potential interest (e.g., Native Americans with traditional lifestyles, people in hot climates, people with high physical activity, people with certain medical conditions) were not large enough to evaluate intake by members of the subgroup, although some people from those groups are included in the overall sample (EPA 2000a).

Tables B-11 to B-14 summarize fluoride intakes that would result from ingestion of community water (for the mean, 90th, 95th, and 99th percentiles of consumption estimated by EPA) at various levels of water fluoride ("optimal" fluoridation levels of 0.7, 1.0, or 1.2 mg/L, and the present secondary MCL [SMCL] and MCL of 2 and 4 mg/L, respectively). The SMCL and MCL are included for purposes of comparison; most people in the Unites States do not drink water with those fluoride levels. An average consumer below the age of 6 months would have an intake of 0.06-0.1 mg/kg/day from fluoridated water (0.7-1.2 mg/L), whereas an adult would ingest approximately 0.01-0.02 mg/kg/day. Individuals at the upper levels of water intake from EPA's estimates (Table B-14) could have fluoride intakes in excess of 1 mg/day at the lowest levels of fluoridation up to about 6 mg/day for some adults, depending on age and level of water fluoridation. Persons in the high-water-intake groups described above could have even higher intakes.

TABLE B-11 Estimated Intake of Fluoride from Community Water for Average Consumers[a]

Population	Water Intake, mL/day	Fluoride Level				
		0.7 mg/L	1 mg/L	1.2 mg/L	2 mg/L	4 mg/L
		Intake, mg/day				
All consumers	1,000	0.70	1.00	1.20	2.00	4.00
<0.5 year	529	0.37	0.53	0.63	1.06	2.12
0.5-0.9 year	502	0.35	0.50	0.60	1.00	2.01
1-3 years	351	0.25	0.35	0.42	0.70	1.40
4-6 years	454	0.32	0.45	0.54	0.91	1.82
7-10 years	485	0.34	0.49	0.58	0.97	1.94
11-14 years	641	0.45	0.64	0.77	1.28	2.56
15-19 years	817	0.57	0.82	0.98	1.63	3.27
20-24 years	1,033	0.72	1.03	1.24	2.07	4.13
25-54 years	1,171	0.82	1.17	1.41	2.34	4.68
55-64 years	1,242	0.87	1.24	1.49	2.48	4.97
≥65 years	1,242	0.87	1.24	1.49	2.48	4.97
	Water Intake, mL/kg/day	Intake, mg per kg body weight/day				
All consumers	17	0.012	0.017	0.020	0.034	0.068
<0.5 year	88	0.062	0.088	0.106	0.176	0.352
0.5-0.9 year	56	0.039	0.056	0.067	0.112	0.224
1-3 years	26	0.018	0.026	0.031	0.052	0.104
4-6 years	23	0.016	0.023	0.028	0.046	0.092
7-10 years	16	0.011	0.016	0.019	0.032	0.064
11-14 years	13	0.009	0.013	0.016	0.026	0.052
15-19 years	12	0.008	0.012	0.014	0.024	0.048
20-24 years	15	0.011	0.015	0.018	0.030	0.060
25-54 years	16	0.011	0.016	0.019	0.032	0.064
55-64 years	17	0.012	0.017	0.020	0.034	0.068
≥65 years	18	0.013	0.018	0.022	0.036	0.072

[a]Based on water consumption rates estimated by EPA (2000a).

EXPOSURES FROM FLUORINATED ANESTHETICS

The sampled data in Table B-15 illustrate wide ranges of reported mean peak serum fluoride concentrations from the use of fluorinated anesthetics under various surgical conditions and for different age groups ranging from 22-day-old infants to people > 70 years old. These data are collected from studies conducted in many countries, including Australia, France, Finland, Germany, Ireland, Japan, the United Kingdom, and the United States. The

TABLE B-12 Estimated Intake of Fluoride from Community Water for 90th Percentile Consumers[a]

Population	Water Intake, mL/day	Fluoride Level				
		0.7 mg/L	1 mg/L	1.2 mg/L	2 mg/L	4 mg/L
		Intake, mg/day				
All consumers	2,069	1.45	2.07	2.48	4.14	8.28
<0.5 year	943	0.66	0.94	1.13	1.89	3.77
0.5-0.9 year	950	0.67	0.95	1.14	1.90	3.80
1-3 years	719	0.50	0.72	0.86	1.44	2.88
4-6 years	940	0.66	0.94	1.13	1.88	3.76
7-10 years	995	0.70	1.00	1.19	1.99	3.98
11-14 years	1,415	0.99	1.42	1.70	2.83	5.66
15-19 years	1,669	1.17	1.67	2.00	3.34	6.68
20-24 years	2,175	1.52	2.18	2.61	4.35	8.70
25-54 years	2,326	1.63	2.33	2.79	4.65	9.30
55-64 years	2,297	1.61	2.30	2.76	4.59	9.19
≥65 years	2,190	1.53	2.19	2.63	4.38	8.76
	Water Intake, mL/kg/day	Intake, mg per kg body weight/day				
All consumers	33	0.023	0.033	0.040	0.066	0.132
<0.5 year	169	0.118	0.169	0.203	0.338	0.676
0.5-0.9 year	116	0.081	0.116	0.139	0.232	0.464
1-3 years	53	0.037	0.053	0.064	0.106	0.212
4-6 years	45	0.032	0.045	0.054	0.090	0.180
7-10 years	33	0.023	0.033	0.040	0.066	0.132
11-14 years	27	0.019	0.027	0.032	0.054	0.108
15-19 years	26	0.018	0.026	0.031	0.052	0.104
20-24 years	31	0.022	0.031	0.037	0.062	0.124
25-54 years	32	0.022	0.032	0.038	0.064	0.128
55-64 years	32	0.022	0.032	0.038	0.064	0.128
≥65 years	32	0.022	0.032	0.038	0.064	0.128

[a]Based on water consumption rates estimated by EPA (2000a).

minimum alveolar concentration per hour (MAC-hr) ranged from short-term (e.g., for cesarean section as reported by Abboud et al. 1989) to prolonged (e.g., >10 hours as reported by Murray et al. 1992 and Obata et al. 2000) surgery and up to 7 days of continuous exposure for critically ill patients (e.g., as reported by Osborne et al. 1996). Test subjects included healthy males who underwent 3-9 hours of anesthesia (Munday et al. 1995), female smokers (Laisalmi et al. 2003), infants and children (age as indicated

TABLE B-13 Estimated Intake of Fluoride from Community Water for 95th Percentile Consumers[a]

Population	Water Intake, mL/day	Fluoride Level				
		0.7 mg/L	1 mg/L	1.2 mg/L	2 mg/L	4 mg/L
		Intake, mg/day				
All consumers	2,600	1.82	2.60	3.12	5.20	10.40
<0.5 year	1,064	0.74	1.06	1.28	2.13	4.26
0.5-0.9 year	1,122	0.79	1.12	1.35	2.24	4.49
1-3 years	952	0.67	0.95	1.14	1.90	3.81
4-6 years	1,213	0.85	1.21	1.46	2.43	4.85
7-10 years	1,241	0.87	1.24	1.49	2.48	4.96
11-14 years	1,742	1.22	1.74	2.09	3.48	6.97
15-19 years	2,159	1.51	2.16	2.59	4.32	8.64
20-24 years	3,082	2.16	3.08	3.70	6.16	12.33
25-54 years	2,926	2.05	2.93	3.51	5.85	11.70
55-64 years	2,721	1.90	2.72	3.27	5.44	10.88
≥65 years	2,604	1.82	2.60	3.12	5.21	10.42
	Water Intake, mL/kg/day	Intake, mg per kg body weight/day				
All consumers	44	0.031	0.044	0.053	0.088	0.176
<0.5 year	204	0.143	0.204	0.245	0.408	0.816
0.5-0.9 year	127	0.089	0.127	0.152	0.254	0.508
1-3 years	68	0.048	0.068	0.082	0.136	0.272
4-6 years	65	0.046	0.065	0.078	0.130	0.260
7-10 years	39	0.027	0.039	0.047	0.078	0.156
11-14 years	36	0.025	0.036	0.043	0.072	0.144
15-19 years	32	0.022	0.032	0.038	0.064	0.128
20-24 years	39	0.027	0.039	0.047	0.078	0.156
25-54 years	40	0.028	0.040	0.048	0.080	0.160
55-64 years	38	0.027	0.038	0.046	0.076	0.152
≥65 years	37	0.026	0.037	0.044	0.074	0.148

[a]Based on water consumption rates estimated by EPA (2000a).

in Table B-15), and patients with renal insufficiency (Conzen et al. 1995). In general, higher MAC-hr resulted in higher peak serum inorganic fluoride concentration. None of the studies presented in Table B-15 shows clear evidence of renal impairment as a result of the increased serum fluoride concentration, except transient reduction in renal function among the elderly (>70 years) reported by Hase et al. (2000). Higher peak serum concentration

TABLE B-14 Estimated Intake of Fluoride from Community Water for 99th Percentile Consumers[a]

Population	Water Intake, mL/day	Fluoride Level.				
		0.7 mg/L	1 mg/L	1.2 mg/L	2 mg/L	4 mg/L
		Intake, mg/day				
All consumers	4,273	2.99	4.27	5.13	8.55	17.09
<0.5 year	1,366	0.96	1.37	1.64	2.73	5.46
0.5-0.9 year	1,529	1.07	1.53	1.83	3.06	6.12
1-3 years	1,387	0.97	1.39	1.66	2.77	5.55
4-6 years	1,985	1.39	1.99	2.38	3.97	7.94
7-10 years	1,999	1.40	2.00	2.40	4.00	8.00
11-14 years	2,564	1.79	2.56	3.08	5.13	10.26
15-19 years	3,863	2.70	3.86	4.64	7.73	15.45
20-24 years	5,356	3.75	5.36	6.43	10.71	21.42
25-54 years	4,735	3.31	4.74	5.68	9.47	18.94
55-64 years	4,222	2.96	4.22	5.07	8.44	16.89
≥65 years	3,668	2.57	3.67	4.40	7.34	14.67
	Water Intake, mL/kg/day	Intake, mg per kg body weight/day				
All consumers	79	0.055	0.079	0.095	0.158	0.316
<0.5 year	240	0.168	0.240	0.288	0.480	0.960
0.5-0.9 year	170	0.119	0.170	0.204	0.340	0.680
1-3 years	112	0.078	0.112	0.134	0.224	0.448
4-6 years	95	0.067	0.095	0.114	0.190	0.380
7-10 years	60	0.042	0.060	0.072	0.120	0.240
11-14 years	54	0.038	0.054	0.065	0.108	0.216
15-19 years	62	0.043	0.062	0.074	0.124	0.248
20-24 years	80	0.056	0.080	0.096	0.160	0.320
25-54 years	65	0.046	0.065	0.078	0.130	0.260
55-64 years	58	0.041	0.058	0.070	0.116	0.232
≥65 years	53	0.037	0.053	0.064	0.106	0.212

[a]Based on water consumption rates estimated by EPA (2000a).

was reported for smokers (Cousins et al. 1976; Laisalmi et al. 2003) and is associated with alcohol, obesity, and multiple drug use (Cousins et al. 1976). Because the reference point for the potential nephrotoxicity in these studies was the peak serum fluoride concentration, data are generally not available for an estimation of the total fluoride load or the area under the curve from the use of these anesthetics.

TABLE B-15 Serum Inorganic Fluoride Concentration from Fluorinated Anesthetic Agents

Age (range)	No. of Subjects	MAC-hour[a]	Mean Serum Inorganic Fluoride, μM		References
			Baseline	Peak	
Isoflurane					
51 years	13	NA	NA	No change	Hara et al. 1998
NA	90	NA	NA	3	Groudine et al. 1999
>70 years	6	3.7	NA	4	Hase et al. 2000
55.5 years	26	NA	about 2.5	5	Goldberg et al. 1996
57 years	24	1.1	3.8	5.4	Newman et al. 1994
28 years	11	9.2	<2	5.5	Higuchi et al. 1995
28 years[b]	20	0.06	5.6	5.6	Abboud et al. 1989
27.7 years[b]	20	0.14	5.9	5.6	Abboud et al. 1989
48.5 years	20	15.9	NA	7.4	Obata et al. 2000
53.7 years	7	4.8	NA	8	Matsumura et al. 1994
26-54 years	5	NA[c]	2.1-2.4	8.4-27.9	Osborne et al. 1996
20-75 years	9	19.2	3.5-3.8	43.2	Murray et al. 1992
Enflurane					
22 days to 11 years	40	0.3-0.7	NA	2-8	Oikkonen and Meretoja 1989
		0.7-1.5	NA	4-10	Oikkonen and Meretoja 1989
		1.5-3.3	NA	6-10	Oikkonen and Meretoja 1989
22 day	1	0.6	NA	3	Oikkonen and Meretoja 1989
29 day	1	1.5	NA	7	Oikkonen and Meretoja 1989
3 months	1	1.6	NA	11	Oikkonen and Meretoja 1989
4 months	1	1.6	NA	11	Oikkonen and Meretoja 1989
9 months	1	2.0	NA	7	Oikkonen and Meretoja 1989
1-9 years	8	NA	1.7	10.5	Hinkle 1989
47-60 years	5	4-6.8	about 2-3	7	Sakai and Takaori 1978
63.9 years	20	1.07	NA	13.3	Conzen et al. 1995
48 years(27-58 years)	16	1	NA	13.8	Laisalmi et al. 2003

435

44 years (35-39 years)[d]	17	1	NA	18.7	Laisalmi et al. 2003
59.3 years	40	2.8	1.2	16.75	Blanco et al. 1995
47.8 years	8	1.24	2-2.5	18	Cousins et al. 1987
40.2 years	10	2.7	1.8	22.2	Cousins et al. 1976
18-35 years	5	6		28.1	Munday et al. 1995
18-35 years	5		NA	27.5	Munday et al. 1995
Halothane					
41.5 years	10	4.9	1.9	1.6	Cousins et al. 1976
6.2 years (1-12 years)	40	2.6	NA	1.8	Sarner et al. 1995
42-57 years	5	2.9-4.9	2-3	3	Sakai and Takaori 1978
50 years	8	2.5	2-2.5	4	Cousins et al. 1987
28.9 years	20	0.07	5.9	5.6	Abboud et al. 1989
9.2 years (5-12 years)	25	2.2	NA	6	Taivainen et al. 1994
20-75 years	10	19.5	3.8	12.6	Murray et al. 1992
Sevoflurane					
12 months (7.7-25 months)	41	4.7	NA	13.8	Lejus et al. 2002
6.2 years (1-12 years)	40	2.6	NA	14.7	Sarner et al. 1995
>70 years	7	5.1	NA	18	Hase et al. 2000
8.8 years	25	2.2	NA	21	Taivainen et al. 1994
50 years	25	0.8	3.8	23	Newman et al. 1994
67.4 years	21	1.01	NA	2.5	Conzen et al. 1995
60.5 years	40	2.9	1.2	27.7	Blanco et al. 1995
52.7 years	24	NA	about 2.5	28	Goldberg et al. 1996
18-35 years	5	3	NA	30.5	Munday et al. 1995
	5	6		31-34	
	5	9		36.6	
29 years	15	9.9	<2	36.8	Higuchi et al. 1995
53 years	13	3.7	NA	about 31	Hara et al. 1998
NA	98	2.9	NA	40	Groudine et al. 1999

continued

TABLE B-15 Continued

Age (range)	No. of Subjects	MAC-hour[a]	Mean Serum Inorganic Fluoride, μM		References
			Baseline	Peak	
26.6 years (19-49 years)	11	10.6	NA	41.9	Higuchi et al. 1994
56.8 years	10	18.0 high flow	NA	47.1	Obata et al. 2000
62.0 years	10	16.7 low flow	NA	53.5	Obata et al. 2000
54.9 years	8	6.1	NA	54	Matsumura et al. 1994
24 years	8	14.0	<2	57.5	Higuchi et al. 1995

[a]MAC is the minimum alveolar concentration, or the mean end-tidal anesthetic concentration. When MAC-hr is not reported, it is estimated as MAC-hr = (mean percent concentration) x (anesthesia time).
[b]Cesarean section patients with induction to delivery time of 7.4-8.4 minutes.
[c]Critically ill patients under anesthesia for 5-7 days at 0.6-1.2% isoflurane.
[d]Smoking > 10 cigarettes a day.

ABBREVIATION: NA, not applicable.

TABLE B-16 Summary of Estimated Safe and Adequate Daily Dietary Intakes[a] of Fluoride

Age, years	Weight, kg[b]	Range, mg/day		Range, mg/kg/day[c]	
0-0.5	6	0.1	0.5	0.017	0.083
0.5-1	9	0.2	1.0	0.022	0.11
1-3	13	0.5	1.5	0.038	0.12
4-6	20	1.0	2.5	0.050	0.13
7-10	28	1.5	2.5	0.054	0.089
Males					
11-14	45	1.5	2.5	0.033	0.056
15-18	66	1.5	2.5[d]	0.023	0.038
19-24	72	1.5	4.0[e]	0.021	0.056
25-50	79	1.5	4.0	0.019	0.051
51+	77	1.5	4.0	0.019	0.052
Females					
11-14	46	1.5	2.5	0.033	0.054
15-18	55	1.5	2.5[d]	0.027	0.045
19-24	58	1.5	4.0[e]	0.026	0.069
25-50	63	1.5	4.0	0.024	0.063
51+	65	1.5	4.0	0.023	0.062

[a]The term "safe and adequate daily dietary intake" was used by the NRC (1989b) "when data were sufficient to estimate a range of requirements, but insufficient for developing [a Recommended Dietary Allowance]." This category was to be accompanied by "the caution that upper levels in the safe and adequate range should not be habitually exceeded because the toxic level for many trace elements may be only several times usual intakes." Use of this term should not be taken to imply that the present committee considers these intakes to be safe or adequate.
[b]Median for age group.
[c]Calculated from range (mg/day) and weight (kg) given for age groups.
[d]Upper limit for children and adolescents (upper age not specified).
[e]Upper limit for adults.

SOURCE: NRC 1989b.

REFERENCE INTAKES OF FLUORIDE

Table B-16 provides the median weight and range of fluoride intake (mg/day; safe and adequate daily dietary intake[5]), by age group, from the National Research Council (NRC 1989b). Table B-17 provides the reference

---

[5]The term "safe and adequate daily dietary intake" was used by the NRC (1989b) "when data were sufficient to estimate a range of requirements, but insufficient for developing [a Recommended Dietary Allowance]." This category was to be accompanied by "the caution that upper levels in the safe and adequate range should not be habitually exceeded because the toxic level for many trace elements may be only several times usual intakes." Use of this

TABLE B-17 Summary of Dietary Reference Intakes of Fluoride

Age, years	Reference Weight, kg	Adequate Intake		Tolerable Upper Intake	
		mg/d	mg/kg/day^a	mg/d	mg/kg/day^a
0-0.5	7	0.01	0.0014	0.7	0.10
0.5-1	9	0.5	0.056	0.9	0.10
1-3	13	0.7	0.054	1.3	0.10
4-8	22	1	0.045	2.2	0.10
9-13	40	2	0.050	10	0.25
Boys 14-18	64	3	0.047	10	0.16
Girls 14-18	57	3	0.053	10	0.18
Males 19+	76	4	0.053	10	0.13
Females 19+	61	3	0.049	10	0.16

^{a}Calculated from intake (mg/day) and weight (kg) given for age groups by IOM (1997) and ADA (2005).

SOURCES: IOM 1997; ADA 2005.

weight and range of fluoride intake (mg/day; dietary reference intake), by age group, from the Institute of Medicine (IOM 1997) and the American Dental Association (ADA 2005). In both tables, the intakes in terms of mg/kg/day were calculated from the cited information as indicated.

---

term should not be taken to imply that the present committee considers these intakes to be safe or adequate.

APPENDIX
C

Ecologic and Partially Ecologic Studies in Epidemiology

Individual-level studies collect information on outcome, exposure, and covariates (potential confounders and effect modifiers) for each individual. Ecologic studies collect information about groups. Partially ecologic studies use a combination of individual-level and group-level variables.

The goal of most ecologic studies is to make inferences about individuals based on aggregated data. Unfortunately, severe bias can occur. (Bias in this context means systematic errors in the results of the analysis; it does not impugn the integrity or intention of the researchers). Ecologic bias has several sources (Greenland 1992; Greenland and Robins 1994; Morgenstern 1998; Webster 2000):

• Nondifferential exposure misclassification within groups (which tends to bias results away from the null)
• Confounding within and between groups
• Effect measure modification within and between groups
• Misspecification error when model is nonlinear
• Inadequate control of covariates
• Magnification of bias by aggregation due to confounding by group and effect measure modification by group
• Failure to weight by population
• Failure to standardize both outcome and exposure in the same way.

Instead of simply dismissing all ecologic studies as unreliable, it is preferable to estimate the direction and magnitude of potential biases. Quantifying bias in ecologic studies is quite difficult in practice. Nevertheless, certain design features tend to reduce ecologic bias, including the following:

1. Studies with outcome variables that can be modeled with weighted or ordinary least-squares regression (e.g., bone fluoride levels) are generally preferable to those with binary outcomes or rates, commonly modeled with logistic or log-linear regression. Nonlinear ecologic models can induce bias due to misspecification.

2. Exposure variables that are continuous on the individual-level before aggregation are generally preferable to those that are dichotomous (aggregation of dichotomous exposures typically produces variables of the form "fraction exposed"). The latter can be subject to nondifferential exposure misclassification within groups, tending to bias ecologic studies away from the null; they also tend to increase the amount of bias magnification. In contrast, using of the average exposure within each group need not cause measurement error on the ecologic level, a special case of the Berkson error model. Errors of this type produce unbiased results in ordinary linear regression; in log-linear regression, bias also depends on variance of the errors.

3. Exposure should be as uniform as possible within groups but as different as possible between groups.

4. Avoid, if possible, confounders with highly nonlinear relationships to outcome, because these can be very difficult to control in ecologic studies.

The following two types of partially ecologic studies are often used in epidemiology.

1. Multilevel models typically supplement individual-level variables with contextual variables. The latter are intrinsically group-level variables that have no real counterpart on the individual-level, (e.g., herd immunity or income inequality).

2. Studies that measure outcome and covariates at the individual level, but exposure at the group level, are commonly used in environmental and occupational epidemiology. This design is sometimes called "semi-individual." For example, fluoride concentrations might be measured in the water system serving a community. Everyone in that group is assigned the same exposure. Exposure is an aggregated variable, not an intrinsically group-level variable. Feasibility is the typical reason for using this design; individual exposure measurements are typically expensive and time-consuming, if they are possible at all.

The semi-individual kind of partially ecologic study can be thought of as individual-level with exposure measurement error. Unfortunately, semi-individual studies are not necessarily free of ecologic bias. Suppose the

APPENDIX C

ecologic exposure variable is the fraction exposed in the group (aggregated from dichotomous exposures at the individual level). Nondifferential exposure misclassification within groups tends to produce bias away from the null as in ecologic studies. Although bias magnification (see list above) can occur, the amount of bias tends to be intermediate between a fully ecologic study and a fully individual study (at least in certain cases that have been analyzed). Because covariate information is collected at the individual level, the ability to control for confounding can be much better than with purely ecologic studies. For more discussions of these issues, see Webster (2000, 2002) and Björk and Strömberg (2002).

In sum, semi-individual studies are generally more trustworthy than fully ecologic studies. Studies using exposure variables based on continuous individual-level exposures are preferable to those based on dichotomous individual-level exposures.

APPENDIX
D

Comparative Pharmacokinetics of Rats and Humans

In healthy young and middle-aged adult humans, fasting plasma fluoride concentrations (expressed as micromoles per liter [µmol/L]) are thought to be approximately equal to concentrations in water (expressed as parts per million [ppm] or milligrams per liter [mg/L]) provided that water is the major source of chronic exposure (NRC 1993; Whitford 1996). Dunipace et al. (1995) exposed weanling male Sprague-Dawley rats to fluoride in water plus a low-fluoride diet for 18 months. Plasma fluoride concentrations increased up to 3 months and remained fairly constant afterward. Plasma levels (µmol/L) were three to seven times less than water concentrations (ppm or mg/L) at several different concentrations and time points. In another chronic experiment with Sprague-Dawley rats, plasma/water fluoride ratios decreased from 4.2 at 2 months to 1.5 at 18 months (Whitford and Birdsong-Whitford 2000; G. Whitford, University of Georgia, personal communication, June 2, 2004). The reason for the difference between the experiments is unclear. Dunipace et al. (1995) concluded that rats require about five times greater water concentrations than humans to reach the same plasma concentration. That factor appears uncertain, in part because the ratio can change with age or length of exposure. In addition, this approach compares water concentrations, not dose. Plasma levels can also vary considerably both between people and in the same person over time (Ekstrand 1978).

Comparing bone fluoride levels in a 16-week rat experiment with human data from Zipkin et al. (1958), Turner et al. (1992) estimated that "humans incorporate fluoride ~18 times more readily than rats when the rats are on a normal calcium diet." The comparison was based on water fluoride concentrations.

Several longer-term animal experiments are compared in Table D-1. The National Toxicology Program (NTP) (Bucher et al. 1991) and Maurer et al. (1990) experiments are well-known long-term fluoride carcinogenicity assays. Of the four studies, Maurer et al. (1990) added fluoride to feed; the others added fluoride to water. Figure D-1 shows results for male rats for the three studies that added fluoride to water. Fluoride bone concentrations for female rats were somewhat higher in the NTP study and somewhat lower in the Maurer et al. study. Femur and vertebra fluoride concentrations were similar in the Dunipace et al. (1995) study. Femur diaphysis fluoride concentrations were similar to concentrations in other sites, except for femur epiphysis, which was higher (Whitford and Birdsong-Whitford 2000; G. Whitford, University of Georgia, personal communication, June 2, 2004). Figure D-1 also shows regression lines through each set of rat data, as well as the crude and adjusted estimates for the human data (Zipkin et al. 1958) discussed earlier. The adjusted line estimates bone concentrations in males with 70 years of residence, but the slope is very similar to the crude model.

Assuming that linear models are realistic in this range and that rats at 18 to 24 months are roughly physiologically comparable to humans at 70 years (Dunipace et al. 1995), the committee compared the slopes for the human and rat studies. The estimates in the left column of Table D-2 (bone versus water) were computed by dividing the slopes for the human data by the slopes estimated for the Dunipace and NTP rat studies. (The commit-

TABLE D-1 Four Chronic Rat Experiments That Measured Fluoride in Bone

	Dunipace et al. 1995	NTP[a]	Maurer et al. 1990	Whitford and Birdsong-Whitford 2000[b]
Strain	Sprague-Dawley	F344/N	Sprague-Dawley	Sprague-Dawley
Sampling	3, 6, 12, 18 months	103 weeks	99 weeks	2, 6, 12, 18 months
Start time	Weanling	Weanling	6 weeks	6 weeks
Sex	M	M, F	M, F	
Water fluoride, mg/L	0, 5, 15, 50	0, 11, 45, 79	—	1, 10, 100
Diet fluoride, ppm	≤1.2	8	Various	
Bone samples	Femur, vertebra	Humerus	Radius, ulna	Femur, radius, calvarium

[a] The NTP results were published by Bucher et al. (1991).
[b] Data are available only in abstract form; unpublished data provided by G. Whitford, University of Georgia, personal communication, June 2, 2004

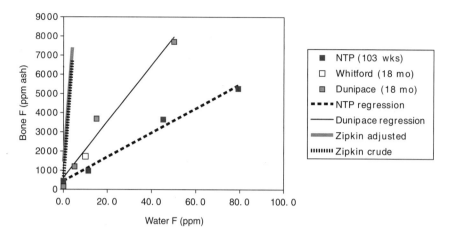

FIGURE D-1 Comparison of bone concentrations in humans and rats on the basis of drinking water concentration

Male rats: NTP (humerus), Whitford (femur diaphysis), Dunipace (femur). Zipkin data: Regression results from crude and adjusted model, the latter assuming males and 70 years residency.

Regression results:
Dunipace: $y = 625 + 147x$ ($r^2 = 0.97$)
NTP: $y = 443 + 63.1x$ ($r^2 = 0.99$)
Human (crude): $y = 517 + 1,549x$
Human (adjusted to male, 70 years residence): $y = 1,300 + 1,527x$

tee also estimated two slopes for the human data, crude and adjusted for length of residency and sex. The crude and adjusted estimates are similar, barely changing the ratios in Table D-2.) These results suggest that rats require water concentrations 10 to 20 times higher than humans to achieve comparable bone fluoride concentrations.

Why are the Dunipace bone concentrations larger than the NTP results? As shown in Table D-1, the NTP study was longer and had higher fluoride concentrations in feed, but both of those factors should increase bone concentrations. The use of different rat strains could contribute to the difference. Type of bone is unlikely to explain the difference. Even if water concentrations are the same, doses might be different. The NTP study provided estimates of average absorbed fluoride doses (assuming 100% from water, 60% from feed) of 0.2, 0.8, 2.5, and 4.1 mg/kg/day for the four experimental groups. Using data provided by Dunipace et al. (1995), the committee estimates average fluoride doses of 0.042, 0.34, 0.96, and 2.83

TABLE D-2 Comparative Uptake of Fluoride Between Humans and Rats

	Bone Versus Water	Bone Versus Dose[a]
Zipkin/NTP	24 to 25[b]	42
Zipkin/Dunipace	10 to 11[b]	20
Zipkin/Maurer	NA	40

[a]Use of the crude and adjusted human models produces very similar results (difference of less than 1).
[b]The lower value uses the adjusted human model (male, 70 years residency); the higher value uses the crude human model.

mg/kg/day for the four experimental groups (divide fluoride intake, μg/day, by body weight for each water concentration and each time interval: 3, 6, 12, and 18 months). At each water concentration, the doses decrease over time. Compute the time-weighted average dose. That does not account for absorption, but feed intake is a small fraction of the total, especially for higher doses. Figure D-2 plots the average doses versus bone fluoride for both studies. Use of average dose reduces the difference in slopes between the Dunipace and NTP studies but not very much. Dunipace et al. found that bone fluoride concentrations increased very rapidly in the first 3 months, followed by a slow increase. As a result, average dose might not be the best metric. On the basis of water consumption rates, exposures appear similar at 3 months (C. Turner, Indiana University, personal communication). Calcium concentrations in feed were higher in the NTP study (0.6 ppm) than in the Dunipace study (0.5 ppm), reducing fluoride absorption (C. Turner, Indiana University, personal communication). The slope estimated for the Maurer data lies between the other two, but the results of this experiment appear to be nonlinear.

To estimate dose for the Zipkin data, the committee assumed the same water consumption (2 L/day) and body weight (70 kg) for every subject, based on standard the U.S. Environmental Protection Agency figures. This assumption multiplies the slope calculated earlier by a constant, 70/2.

The right-hand column of Table D-2 compares human and rat fluoride uptake on an average dose basis. The ratio of the slopes has increased to 20 to 40. The ratios would be higher if a smaller water consumption rate for humans had been assumed. The very high bone concentration predicted by Rao et al. (1995) for women exposed to fluoride in drinking water at 4 mg/L for 70 years suggests an even higher ratio.

Because many assumptions were involved in estimating the values presented in Table D-2, they should be used with caution. But values support a rat-to-human conversion factor for bone fluoride uptake of at least an order of magnitude.

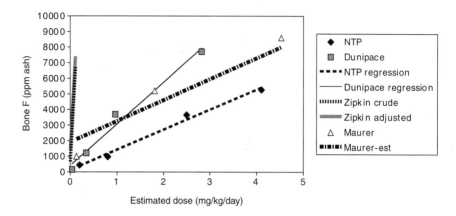

FIGURE D-2 Comparison of bone concentrations in humans and rats on the basis of estimated dose.

To keep the results visible, the figure omits the high data point from Maurer et al. (11.3 mg of fluoride/kg/day, 16,760 mg/kg ash).

Male rats: NTP (humerus), Dunipace (femur), Maurer (radius and ulna).
Zipkin data: Regression results from crude and adjusted model, the latter assuming males and 70 years residency.

Regression results:
Dunipace: $y = 415 + 2{,}664x$ ($r^2 = 0.98$)
NTP: $y = 145 + 1{,}283x$ ($r^2 = 0.99$)
Maurer: $y = 1{,}911 + 1{,}345x$ ($r^2 = 0.98$)
Human (crude): $y = 517 + 1{,}549(70/2)x$
Human (adjusted to male,
 70 years residence): $y = 1{,}300 + 1{,}527(70/2)x$

APPENDIX
E

Detailed Information on Endocrine Studies of Fluoride

The tables that follow contain detailed information on the endocrine studies discussed in Chapter 8, including study design, exposure information, and reported effects. Exposure conditions and duration and fluoride concentrations are provided as given in the published articles. Many of the tables include estimates of exposure in units of mg/kg/day to aid in comparing studies. When possible, these estimates were made from information (e.g., intake rate of drinking water, body weight) given in the articles. Where such information was not available in a published article, the assumptions used to make the estimates are listed in footnotes to the tables. Note that for most of the human studies, the exposure estimates (mg/kg/day) are for typical or average values for the groups and do not reflect the full range of likely exposures.

TABLE E-1 Effects of Fluoride on Thyroid Follicular Cell Function in Experimental Animals

Species and Strain	Exposure Conditions	Fluoride Concentration or Dose[a]	Exposure Duration	Effects	Reference
Rats (Hebrew University albino, males; infants at start, 30-32 g) See also Table E-16	Drinking water	0.55, 1, or 10 mg/L (0.055, 0.1, and 1 mg/kg/day)[b]	9 months	No significant differences in basal metabolic ratio, thyroid weight, radioiodine uptake, total blood iodine, protein-bound iodine, or urinary excretion. TSH not measured.	Gedalia et al. 1960
Rats (females, 180-230 g)	Gastric tube 0.2 or 2.2 μg/day iodine in diet	750 μg/day in 1 mL water (3.3-4.2 mg/kg/day)	2 months	No effect of fluoride on body weight, weight of thyroid, total composition of iodinated amino acids, or amount of iodide present in the thyroid. No effect of fluoride on iodine excretion in the higher-iodine group. Decreased protein-bound iodine, T3, and T4 (low-iodine group). Decreased biogenesis of T3 and T4 following administration of ^{131}I (low- and high-iodine groups). TSH not measured.	Stolc and Podoba 1960
Rats (Wistar, males; initial weight 170-230 g; 13 per group)	Drinking water Dietary iodine, 0.45 μg/g feed (0.45 ppm)	0, 0.1, or 1 mg/day (0, 0.43-0.59, or 4.3-5.9 mg/kg/day)	60 days	Decreased plasma T3 and T4, decreased free T4 index, increased T3-resin uptake (all changes statistically significant except for the decrease in T3 for the group receiving 0.1 mg/day)[c] TSH not measured.	Bobek et al. 1976

Species	Exposure	Dose	Duration	Effects	Reference
Cows (Holstein; various states of lactation, 9-13 cows from each of 9 herds) See also Table E-3	Feed supplements	1-22 mg/kg F in feed (estimated) (approximate doses, 0.03-0.7 mg/kg/day)[d]	Chronic	Urinary fluoride ≥ 2.9 mg/L (range 1.04-15.7 mg/L, average 5.13 mg/L). Decreased T3, T4, cholesterol and increased eosinophils with increasing urinary fluoride (adjusted for stage of lactation); serum calcium correlated with T3 and T4. Fluorosis herds (S1, C4, V3, B2) had lower T4 than herds W, B, M, G ($P < 0.05$). Feeding of iodinated casein to herd B2 for 3 weeks resulted in 100% increase in milk production, increased hematopoiesis, reduced eosinophils, increased serum calcium, decreased serum phosphorus, and increase in serum T4 from 3.4 to 14.1 µg/dL. TSH not measured. Bone fluoride: mean, 2,400 ppm in ash (range, 850-6,935, 22 specimens from 8 herds).	Hillman et al. 1979
Rats (Wistar) See also Table E-16	Drinking water and diet	Water: 0, 1, 5, 10, 50, 100, or 200 mg/L Diet: 0.31 or 34.5 ppm (0, 0.1, 0.5, 1, 5, 10, or 20 mg/kg/day from water and 0.025 or 2.8 mg/kg/day from feed)[e]	54-58 days	Elevated T3 and T4 in rats on 1 mg/L in drinking water and low-fluoride diet. Low T3 and normal T4 in rats on 1, 5, or 10 mg/L in drinking water and high-fluoride diet. Decreased TSH and GH in animals receiving 100 or 200 mg/L in drinking water. Full details not available.	Hara 1980

continued

TABLE E-1 Continued

Species and Strain	Exposure Conditions	Fluoride Concentration or Dose[a]	Exposure Duration	Effects	Reference
Rats (Wistar, 3-month-old, 200-400 g)	Drinking water Animals were kept 21 days on a diet containing 0.15% PTU to deplete their thyroid glands of iodine and thyro-globulin. For the next 2 days, a low-iodine diet (0.04 µg/g) was fed, but no more PTU. During the next 6 days the rats were given sufficient iodine (1.5 µg of iodide/mL of drinking water, labeled with 0.1 µCi of 125I). Then fluoride was given as indicated.	60 or 200 mg/L (6-20 mg/kg/day)[b]	6 days	Serum fluoride at end of experiment (µg/mL): 0.165 (controls), 0.246 (60 mg/L), and 0.576 (200 mg/L). No significant differences from control values for relative thyroid weight, iodine content of thyroglobulin, thyroidal content of organic iodine, or amounts of monoiodotyrosine, diiodotyrosine, T3, and T4. TSH not measured.	Siebenhüner et al. 1984

Cows (Holstein, females; age 5-6 months at start, 30 animals total)	NaF added to feed Iodine intake not stated, presumably adequate	30 or 50 ppm in feed Approx. 0.8 or 1.4 mg/kg/day at 30 weeks of age; approx. 0.5 or 0.8 mg/kg/day at 100 weeks of age	Data reported through age 100 weeks	Serum fluoride at age 88 weeks (mg/L): 0.06 (controls), 0.20 (30 ppm in feed), and 0.28 (50 ppm in feed). Urinary fluoride at age 88 weeks (mg/L): 1 (controls), 13 (30 ppm in feed), and 20 (50 ppm in feed). Bone fluoride at age 17 months (ppm in tail vertebra, means of groups of 5 animals): 352 and 453 (controls), 2,306 and 2,712 (30 ppm in feed), and 3,539 and 3,946 (50 ppm in feed). No significant differences from control values for T4 concentration and T3 uptake at ages 40, 56, 72, and 88 weeks.[f] TSH not measured.	Clay and Suttie 1987
Rats (Wistar, males and females; 120 ± 19 g at start, 212 animals total) See also Table E-2	Drinking water Low or normal iodine	10 or 30 mg/L in drinking water (1 or 3 mg/kg/day)[b]	7 months	10 mg/L and normal iodine: no significant effect (some decrease in serum T4 and T3). 30 mg/L and normal iodine: statistically significant decreases in T4, T3, thyroid peroxidase, ^{131}I uptake, [^{3}H]-leucine uptake, and thyroid weight. 10 mg/L and low iodine: abnormalities in thyroid function beyond those attributable to low iodine; reduced thyroid peroxidase; low T4, without compensatory transformation of T4 to T3. TSH not measured.	Guan et al. 1988

continued

TABLE E-1 Continued

Species and Strain	Exposure Conditions	Fluoride Concentration or Dose[a]	Exposure Duration	Effects	Reference
Mice (Kunmin, males; 288 animals in 9 groups of 32 each; 13-15 g at start)	Drinking water (NaF) Iodine: low (0 µg/L); normal (20 µg/L); excess 2500 µg/L) Low-iodine, low-fluoride chow fed to all groups.	Low, 0 mg/L; normal, 0.6 mg/L; excess, 30 mg/L (0, 0.06, and 3 mg/kg/day)[b]	100 or 150 days	For iodine-excess groups, thyroid weight relative to body weight decreased significantly with increasing fluoride intake. For iodine-deficient groups, goiter incidence at 100 days was 18%, 40%, and 66% for low-, normal-, and high-fluoride groups, respectively; at 150 days, goiter incidence was 81-100%. Fluoride-excess groups at 100 days had elevated T4 with all concentrations of iodine intake and elevated T3 for iodine-deficient animals. Fluoride excess significantly inhibited radioiodine uptake in iodine-deficient and iodine-normal groups. Incisor fluorosis occurred only in the fluoride excess groups; severity was greater in the iodine-deficient animals. Bone fluoride in fluoride-excess animals was greater in iodine-deficient (means, 2,560-2,880 ppm ash) or iodine-excess animals (means, 2,140-2,380 ppm ash) than in iodine-normal animals (means, 1,830-2,100 ppm ash). TSH not measured.	Zhao et al. 1998

Subject	Exposure	Duration	Effects	Reference
Cattle near aluminum smelter in India	Contaminated pasture from smelter emissions No information on iodine intake	Not available	Skeletal and enamel fluorosis (58% of animals within 3 km of plant were affected). Significantly decreased concentrations of T3. Significantly increased concentrations of alkaline phosphatase, inorganic phosphorus, and creatinine. Urinary fluoride averaged 26.5 mg/L close to smelter. Full details not available.	Swarup et al. 1998
Cattle, buffaloes, sheep, and goats in 21 villages in India (286 calves, 1,675 adult cattle, 290 adult buffaloes, 780 goats, 564 sheep)	Drinking water No information on iodine intake	1.5–4 mg/L in drinking water	Native livestock present in relevant area since birth	Prevalence of enamel fluorosis up to 75% (adult buffalo), 70% (adult cattle), or 100% (calves), depending on location; prevalence of skeletal fluorosis up to 37.5% (buffalo) or 29% (cattle), depending on location; no evidence of enamel or skeletal fluorosis in goats or sheep. No clinical evidence of goiter in any fluorotic animals. Animals not showing clinical signs of fluorosis were not examined for goiter. No measurements of any thyroid hormone parameters or TSH.	Choubisa 1999
Mice (Wistar, adult females; about 30 g at beginning; fluoride was administered during pregnancy and lactation)[g]	Drinking water (Iodine intake 0.720 ± 0.12 µg/g in diet)	500 mg/L in drinking water (50 mg/kg/day to the mothers)[b]	From day 15 of pregnancy to day 14 of lactation	Body weight of pups at 14 days old was reduced 35%; 75% decrease in plasma T4 in pups; 17% decrease in cerebral protein in pups; histological changes in cerebellum in pups. TSH not measured.	Trabelsi et al. 2001

continued

453

TABLE E-1 Continued

Species and Strain	Exposure Conditions	Fluoride Concentration or Dose[a]	Exposure Duration	Effects	Reference
Cows (3 years old with chronic fluorosis, 10 controls without fluorosis, from different regions of Turkey)	Drinking water Iodine intakes not specifically stated	5.7-15.2 mg/L in drinking water (approximate doses, 0.7-1.8 mg/kg/day)[b]	Lifelong	Mean values of T4, T3, and PBI in fluorotic animals were below the normal ranges and also significantly less than in controls. Low concentrations of bioavailable iodine in fluorosis region might be a factor. TSH not measured.	Cinar and Selcuk 2005

[a] Information in parentheses was calculated from information given in the papers or as otherwise noted.
[b] Based on water consumption of about 10% of body weight.
[c] ATSDR (2003) stated that an intermediate-duration minimal risk level (MRL) derived from this study of thyroid effects in rats would have been lower (more protective) than the chronic-duration MRL derived from a human study of bone effects (0.05 mg/kg/day).
[d] Based on feed consumption of 16 kg/day (dry weight) and body weight of 500 kg.
[e] Based on water consumption of about 10% of body weight and feed consumption of about 8% of body weight.
[f] Text says "triiodothyronine uptake" and table says "thyroxine uptake." Data for different treatment groups were not given.
[g] In many mammalian species, maternal fluoride exposures are not well reflected by fluoride concentrations in milk; therefore, the impacts of fetal exposure and of reduced milk production by the mothers must also be considered.
[h] Based on water consumption of 60 L/day and body weight of 500 kg.

ABBREVIATIONS: GH, growth hormone; PBI, protein-bound iodine; TSH, thyroid-stimulating hormone.

TABLE E-2 Summary of Effects of Fluoride Exposure for Rats with Different Amounts of Iodine Intake (Means ± SD)

Group	Body Weight, g	Urinary Fluoride, mg/L	Urinary Iodine, μg/24 hours	131I Uptake, % at 24 hours	Serum T4, μg/dL	Serum T3, ng/dL	TPO, G.U./100 g of body weight	[3H] Leucine Uptake, cpm/10 mg	Thyroid Weight, mg/g
1 (control; normal iodine,[a] normal fluoride[b])	293 ± 57	1.23 ± 0.22	1.110 ± 0.226	47.37 ± 5.66	3.64 ± 1.45	70.65 ± 30.29	2.04 ± 0.22	1,808 ± 358	9.97 ± 3.52
2 (normal iodine,[a] fluoride, 10 mg/L in drinking water)	294 ± 85	6.65 ± 0.91[c]	1.215 ± 0.357	44.74 ± 5.14	3.02 ± 1.48	61.96 ± 26.02	1.98 ± 0.51	1,728 ± 790	9.58 ± 2.40
3 (normal iodine,[a] fluoride, 30 mg/L in drinking water)	254 ± 68[c]	8.16 ± 0.89[c]	1.150 ± 0.87	42.73 ± 4.31[c]	1.44 ± 0.39[c]	43.00 ± 11.31[c]	1.73 ± 0.24[c]	1,258 ± 293[c]	7.90 ± 2.37[c]
4 (low iodine,[d] normal fluoride[f])	289 ± 72	1.23 ± 0.26	0.095 ± 0.029[c]	58.40 ± 9.54[c,e]	0.76 ± 0.70[c]	95.81 ± 25.18[c]	2.57 ± 0.44[c]	2,252 ± 683[c]	19.91 ± 11.23[c]
5 (low iodine,[d] fluoride, 10 mg/L in drinking water)	308 ± 63	6.23 ± 0.88[c]	0.099 ± 0.017[c]	59.05 ± 7.59[c,e]	0.65 ± 0.57[c]	68.05 ± 21.96	1.75 ± 0.21[c]	1,804 ± 459	20.13 ± 22.10[c]

[a]Normal iodine: 310 ng/g in diet; 8.2 ng/mL in drinking water.
[b]Fluoride: 1.856 ppm in diet; 0.4 mg/L in drinking water.
[c]$P < 0.01$, compared with group 1 (control).
[d]Low iodine: 20–62.5 ng/g in diet; deionized drinking water.
[e]Also statistically significant at 2 hours and 6 hours ($P < 0.01$, compared with group 1).
[f]Fluoride: 1.743 ppm in diet; deionized water.

ABBREVIATIONS: cpm, counts per minute; G.U., guaiacol unit; TPO, thyroid peroxidase.
SOURCE: Guan et al. 1988. Reprinted with permission; copyright 1988, Chinese Medical Association.

TABLE E-3 Summary of Selected Findings for Fluoride-Exposed Dairy Cows

Herd[a]	Number Observed	Urinary Fluoride, mg/L[b]	Serum T4, µg/dL[c]	Serum T3, ng/dL[d]	Plasma Calcium, mg/dL[e]
W	12	2.92 ± 0.52	4.60 ± 0.34	175 ± 7.2	10.1 ± 0.15
B	12	5.37 ± 0.43	4.83 ± 0.19	168 ± 5.8	9.5 ± 0.11
M	12	6.39 ± 0.92	5.30 ± 0.38	177 ± 8.4	9.6 ± 0.11
G	12	6.33 ± 0.74	4.82 ± 0.28	159 ± 7.7	9.4 ± 0.15
P	12	3.47 ± 0.47	—	—	9.3 ± 0.12
S1	12	6.29 ± 1.08	3.59 ± 0.26	126 ± 8.4	9.1 ± 0.17
C4	9	—[f]	2.21 ± 0.54	—	9.5 ± 0.14
V3	10	—	3.35 ± 0.47	—	9.5 ± 0.13
B2	13	—	3.39 ± 0.42	—	8.9 ± 0.12

[a]Herd identification as reported by Hillman et al. (1979). Enamel fluorosis and elevated bone fluoride were confirmed in herds S1, C4, V3, and B2. Cows were uniformly distributed throughout lactation in all herds.
[b]W < all others ($P < 0.05$).
[c]C4 < all others; S1, V3, B2 < W, B, M, G ($P < 0.05$).
[d]S1 < W, B, M, G ($P < 0.05$).
[e]B2 < M, W; S1, P, G < W ($P < 0.05$).
[f]—indicates not measured or not reported.

SOURCE: Hillman et al. 1979. Reprinted with permission; copyright 1979, Journal of Dairy Science.

TABLE E-4 Effects of Fluoride in Drinking Water on Thyroid Follicular Cell Function in Humans

Study Population(s) and Type	Fluoride Concentration[a] and Exposure Duration/Conditions	Iodine Status and Other Information	Effects	Reference
India, 3 villages, 2,008 persons, all ages Ecologic study; cross-sectional; entire population of each village included	5.4, 6.1, and 10.7 mg/L (means for the villages) Lifelong	Iodine in drinking water: 14.4-175.3 µg/L (inverse relationship to fluoride concentration). Iodine from salt: 86 µg/day. Calcium in diet: 480 mg/day. Diet considered deficient in proteins, fats, calcium, vitamins A and C.	Transient goiters in persons aged 14-17; associated with increased fluoride in water and with decreased iodine in water.	Siddiqui 1960
Israel, 2,685 girls, ages 7-18 Ecologic study; cross-sectional; may have included all eligible subjects, but not specifically stated	<0.1-0.9 mg/L Lifelong	Iodine in drinking water: <2-100 µg/L.	Endemic goiter associated with low iodine content of water, but not with fluoride content of water.	Gedalia and Brand 1963
U.S., adults ages 18-60; 106 from Crisfield, Maryland (42% female); 109 from New York City (29% female) Ecologic exposure measure; cross-sectional; no information on subject selection	0.09 mg/L in New York City 3.48 mg/L in Crisfield, Maryland ≥10 years exposure	General iodine status not given. Crisfield: the 3 individuals with the highest PBI concentrations were all on iodine medication for non-thyroidal disease, and one of the individuals with the lowest PBI had had a partial thyroidectomy for a thyroid cyst.[b] New York City: the individual with the highest PBI was taking 3 grains of thyroid daily.[b]	No differences in PBI. No gross thyroid abnormalities or gross evidence for thyroid disease. Mild or moderate enamel fluorosis in 75% of individuals from Crisfield.	Leone et al. 1964

continued

TABLE E-4 Continued

Study Population(s) and Type	Fluoride Concentration[a] and Exposure Duration/Conditions	Iodine Status and Other Information	Effects	Reference
Nepal, 648 persons in 13 villages with similar iodine concentrations in water, all ages Ecologic study; cross-sectional; samples represented about one-third of the population in each village (children presenting for inoculations plus accompanying adults)	< 0.1 to 0.36 mg/L Lifelong	Iodine in drinking water: ≤1 µg/L. Diet low in iodine; iodized salt not available. Calcium in water, 3-148 mg/L. Magnesium in water, 0.5-77 mg/L. Water hardness, 10-670 ppm (as $CaCO_3$).	Goiter prevalence (5-69%) positively associated with fluoride concentration ($\rho = 0.74$, $P < 0.01$). Goiter prevalence of at least 20% associated with fluoride concentrations ≥ 0.19 mg/L. Goiter prevalence also associated positively with water hardness ($\rho = 0.77$, $P < 0.01$), calcium ($\rho = 0.78$, $P < 0.01$) and magnesium ($\rho = 0.83$, $P < 0.01$). Effect of fluoride was independent of that of hardness.	Day and Powell-Jackson 1972
India, 9 patients with moderate to severe skeletal fluorosis (6 males, 3 females), mean age 29 years; 5 control individuals (3 males, 2 females), mean age 31 years Case-control study; individual estimates of current fluoride intake, measurements of fasting plasma fluoride and urinary fluoride; incomplete information on selection of subjects and controls	7.8-8.0 or 24.5-25.0 mg/L Current exposure to 0.8 and 1.8 mg/L in water for the 2 persons who had moved Lifelong 2 persons had moved to nonendemic areas 2 or 5 years previously Symptomatic for 10-15 years	Iodine status not given	PBI values all normal (4.2-5.8 µg/100 mL). No evidence of goiter or thyroid dysfunction. See also Tables E-9, E-10, and E-12	Teotia et al. 1978

Germany, 13-15 years old, males and females, 17 in low-fluoride group and 26 in high-fluoride group Ecologic exposure measure; cross-sectional; no information on subject selection; 2 of the original 19 in low-fluoride group excluded upon discovery of hyperthyroidism	0.1-0.2 and 3 mg/L Lifelong	Iodine status not given	No significant differences in T3 uptake, T4, free T4 index, T3, reverse T3, thyroglobulin, TSH, thyroglobulin antibodies, or microsomal thyroid antibodies. Unexplained decrease in thyroglobulin in girls (31.3 ± 12.9 ng/mL in the low-fluoride group and 13.8 ± 4.3 ng/mL in the high-fluoride group); this difference is also reflected in the means for boys and girls combined.	Baum et al. 1981
Ukraine, 2 cities with different water fluoride concentrations Ecologic exposure measure; cross-sectional; no information on subject selection	Values not given	Iodine status not given	Iodine deficiency and "adaptive amplification of the hypophyseal-thyroid system" (increased TSH?) in residents with high fluoride in drinking water; increased incidence of "functional disturbance" of the thyroid, but no structural changes. Full details not available.	Sidora et al. 1983

continued

TABLE E-4 Continued

Study Population(s) and Type	Fluoride Concentration[a] and Exposure Duration/Conditions	Iodine Status and Other Information	Effects	Reference
Ukraine, 47 healthy persons (ages 19-59), 43 persons with hyperthyroidism (ages 18-58), and 33 persons with hypothyroidism (ages 20-55) Ecologic exposure measure; cross-sectional; no information on subject selection other than by thyroid status See also Table 8-6	Region I: 0.5-1.4 mg/L (mean, 1.0) Region II: 1.6-3.5 mg/L (mean, 2.3) Lifelong (permanent residents)	Iodine status not given	Among normal individuals, significantly increased serum TSH and thyroidal ^{131}I uptake and significantly decreased serum T3 in Region II, although values still within normal ranges. Differences between Regions I and II not seen among thyroidopathy patients. No information on the prevalence of thyroid disease in the two regions.	Bachinskii et al. 1985
China, children ages 7-14, 250 in Area A and 256 in Area B Ecologic exposure measure; cross-sectional; no information on subject selection	Area A, 0.88 mg/L (enamel fluorosis, 20.80%) Area B, 0.34 mg/L (enamel fluorosis, 16.00%) Lifelong	Iodine in drinking water (µg/L): Area A, 5.21; Area B, 0.96 Goiter prevalence: Area A, 91%; Area B, 82%	Area A had higher TSH, slightly higher ^{131}I uptake, and lower mean IQ than Area B. Area A also had reduced T3 and elevated reverse T3, compared with Area B. Urine fluoride (mg/L): Area A, 2.56; Area B, 1.34-1.61.	Lin et al. 1991

Study population	Exposure	Results	Reference
India, 22,276 individuals in a single district, all ages Ecologic study; cross-sectional; subjects included 1% of total population and 5% of school children of randomly selected villages	≥1 mg/L Enamel fluorosis prevalence ranged from 6.0% to 59.0% (12.2% overall) Lifelong	Iodine in drinking water ≥ 10 µg/L Goiter prevalence ranged from 9.5% to 37.5% (14.0% overall)	Significant positive correlation between prevalence of goiter and enamel fluorosis ($r = 0.4926$, $P < 0.001$). No significant correlation between water iodine concentration and goiter prevalence ($r = 0.1443$, $P > 0.05$). In regions with water iodine concentrations > 20 µg/L, goiter prevalence was significantly higher in regions with fluoride > 2 mg/L (27.8%) than in regions with fluoride < 2 mg/L (17.1%). No evidence for functional changes in thyroid activity associated with the presence of goiter.	Desai et al. 1993
China, no details available Ecologic study; probably cross-sectional; no information on subject selection	High fluoride, values not given (Enamel fluorosis in children, 72.9%) Lifelong	High iodine, values not given	Urinary fluoride: 2.08 ± 1.03 mg/L Urinary iodine: 816.25 ± 1.80 µg/L. Reduced ^{131}I uptake rate, elevated serum TSH, with respect to controls. Prevalence of thyroid enlargement was 3.8% in adults and 29.8% in children, and of enamel fluorosis, 35.5% and 72.9%, respectively.	Yang et al. 1994

continued

TABLE E-4 Continued

Study Population(s) and Type	Fluoride Concentration[a] and Exposure Duration/Conditions	Iodine Status and Other Information	Effects	Reference
India, 500 individuals from 52 villages in 2 districts; blood samples from randomly selected subset of control and fluorotic individuals Ecologic exposure measure; cross-sectional; no information on selection of original set of subjects	1.0-6.53 mg/L (18 villages, <2 mg/L; 26 villages, 2-4 mg/L; 8 villages, >4 mg/L; 74% with slight to severe mottling of teeth) Control, 0.56-0.72 mg/L Lifelong	Iodine status not given	Serum fluoride (mg/L): 38%, <0.2; 47%, 0.2-0.4; 15%, >0.4. Significant increase in serum T4 ($P < 0.001$): 14.77 ± 0.512 µg/dL versus 9.16 ± 0.63 µg/dL[c] (ranges, 7.2-20.0 versus 5.4-13.0). No significant differences in concentrations of serum T3 and TSH.	Michael et al. 1996
South Africa, 671 children, ages 6, 12, and 15, from six towns selected by fluoride concentration of drinking water Ecologic exposure measure; cross-sectional; study population included all children of the designated ages who spent their entire lives in the study towns See also Table E-5	Low: 0.3 and 0.5 mg/L Medium: 0.9 and 1.1 mg/L High: 1.7 and 2.6 mg/L Severe mottling of teeth in most children in the high-fluoride towns, not seen in the other towns Lifelong	Iodine in water, 105 to >201 µg/L[d] Iodine in urine, 193 to >201 µg/L[d] (median values) Iodine status considered sufficient (possibly even high)	Goiter prevalence ranged from 5.2% to 29.0% (15.3-29.0% for 5 of the 6 towns). The two towns with the highest fluoride had the highest goiter rates (27.7 and 29.0%). The town with 5.2% goiter prevalence had substantially less undernutrition than the other 5 towns.	Jooste et al. 1999

| India, 90 children, ages 7-18 with enamel fluorosis; 21 controls, ages 8-20 without enamel fluorosis Case-control study, subjects with and without enamel fluorosis, also selected by water fluoride concentration; cross-sectional; ecologic exposure measure (water fluoride concentration) but urine and serum fluoride also measured | Children with dental fluorosis: 1.1-14.3 mg/L (mean, 4.37 mg/L) Children without fluorosis: Group I, 0.14-0.81 mg/L (mean, 0.23 mg/L); Group II, 0.14-0.73 mg/L (mean, 0.41 mg/L) Lifelong | Iodine supplementation via iodized salt for more than a decade previously, considered satisfactory | 49 of 90 children with fluorosis had "well-defined hormonal derangements"; findings were borderline in the remainig 41 children. Five distinct categories of hormonal deviations: normal FT4 and FT3, elevated TSH (subclinical hypothyroidism, 23 of 90) normal FT4 and TSH, low FT3 (low T3 syndrome, 16 of 90); borderline low T3 in many of the other children normal FT4, elevated FT3 and TSH (7 of 90); T4 on low end of normal range, possible T3 toxiccsis normal FT3, low FT4, elevated TSH (2 of 90) normal FT4, low FT3, elevated TSH (1 of 90) Categories 2-5 all associated with or can be caused by abnormal deiodinase activity. Only 4 control children had serum fluoride concentrations below the normal upper limit; approximately 50% of the control children also had "hormonal deviatiors"; children with "safe" water (< 1 mg/L fluoride) were taking in too much fluoride, presumably from nonwater sources. | Susheela et al. 2005 |

continued

TABLE E-4 Continued

Study Population(s) and Type	Fluoride Concentration[a] and Exposure Duration/Conditions	Iodine Status and Other Information	Effects	Reference
			Urinary fluoride concentrations (normal upper limit, 0.1 mg/L): Children with fluorosis, 0.41-12.8 mg/L (mean, 3.96 mg/L) Controls, 0.09-4.2 mg/L Serum fluoride concentrations (normal upper limit, 0.02 mg/L): Children with fluorosis, 0.02-0.41 mg/L (mean, 0.14 mg/L) Controls, 0.02-0.29 mg/L	

[a]Due to the great range of ages included in the various studies, and because the reports do not include dose estimates (mg/kg/day), comparisons in this table are best made in terms of fluoride concentrations in drinking water. Approximations of representative doses have been made as follows: Day and Powell Jackson (1972) (iodine deficiency present): [F] = ≥ 0.2 mg/L; intake of 1 L/day for a 20-kg child; approximate dose ≥ 0.01 mg/kg/day.
Bachinskii et al. (1985): [F] = 1.6-3.5 mg/L; intake of 2 L/day for a 70-kg adult; approximate dose, 0.05-0.1 mg/kg/day.
Lin et al. (1991) (iodine deficiency present): [F] = 0.88 mg/L; intake of 1 L/day for a 30-kg child; approximate dose, 0.03 mg/kg/day.
Michael et al. (1996): [F] = 1.0-6.5 mg/L; intake of 2 L/day for a 60-kg adult; approximate dose, 0.03-0.22 mg/kg/day.
Jooste et al. (1999): [F] = 1.7 and 2.6 mg/L; intake of 1 L/day for a 20-kg child or 2 L/day for a 50-kg teenager; approximate doses, 0.09-1.3 mg/kg/day for the child and 0.07-0.1 mg/kg/day for the teenager.
Susheela et al. (2005): [F] = 1.1-14.3 mg/L; intake of 2 L/day for a 50-kg teenager; approximate dose, 0.04-0.6 mg/kg/day.
[b]McLaren (1976) suggested that these individuals should not have been included in the samples or else that further research on the etiology should have been carried out.
[c]The units for serum T4 given by Michael et al. 1996 are ng/mL, but most likely μg/dL was meant. In units of μg/dL, these mean values are in the normal range for the controls and slightly above the normal range for the endemic fluorosis population. If the values are in ng/mL, then both means are below the normal range for serum T4.
[d]Iodine concentrations reported as 0.83 to > 1.58 μmol/L in water and 1.52 to > 1.58 μmol/L in urine.

ABBREVIATIONS: FT3, free T3; FT4, free T4; PBI, protein-bound iodine.

TABLE E-5 Summary of Selected Parameters for Six South African Towns

Town	Sample Size	Fluoride in Drinking Water, mg/L	Goiter Prevalence, %	Median Urinary Iodine, µg/L[a]	Iodine in Drinking Water, µg/L[b]	Iodine in Iodized Salt, ppm
Williston	85	0.3	15.3	> 201	105	28
Victoria West	127	0.5	17.3	> 201	> 201	5
Frazerburg	87	0.9	18.4	193	127	11
Carnarvon	95	1.1	5.2	> 201	—[c]	9
Brandvlei	94	1.7	27.7	> 201	> 201	5
Kenhardt	183	2.6	29.0	> 201	143	4

[a]Reported as > 1.58, > 1.58, 1.52, > 1.58, > 1.58, and > 1.58 µmol/L, respectively.
[b]Reported as 0.83, > 1.58, 1.00, > 1.58, and 1.13 µmol/L, respectively.
[c]No water sample.

SOURCE: Jooste et al. 1999. Reprinted with permission; copyright 1999, Macmillian Publishers Ltd.

TABLE E-6 Summary of Findings in Healthy Persons and Persons with Thyroid Disease

Group	Region	No.	Fluoride in Drinking Water, mg/L	Fluoride in Urine, mg/L	Fluoride in Urine, mg/day	Fluoride in Serum, mg/L	Fluoride in Erythrocytes, mg/L	^{131}I Uptake, 24 hours, %	T4, µg/dL[a]	T3, ng/dL[b]	TSH, milliunits/L
Hyperthyroid	I	21	1.2 ± 0.2	1.5 ± 0.2	2.1 ± 0.4	0.18 ± 0.01	0.46 ± 0.03	61 ± 7[c]	19 ± 1.2[c]	340 ± 46[c]	0.8 ± 0.12[c]
	II	22	2.2 ± 0.2[d]	2.9 ± 0.5[d]	3.9 ± 0.9[d]	0.19 ± 0.01[e]	0.51 ± 0.10	72 ± 13[e,c]	20 ± 1.8[e]	460 ± 120[e]	0.6 ± 0.08[e]
Hypothyroid	I	14	1.1 ± 0.1	1.4 ± 0.2	1.6 ± 0.2	0.23 ± 0.02	0.55 ± 0.10	8.5 ± 2.7[c]	2.0 ± 0.54[c]	72 ± 26[c]	51 ± 11[c]
	II	19	2.5 ± 0.5[d]	2.8 ± 0.4[d]	3.7 ± 0.7[d]	0.29 ± 0.02	0.61 ± 0.02	9.8 ± 1.3[e]	2.3 ± 0.16[e]	65 ± 6.5[e]	58 ± 17[e]
Controls	I	17	1.0 ± 0.1	1.5 ± 0.2	1.9 ± 0.3	0.21 ± 0.01	0.55 ± 0.10	24 ± 3	7.5 ± 0.62	180 ± 20	2.4 ± 0.2
	II	30	2.3 ± 0.1[c]	2.4 ± 0.2[c]	2.7 ± 0.2[c]	0.25 ± 0.01[c]	0.61 ± 0.03	33 ± 4[c]	7.3 ± 0.47	130 ± 13[c]	4.3 ± 0.6[c]

[a] Reported as 250 ± 16, 261 ± 23, 26 ± 7, 29 ± 2, 97 ± 8, and 94 ± 6 nmol/L, respectively.
[b] Reported as 5.2 ± 0.7, 7.1 ± 1.8, 1.1 ± 0.4, 1.0 ± 0.1, 2.8 ± 0.3, and 2.0 ± 0.2 nmol/L, respectively.
[c] $P < 0.05$ compared with controls residing in Region I.
[d] $P < 0.05$ compared with patients with corresponding thyropathies residing in Region I.
[e] $P < 0.05$ compared with controls residing in Region II.

SOURCE: Adapted from Bachinskii et al. (1985).

TABLE E-7 Effects of Clinical Fluoride Exposure on Thyroid Follicular Cell Function in Humans

Study Population(s) and Type	Exposure Conditions and Duration	Fluoride Concentration or Dose	Effects	Reference
Switzerland, patients with hyperthyroidism, males and females, 15 total Clinical trial; nonblinded; comparison with before-treatment values; mechanistic rather than therapeutic study	NaF, orally (3 times per day) or intravenously (once per day) Iodine status not given 20-245 days	2-10 mg/day [0.029-0.14 mg/kg/day][a]	Clinical improvement in 6 of 15 patients (symptoms of hyperthyroidism relieved, both BMR and plasma PBI reduced to normal concentrations); BMR or PBI was often improved in the other 9. Greatest improvement in women between 40 and 60 years old with a moderate degree of thyrotoxicosis.	Galletti and Joyet 1958
Germany, women with osteoporosis, 26 total completed 6 months of treatment (median age 62.1 years) Clinical therapeutic trial; nonblinded; comparison with before-treatment values; 38 patients originally enrolled, 3 excluded for disturbance of thyroid function	NaF, orally (twice per day) Iodine status not given Only 10 patients took their medicine regularly (as indicated by measurements of plasma fluoride) 6 months	36 mg/day or less Reduction to half dose necessary for 6 patients [0.3 or 0.6 mg/kg/day][b]	Tested for T3 uptake, T4, free T4 index, T3, and TSH; tested before start of trial and after 3 and 6 months. No changes observed in thyroid function or size.	Eichner et al. 1981

continued

TABLE E-7 Continued

Study Population(s) and Type	Exposure Conditions and Duration	Fluoride Concentration or Dose	Effects	Reference
Denmark, osteoporosis patients, 140 females, 23 males, aged 16-84 years, mean 63.7 years Clinical therapeutic trial; non-blinded; 163 consecutive patients (1975-1983) presenting with osteoporosis and at least one atraumatic spinal fracture and who started treatment with fluoride, calcium and vitamin D; comparison with before-treatment values	NaF, orally (3 times per day with meals) Iodine status not given Calcium phosphate and vitamin D were supplemented Mean duration 2.8 years (5 years for 43 patients)	27 mg/day during first year Later adjusted to maintain serum fluoride between 0.095 and 0.19 mg/L (5 and 10 µmol/L) [0.45 mg/kg/day][b]	No changes in thyroid function (T4, T3, T3 uptake, TSH). Joint-related (51%) and gastrointestinal (25%) side effects at some point during treatment; 6% withdrew due to side effects; side effects rare when doses reduced to 14-18 mg/day.	Hasling et al. 1987

[a]Based on 70-kg body weight.
[b]Based on 60-kg body weight.

ABBREVIATIONS: BMR, basal metabolic rate; PBI, protein-bound iodine

TABLE E-8 Effects of Fluoride on Thyroid Parafollicular Cell Function in Experimental Animals

Species and Strain	Exposure Conditions	Concentration or Dose[a]	Exposure Duration	Effects	Reference
Rats (Sprague-Dawley, albino, 200 g at start, 16 total, both sexes)	A: Drinking water (8 animals) B: Intraperitoneal (4 animals) C: Controls (4 animals)	A: 40 mg/L [4 mg/kg/day][b] B: 20 mg/kg/day	A: 2 months B: 4 days (lived with controls for 2 months, ip injections on last 4 days)	No morphological differences in parafollicular cells. No evidence for short-term release of calcitonin, but calcitonin not directly measured.	Sundström 1971
Pigs (females, 20 with thyroidectomy at 10 weeks old, 20 intact; 8 months old at start of experiment; bred at 8 1/2 months old)	Basal ration (Ca deficient); basal ration plus Ca and P; basal ration plus NaF; basal ration plus Ca, P, and NaF Iodinated casein (0.2 g/day) fed to thyroidectomized animals	2 mg/kg/day (fluoride in ration adjusted periodically to maintain this dose)	Approximately 6 months Experiment terminated when litters were 7 weeks old (maternal age approximately 14 months)	Retarding effect on cortical bone remodeling; intact thyroid gland necessary for this effect (effect not seen in thyroidectomized animals with replacement of thyroid hormone but not calcitonin). Bone fluoride in intact animals (μg/g): basal, 285; basal plus Ca and P, 181; basal plus NaF, 3,495; basal plus Ca, P, and NaF, 3,249. Bone fluoride in thyroidectomized animals (ppm): basal, 280; basal plus Ca and P, 252; basal plus NaF, 3323; basal plus Ca, P, and NaF, 3197.	Rantanen et al. 1972

[a] Information in brackets was calculated from information given in the papers or as otherwise noted.
[b] Based on water consumption of about 10% of body weight.

TABLE E-9 Effects of Fluoride on Thyroid Parafollicular Cell Function in Humans

Study Population(s) and Type	Exposure Conditions and Duration	Concentration or Dose[a]	Effects	Reference
India, 9 patients with moderate to severe skeletal fluorosis (6 males, 3 females), mean age 29 years; 5 controls (3 males, 2 females) mean age 31 years. Case-control study; individual estimates of current fluoride intake, measurements of fasting plasma and urinary fluoride; incomplete information on selection of subjects and controls. See also Tables E-4, E-10, and E-12	Drinking water, area with endemic skeletal fluorosis 2 persons had moved to nonendemic areas 5 or 2 years previously. Exposed since birth Symptomatic for 10-15 years	A) 8.7-9.2 mg/day for 3 persons (7.8-8.0 mg/L in water) [0.145-0.15 mg/kg/day][b] B) 21.0-52.0 mg/day for 4 persons (24.5-25.0 mg/L in water) [0.35-0.87 mg/kg/day][b] C) 2.5 and 3.8 mg/day for 2 persons (0.8 and 1.8 mg/L in water) [0.04-0.06 mg/kg/day][b] D) 1.2-2.2 mg/day for 5 controls (0.7-1.0 mg/L in water) [0.02-0.04 mg/kg/day][b]	Elevated calcitonin concentrations: A, 3 of 3; B, 4 of 4; C, 1 of 2 (8 of 8 individuals with intake ≥ 3.8 mg/day; plasma fluoride ≥ 0.11 mg/L (5.7 µmol/L); urinary fluoride ≥ 2.2 mg/day).	Teotia et al. 1978
Russia, description of subjects not available. Occupational study; probably cross-sectional; full details not available	Occupational exposure (fluorine production) Duration not available	Not available	Elevated concentrations of calcitonin in blood.	Tokar' et al. 1989
Review of epidemiological studies from 1963-1997 (45,725 children) See also Table E-12	Drinking water Comparison of groups with adequate (>800 mg/day) and inadequate (<300 mg/day) dietary calcium intake Exposed since birth	1.5-25 mg/L	Normal or elevated plasma calcitonin.	Teotia et al. 1998
China, 50 male fluoride workers and 50 controls. Occupational cohort study; cross-sectional; measurements of fluoride in serum and urine; full details not available	Occupational exposure Duration not available	Not available	Elevated concentrations of serum calcitonin and parathyroid hormone.	Huang et al. 2002

[a]Doses in brackets were calculated from information given in the papers; other information is as reported.
[b]Based on 60-kg body weight.

TABLE E-10 Summary of Selected Findings for Nine Patients with Endemic Skeletal Fluorosis and Five Controls

Case Number[a]	Age	Sex	Fluoride in Drinking Water, mg/L	Fluoride Intake, mg/day	Urinary Fluoride, mg/day	Plasma Fluoride, mg/L[b]	Urinary Calcium, mg/day	Plasma Calcium, mg/dL	Calcitonin, µg/L	IPTH[c], µg/mL
1_control	35	F	1.0	1.2	0.8	0.023	120	9.5	<0.08	<0.35
3_control	22	M	0.8	1.6	0.2	0.021	115	10.0	<0.08	0.40
2_control	25	M	0.8	1.8	0.6	0.030	95	10.2	<0.08	0.50
4_control	32	M	0.7	2.0	1.0	0.020	170	9.6	<0.08	<0.35
5_control	34	F	1.0	2.2	1.2	0.038	130	9.8	<0.08	0.35
2*	25	M	0.8	2.5 (38)[d]	1.2	0.036	85	10.1	<0.08	0.55
4*	18	M	1.8	3.8 (30)[e]	2.2	0.12	80	9.7	0.14[f]	0.40
8	36	M	7.8	8.7	3.2	0.15	65	8.9	0.10[f,g]	0.70[h]
7	25	F	8.0	9.2	4.2	0.15	60	8.3	0.10[f]	0.50
6	22	M	8.0	9.2	5.8	0.18	70	8.8	0.12[f]	0.35
1	36	F	24.5	21.0	10.0	0.11	75	9.8	0.18[f]	0.40
3[i]	34	F	25.0	28.0	11.0	0.17	70	9.65	0.18[f]	1.10[h]
5	35	M	25.0	48.0	15.0	0.14	65	9.8	0.10[f]	0.80[h]
9[i]	58	M	25.0	52.0	18.5	0.26	78	10.6	0.10[f]	1.50[h]

[a]Case number as reported by Teotia et al. (1978), arranged in order of increasing fluoride intake. Control subjects are indicated. Asterisks by the case numbers indicate patients no longer living in the high-fluoride area; case 2 had moved 5 years previously and case 4 had moved 2 years previously.
[b]Plasma fluoride reported in µmol/L as follows: 1.2, 1.12, 1.6, 1.05, 2.0, 1.9, 6.1, 7.8, 8.0, 9.7, 5.7, 9.2, 7.5, and 13.6.
[c]Plasma immunoreactive parathyroid hormone.
[d]Fluoride intake before moving had been 38 mg/day.
[e]Fluoride intake before moving had been 30 mg/day.
[f]Considered elevated above calcitonin concentrations found in normal controls.
[g]Listed as "<0.10" in Table 1 of Teotia et al. (1978) but assumed to be a misprint of "0.10" based on information in the text of that paper.
[h]Considered elevated above IPTH concentrations found in normal controls.
[i]Patient had radiographic findings suggestive of secondary hyperparathyroidism.

SOURCE: Adapted from Teotia et al. (1978).

TABLE E-11 Effects of Fluoride on Parathyroid Function in Experimental Animals

Species and Strain	Exposure Conditions	Concentration or Dose[a]	Exposure Duration	Effects	Reference
Sheep (4 pairs of twin lambs)	Drinking water No information on dietary calcium	200 mg/L (NaF) [90 mg/L] [9 mg/kg/day][b]	1 week or 1 month	After 1 week, only slight changes in parathyroid ultrastructure; after 1 month, hypertrophy and ultrastructural changes considered to be indicative of increased activity in most cells. Fivefold increase in blood PTH as early as 1 week, remained raised through 1 month. Severely reduced skeletal growth, no evidence of increased resorption, no definite pathology of kidney.	Faccini and Care 1965
Rabbits (strain and sex not stated, 48-42 days old at start)	Oral supplement No information on dietary calcium	10 mg/kg/day	14 weeks; some animals followed for another 24 weeks after withdrawal of fluoride	No significant differences in serum calcium or magnesium; no significant differences in histological, morphometric, or ultrastructural features; no evidence for increased production of PTH or secondary hyperparathyroidism. PTH concentrations not measured.	Rosenquist and Boquist 1973
Rats (Sprague-Dawley, weanling male, 45 g; either thyroid-parathyroidectomized or sham-operated; 17-21 animals per group)	Drinking water 0.6% calcium in diet	90 mg/L [9 mg/kg/day][b] Controls, <1 mg/L	15 days	No effect of fluoride on serum calcium, serum phosphorus, or body weight in either group. No effect of fluoride on serum immunoreactive PTH in sham-operated group. Significantly increased periosteal bone formation, significantly decreased endosteal bone formation, increased endosteal bone resorption; effects on bone were thought not to be due to increased PTH activity.	Liu and Baylink 1977

Rats (Sprague-Dawley, males, 290-300 g; 12 animals per group)	Drinking water Dietary calcium not given	150 mg/L [15 mg/kg/day][b]	10 weeks	Ultrastructural evidence (from transmission electron microscopy) of increased parathyroid activity: higher percentage of active chief cells (90% versus 6%), increased numbers of secretory granules, accumulation of glycogen granules. Results considered indicative of a type of secondary hyperparathyroidism.	Ream and Principato 1981a; 1981b; 1981c
Rats (Wistar albino, males, 95-105 g; 5 animals per group)	Intraperitoneal	15.8 mg/kg (35 mg/kg NaF)	Single dose, killed 0-24 hours later	Increased serum phosphorus; decreased urinary phosphorus; no change in serum calcium; increased urinary calcium; increased calcium, magnesium, and cAMP in renal cells (increase in cAMP was temporary); increased activity of Ca^{2+}-ATPase in kidney. Effects were suppressed in thyroid-parathyroidectomized animals. PTH concentrations not measured.	Suketa and Kanamoto 1983
Rats (Wistar, male, age 5 weeks, 80 g; 40 animals total)	Drinking water and feed	Drinking water: 50 mg/L in treated group, 0.5 mg/L in controls Feed: 5 mg/kg feed (0.26 mM/kg feed) [Approximate doses: treated group, 5.4 mg/kg/day; controls, 0.45 mg/kg/day][c]	46 weeks Calcium-deficient diet for last 16 weeks (from age 35 weeks, approximately 500 g) for half of the animals	Average serum immunoreactive PTH reduced in fluoride-treated animals (not significantly) at 35 weeks. At 51 weeks, normal increase in PTH in response to a dietary calcium deficiency did not occur in fluoride-treated animals (inhibition of normal parathyroid function). Small but significant increase in calculated cytoplasmic volume was observed in calcium-deficient animals given fluoride. Normal serum calcium concentrations in all groups.	Rosenquist et al. 1983

473

continued

TABLE E-11 Continued

Species and Strain	Exposure Conditions	Concentration or Dose[a]	Exposure Duration	Effects	Reference
Pigs (female, 8 months old, average weight 112 kg; 8 animals per group)	Daily oral supplement High calcium and vitamin D in diet	2 mg/kg/day (Fluoride in feed and water approximately 0.05 mg/kg/day)	6 months (average weight, 166 kg)	Plasma fluoride (mg/L): controls, 0.013; treated, 0.24; peak (40-100 minutes after dose), >1.9. Skeletal fluorosis without changes in plasma calcium, parathyroid activity, or vitamin D concentrations. No effect on PTH (measured after 4 months).	Andersen et al. 1986
Sheep (females, 3 breeds, average age 6.0 ± 2.8 years, 55-60 kg; 2 groups of 7 animals)	Oral with dry feed Normal dietary calcium without calcium supplementation	0.45 or 2.3 mg/kg/day (NaF 1 or 5 mg/kg/day)[d]	45 days	Significant decrease in serum calcium and phosphorus in both groups; significant increase in osteocalcin in second group. Variable increase in serum PTH in both groups, not statistically significant due to wide variation, but mean serum PTH in both groups at least twice as high at 45 days (4.9 ± 3.5 and 3.9 ± 0.9 milliunits/mL) as at beginning of experiment (1.9 ± 0.3 milliunits/mL in both groups). Effects on osteoblast birth rate and life span; increased bone formation and resorption, but formation greater than resorption (net increase in bone mass); possible secondary hyperparathyroidism. Serum fluoride (means, mg/L): initial (both groups), 0.10-0.11; final (45 days), first group, 0.24, second group, 0.82; peak > 0.5 at 3 hours after single dose of NaF at 3.5 mg/kg (fluoride, 1.6 mg/kg). Bone fluoride (means, ppm in ash): initial, 2,200-2,500; final, 2,700-3,200.	Chavassieux et al. 1991

Subjects	Exposure	Dose	Duration	Effects	Reference
Rats (Sprague-Dawley, male, 40-50 g weanlings at start, 68-77 animals per group)	Drinking water	5, 15, or 50 mg/L (0.26-0.45, 0.69-1.31, and 2.08-3.46 mg/kg/day, decreasing with increasing body weight)	3, 6, 12, or 18 months	"No significant effect" on plasma calcium or alkaline phosphatase; specific data by treatment group not reported. PTH concentrations not measured.	Dunipace et al. 1995
Rabbits (Dutch-Belted, female, 3 1/2 months old at start, 1.55 kg; 2 groups of 12 animals) See also Table E-16	Drinking water	0 and 100 mg/L [7-10.5 mg/kg/day][e]	6 months	Decreased serum calcium (3%, possibly in the protein-bound fraction). No statistically significant changes in PTH, vitamin D metabolites, or serum phosphorus; mean PTH elevated 3%. Increased bone-specific alkaline phosphatase and tartrate-resistant acid phosphatase, indicative of increased bone turnover. Increased bone mass, but decreased bone strength. Increased serum fluoride (0.73 mg/L versus 0.044 mg/L) and bone fluoride (6,650-7,890 ppm in ash versus 850-1,150 ppm in ash). High intake of calcium and vitamin D from rabbit chow, probable explanation for absence of secondary hyperparathyroidism.	Turner et al. 1997
Rats (strain not available)	Drinking water Dietary calcium adequate or low	100 mg/L [10 mg/kg/day][b]	2 months	Animals on low-calcium diet: osteomalacia, osteoporosis, accelerated bone turnover, increased serum alkaline phosphatase, increased osteocalcin, increased PTH. Animals on adequate calcium diet: slightly increased osteoblastic activity (elevated serum alkaline phosphatase activity and increased average width of trabecular bone after 1 year).	Li and Ren 1997

continued

TABLE E-11 Continued

Species and Strain	Exposure Conditions	Concentration or Dose[a]	Exposure Duration	Effects	Reference
Rats (Sprague-Dawley, male, 30 to 40 g weanlings at start, 432 animals total)	Drinking water Either calcium-deficient diet or diet deficient in protein, energy, or total nutrients	5, 15, or 50 mg/L [0.5, 1.5, or 5 mg/kg/day][b]	16 or 48 weeks	No significant effect on plasma calcium or alkaline phosphatase; specific data by fluoride treatment group not reported. PTH concentrations not measured. Calcium-deficient animals absorbed and retained more fluoride than controls and, in highest fluoride group, gained significantly less weight. Combination of general malnutrition and calcium deficiency was not examined.	Dunipace et al. 1998
Monkeys (cynomolgus, females, 2.5-3.5 kg)	Isoflurane anesthesia	Not available	2 hours	Increased serum inorganic fluoride; decreased ionized calcium; increased PTH and osteocalcin in response to decreased calcium. Serum fluoride 0.070 mg/L versus 0.046 mg/L with ketamine/atropine anesthesia.	Hotchkiss et al. 1998
Rats (Wistar, females, 4-5 months old, 130-150 g)	Drinking water	500 mg/L (50 mg/kg/day)[b,f]	60 days	Hypocalcemia, attributed to suppressed gastrointestinal absorption of calcium. Decreased weight gain; inhibition of acetylcholinesterase and total cholinesterase in brain and serum; decreased spontaneous motor activity and endurance time. PTH not measured.	Ekambaram and Paul 2001
Rats (Wistar, adult females, 150-170 g at start; fluoride administered during pregnancy and lactation)[g]	NaF orally by feeding tube	40 mg/kg/day NaF (18 mg/kg/day fluoride to the mothers)	Day 6 of gestation through day 21 of lactation	Hypocalcemia in mothers and offspring. PTH not measured. Significant changes in other serum cations (sodium, potassium) and phosphorus. Significant recovery on withdrawal of NaF.	Verma and Guna Sherlin 2002b

| Rats (Sprague Dawley weanlings) | Drinking water to dams and then to weanling pups Some groups with calcium deficient diet (dams and pups) | 50 mg/L (5 mg/kg/day)[b] | Day 11 of gestation through 9 weeks old; continued until 15 weeks old with restored calcium, low fluoride, or both | Decreased serum calcium, increased serum alkaline phosphatase, increased concentrations of vitamin D metabolites (both 25(OH)D_3 and 1,25(OH)$_2D_3$). Decreased transcription of genes for vitamin D receptor and calbindin D 9 k; increased transcription of calcium-sensing receptor gene. Continued fluoride excess even with calcium supplementation continued to be detrimental. PTH not measured. | Tiwari et al. 2004 |

[a]Information in brackets was calculated from information given in the papers or as otherwise noted.
[b]Based on water consumption of about 10% of body weight.
[c]Based on water consumption of about 10% of body weight and feed consumption of about 8% of body weight; ATSDR (2003) gives a fluoride dose of 3.3 mg/kg (presumably per day) for these animals.
[d]Choice of doses based on a therapeutic dose of NaF (1 mg/kg/day) and a toxic dose of fluoride (5 mg/kg/day) (Chavassieux et al. 1991).
[e]Based on average daily water consumption of 163 mL, mean initial weight of 1.55 kg, and mean final weight of 2.33 kg for the fluoride-treated group.
[f]The dose was selected to produce toxic effects in a short time, without lethality (Ekambaram and Paul 2001).
[g]In many mammalian species, maternal fluoride exposures are not well reflected by fluoride concentrations in milk; therefore, the impacts of fetal exposure and of reduced milk production by the mothers must also be considered.

TABLE E-12 Effects of Fluoride on Parathyroid Function in Humans (Clinical, Occupational, and Population Studies)

Study Population(s) and Type	Exposure Conditions	Concentration or Dose[a] and Exposure Duration	Effects	Reference
Denmark, 14 normal subjects (5 fasting, 9 nonfasting, ages 22-38 years) Experimental study	Oral dose of NaF	27 mg of fluoride (60 mg NaF) [0.4 mg/kg][b] Single dose Measurements made at 1, 2, 3, and 24 hours	Decreased serum calcium and phosphorus; increased immunoreactive PTH. Measured serum fluoride peak 0.8-0.9 mg/L. Uncertainty as to peak fluoride and PTH, minimum Ca and phosphorus concentrations. No differences between fasting and nonfasting subjects except for a higher increase in serum fluoride concentration in fasting subjects.	Larsen et al. 1978
France, 21 surgery patients (12 males and 9 females; ages 20-60 years) Experimental study; subjects had orthopedic (16), opthalmologic (3), or plastic (2) surgery; study excluded patients who were obese, had altered renal function or previously recognized diseases, or received blood transfusions or undescribed medications; initial values used as controls	Enflurane anesthesia	Not available 60-165 min. (mean, 95.5 ± 26 minutes)	Variations in phosphorus clearance suggestive of a transitory hypersecretion of PTH; initial fall in serum calcium, return to preoperative concentration after 24 hours (variations in calcium balance were not highly significant). PTH not measured. Maximum serum inorganic fluoride: 0.12 mg/L (versus 0.039 mg/L in controls).	Duchassaing et al. 1982

Subjects	Treatment	Dose/Duration	Results	Reference
The Netherlands, 91 osteoporosis patients (61 females, 30 males; mean ages by type of treatment were 57.6-67.3 years). Clinical therapeutic trial; non-blinded; subjects had osteoporosis with one or more vertebral fractures before participation in the study, had normal concentrations for serum creatinine and liver enzymes, were treated as outpatients, were mobile and advised to exercise; pretreatment values used as controls	Oral sodium fluoride (capsules, enteric coated tablets, or enteric coated, slow release tablets) Calcium supplementation of 1,000 mg/day	Mean fluoride dosages by group between 18 and 36 mg/day (NaF, 40-80 mg/day) [fluoride, 0.57-1.1 mg/kg/day]^b 2 years	Patients divided into "responders" and "nonresponders" (NR) by (1) degree of increase in serum alkaline phosphatase concentration (20% NR); (2) changes in bone mineral content (26% NR); (3) occurrence of femoral neck fracture (6.6% NR). Patients with a fracture had lower serum alkaline phosphatase changes and higher increases in PTH.	Duursma et al. 1987
England (7 healthy males; ages 24-43 years) Experimental study	Oral NaF tablets Calcium intakes 400-800 mg/day	27 mg/day (NaF, 60 mg/day) [fluoride, 0.39 mg/kg/day]^b 3 weeks, followed up 6 weeks later	No significant changes in plasma alkaline phosphatase, 25-hydroxy vitamin D, PTH, total and ionized calcium, phosphorus, or albumin. Significant increase in serum osteocalcin. PTH elevated slightly but not significantly (50 ± 17.6 pM/L after versus 43 ± 5.3 pM/L before); large standard deviation indicates variable response (not seen with other parameters).	Dandona et al. 1988

continued

TABLE E-12 Continued

Study Population(s) and Type	Exposure Conditions	Concentration or Dose[a] and Exposure Duration	Effects	Reference
England, osteoporosis patients (34 females aged 49-74 years; 7 males aged 45-69 years; all with postmenopausal or idiopathic osteoporosis; all had normal renal function; 6 females were on hormone replacement therapy) Experimental study	NaF orally in gelatin capsules Calcium supplementation was started at least 6 weeks (median, 20 weeks) prior to study	27 mg/day (NaF, 60 mg/day) [fluoride, 0.39 mg/kg/day][b] 8 days	Decreased serum calcium (total and ionized); decreased serum phosphorus; increased concentrations of biologically active PTH (more than 5-fold); major changes occurred within 48 hours, some return toward normal after that. Patients divided into 2 groups by stability of serum calcium and phosphorus concentrations; the groups varied in their response to NaF with respect to mineral absorption and balance.	Stamp et al. 1988
England, osteoporosis patients (22 controls; 2 males and 20 females, mean age 67 ± 8 years, range 51-83 years; 18 treated patients, 5 males and 13 females, mean age 61 ± 12 years, range 41-78 years; 10 patients were common to both groups [before and after treatment]; 8 females were on hormone replacement therapy) Experimental study; longitudinal for 10 patients	NaF orally in gelatin capsules Calcium supplementation was started prior to study	27 mg/day (NaF, 60 mg/day) [fluoride, 0.39 mg/kg/day][b] 15 ± 10 months	Increased concentrations of biologically active PTH (bio-PTH) in treated group (log-transformed means, 10.6 versus 2.5 pg/mL; ranges, 1.6-126 versus 0.25-10.9 pg/mL). Significantly higher serum alkaline phosphatase (SAP) in treated group. Fluoride-treated patients with elevated concentrations of bio-PTH (> 18 pg/mL) had significantly lower concentrations of SAP than other treated patients, indistinguishable from controls; elevated bio-PTH also associated with relative hypophosphatemia and relative hypocalciuria; possibly excessive PTH accounts for "refractory" state of some patients—nonresponsiveness to fluoride therapy.	Stamp et al. 1990

Subjects/Study Design	Treatment	Dose/Duration	Results	Reference
U.S., female osteoporosis patients (patients with previous history of hyperparathyroidism and several other conditions were excluded) Initial recruitment included 203 in-state patients from previous fluoride trials and 95 controls who had not taken fluoride; of these, 40 fluoride patients and 43 controls were scheduled for appointments; 15 fluoride patients were no longer taking fluoride or failed the appointments; 5 controls failed the appointments; final study included 25 fluoride patients and 38 controls (mean ages, 70.1 for fluoride group, 69.5 for controls) Cross-sectional study; fluoride-treated patients and non-treated controls recruited from database of osteoporosis patients of one investigator; fasting samples; analyses of drinking water, blood, and urine performed blindly; results reported as means of groups and as number outside the normal range for the parameter; urine and plasma fluoride were clearly different between groups; no significant difference in mean water fluoride concentrations See also Table E-17	Slow-release sodium monofluoro-phosphate plus calcium carbonate at 1,500 mg/day Most controls (n = 38) had calcium supplementation	23 mg/day (mean dose) [fluoride, 0.33 mg/kg/day][b] 1.4-12.6 years (mean, 4.2 years)	No significant difference in mean calcium concentrations between groups; 2 of 25 individuals outside normal range (versus 0 of 38 controls). Significant difference (elevation) in mean alkaline phosphatase concentrations between groups; 8 of 25 individuals outside normal range (versus 0 of 38 controls); for those 8, a significant elevation in bone isoenzymes was found. For 24 of the 25 patients, calcium was significantly lower than baseline (pretreatment) values and alkaline phosphatase was significantly higher. PTH not measured. Urine fluoride (mg/L, mean and SD): fluoride group, 9.7 (4.1); controls, 0.8 (0.5); plasma fluoride (mg/L, mean and SD)[c]: fluoride group, 0.17 (0.068); controls, 0.019 (0.0076)	Jackson et al. 1994

continued

TABLE E-12 Continued

Study Population(s) and Type	Exposure Conditions	Concentration or Dose[a] and Exposure Duration	Effects	Reference
China, healthy adults (approximately 120 per group, with either normal or inadequate nutritional intakes; mean ages of groups, 44.9-47.7 years) Cross-sectional cohort study; subjects grouped by location (water fluoride concentration) and nutritional status; populations generally similar (e.g., socially and economically; estimated fluoride intakes and measurements of urine and plasma fluoride and other parameters were made for individuals but results reported only for groups; probable overlap between low (< 0.3 mg/L) and middle (around 1 mg/L) fluoride exposure groups for each nutritional category; no mention of whether analyses were performed blindly See also Table E-17	Drinking water Normal nutrition defined as > 75 g of protein and >600 mg of Ca per day Inadequate nutrition defined as <60 g of protein and <400 mg of Ca per day	0.23, 1.02, and 5.03 mg/L (normal nutrition) 0.11, 0.90, and 4.75 mg/L (inadequate nutrition) Estimated intakes: 1.70, 3.49, and 14.8 mg/day (normal nutrition); 1.20, 2.64, 15.32 mg/day (inadequate nutrition) At least 35 years of continuous residency in the study area	Significant decrease in plasma calcium concentration associated with an increase in fluoride exposure in the populations with inadequate nutrition; not detected in subjects with normal nutrition. Elevated alkaline phosphatase activity with increased fluoride exposure in all populations, with higher values in subjects with inadequate nutrition. All values[d] within the normal range regardless of fluoride exposure and nutritional condition. PTH concentrations not measured.	Li et al. 1995

U.S., osteoporosis patients (Group I, "good responders," 13 postmenopausal females and 3 males; Group II, "poor responders," 7 postmenopausal females and 3 males; Group III, untreated controls, 10 age-matched postmenopausal females) Cross-sectional study of fluoride-treated osteoporosis patients; non-fluoride-treated osteoporosis patients as controls	Oral doses of NaF or sodium monofluoro-phosphate Calcium intake at least 1,500 mg/day	30.6 ± 6.6 mg/day (range, 17.4-40.0 mg/day) [0.44 ± 0.9 mg/kg/day; range, 0.25-0.57 mg/kg/day]^b 32 ± 19 months (range, 13-89 months)	Patients who showed a rapid increase in spinal bone density also showed a general state of calcium deficiency and secondary hyperparathyroidism. Serum PTH elevated in 4 "good responders" and 1 "poor responder" but no controls; all 5 with elevated PTH were calcium deficient; mean PTH concentrations were similar for all 3 groups. Degree of calcium deficiency in fluoride-treated patients was proportional to serum concentrations of PTH, alkaline phosphatase, procollagen peptide, and osteocalcin and to urine hydroxyproline concentrations. Fluoride therapy can cause calcium deficiency, even in patients with a high calcium intake; osteogenic response to fluoride can increase the skeletal requirement for calcium.	Dure-Smith et al. 1996
U.S., 199 adult volunteers (mean ages of groups, 62.3, 58.6, 57.2 years) Ecological study; cross-sectional; subjects grouped by location (water fluoride concentration); subjects not randomly selected; nonfasting samples; urine and plasma fluoride concentrations significantly different for groups; study parameters reported by groups; no information on whether analyses were performed blindly See also Table E-17	Drinking water, natural fluoride Dietary calcium and calcium concentrations in drinking water were not discussed	0.2, 1.0, 4.0 mg/L [0.003, 0.01, 0.06 mg/kg/day]^b At least 30 years of continuous residency in their communities	Some differences in mean plasma calcium and phosphorus concentrations among groups were statistically significant (lower calcium at 0.2 mg/L than 1.0 or 4.0; higher phosphorus at 4.0 mg/L than 0.2 or 1.0); no significant differences among mean alkaline phosphatase concentrations; all mean values were within normal ranges. PTH not measured.	Jackson et al. 1997

continued

TABLE E-12 Continued

Study Population(s) and Type	Exposure Conditions	Concentration or Dose[a] and Exposure Duration	Effects	Reference
U.S., 75 osteoporosis patients (36 with placebo and 39 with fluoride) Placebo-controlled therapeutic study; subjects randomly assigned to treatment groups; no information on whether analyses were performed blindly	Oral doses of slow-release NaF Both groups given calcium at 800 mg/day as calcium citrate	23 mg/day (NaF, 50 mg/day) [approximate fluoride dose, 0.33 mg/kg/day][b] 2 cycles of 12 months of treatment, 2 months off; analyses at 0, 6, 12, and 14 months for each cycle Calcium supplemented continuously throughout	No significant changes in most parameters. Decrease in immunoreactive PTH from beginning values (due to increased calcium intake); fluoride-treated group slightly and consistently (but not significantly) higher than placebo group. Decrease in serum 1,25-dihydroxy vitamin D in placebo group but not in fluoride-treated group.	Zerwekh et al. 1997b
China, 50 male fluoride workers and 50 controls Occupational cohort study; cross-sectional; measurements of fluoride in serum and urine; full details not available	Occupational exposure	Not available	Elevated concentrations of serum calcitonin and PTH.	Huang et al. 2002

[a]Information in brackets was calculated from information given in the papers or as otherwise noted.
[b]Based on 70-kg body weight.
[c]Reported as 9.0 (3.6) μmol/L for the fluoride group and 1.0 (0.4) μmol/L for the controls.
[d]Not stated whether this refers to mean values or all individual values.

TABLE E-13 Effects of Fluoride on Parathyroid Function in Humans (Studies of Endemic Fluorosis Patients)

Study Population(s)	Exposure Conditions	Concentration or Dose[a] and Exposure Duration	Effects	Reference
India, 25 cases of skeletal fluorosis (21 males, 4 females, aged 30-76, with radiologically proved skeletal fluorosis) 25 adult controls (19 males, 6 females, aged 25-75, not from endemic fluorosis area, and with no evidence of enamel or skeletal fluorosis or of bone or renal disease) Case-control study	Drinking water (endemic fluorosis areas)	Not given Probably lifelong	No significant differences between cases and controls in serum calcium, serum inorganic phosphate, phosphate clearance, or 24-hour urinary calcium excretion (the latter either on a normal diet or on days 4-6 of a low-calcium diet); mean phosphate clearance was reduced, but not significantly. Significantly higher serum alkaline phosphatase values in individuals with fluorosis. No measurements of PTH.	Singh et al. 1966
United States, 18-year-old boy, 57.4 kg, with renal insufficiency Case report See also Table 2-3	"High" intake of well water containing fluoride; current intake, 7.6 L/day (2 gallons per day)	2.6 mg/L [0.34 mg/kg/day] Since early childhood	Elevated serum immunoreactive PTH (more than 3 times normal value), slightly elevated serum calcium. Enamel fluorosis and roentgenographic bone changes consistent with "systemic fluorosis."[b]	Juncos and Donadio 1972

continued

TABLE E-13 Continued

Study Population(s)	Exposure Conditions	Concentration or Dose[a] and Exposure Duration	Effects	Reference
India, 20 patients with skeletal fluorosis (17 males, 3 females, age 42-68 years) Detailed studies on 5 of these patients (all males, age 42-60 years, duration of symptoms 5-11 years, no evidence of renal disease or intestinal malabsorption) Case reports; individual measurements of plasma and urine parameters and bone samples; comparison with values obtained from persons in nonfluorotic areas	Drinking water (endemic fluorosis areas) Dietary calcium and vitamin D considered adequate	> 25 mg/day [> 0.4 mg/kg/day][c] Lifelong	Clear evidence of secondary hyperparathyroidism in the 5 patients studied in detail; radiological findings consistent with hyperparathyroidism. Increased plasma alkaline phosphatase, increased phosphate clearance, decreased tubular reabsorption of phosphate, increased urinary fluoride, decreased urinary calcium. Normal plasma calcium and phosphate in 4 persons; elevated plasma calcium and decreased plasma phosphate in 1 person. Elevated serum immunoreactive PTH in all 5, especially in the person with elevated plasma calcium and decreased plasma phosphate (a parathyroid adenoma was later found in that individual, possibly attributable to long-standing hyperplasia as a result of excessive fluoride intake). Excess calcium and fluoride in bone in all 5 (11.8-13.2 versus 10.8 g of calcium per 100 g of dry fat-free bone ash; 265-585 versus 30 mg of fluoride per 100 g of dry fat-free bone ash). Urinary fluoride: 3.0-4.8 mg/L/day.	Teotia and Teotia 1973

Study population	Exposure source	Dose/Duration	Results	Reference
India, 9 patients with moderate to severe skeletal fluorosis (6 males, 3 females, mean age 29 years) 5 controls (3 males, 2 females; mean age 31 years) Case-control study; individual estimates of current fluoride intake, measurements of fasting plasma fluoride and urinary fluoride; incomplete information on selection of subjects and controls See also Tables E-4, E-9, and E-10	Drinking water, area with endemic skeletal fluorosis 2 persons had moved to non-endemic areas 5 or 2 years previously	A) 8.7-9.2 mg/day for 3 persons (7.8-8.0 mg/L in water) [0.145-0.15 mg/kg/day]^d B) 21.0-52.0 mg/day for 4 persons (24.5-25.0 mg/L in water) [0.35-0.87 mg/kg/day]^d C) 2.5 and 3.8 mg/day for 2 persons (0.8 and 1.8 mg/L in water) [0.04-0.06 mg/kg/day]^d D) 1.2-2.2 mg/day for 5 controls (0.7-1.0 mg/L in water) [0.02-0.04 mg/kg/day]^d Since birth Symptomatic for 10-15 years	Increased PTH concentrations: A, 1 of 3; B, 3 of 4 [4 of 6 individuals with plasma fluoride ≥ 0.15 mg/L (7.8 μmol/L)]. Radiographs of 2 of the 4 persons were consistent with secondary hyperparathyroidism.	Teotia et al. 1978
India, 4 siblings (aged 8-18; 2 males, 2 females) and their mother (age 40), all with skeletal fluorosis Case reports; individual estimates of fluoride intake from water, measurements of serum fluoride and other parameters; age-matched Indian controls	Drinking water source, 16.2 mg/L Calcium intakes considered normal (500-820 mg/day)	16-49 mg/day from water, plus any contribution from food [0.5 mg/kg/day for the younger children; 0.5-1 mg/kg/day for the older children and mother]^e Symptomatic for at least 2 years	Normal total and ionized calcium concentrations; normal vitamin D concentrations in children; subnormal total and ionized calcium and subnormal vitamin D in the mother. Significantly elevated PTH, elevated osteocalcin, and elevated alkaline phosphatase in all 5. Findings consistent with secondary hyperparathyroidism. Skeletal changes, biochemical hyperparathyroidism, and elevated osteocalcin were similar in all 5, regardless of nutritional status (low in calories and protein for the mother, more nearly adequate for the children) and vitamin D status. Serum fluoride: 0.29-0.45 mg/L in the children (not measured in the mother).	Srivastava et al. 1989

continued

TABLE E-13 Continued

Study Population(s)	Exposure Conditions	Concentration or Dose[a] and Exposure Duration	Effects	Reference
South Africa (260 children, 119 boys, 141 girls; ages 6-16, in an area with endemic skeletal fluorosis) 9 children (8 boys, 1 girl) studied individually; mean age, 13.7 ± 4.4 years; from the same area Prevalence (cross-sectional) study with ecologic measure of exposure; random selection of participants Case reports of 9 hospitalized individuals	Drinking water	8-12 mg/L [0.2-1.2 mg/kg/day][f] Probably lifelong for most For the 9 children, at least 8 years	Hypocalcemia present in 23% of the children; hypophosphatemia in 15%; elevated alkaline phosphatase in about 25%. Normal serum 25(OH)D concentrations in the 40 children in whom it was measured. Hypocalcemia in 6 of 9 studied individually; low concentrations of 25(OH)D in 2; elevated 1,25(OH)2D in 7. Bone fluoride elevated about 10-fold in the 7 children measured: 4,430-6,790 ppm in ash, mean 5,580 ppm in ash. Reduced phosphaturic response during a PTH-stimulation test (suggestive of pseudohypoparathyroidism Type II), directly related to presence of hypocalcemia, corrected by correcting the hypocalcemia. PTH concentrations not measured. Severe hyperosteoidosis associated with secondary hyperparathyroidism and a mineralization defect. Fluoride ingestion may increase calcium requirements and exacerbate the prevalence of hypocalcemia.	Pettifor et al. 1989

Review of epidemiological studies from 1963-1997 (45,725 children) See also Table E-9	Drinking water Comparison of groups with adequate (> 800 mg/day) and inadequate (< 300 mg/day) dietary calcium intake	1.5-25 mg/L Since birth	High plasma fluoride, alkaline phosphatase, osteocalcin, PTH, and 1,25(OH2)D3; normal or elevated plasma calcitonin; normal plasma calcium, magnesium, phosphorus, and 25-(OH)D. Combination of fluoride exposure and calcium deficiency led to more severe effects of fluoride, metabolic bone diseases, and bone deformities. Toxic effects of fluoride occur at a lower concentration of fluoride intake (>2.5 mg/day) when there is a calcium deficiency; fluoride exaggerates the metabolic effects of calcium deficiency on bone.	Teotia et al. 1998
India, children aged 6-12 in four regions (18-30 kg, 50 children per village) Cross-sectional cohort study; random selection of subjects; subjects grouped by location (water fluoride concentration); individual estimates of fluoride intake, measurements of serum and urinary fluoride, other end points; results reported by group See also Table E-14	Drinking water Calcium intake considered adequate (S.K. Gupta, Satellite Hospital, Banipark, Jaipur, personal communication, December 11, 2003)	2.4, 4.6, 5.6, and 13.5 mg/L [0.25-0.41, 0.40-0.67, 0.48-0.80, and 1.1-1.8 mg/kg/day]g Lifelong	Serum calcium concentrations within normal range for all groups; serum PTH concentrations elevated in two highest groups; serum PTH correlated with fluoride intake and with severity of clinical and skeletal fluorosis.	Gupta et al. 2001

continued

TABLE E-13 Continued

Study Population(s)	Exposure Conditions	Concentration or Dose[a] and Exposure Duration	Effects	Reference
India, 1 adult female Case report	Drinking water "8.4 times above the normal"	Chronic	Fluorosis, leading to secondary hyperparathyroidism manifesting as osteomalacia and a resorptive cavity in the head and neck of the femur; low serum calcium, elevated serum alkaline phosphatase; serum and urine fluoride "86 and 63 times above the normal."	Chadha and Kumar 2004

[a]Information in brackets was calculated from information given in the papers or as otherwise noted.
[b]Juncos and Donadio (1972) described two patients with renal insufficiency and systemic fluorosis; PTH was not reported for the second patient.
[c]Based on consumption of 2 L of drinking water per day by a 60-kg adult.
[d]Based on 60-kg body weight.
[e]Based on 30- to 35-kg body weight for the younger children and 50- to 60-kg weight for the older children and mother.
[f]Based on consumption of 1-2 L of drinking water per day by a 20-to 40-kg child.
[g]Based on mean intakes (mg/day) for 18- to 30-kg children.

ABBREVIATIONS: 25(OH)D, 25-hydroxy vitamin D; 1,25(OH)$_2$D, 1,25-dihydroxy vitamin D.

TABLE E-14 Summary of Selected Findings for Children in Four Villages[a]

Village	Fluoride in Drinking Water, mg/L	Fluoride Intake, mg/day[b]	Serum Fluoride, mg/L	Urinary Fluoride, mg/L	Serum Calcium, mg/dL	IPTH[c], pM/L	Enamel Fluorosis Score[d]	Clinical Fluorosis[e]	Skeletal Fluorosis[f]
Ramsagar ki Dhani	2.4	7.35 (1.72)	0.79 (0.21)	9.45 (4.11)	9.23 (1.89)	31.64 (2.82)	2.71 (1.09)	0.95 (0.22)	0.68 (0.67)
Rampura	4.6	11.97 (1.8)	1.10 (0.58)	15.9 (9.98)	10.75 (1.66)	40.98 (26.9)	1.73 (1.09)	1.00 (0.00)	0.50 (0.61)
Shivdaspura	5.6	14.45 (3.19)	1.10 (0.17)	17.78 (7.77)	9.68 (0.99)	75.07 (31.75)	2.44 (1.32)	1.00 (0.00)	0.79 (0.91)
Raipuria	13.6	32.56 (9.33)	1.07 (0.17)	14.56 (7.88)	10.39 (1.44)	125.10 (131.14)	3.43 (1.70)	1.51 (0.51)	0.95 (1.12)

[a]Mean (standard deviation) of 50 children per village, ages 6-12, body weight 18-30 kg.
[b]Total from food and water.
[c]PTH, midmolecule fragment; normal range, 48.1 ± 11.9 pM/L.
[d]Grading of enamel fluorosis: 0, normal; 0.5 questionable fluorosis; 1, very mild fluorosis; 2, mild fluorosis; 3, moderate fluorosis; 4 severe fluorosis (defined in more detail by Gupta et al. 2001).
[e]Clinical (nonskeletal) fluorosis grading: 1, mild; 2, moderate; 3, severe (defined in more detail by Gupta et al. 2001).
[f]Skeletal (radiological) fluorosis grading: 1, mild; 2, moderate; 3, severe (defined in more detail by Gupta et al. 2001).

SOURCE: Gupta et al. 2001. Reprinted with permission; copyright 2001, Indian Pediatrics.

TABLE E-15 Effects of Fluoride on Pineal Function in Animal and Human Studies

Species	Exposure Conditions	Concentration or Dose[a]	Exposure Duration
Mongolian gerbil (*Meriones unguiculatus*; males and females, from birth)	Fluoride in feed (primarily); oral administration of fluoride through 24 days for high-fluoride group	Low-fluoride group, 7 mg/kgfeed after age 24 days [0.7 mg/kg/day][b] High-fluoride group, 2.3 mg/kg/day orally, 5 days/week through age 24 days; 37 mg/kgfeed thereafter [3.7 mg/kg/day][b,c]	Birth through 28 weeks 24-hour urinary 6-sulfatoxymelatonin measured at 7, 9, 11.5, 16, 28 weeks
Humans (female; 233 in Newburgh, NY; 172 in Kingston, NY) Ecologic study; most of the eligible children in both cities; nonblinded	Fluoride in drinking water	Newburgh, 1.2 mg/L [0.01-0.2 mg/kg/day][d] Kingston, "essentially fluoride-free" [0.001-0.02 mg/kg/day][e]	Up to 10 years (ages 7-18 at time of study; ages at beginning of exposure varied from prenatal to 9 years)
Humans (female; 337 in Kunszentmárton and 467 in Kiskunmajsa, ages 10-19.5 at time of study) Ecologic study; probably included most of the eligible children in both cities; nonblinded	Fluoride in drinking water (probably natural fluoride)	Kunszentmárton, 1.09 mg/L Kiskunmajsa, 0.17 mg/L [0.01-0.2 mg/kg/day versus 0.001-0.02 mg/kg/day][f]	Lifelong

[a]Information in brackets was calculated from information given in the papers or as otherwise noted.

[b]Based on estimated feed consumption of about 10% of body weight per day.

[c]High-fluoride group was given 50 mg/L in drinking water during 24-hour metabolism studies when usual feed was not given.

[d]Estimated fluoride intakes based on ranges of weight and water consumption for children aged 0-18 and fluoride concentration of 1.2 mg/L in drinking water; higher fluoride intakes are associated with the smallest children or the highest water intakes. Some individual intakes could have been lower or higher than the range shown.

Effects	Reference
Altered rhythms and peaks of melatonin production; significantly lower pineal melatonin production in prepubescent gerbils in high-fluoride than in low-fluoride group. Sexual maturation in females occurred earlier in high-fluoride group (79% versus 42% showing vaginal opening at 7 weeks and 70% versus 16% showing differentiated ventral glands at 11.5 weeks). Lower testicular weight at 16 weeks in males. At 28 weeks, fluoride concentration in trabecular bone ash was 600-700 mg/kg in low-fluoride animals and 2,800 mg/kg in high-fluoride animals.	Luke 1997
Average age at menarche 12 years in Newburgh, versus 12 years 5 months in Kingston; described as not statistically significant. At time of study, 35.2% in Newburgh and 35.0% in Kingston were past menarche (adjusted for age distribution). Distributions of actual menarcheal age not available. Girls exposed since birth or before had not yet reached menarche.	Schlesinger et al. 1956
Median value of menarcheal age; 12.779 years in Kunszentmárton and 12.79 years in Kiskunmajsa; distributions of actual menarcheal age not available. Distributions of the frequency of girls having reached menarche by the time of the study show, for most age groups below 15 years, higher likelihood of having reached menarche for Kunszentmárton than for Kiskunmajsa (data were not adjusted for different age distributions in the two towns). Of those reporting having reached menarche by the time of the study (159 in Kunszentmárton and 270 in Kiskunmajsa), the youngest were 10 (1 girl), 11 (2 girls), and 11.5 (6 girls) in Kunszentmárton (8.0% of the total in the 10-11.5 age groups, 5.7% of all postmenarcheal girls) and 11.5 (5 girls) in Kiskunmajsa (4.7% of the total in the 10-11.5 age groups, 1.9% of all postmenarcheal girls).	Farkas et al. 1983

^{e}Estimated as a factor of 10 lower than for a fluoride concentration of 1.2 mg/L. Some individual intakes could have been lower or higher than the range shown.

^{f}Ranges assumed to be close to those given for Schlesinger et al. (1956) above. Some individual intakes could have been lower or higher than the ranges shown.

TABLE E-16 Effects of Fluoride on Other Endocrine Organs in Experimental Animals

Species and Strain	Exposure Conditions	Concentration or Dose[a]	Exposure Duration
Rabbits (young adult)	Intravenous	3 mg/kg/day	2 months
Rats (Long-Evans; 2 groups, each with 10 experimental and 5 control; age 49 or 52 days at start, 160-180 g)	Intraperitoneal (controls injected with NaCl)	Acute, 406.47 mg, NaF total [average dose, 68 mg/kg/day][b] Chronic, 1131.65 mg of NaF total [average dose, 18 mg/kg/day][b]	Acute, 15 days Chronic, 100 days
Rats (Hebrew University albino, males; infants at start, 30-32 g) See also Table E-1	Drinking water	0.55, 1, or 10 mg/L [0.055, 0.1, and 1 mg/kg/day][c]	9 months
Rats (Sprague-Dawley, males, 325-350 g)	Intravenous	6 mg/kg/hour	3 hours
Rats (Wistar) See also Table E-1	Drinking water and diet	Water: 0, 1, 5, 10, 50, 100, or 200 mg/L Diet: 0.31 or 34.5 ppm [0, 0.1, 0.5, 1, 5, 10, or 20 mg/kg/day from water and 0.025 or 2.8 mg/kg/day from feed][d]	54-58 days
Rats (Wistar albino, males, 95-105 g)	Intraperitoneal (controls injected with NaCl)	15.8 mg/kg (35 mg/kg of NaF)	Single dose
Rats (inbred strain IIM, females, 180-220 g)	Oral administration of NaF by gastric tube	7.6 mg/kg	Single dose, after fasting for 24 hours

Effects	Reference
Adrenal weights averaged 20% greater than in controls. Body weight increase was 17% lower than in controls.	Stormont et al. 1931
Acute: 7 of 10 survived, 6 were analyzed (1 "exhibited such bizarre overall changes" that it was omitted from the study). Chronic: 5 of 10 survived. Increased adrenal weight (about 30%) in both groups; enlarged adrenal cortex; normal cortical and medullary cytology. Increased width of connective tissue and increased mitotic activity in pancreases of most animals.	Ogilvie 1953
No histological changes or weight differences in adrenals or pancreases; increase in pituitary weight (not significant for 1 mg/L, significant for 10 mg/L).	Gedalia et al. 1960
Depression of glucose utilization, measured in terms of the output of $14CO_2$; serum glucose was not measured but presumably was elevated in accordance with decreased utilization.	Dost et al. 1977
Decrease in pituitary weight in animals receiving 200 mg/L in drinking water. Decreased TSH and growth hormone in animals receiving 100 or 200 mg/L in drinking water. Full details not available.	Hara 1980
Elevated serum glucose and enhanced glucose-6-phosphate dehydrogenase (G6PD) activities in liver and kidney; attributed to stimulation of adrenal function, both medullary and cortical; changes in glucose concentrations and G6PD activities suppressed by adrenalectomy but not by thyroid-parathyroidectomy.	Suketa et al. 1985
Immediate fall in insulin concentrations (to 50% of basal concentration after 15 minutes) and consequent increase in glycemia (peak at about 1 1/2 hours), returned to normal in 4-5 hours. Decreased insulin response to glucose challenge when fluoride administered 15 minutes before glucose challenge (versus together with or immediately after). Appeared to be direct effect on insulin secretion, not on insulin receptors; hypoglycemic response to exogenous insulin was not impaired by pretreatment with fluoride. Plasma fluoride: 0.1-0.3 mg/L (5-15 µmol/L).	Rigalli et al. 1990

continued

TABLE E-16 Continued

Species and Strain	Exposure Conditions	Concentration or Dose[a]	Exposure Duration
Rats (female, IIM line, age 21 days at start)	Drinking water (NaF)	95 mg/L (5 mmol/L) [10 mg/kg/day][c]	100 days
Rats (Sprague-Dawley, male, 40-50 g weanlings at start, 68-77 animals per group)	Drinking water	5, 15, or 50 mg/L [0.26-0.45, 0.69-1.31, and 2.08-3.46 mg/kg/day] (changing with increasing body weight)	3, 6, 12, or 18 months
Rats (female, IIM line, age 21 days at start)	Drinking water (NaF)	95 mg/L (5 mmol/L) [10 mg/kg/day][c]	3 months
Rats (Zucker, males, normal and fatty diabetic, age-matched, 8 weeks old at start of study, initial weights 282 g for controls and 351 g for diabetics)	Drinking water (NaF) (minimal contribution from feed)	0, 5, 15, or 50 mg/L in drinking water (<1.2 ppm in feed) [Control: 0.05, 0.31, 0.85, and 2.8 mg/kg/day Diabetic: 0.09, 2.0, 6.0, and 15.5 mg/kg/day][e] Reported doses for control rats (mg/kg/day): 0.33 for 5 mg/L and 3.04 or 50 mg/L; for diabetic rats, 1.99 for 5 mg/L and 16.26 for 50 mg/L	3 or 6 months

Effects	Reference
Subtle disturbance of glucose tolerance as shown by glucose tolerance tests, associated with period of elevated fluoride concentrations in plasma and soft tissue (deterioration of glucose tolerance for about 50 days and then normalization by 100 days, when maximum bone mass was achieved and plasma fluoride returned to normal concentrations). Bone mass higher 6-12% greater in fluoride-treated animals (depending on portion of skeleton considered). Bone fluoride (ppm in ash): controls, 1,160-1,410; treated, 6,880-8,550 (depending on portion of skeleton considered).	Rigalli et al. 1992
"No significant effect" on fasting plasma glucose concentrations; specific data by treatment group not reported.	Dunipace et al. 1995
Abnormal glucose tolerance tests when plasma diffusible fluoride exceeds 0.1 mg/L (5 µmol/L). Effects on glucose homeostasis not seen with equivalent (5 mmol/L) amount of sodium monofluorophosphate (MFP); plasma diffusable fluoride always below 0.04 mg/L (2 µmol/L); protein-bound MFP did not affect glucose homeostasis.	Rigalli et al. 1995
Water intake and fluoride intake approximately 6 times higher in diabetics than in controls for a given fluoride concentration; fluoride absorption about 75% in diabetics versus 63% in controls; fluoride retention about 40% (39-42%) in diabetics versus increasing with fluoride dose (27-45%) in controls. Plasma and tissue fluoride concentrations increased with fluoride dose, significantly higher for diabetics than for controls. Plasma fluoride (mg/L) in controls: 0.008-0.010, 0.015-0.017, 0.029, and 0.072-0.082; in diabetics: 0.0097-0.012, 0.036-0.046, 0.10-0.12, and 0.26-0.36.^{f} Bone fluoride (ppm in ash) in controls: 171-194, 410-560, 872-1,330, and 2,500-3,600; in diabetics: 200-310, 1,000-2,000, 2,700-4,700, and 6,800-9,500. Same mean blood glucose value (453.5 ± 8.2 mg/dL) given for initial and final values in diabetic rats—one of them is probably not correct; for controls, initial value of 121.9 ± 1.7 mg/dL and final value of 129.6 ± 1.7 mg/dL. Markers examined: plasma urea, glucose (nonfasting), creatinine, calcium, phosphorus, uric acid, cholesterol, total protein, albumin, total bilirubin, alkaline phosphatase, glutamate oxaloacetate transaminase; urine urea, creatinine; creatinine clearance; histological evaluations; bone marrow sister chromatid exchanges. Significant differences in many parameters between normal and diabetic animals; with respect to fluoride intake, significant differences only for diabetic rats with fluoride at 50 mg/L (lower plasma cholesterol, higher total protein in plasma, increased width of tibial cortex).	Dunipace et al. 1996

continued

TABLE E-16 Continued

Species and Strain	Exposure Conditions	Concentration or Dose[a]	Exposure Duration
Rabbits (Dutch-Belted, female, 3 1/2 months old at start, 1.55 kg) See also Table E-11	Drinking water	0 and 100 mg/L [7-10.5 mg/kg/day][g]	6 months
Rats (Sprague-Dawley, male, 30-40 g weanlings at start, 432 animals total)	Drinking water Either calcium-deficient diet or diet deficient in protein, energy, or total nutrients	5, 15, or 50 mg/L [0.5, 1.5, or 5 mg/kg/day][c]	16 or 48 weeks
Rats (Charles River, Wistar, females, normal and with streptozotocin-induced diabetes, 8 per group) C: normal, no fluoride in water F_{10}: normal, fluoride in water D: diabetic, no fluoride in water DF_{10}: diabetic, fluoride in water FF: normal, with fluoride intake adjusted to match that of DF_{10} (1.6-3 mg/day per rat)	Drinking water and feed (NaF in drinking water)	Drinking water: Groups C and D, 0 mg/L Groups F_{10} and DF_{10}, 10 mg/L Group FF, adjusted to match fluoride intake of DF_{10} Feed: 13 ppm (all groups) [C: 1.0-1.5 mg/kg/day F_{10}: 2.1-2.9 mg/kg/day D: 2.2-2.5 mg/kg/day DF_{10}: 8.4-18.6 mg/kg/day FF: 8.3-11.8 mg/kg/day][i]	3 weeks
Horses (6 total, thoroughbreds, average age 5 years, average weight 509 kg, euthanized at end of experiment)	Sevoflurane anesthesia	Not available	Mean, 18.5 hours

APPENDIX E

Effects	Reference
Statistically significant ($P < 0.05$) increase in serum glucose (17%). Increased IGF-1 (40%). Insulin or other regulators of serum glucose were not measured. No effect of fluoride on serum urea, creatinine, phosphorus, total protein, albumin, or bilirubin; serum glutamate oxaloacetate transaminase; or total alkaline phosphatase. Increased serum fluoride (0.728 versus 0.0441 mg/L)[h] and bone fluoride (6,650-7,890 versus 850-1,150 ppm in ash).	Turner et al. 1997
No significant effect on fasting plasma glucose; specific data by fluoride treatment group not reported. Combination of general malnutrition and calcium deficiency was not examined.	Dunipace et al. 1998
Normal rats had similar intakes of feed and water regardless of fluoride intake; final body weights were similar. Diabetic rats had 3-5 times higher water intake than normal rats and almost twice the feed intake; final body weights for group D were lower than for normal rats; final body weights for group DF_{10} were lower than initial body weights. Increase in overall severity of diabetes and higher fasting blood glucose concentrations in fluoride-treated diabetic rats; about 400 mg/dL (22 mM/L) in DF_{10} versus 250 mg/dL (14 mM/L) in D and 90 mg/dL (5 mmol/L) in C, F_{10}, and FF. Plasma fluoride (approximate, mg/L): C, 0.029; F_{10}, 0.038; D, 0.038; DF_{10}, 0.095; FF, 0.057.[j] Bone (femoral) fluoride (approximate, ppm in ash): C, 400; F_{10}, 600; D, 400; DF_{10}, 1000; FF, 1900). Fluoride treatment in nondiabetic rats did not cause significant alteration of blood glucose concentrations.	Boros et al. 1998
Mean plasma fluoride after 8 hours was 0.7-0.9 mg/L (38-45 µmol/L). Total and ionized calcium decreased over time; ionized calcium remained within normal limits; total calcium below normal values after 2 hours. Serum glucose concentrations increased throughout, exceeding normal concentrations at 6 hours and thereafter, but within the values commonly observed during general inhalation anesthesia in horses; glucosuria also present after 10 hours.	Driessen et al. 2002

continued

TABLE E-16 Continued

Species and Strain	Exposure Conditions	Concentration or Dose[a]	Exposure Duration
Rats (Wistar, adult females, 150-170 g at start; fluoride administered during pregnancy and lactation)[k]	NaF orally by feeding tube	40 mg/kg/day NaF (18 mg/kg/day fluoride to the mothers)	Day 6 of gestation through day 21 of lactation
Rats (Wistar FL, males, 14 weeks old, 8 treated, 10 controls)	Intraperitoneal injection	35 mg/kg NaF (15.8 mg/kg fluoride) in physiological saline Controls, saline only	Single dose, sacrificed 90 minutes later

[a]Information in brackets was calculated from information given in the papers or as otherwise noted.

[b]Based on average of initial and final mean body weights.

[c]Based on water consumption of about 10% of body weight, with no significant differences in body weight with fluoride intake.

[d]Based on water consumption of about 10% of body weight and feed consumption of about 8% of body weight, with no significant differences in body weight with fluoride intake.

[e]Based on final (6-month) mean body weights of 508.8 g for controls and 445.4 g for diabetics, with pretermination (3- and 6-month combined) metabolic data for fluoride intake.

[f]Plasma fluoride (µmol/L) in controls: 0.42-0.54, 0.8-0.9, 1.5, and 3.8-4.3; in diabetics: 0.51-0.65, 1.9-2.4, 5.5-6.1, and 13.6-19.2.

[g]Based on average daily water consumption of 163 mL, mean initial weight of 1.55 kg, and mean final weight of 2.33 kg for the fluoride-treated group.

[h]Serum fluoride: 38.31 versus 2.32 µmol/L.

[i]Based on average daily fluoride intake for days 1-4 with average initial body weight for all groups and average daily intake for days 15-21 with average final body weight for the group.

[j]Plasma fluoride (approximate, µmol/L): C, 1.5; F_{10}, 2; D, 2; DF_{10}, 5; FF, 3.

[k]In many mammalian species, maternal fluoride exposures are not well reflected by fluoride concentrations in milk; therefore, the impacts of fetal exposure and of reduced milk production by the mothers must also be considered.

Effects	Reference
Marked hypoglycemia in mothers and offspring, attributed to reduced feed consumption. Reduced serum protein content, significant increases in serum sodium and potassium. Significant recovery on withdrawal of NaF or supplementation with vitamins C, D, and E.	Verma and Guna Sherlin 2002a
Hyperglycemia (47% increase), accompanied by impairment in renal function, decreased calcium concentrations (13%).	Grucka-Mamczar et al. 2005

TABLE E-17 Effects of Fluoride on Other Endocrine Organs in Humans

Study Population(s)	Exposure Conditions	Concentration or Dose[a] and Exposure Duration
76 male and female inmates of Japanese mental hospital Observational study; summary of cases; cross-sectional	Thought to be from pesticide use	Not available Chronic
41 Russian males with fluorosis, ages 33-45, 19 controls (no contact with fluorine compounds) Case-control study; cross-sectional; full details not available	Occupational exposure	Not available >15 years for some
Volunteers in Argentina, 6 adults Experimental study; subjects included the authors of the report and members of their laboratory	Oral administration to fasting persons	27 mg of fluoride (60 mg of NaF) [0.4 mg/kg][b] Single dose
25 young adults (14 males, 11 females) in India with endemic fluorosis (skeletal and enamel), ages 15-30 years (nonobese, nonsmokers, no personal or family history of diabetes mellitus or hypertension) 25 controls with normal fluoride intake (age, sex, and body mass index matched; comparable social and working conditions) Case-control study; cross-sectional for all; longitudinal for subjects initially found to have impaired glucose tolerance; tests were repeated after 6 months on a low-fluoride water source	Drinking water	2-13 mg/L in drinking water [0.067-0.43 mg/kg/day][c] Controls: < 1 mg/L [< 0.03 mg/kg/day][c] Since birth
Poland, residents of Skawina (living in the vicinity of an aluminum smelter) and Chorzów (employed in any of 3 industries); approximately 50 individuals per group (approximately 200 total) Ecologic measure of exposure (exposure to environmental fluorides from industrial pollution)	Airborne fluorides Skawina: chronic exposure to fluorine compounds Chorzów: chronic exposure to environmental fluorides and other toxic compounds	8-10 times the Maximum Allowable Concentration for fluoride of 1.6 µg/m^3 (12.8-16 µg/m^3)

Effects	Reference
Endocrine disturbances including melanosis in 20 of 76 patients; attributed to dysfunction of parathyroids and adrenals, reversed upon treatment for chronic fluorine poisoning.	Spira 1962
Elevated follicle-stimulating hormone and decreased testosterone in blood in all men with fluorosis; elevated blood luteinizing hormone in men with long-term exposure (>15 years).	Tokar' and Savchenko 1977
After 1 hour, significant fall of plasma insulin concentrations and increased fluoride; reduced insulin response to glucose challenge. Plasma fluoride: 0.1-0.3 mg/L (5-15 µmol/L).	Rigalli et al. 1990
Impaired glucose tolerance (IGT) in 40% (6 males, 4 females); fasting serum fluoride concentrations positively correlated ($P < 0.01$) with area under glucose curve in those 10; effect appeared to be reversible on provision of drinking water with "acceptable" fluoride concentrations (<1 mg/L). For all 25 endemic fluorosis patients, significant positive correlation between serum fluoride and fasting serum immunoreactive insulin; significant negative correlation between serum fluoride and fasting glucose:insulin ratio. Normal serum calcium, inorganic phosphorus, and vitamin D; elevated serum alkaline phosphatase in patients with endemic fluorosis. Urine fluoride (mg/L): fluorosis patients, 2-8; controls, 0.2-0.5. Serum fluoride (mg/L): patients with IGT, 0.08 ± 0.04; patients with normal glucose tolerance, 0.02 ± 0.01; controls, 0.01 ± 0.009; IGT patients after 6 months on low-fluoride water, 0.02 ± 0.01.	Trivedi et al. 1993
Excessive excretion of fluorides in urine (53-100% with urine fluoride > 2.3 mg/L; for Skawina, mean = 5.6 mg/L; SD = 2.5, n = 46), associated with a decrease in urine and erythrocyte magnesium concentrations (36-65% with urine magnesium < 5.4 mg/L); increased blood glucose and lactate concentrations, which were normalized by magnesium supplementation. For Skawina, 74% had blood glucose results above the norm (70-100 mg/dL or 3.89-5.55 mmol/L; n = 42).	Kedryna et al. 1993

continued

TABLE E-17 Continued

Study Population(s)	Exposure Conditions	Concentration or Dose[a] and Exposure Duration
U.S., female osteoporosis patients (patients with previous history of hyperparathyroidism and several other conditions were excluded) Initial recruitment included 203 in-state patients from previous fluoride trials and 95 controls who had not taken fluoride; of these, 40 fluoride patients and 43 controls were scheduled for appointments; 15 fluoride patients were no longer taking fluoride or failed the appointments; 5 controls failed the appointments; final study included 25 fluoride patients and 38 controls (mean ages, 70.1 for fluoride group, 69.5 for controls) Cross-sectional study; fluoride-treated patients and non-fluoride-treated controls recruited from database of osteoporosis patients of one investigator; fasting samples; analyses of drinking water, blood, and urine performed blindly; results reported as means of groups and as number outside the normal range for the parameter; urine and plasma fluoride clearly different between groups; no significant difference in mean water fluoride concentrationsSee also Table E-12	Slow-release sodium monofluoro-phosphate plus 1,500 mg/day calcium carbonate Most controls (n = 38) had calcium supplementation	23 mg/day (mean dose) [0.33 mg/kg/day][b] 1.4-12.6 years (mean, 4.2 years)
China, healthy adults (approximately 120 per group, with either normal or inadequate nutritional intakes; mean ages of groups, 44.9-47.7 years) Cross-sectional cohort study; subjects grouped by location (water fluoride concentration) and nutritional status; populations generally similar (e.g., socially and economically); estimated fluoride intakes and measurements of urine and plasma fluoride and other parameters were made for individuals but results reported only for groups; probably overlap between low (<0.3 mg/L) and middle (around 1 mg/L) fluoride exposure groups for each nutritional category; no mention of whether analyses were performed blindly See also Table E-12	Drinking water Normal nutrition defined as > 75 g/day protein and Ca >600 mg/day Inadequate nutrition defined as <60 g/day protein and Ca <400 mg/day	0.23, 1.02, and 5.03 mg/L (normal nutrition) 0.11, 0.90, and 4.75 mg/L (inadequate nutrition) Estimated intakes: 1.70, 3.49, and 14.8 mg/day (normal nutrition); 1.20, 2.64, 15.32 mg/day (inadequate nutrition) At least 35 years of continuous residency in the study area

Effects	Reference
Mean fasting blood glucose concentrations 104.7 (SD = 53.0) for fluoride-treated group and 95.2 (SD = 10.3) for controls (difference not considered significant); 3 of 25 fluoride-treated individuals outside normal range (versus 1 of 38 controls). Urine fluoride (mg/L, mean and SD): fluoride group, 9,7 (4.1); controls, 0.8 (0.5); plasma fluoride (mg/L, mean and SD)[d]: fluoride group, 0.17 (0.068); controls, 0.019 (0.0076).	Jackson et al. 1994
No significant differences in mean blood glucose concentrations among groups. Not clear whether samples were fasting or nonfasting.	Li et al. 1995

continued

TABLE E-17 Continued

Study Population(s)	Exposure Conditions	Concentration or Dose[a] and Exposure Duration
2 postmenopausal women in Argentina Experimental study; subjects were members of the authors' department who were receiving NaF as treatment for osteoporosis and who volunteered to undergo glucose tolerance tests; tests were administered in the fasting state	Treatment for osteoporosis	13.6 mg/day (30 mg/day NaF) [0.23 mg/kg/day][e] 9 and 24 months
24 women and 2 men, ages 44-66, former residents of an area of endemic fluorosis in Argentina Ecologic exposure measure; cross-sectional study; fasting blood samples	Drinking water	Not stated Chronic
U.S., 199 adult volunteers (mean ages of groups, 62.3, 58.6, 57.2 years) Ecological study; cross-sectional; subjects grouped by location (water fluoride concentration); subjects not randomly selected; nonfasting samples; urine and plasma fluoride concentrations significantly different for groups; study parameters reported by groups; no information on whether analyses were performed blindly See also Table E-12	Drinking water, natural fluoride Dietary calcium and calcium concentrations in drinking water were not discussed	0.2, 1.0, 4.0 mg/L [0.003, 0.01, 0.06 mg/kg/day][b] At least 30 years of continuous residency in their communities
160 males ages 20-50 years, in Mexico Ecologic exposure measure based on occupation; exposure groups overlapped; no information on selection of subjects	Drinking water alone for 27 men (low group) Occupational exposure and drinking water for 133 men (high group)	3.0 mg/L in drinking water 2-13 mg/day estimated for low group [0.03-0.19 mg/kg/day][b] 3.4-27.4 mg/day estimated for high group [0.05-0.39 mg/kg/day][b] Chronic (at least 1 year for occupational exposure)

[a]Information in brackets was calculated from information given in the papers or as otherwise noted.
[b]Based on 70-kg per person.
[c]Based on consumption of 2 L of drinking water per day by a 60-kg adult.
[d]Reported as 9.0 (3.6) μmol/L for the fluoride group and 1.0 (0.4) μmol/L for the controls.
[e]Based on 60-kg per person.

Effects	Reference
Disturbed glucose homeostasis when given glucose tolerance test. Plasma F: 0.11 and 0.13 mg/L (5.6 and 6.7 µM/L).	Rigalli et al. 1995
Inverse relationship between plasma fluoride and area under curve of insulin during a standard glucose tolerance test. Plasma F: 0.01-0.18 mg/L (0.5-9.2 µM/L). Urine F: > 1.1 mg/day.	de la Sota et al. 1997
No significant differences among mean glucose concentrations (nonfasting); all mean values were within normal ranges.	Jackson et al. 1997
Elevated follicle stimulating hormone; decreased testosterone, inhibin B, and prolactin; apparent reduction in sensitivity of the hypothalamic-pituitary axis to negative feedback action from inhibin B. Fluoride exposures of the two groups overlapped, and occupational exposures included other chemicals besides fluoride.	Ortiz-Perez et al. 2003